Current Progress in Aerospace Engineering

Current Progress in Aerospace Engineering

Edited by Natalie Spagner

www.clanryeinternational.com

Clanrye International,
750 Third Avenue, 9th Floor,
New York, NY 10017, USA

ISBN: 978-1-63240-934-8

Cataloging-in-Publication Data

Current progress in aerospace engineering / edited by Natalie Spagner.
p. cm.
Includes bibliographical references and index.
ISBN 978-1-63240-934-8
1. Aerospace engineering. 2. Aeronautics. 3. Astronautics. 4. Engineering. I. Spagner, Natalie.
TL545 .C87 2020
629.1--dc23

For information on all Clanrye International publications visit our website at www.clanryeinternational.com

Contents

Preface

Aerospace engineering is a discipline which is concerned with the manufacturing of aircraft and spacecraft. It is categorised into two main branches, aeronautical and astronautical engineering. It focuses on the designing of the aircraft structure with respect to atmospheric temperature and pressure. It encompasses several distinct fields such as astrodynamics, aeroacaustics, aeroelasticity, propulsion and others. Modern aerospace engineering also implements computer software for aerospace applications such as flight software, ground control software, test and evaluation software, etc. This book discusses the fundamentals as well as modern approaches of aerospace engineering. The various studies that are contributing towards the advancement of technologies and evolution of this field are examined in detail. This book will serve as a resource guide for students and experts both and contribute to the growth of the discipline.

This book is the end result of constructive efforts and intensive research done by experts in this field. The aim of this book is to enlighten the readers with recent information in this area of research. The information provided in this profound book would serve as a valuable reference to students and researchers in this field.

At the end, I would like to thank all the authors for devoting their precious time and providing their valuable contribution to this book. I would also like to express my gratitude to my fellow colleagues who encouraged me throughout the process.

Editor

The Aerodynamic Behavioral Study of Tandem Fan Wing Configuration

Du Siliang[1] **and Tang Zhengfei**[2]

[1]*Faculty of Mechanical & Material Engineering, Huaiyin Institute of Technology, Huaian 223003, China*
[2]*National Key Laboratory of Rotorcraft Aeromechanics, Nanjing University of Aeronautics and Astronautics, Nanjing 210016, China*

Correspondence should be addressed to Du Siliang; kjofchina@qq.com

Academic Editor: Kenneth M. Sobel

The fan wing aircraft is a new concept based on a new principle, especially its wing which is based on a unique aerodynamic principle. A fan wing can simultaneously generate lift and thrust. In order to further improve its aerodynamic characteristics without changing its basic geometric parameters, two fan wings are installed along the longitudinal body, which is the composition of a tandem fan wing aircraft. Through numerical simulation, the lift and thrust of the fan wings were calculated with the distance, height, and installation angle of the front and rear fan wings changed, and the aerodynamic characteristic interaction rule between the front and rear fan wings was analyzed. In addition, the wind test model of a tandem fan wing was designed, and the results of the wind tunnel test and numerical calculation results were compared to verify the preliminary setup. The results show that at a certain height, distance, and installation angle, aerodynamic characteristics of a tandem fan wing have more advantages compared to the single fan wing. Therefore, the tandem fan wing aircraft's advantages have good prospects for development and application.

1. Introduction

A fan wing aircraft [1–4] has a crossflow fan at the leading edge of each wing. The fan, powered by a conventional engine, pulls the air in from the front and accelerates the air over the trailing edge of the wing. By transferring the power of the engine to the crossflow fan, which spans the whole wing, the fan wing accelerates a large volume of air and produces lift and thrust simultaneously. It cannot be classified as either a fixed wing or rotary aircraft but shares certain fundamental features of both. The fan wing concept with distributed propulsion is described as a simple, stable, and very efficient high lift aircraft. The implementation of a crossflow fan in a wing was first proposed by Nikolaus Laing in 1964 [5], who together with Bruno Eck are responsible for increasing the crossflow fan efficiency. Hancock [6] worked on raising the efficiency of the crossflow fan implementation at Lockheed up to the value of 82%. However, the fan was reported to be so noisy that the test area had to be evacuated during the tests and was never integrated into a wing. The subsequent research studied the flow field through the crossflow fan with high- and low-pressure cavities [7]. Recently, the investigations on the fan wing technology integrated in airfoils showed the high lift potential of the embedded propulsion system and moved the research from experimentation to prototyping [8]. It exhibits great potential for exploitation in a wider range of commercial or military applications.

The implementation of a single fan wing was numerically and experimentally investigated [9–15]. The results confirmed the high lift, fuel consumption reduction, and noise decreasing characteristic. However, the studies above are all conducted on a single fan wing; with reference to the tandem wing aircraft [16] and tandem rotor helicopter [17], we proposed the tandem fan wing aircraft (two fan wings installed along the longitudinal body). This kind of aircraft realizes the flight capability to take off and land at ultrashort distances or vertically with the heavy load, which satisfies the demand for an aircraft at low speed with a heavy load in some particular situations. This paper is aimed at studying a tandem fan wing configuration by CFD method, including the force measurement experiments in the wind tunnel.

2. Model and Calculation Method

2.1. Tandem Fan Wing Model. The layout of the tandem fan wing is shown in Figure 1. We defined the geometric parameters of the single fan wing model (Figure 2, Table 1) and blade geometric parameters (Figure 2(b), Table 2). The geometric parameters of the front and rear wings are the same.

2.2. Calculation Method. In this paper, the maximum rotation speed of the crossflow is 2000 r/min, and the blade tip's rotation speed of the crossflow fan is less than 0.3 Ma. Therefore, the compressibility of airflow can be ignored in the numerical calculation. The whole flow field around the fan wings is in an unsteady state, and it is necessary to take the influence of the Reynolds number into account [18]. The numerical simulations are performed using the commercial general-purpose CFD code FLUENT 14.5 by Fluent Inc. The ANSYS ICEM software was used for grid division. For the CFD analysis, free stream velocity and angle of attack are constant for all rotation velocities. Renormalization group (RNG) $k-\varepsilon$ model was used for turbulence. The pressure-velocity coupling was calculated using the SIMPLEC algorithm. Second-order upwind discretization was considered for the convection terms. The finite volume method with rectangular elements was used for the whole solution domain. The rotating and stationary domains connected each other with a fluid-fluid interface, where the flow continuity is satisfied. To simulate the fan rotating, the area surrounding the blades was designed as a sliding mesh region (Figure 3). A uniform velocity is imposed at the inlet, while a zero relative static pressure is prescribed at the outlet. Unsteady simulations require a proper setting of both the time step size and the convergence criteria within each time step. For this simulation, a time step size equal to 1/20th of the blade passing period captured the unsteady flow well. Within each time step, iterations were performed until the solution no longer changed. It was found necessary to reduce all residuals to at least 10^{-5}. With these, we set up the numerical calculation model for analysis of different distances, heights, and installation angles between the front and the rear wing.

2.3. Example Verification. We took the wind tunnel test results from the literature [19] to verify the accuracy of the numerical simulation method. Given the coming flow, speed of 10 m/s, angle of attack of 0°, and range of crossflow fan rotation speed changing within 400–1200 r/min, we made a contrastive verification of the lift and the thrust of the single fan wing. Figures 4(a) and 4(b) show the test results of the lift and the thrust of the single fan wing at the different rotation speeds of the crossflow fan. As we can see from the pictures, with the increase of crossflow fan rotation speed, the lift and the thrust gradually increased as well, and the calculation results coincide with the trend of test results well with the maximum error less than 10%. Therefore, the numerical method mentioned above can be used in the calculation and analysis of the tandem fan wing aerodynamic characteristics. It can be seen from the curve that when the speed is 1000 r/min, the lift and thrust errors have a minimum value of 0.17 N and 1.04 N, respectively. We use the rotational speed at this time as the reference speed for calculation and analysis.

3. Calculation Results and Analysis

3.1. The Relationship between the Aerodynamic Force and the Distance. Figure 5 shows the definition of the tandem fan wing model with different distances. The height and the angle between the front and the rear wings are zero; we calculated the aerodynamic force with different distances of 500 mm, 600 mm, 700 mm, 800 mm, 900 mm, and 1000 mm (such six conditions).

Figures 6(a) and 6(b) show, respectively, the changing curve of the lift and the thrust's variation following the distance between the front and the rear wings. As you can see from Figure 6(a), the lift force increases with distance when under 700 mm. On the contrary, when over 700 mm, the lift tends to be stable when the distance changes. In addition, the front wing's lift is always greater than the rear wing's, and compared with the single fan wing, the increase of the rear wing's lift is relatively small, even less than the single fan wing's lift when the distance decreases to some extent. On the whole, the average aerolift effect of the tandem fan wing is better than that of a single fan. According to the figures, it has increased by about 27%. After analysis of the force curve in Figure 6(b), we can find that tandem fan wings' added value of thrust is quite smaller than the single wing's when the distance is big. Only when the distance is minimum does the tandem fan wing have its advantage, increasing the thrust by about 51%.

In order to further illustrate the principle of the aerodynamic increase of the tandem fan wing, we drew the velocity contours and streamlines with different distances between the front and the rear wings in Figure 7. As we can see from these figures, due to the interactive influence of the front and the rear airflow, the intensity and position of the low-pressure vortices inside the fan wing are different. With reference to the single flow chart of the flow velocity, when spacing is small, most of the airflow accelerated by the front wing flows into the rear wing, which will be accelerated again by the crossflow fan in the rear fan wing. Therefore, the airflow speed of the rear fan wing is always larger than the front one. When the distance is over or equal to 700 mm, the rear wing's acceleration of airflow from the front wing decreases, most of which escape the rear fan wing's acceleration, which means the flow into the rear wing decreases. Therefore, the rear fan wing's lift is always less than the front fan wing's.

Now we talk about the front and the rear wings' impact on the change of thrust. The thrust produced by the fan wing derives from the acceleration of airflow by the crossflow fan's rotation. When the spacing is small, all the airflow out of the front wing flows into the rear wing, and the flow rate of accelerated flow in the crossflow fan increases in unit time. According to Newton's third law, the crossflow fan's blade gets more reaction so that the rear wing's thrust gets larger

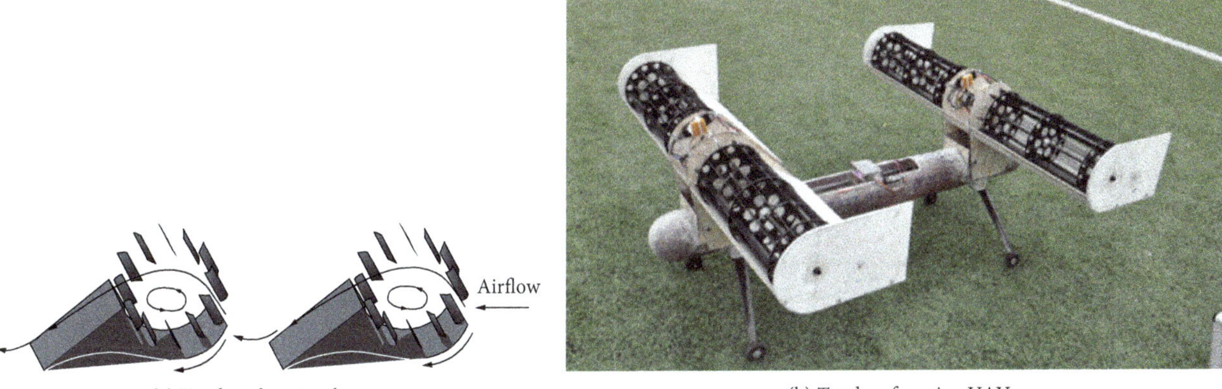

(a) Tandem fan wing layout

(b) Tandem fan wing UAV

FIGURE 1: Tandem fan wing.

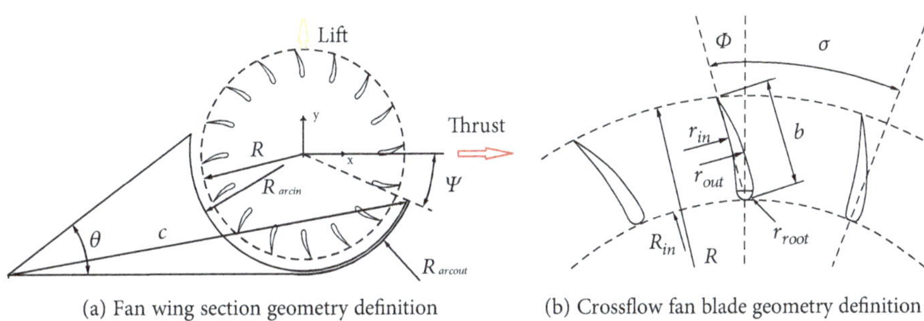

(a) Fan wing section geometry definition

(b) Crossflow fan blade geometry definition

FIGURE 2: Fan wing section.

TABLE 1: Definition of fan wing geometric parameters.

Definition	Value
Fan wing span L (mm)	500
Radius of crossflow fan R (mm)	150
Inner radius of semicircular cavity R_{arcin} (mm)	155
Outer radius of semicircular cavity R_{arcout} (mm)	160
Fan wing chord c (mm)	561
Trailing edge angle θ (°)	36.5°
Opening angle of the leading edge ψ (°)	24°

TABLE 2: Crossflow fan blade geometric parameters.

Parameters	Value
Blade width b (mm)	36
Outer radius of crossflow fan R (mm)	150
Inner radius of crossflow fan R_{in} (mm)	98
Blade outer arc radius r_{out} (mm)	96
Blade inner arc radius r_{in} (mm)	68
Blade root arc radius r_{root} (mm)	3
Blade installation angle ϕ (°)	18
Contiguous blade angle σ (°)	22.5

when the distance narrows. When the distance increases to 700 mm, the thrust of the front wing is the minimum and gradually increases at the same pace of distance. It may be caused by the flow velocity of the edge inclined plane effected by the rear fan wing.

3.2. The Relationship between the Aerodynamic Force and the Height. The definition of the tandem fan wing with different heights is shown in Figure 8. We defined the distance as 800 mm, and the angle between the front and the rear wings does not change. The height difference is $h/2R = 1, 2, 0, 0.5, 0.5$ (five conditions).

Figures 9(a) and 9(b) show the curves of the lift and the thrust of the front and the rear fan wings with different heights. It can be found that when the front fan wing is relatively higher than the rear one, the effect on the lift of the front fan wing is significant, but the rear fan wing is not significant. When $h/2R = 0$, the lift increases to the maximum. The changes of the height difference between the front and the rear wings have little influence on the overall thrust, and the relationship between the thrust of the front and the rear wing shows the opposite trend. Figure 10 shows the velocity contours and streamlines of the height change between the front and rear wings of the tandem fan wing. As we can see, when $h/2R > 0$, airflow accelerated by the front fan wing from its trailing edge slope to the rotating crossflow fan of the rear fan wing is quite stable. When $h/2R = 0.5$, the lift of the front wing is small, which is caused by the

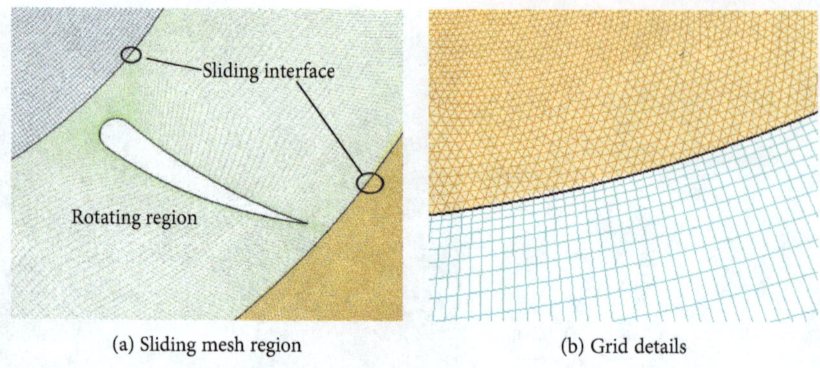

(a) Sliding mesh region (b) Grid details

FIGURE 3: Interface and grid definition.

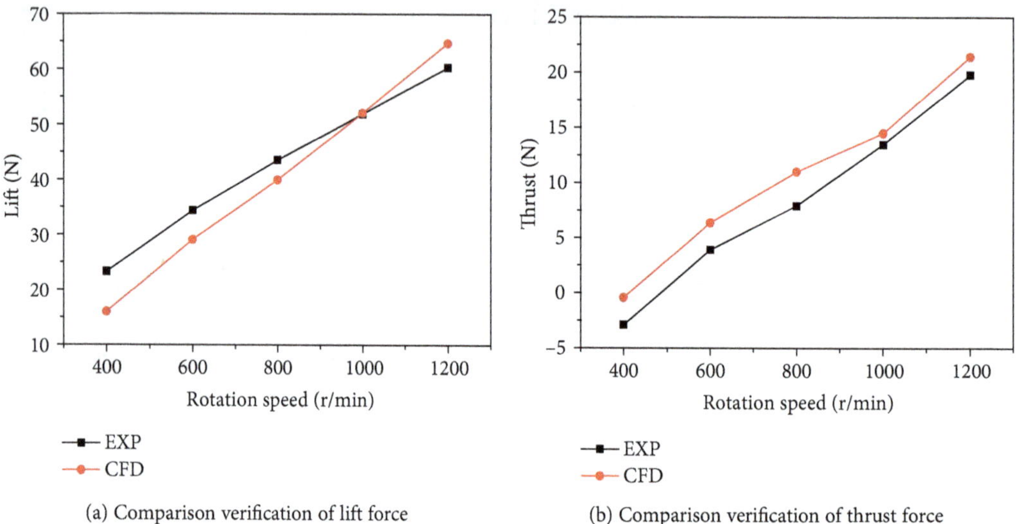

(a) Comparison verification of lift force (b) Comparison verification of thrust force

FIGURE 4: The numerical calculation results and test results with the crossflow fan speed change curve.

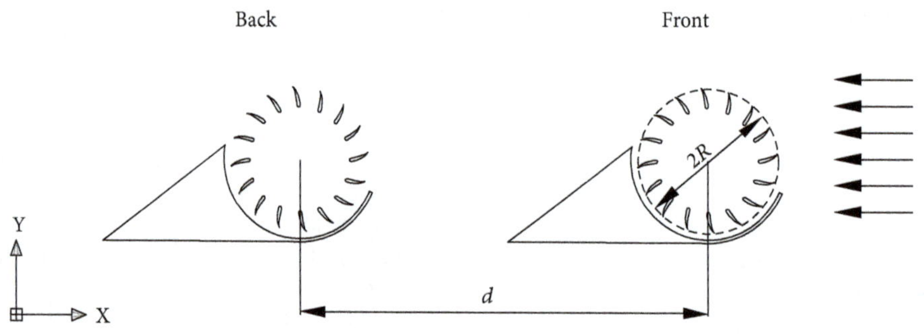

FIGURE 5: Definition of distance between front and rear wings of the tandem fan wing.

decreased pressure on the front wing's surface with the airflow acceleration from the rear wing. The $h/2R = 0.5$ value results in rather "odd" changes in lift and thrust behavior. This is due to the high energy fluid leaving the front fan wing entering the rear fan wing.

3.3. The Relationship between the Aerodynamic Force and the Installation Angle. Figure 11 shows the definition of the tandem fan wing with different installation angles between the front and rear wings. We defined the distance as 800 mm, the height difference between the front and the rear wings as $h/2R = 0$, and the installation angle of the front wing as 0°. Then we calculated the aerodynamic force when the installation angle of the rear fan wing is −20°, −10°, 0°, 10°, and 20°. Next, we defined the installation angle of the rear wing as 0° and calculated the aerodynamic force when

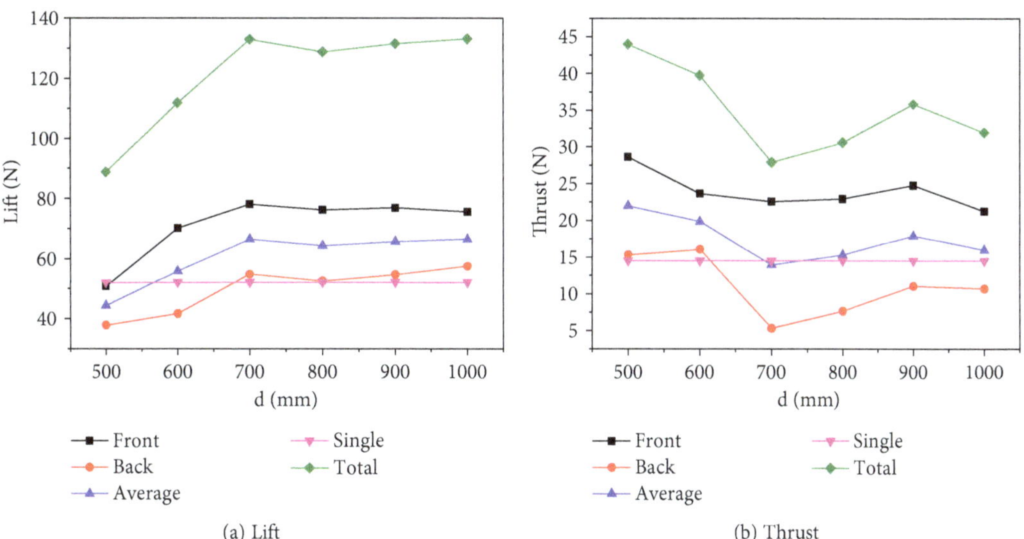

FIGURE 6: Tandem fan wing aerodynamic force with the changing curve following the distance between the front and rear wings.

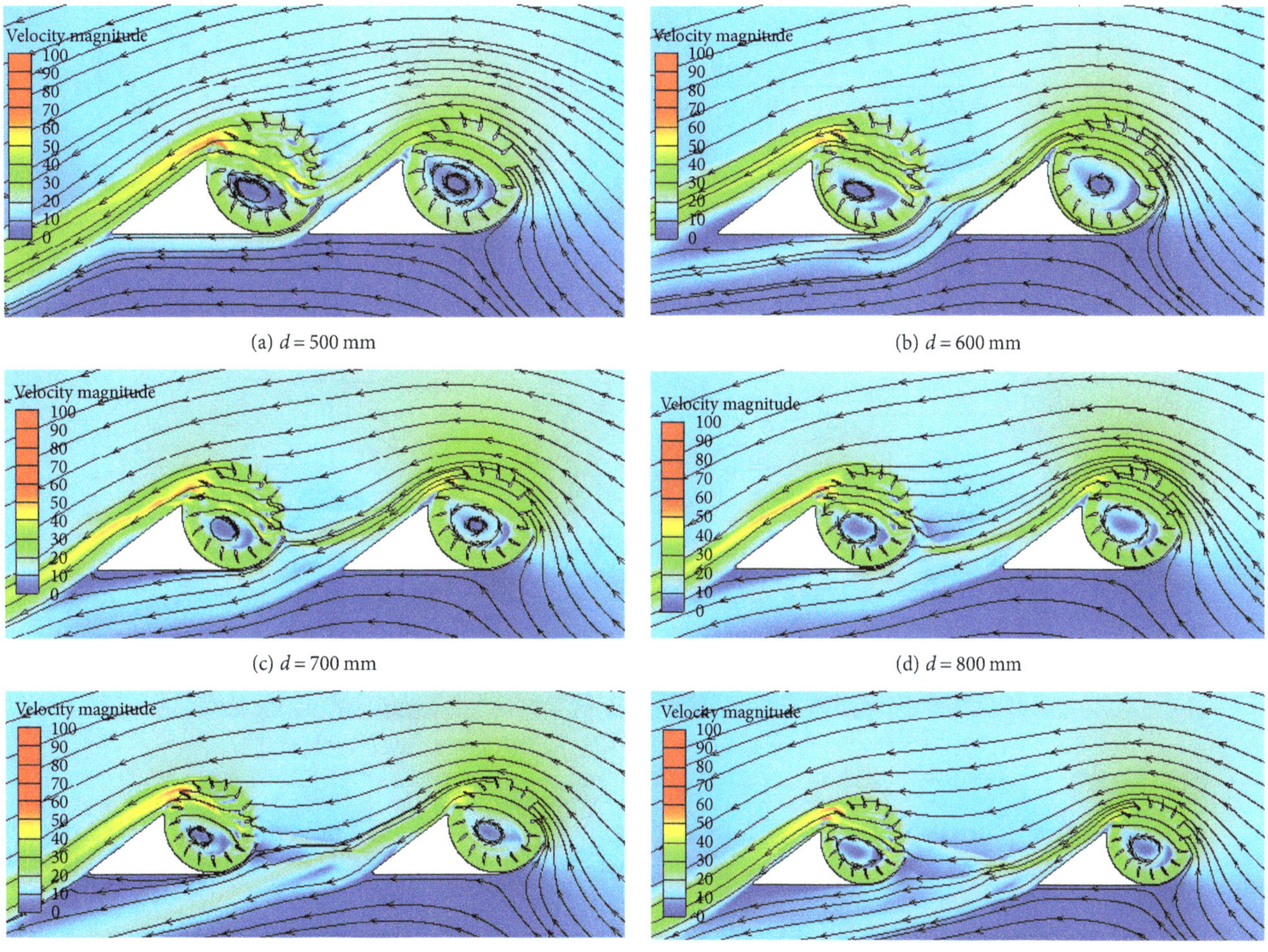

(a) $d = 500$ mm

(b) $d = 600$ mm

(c) $d = 700$ mm

(d) $d = 800$ mm

(e) $d = 900$ mm

(f) $d = 1000$ mm

FIGURE 7: Velocity contours and streamlines of the distance change between the front and rear wings of the tandem fan wing.

the installation of the front wing is -20°, -10°, 0°, 10°, and 20°. Later, we calculated the aerodynamic force when the installation angles of the front and the rear wings are the same. Finally, we calculated the aerodynamic force when the installation angle of the front wing is negative to that of the rear wing.

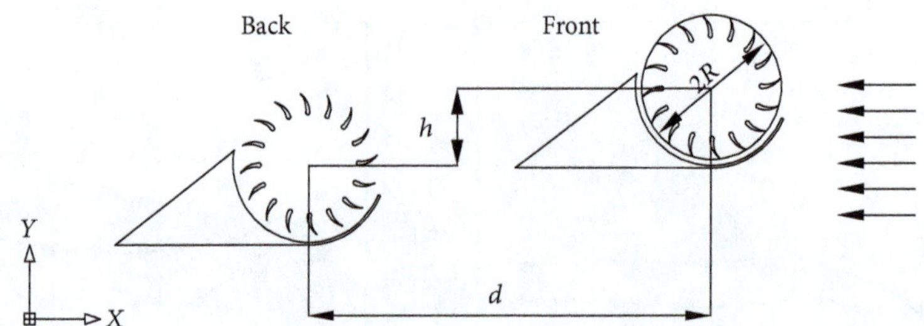

FIGURE 8: Definition of height between the front and rear wings of the tandem fan wing.

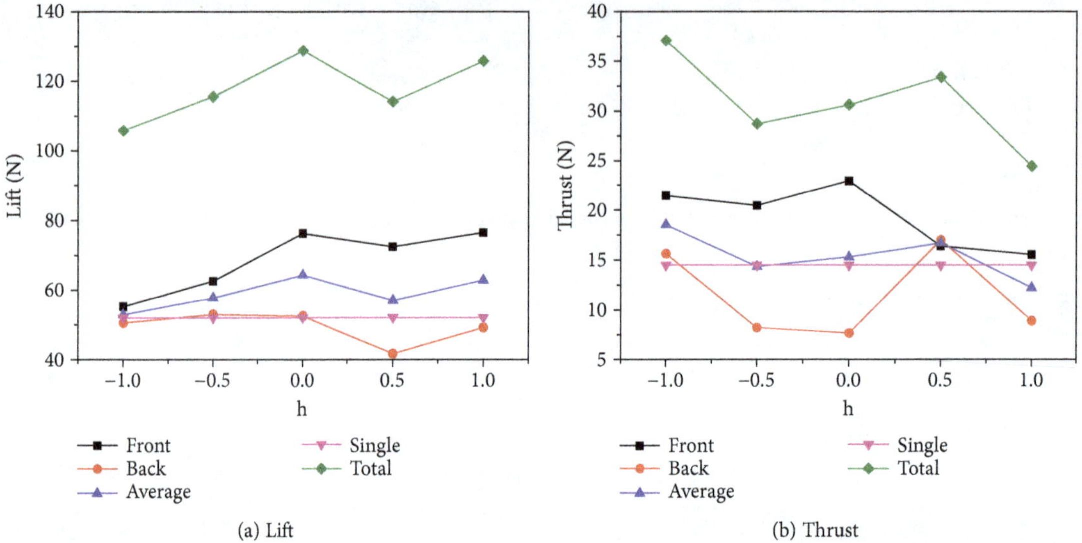

FIGURE 9: Tandem fan wing aerodynamic force with the changing curve following different heights between the front and rear wings.

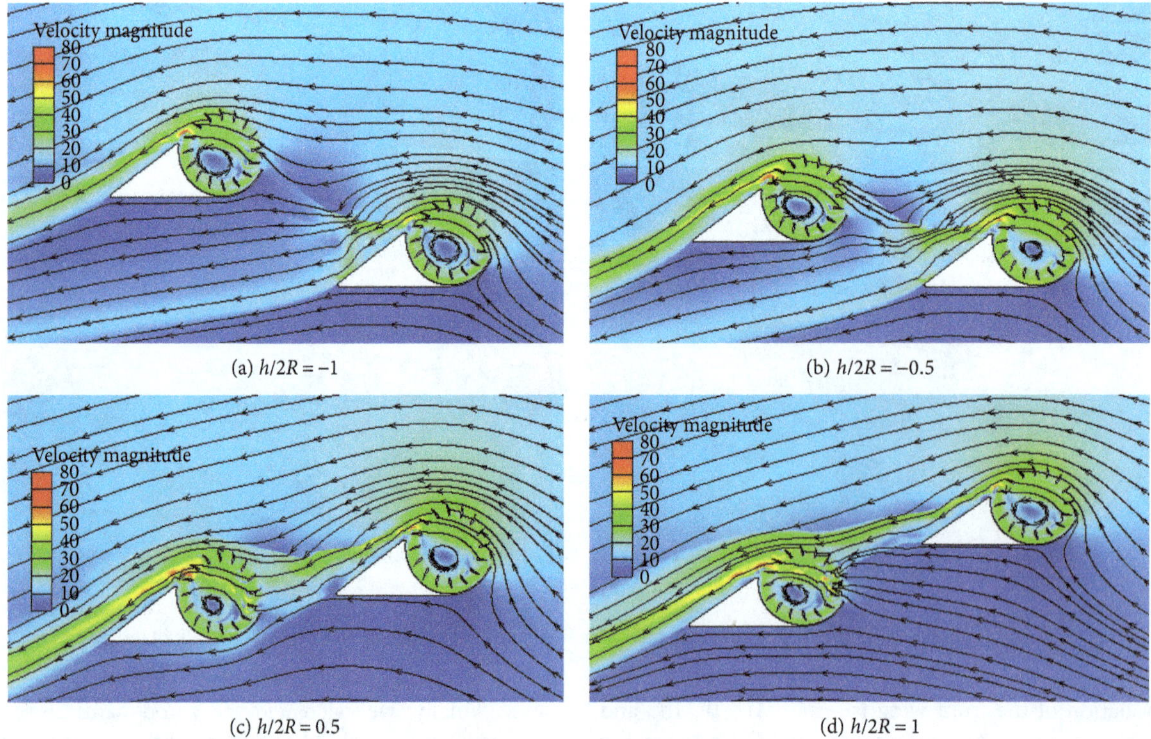

FIGURE 10: Velocity contours and streamlines of the height change between the front and rear wings of the tandem fan wing.

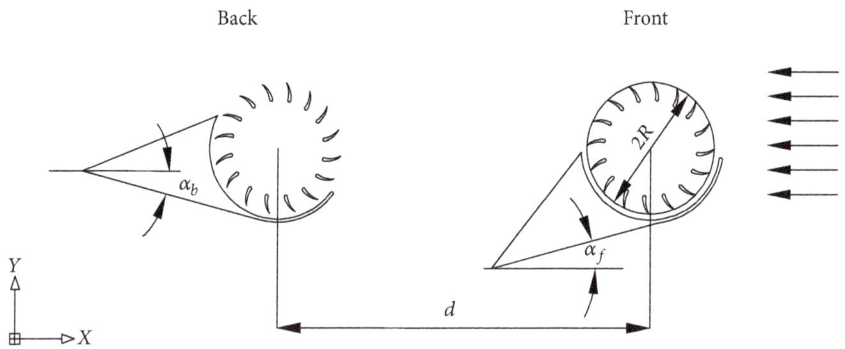

Figure 11: Definition of the angle between the front and rear wings of the tandem fan wing.

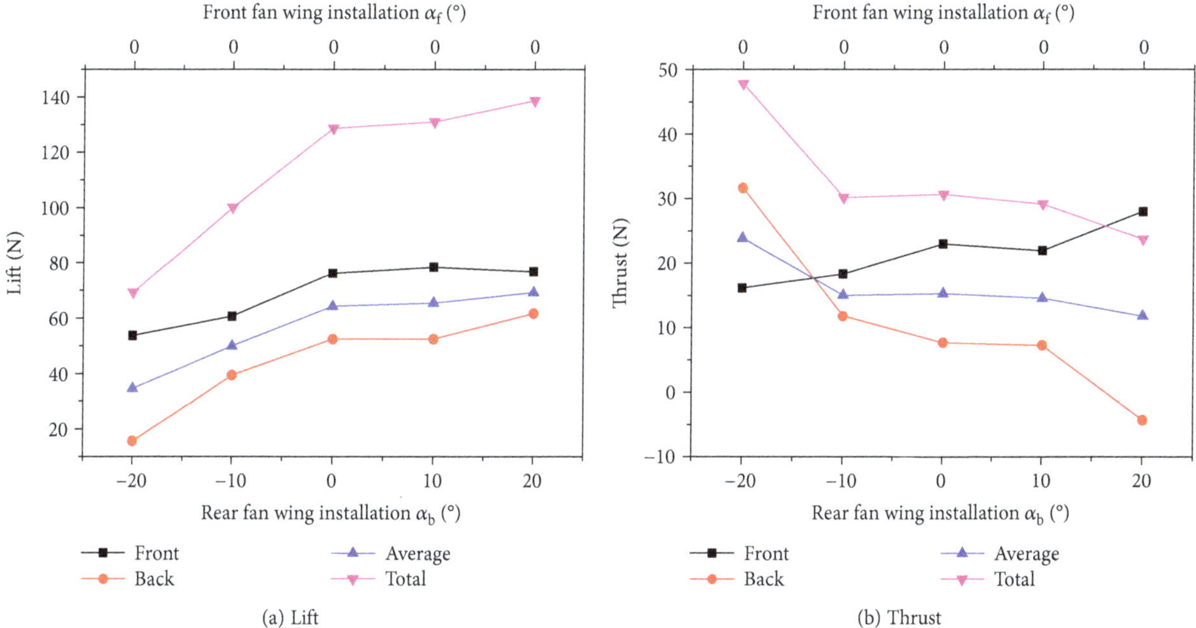

(a) Lift

(b) Thrust

Figure 12: Curve of the lift and thrust of the front and rear fan wings when the front wing is fixed.

3.3.1. The Installation Angle of the Front Fan Wing Is Unchanged. As we can see from Figure 12(a), as the installation angle of the rear wing increases, the lift of both the front and the rear wings keeps pace with it, while the lift of the rear wing has a relatively great effect on that of the front wing, and it is about 20% more than the front wing's (the installation angle of the rear wing is 20°). However, when the installation angle of the rear wing is greater than or equal to 0°, the added value of the rear wing's lift hardly changes. From the average value of the front and the rear wing's lift, we can see that it is increasing all the time, and the overall aerodynamic increase is good. As we can see from Figure 12(b), the rangeability of the average thrust is not very large, and the thrust of the front and the rear wings shows the opposite trend. As is shown in Figure 13, when the installation angle of the rear wing is negative, the airflow of the front wing's trailing edge slope flows in an S-shaped path when across the rear wing's lower surface. When the installation angle of the rear wing is positive, basically, it flows along the airflow of the trailing edge slope, which illustrates that the rear wing has a good rectification effect on the front wing under this condition. The result at the rear angle > 0° is that the lift does not change much as most of the high energy fluid leaving the front fan wing does not enter the rear fan wing.

3.3.2. The Installation Angle of the Rear Fan Wing Is Unchanged. As we can see from Figure 14, when the installation angle of the front wing is greater than 0°, the lift of both the front and the rear wings tends to be stable. When the installation angle of the front wing ranges from −20° to −10°, lift of the rear wing is less than that of the front wing. With analysis of Figure 15, we can find that the accelerated airflow from the front wing just flows into the rear wing, due to the suction effect of the rear wing on the front wing's airflow, which increases the flow rate on the trailing edge wing surface of the front wing and decreases the relative pressure; all these contribute to the increase of the front wing's lift. From the average value of the front and the rear wings' lift, we can find that the average lift hardly changes, and the changes of the front wing's installation angle are insignificant

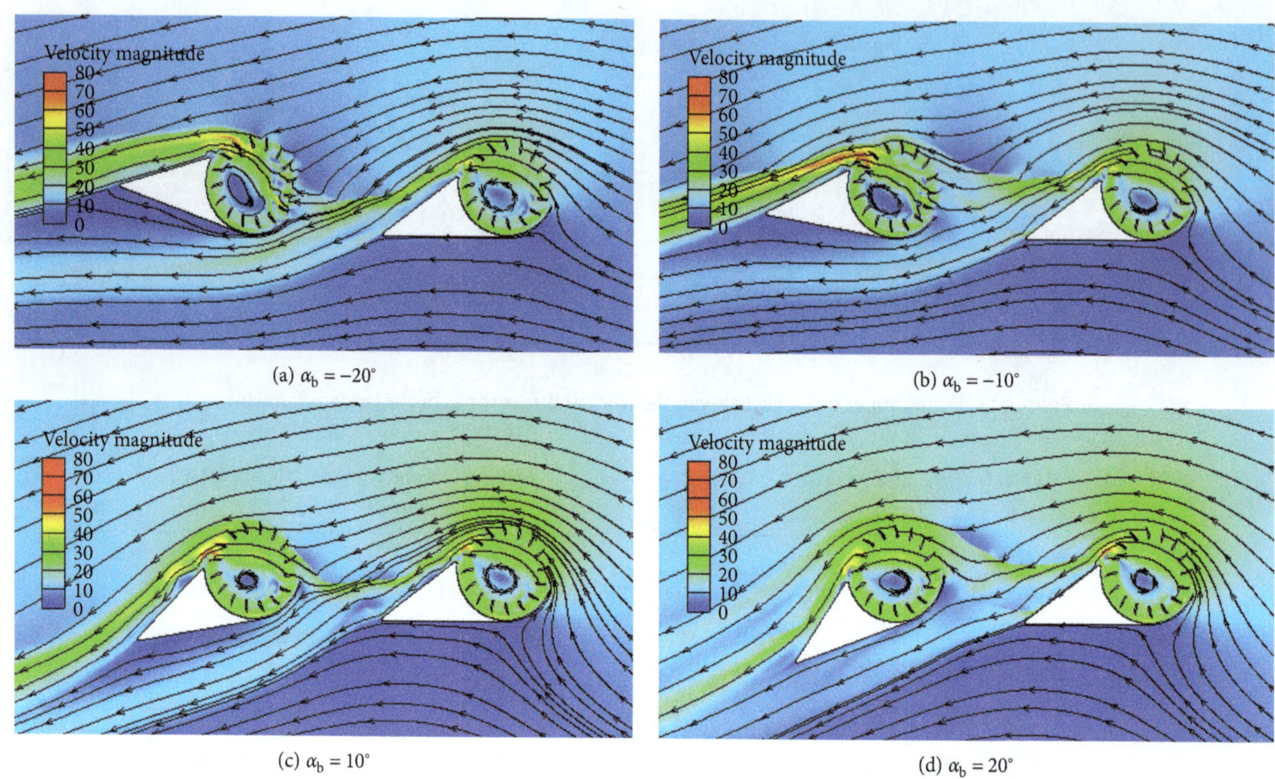

(a) $\alpha_b = -20°$

(b) $\alpha_b = -10°$

(c) $\alpha_b = 10°$

(d) $\alpha_b = 20°$

FIGURE 13: Velocity contours and streamlines of the lift and thrust of the front and rear fan wings when the front wing is fixed.

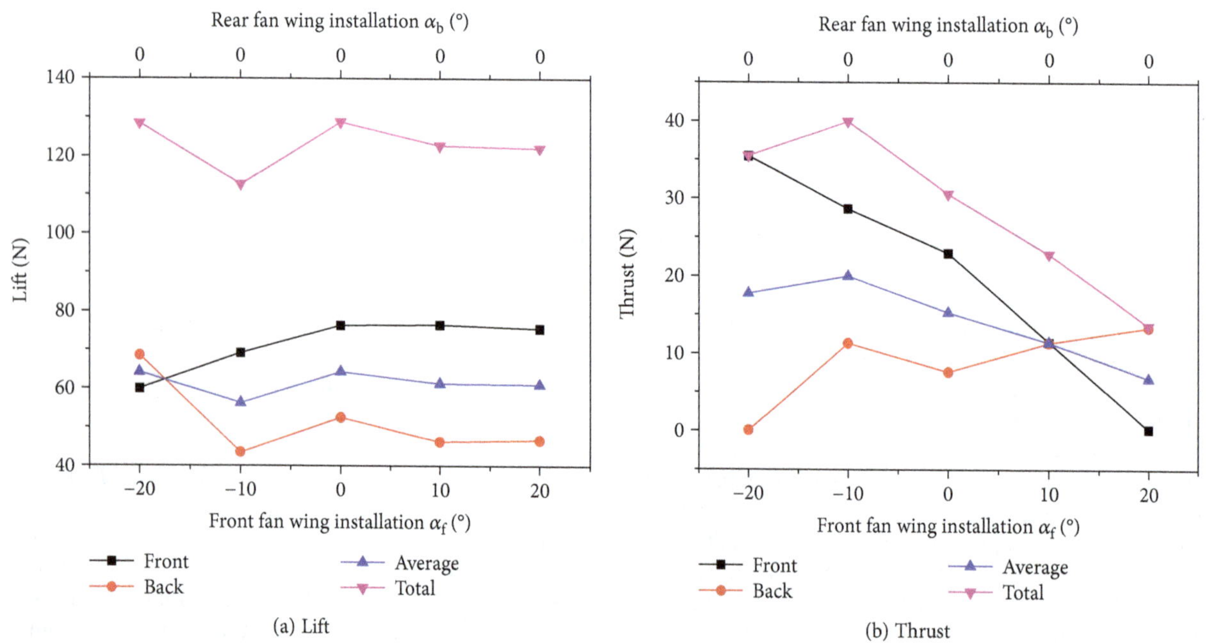

(a) Lift

(b) Thrust

FIGURE 14: Curve of the lift and thrust of the front and rear fan wings when the rear wing is fixed.

in improving the overall aerodynamic force increase. As we can see from Figure 14(b), the average value of the thrust decreases with the increase of the front wing's installation angle. From the current flowchart, we can find that the streamline of the front wing's airflow is disturbed by the rear wing's airflow, and the airflow flowing into the rear wing decreases. Therefore, the rear wing's thrust continues to decrease. The rear angle $< 0°$ value results in rather "odd" changes in lift and thrust behavior. This is due to the high energy fluid leaving the front fan wing entering the rear fan wing.

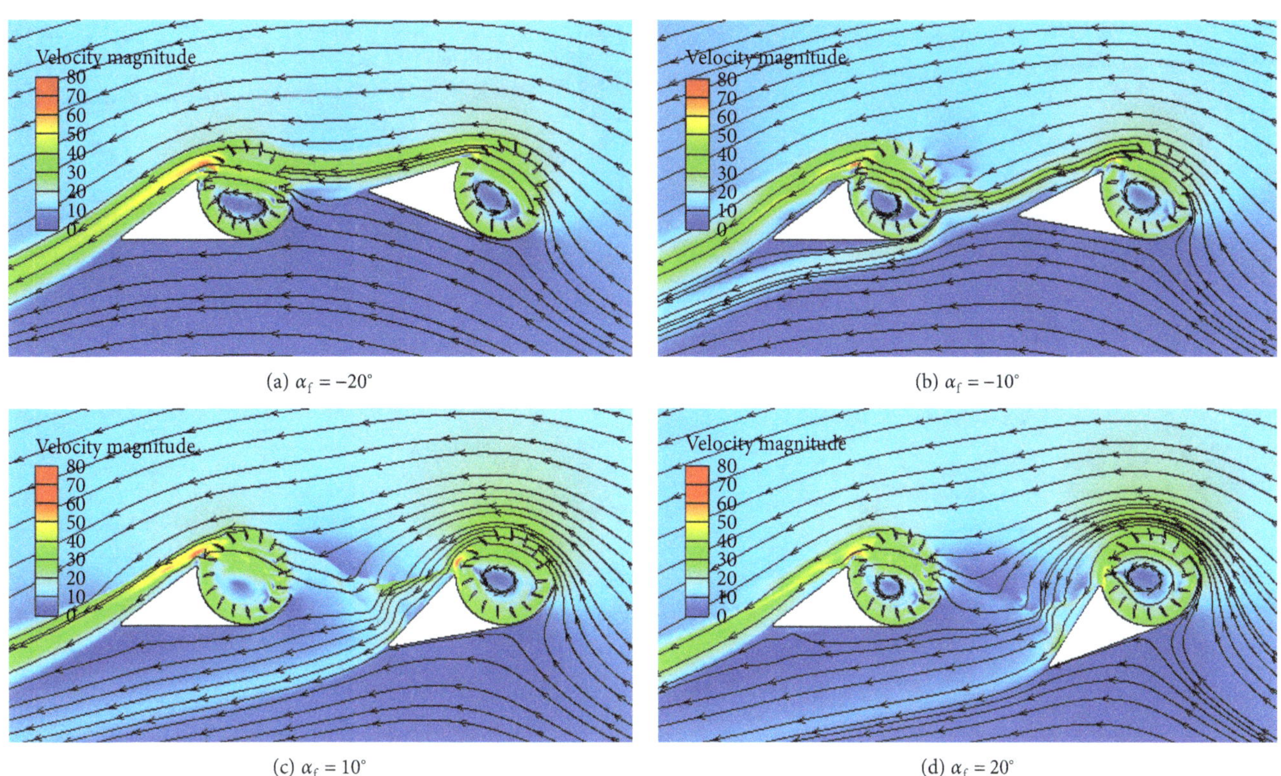

(a) $\alpha_f = -20°$ (b) $\alpha_f = -10°$ (c) $\alpha_f = 10°$ (d) $\alpha_f = 20°$

FIGURE 15: Velocity contours and streamlines of the lift and thrust of the front and rear fan wings when the rear wing is fixed.

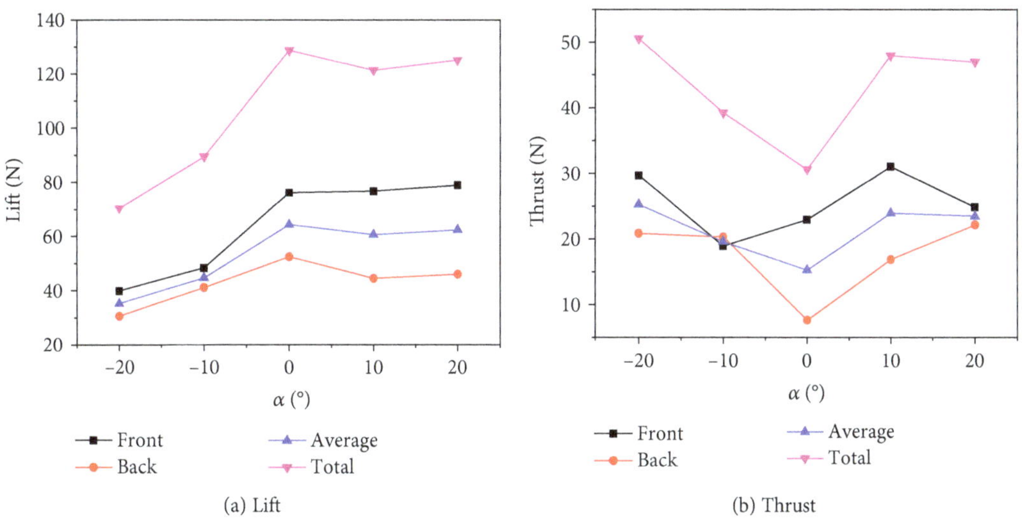

(a) Lift (b) Thrust

FIGURE 16: Curve of the lift and thrust of the front and rear fan wings when the front and rear wings have the same installation angle.

3.3.3. The Installation Angles of the Front and the Rear Fan Wings Are the Same. Under this situation, the installation angle of the front and the rear wings changes in the same way at the same time. As we can see from Figure 16(a), when the installation angles of the front and the rear wings are negative, the aerodynamic force increase effect keeps pace with the increase of the installation angles. When the installation angles of the front and the rear wings are positive, such an effect is not obvious. As is shown in Figure 16(b), the average thrust of both the front and rear wings hardly changes, which illustrates that the change of the installation angles of the front and the rear wings has little influence on the thrust. With analysis of Figure 17, flow separation does not occur when the installation angle of the front and the rear wings is large, which illustrates that the tandem fan wing possesses such aerodynamic characteristic, to keep the velocity streamline stable with the large installation angle. The phenomenon caused by the flow field is similar to the upper section.

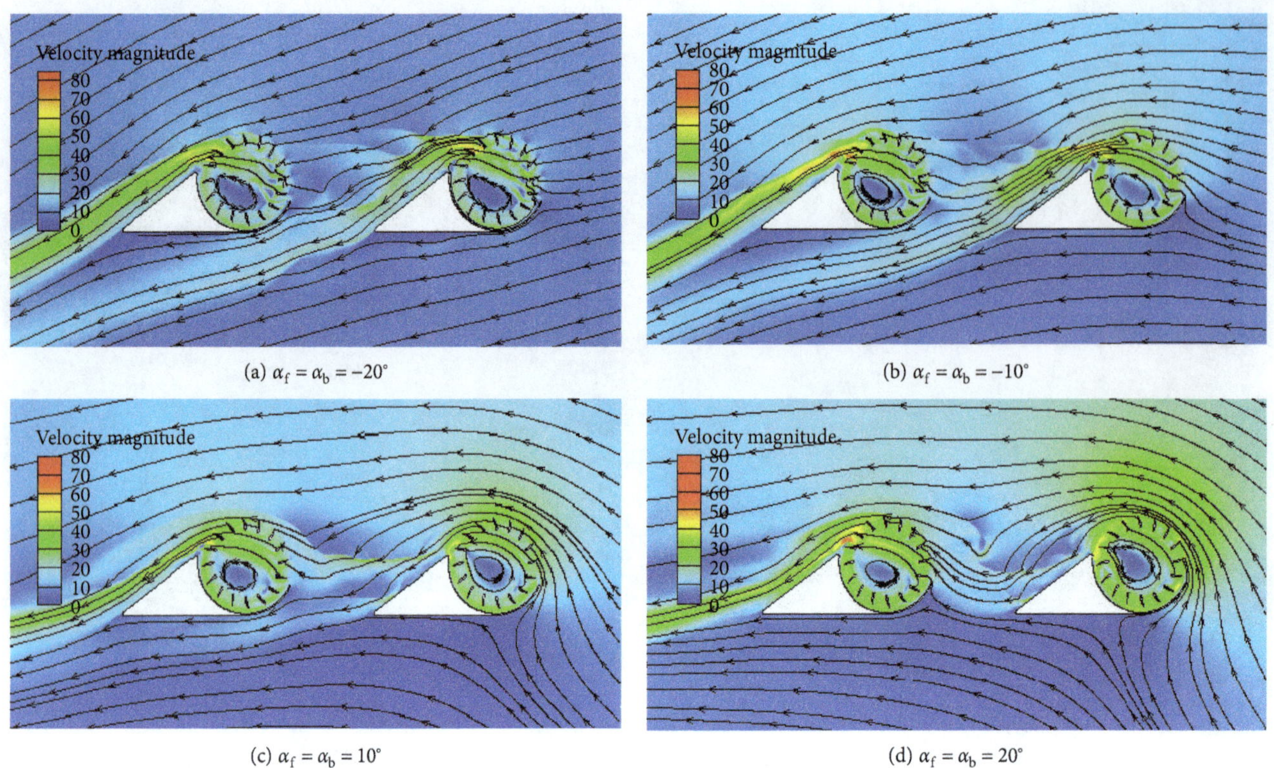

(a) $\alpha_f = \alpha_b = -20°$ (b) $\alpha_f = \alpha_b = -10°$

(c) $\alpha_f = \alpha_b = 10°$ (d) $\alpha_f = \alpha_b = 20°$

FIGURE 17: Velocity contours and streamlines of the front and rear fan wings when the front and rear wings have the same installation angle.

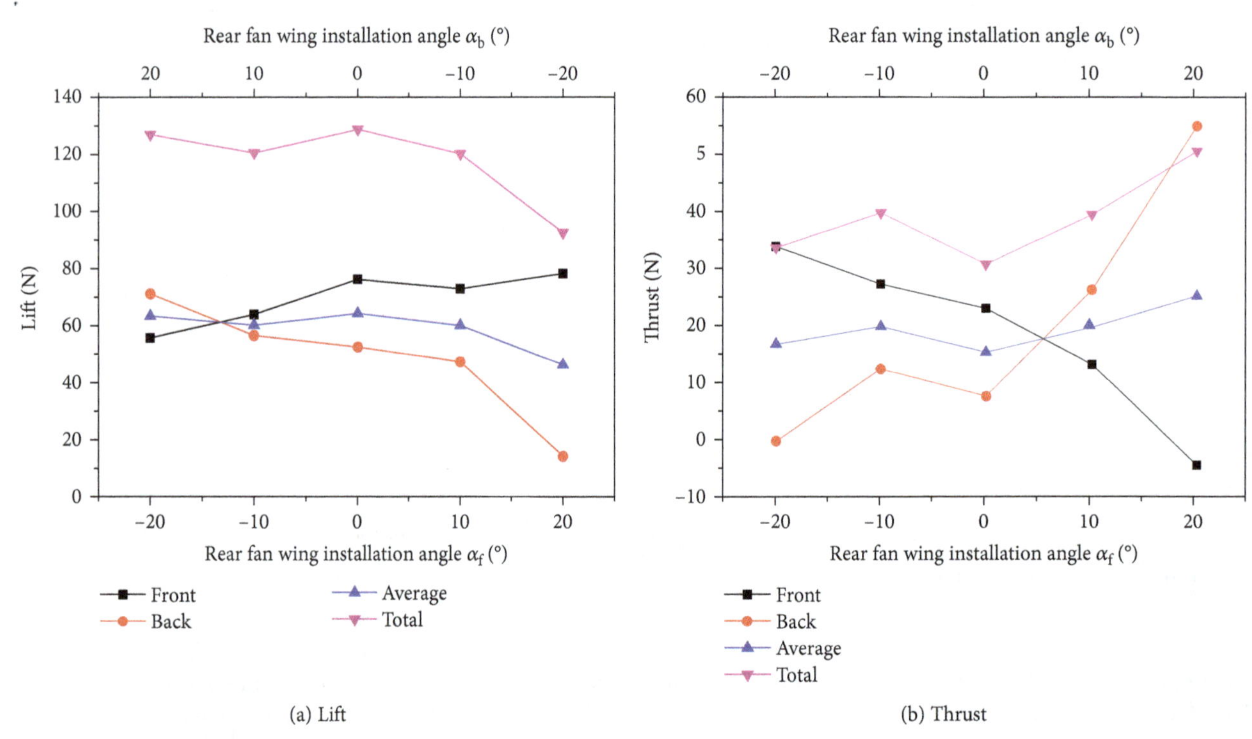

(a) Lift (b) Thrust

FIGURE 18: Curve of the lift and thrust of the front and rear fan wings when the front and rear wings have the opposite installation angle.

3.3.4. The Installation Angles of the Front and the Rear Fan Wings Are Different. The front and the rear wings change in a different direction at the same time with the front wing as the benchmark. Figure 18(a) shows that the lift of the rear wing is smaller than that of the front wing when the installation angle of the rear wing ranges from −20° to −10°. As we can see from the average value of the front and the rear wings' lift, the change of the average lift gradually decreases. As is

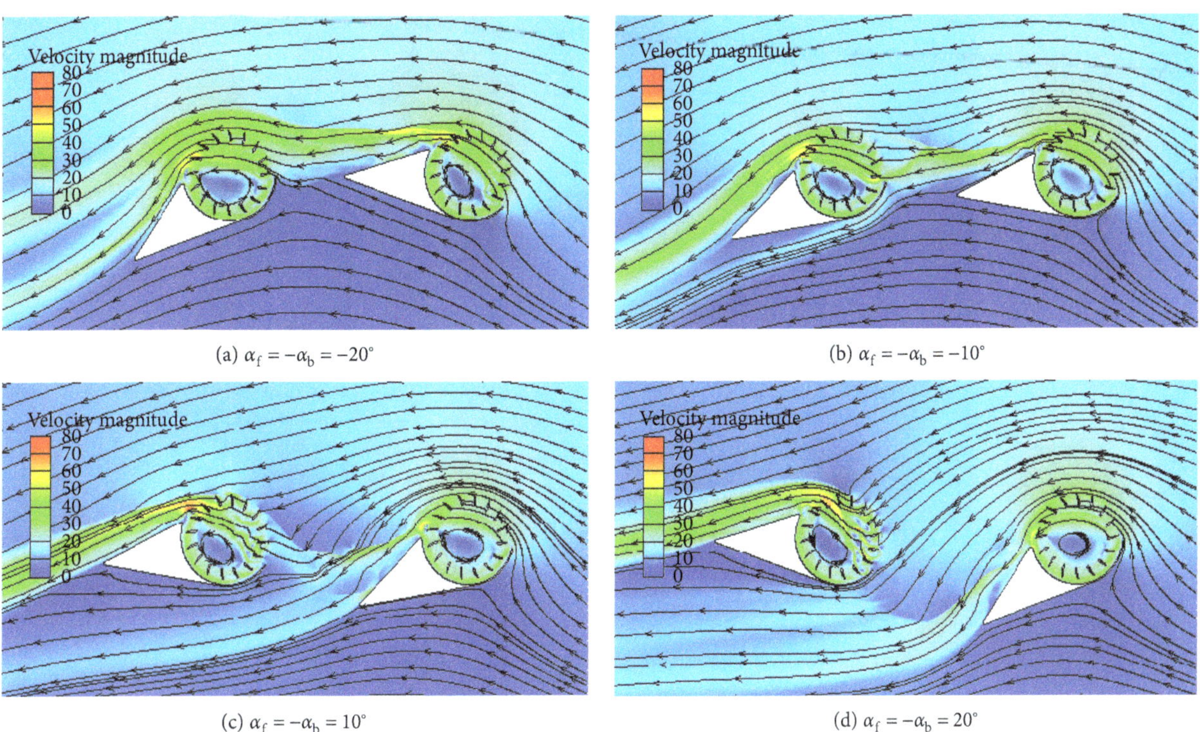

(a) $\alpha_f = -\alpha_b = -20^\circ$ (b) $\alpha_f = -\alpha_b = -10^\circ$

(c) $\alpha_f = -\alpha_b = 10^\circ$ (d) $\alpha_f = -\alpha_b = 20^\circ$

FIGURE 19: Velocity contours and streamlines of the front and rear fan wings when the front and rear wings have the opposite installation angle.

(a) Wind tunnel (b) Experimental model

(c) The position of the experimental model in the wind tunnel (d) The way of changing distance, height, and installation angle

FIGURE 20: Experimental model and equipment.

TABLE 3: Parameters of the wind tunnel.

Parameters	Value
Size of test area (m × m)	3.4 × 2.4
Maximum wind speed (m/s)	40
Minimum stable wind speed (m/s)	5
Shrinkage ratio	4

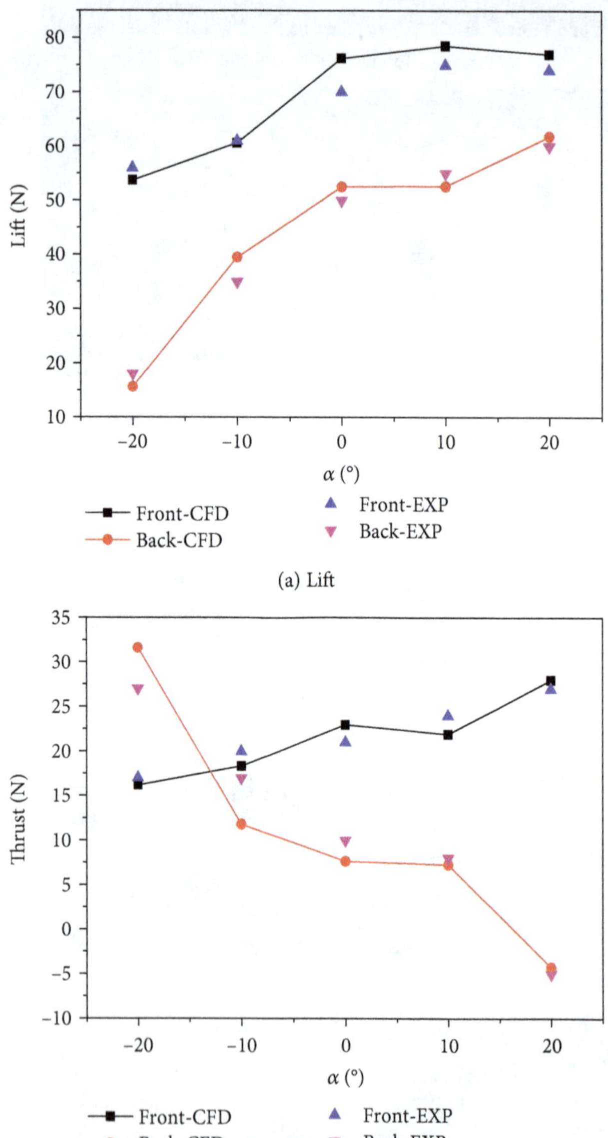

(a) Lift

(b) Thrust

FIGURE 21: Comparison curves between numerical results and experimental results when the front wing is fixed.

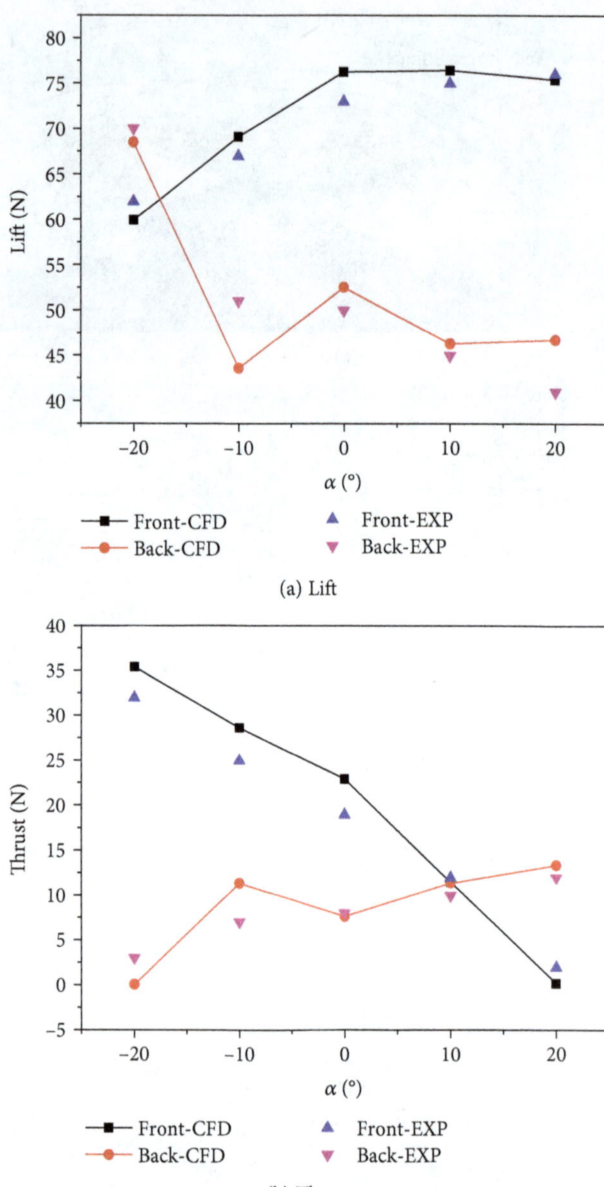

(a) Lift

(b) Thrust

FIGURE 22: Comparison curves between numerical results and experimental results when the rear wing is fixed.

shown in Figure 18(b), the thrust change still demonstrates the trend of substitution between the front and the rear wings. What is more, the average lift does not change too much. As we can see from Figure 19, when the installation angles of the front and the rear wings are the same and negative, the rear wing can get accelerated airflow from the front wing; at this time, the change of the front and the rear wings' lift and thrust is not obvious. When they are the same and positive, the rear wing cannot make use of the accelerated airflow from the front wing, which is indicated in the graph as the lift and thrust fluctuate greatly with the change of installation angle. The interference between the front and rear flow fields causes the abrupt flow field of the trailing edge of the front wing and the streamline bending.

4. Test Verification

4.1. Wind Tunnel and Model Arrangements. This experiment is based on the low-speed open return flow wind tunnel (Figure 20(a)), which is in the National Key Laboratory of Rotorcraft Aeromechanics in Nanjing University

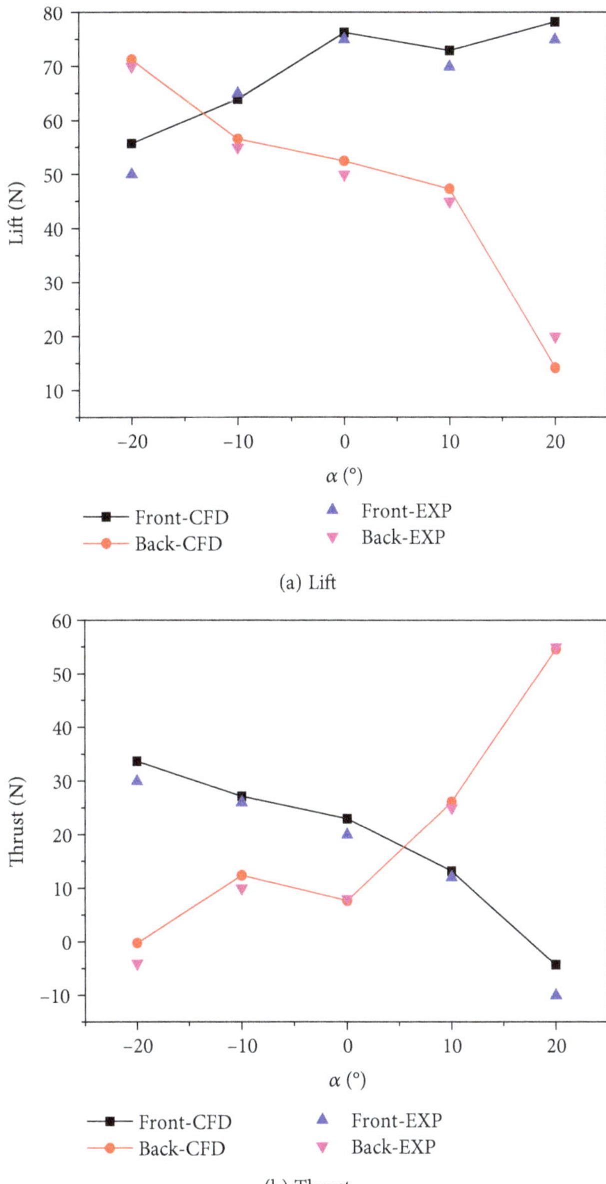

(a) Lift

(b) Thrust

Figure 23: Comparison curves between numerical results and experimental results when the front and rear wings have the same installation angle.

of Aeronautics and Astronautics. The basic parameters of the experimental wind tunnel are shown in Table 3. The fan wing model (Figure 20(b)) used in this experiment is the same size as the numerical calculation model. The model airfoil is made of glass fiber, and the crossflow fan is made of carbon fiber material. The experimental bench is mainly composed of assembled aluminum. The whole experimental bench is placed in the relative position of the wind tunnel as shown in Figure 20(c). The way to change the distance between the front and rear wings is to move the front and rear wings along the horizontal guide rail. Pairs of bolts through the positioning holes on the wing plate change the installation angle of the wing. The height of the wing is changed by replacing different height brackets. Figure 20(d) shows the location of the adjusting element.

4.2. Experimental Results and Comparison. The comparison curves of the test results and numerical calculation results on the lift and the thrust of the front and the rear wings with different installation angles and under three conditions, namely, when the installation angle of the front wing is unchanged, when the installation angle of the rear wing is unchanged, and when the installation angles of the front and the rear wings are kept the same, are shown, respectively, in Figures 21–23. Although there exists some difference, the whole trend of change is the same. The relative error between most of the numerical calculation results and test results is less than 10%, which illustrates that the numerical calculation method is feasible in this paper, and it can be applied in more analyses of aerodynamic characteristics of the tandem fan wing.

5. Conclusion

A tandem fan wing configuration is studied by the CFD method. The numerical setup is validated by experiments performed on the model with a single wing. The aerodynamic performance of this configuration is studied for different relative positions of the wings. It is shown that the lift of the model increases up to a certain distance between the two wings in the free stream direction and then remains steady, but its thrust improves only when this distance is kept low. Also, the impact of the distance between the wings normal to the free stream direction on the aerodynamic performance of the configuration is studied, and it is shown that the best performance results when this distance is about zero. Moreover, the performance of the configuration is studied for different installation angles. It is shown that changing the installation angle does not improve the lift but it has some minor effects on the thrust.

Conflicts of Interest

The authors declare that there is no conflict of interest regarding the publication of this paper.

Acknowledgments

The authors would like to acknowledge the financial support of the Priority Academic Program Development of Jiangsu Higher Education Institutions and useful information obtained from the FanWing company. The received funding did not lead to any conflict of interests.

References

[1] L. Meng, Y. Ye, and N. Li, “Research progress and application prospects of fan-wing aircraft,” *Acta Aeronautica et Astronautica Sinica*, vol. 36, no. 8, pp. 2651–2661, 2015.

[2] G. R. Seyfang, "Fanwing-developments and applications," in *28th Congress of International Council of the Aeronautical Sciences, ICAS*, pp. 1-9, Brisbane, Australia, 2012.

[3] G. R. Seyfang, "Recent developments of the FanWing aircraft," in *The International Conference of the European Aerospace Societies, CEAS*, pp. 1-7, Venice, Italy, 2011.

[4] O. Ahad and J. M. R. Graham, "Flight simulation and testing of the fanwing experimental aircraft," *Aircraft Engineering and Aerospace Technology*, vol. 79, no. 2, pp. 131–136, 2007.

[5] N. Laing, "Fluid flow machine with parallel rotors," 1962, US Patent 3150821A.

[6] J. P. Hancock, "Test of a high efficiency transverse fan," in *16th Joint Propulsion Conference*, Hartford, CT, USA, 1980.

[7] G. J. Harloff, *Cross-Flow Fan Experimental Development and Finite-Element Modeling, Ph. D. Dissertation*, University of Texas at Arlington, Arlington, TX, USA, 1979.

[8] J. D. Kummer and T. Q. Dang, "High-lift propulsive airfoil with integrated crossflow fan," *Journal of Aircraft*, vol. 43, no. 4, pp. 1059–1068, 2006.

[9] T. Q. Dang and P. R. Bushnell, "Aerodynamics of cross-flow fans and their application to aircraft propulsion and flow control," *Progress in Aerospace Sciences*, vol. 45, no. 1-3, pp. 1–29, 2009.

[10] S. Askari and M. H. Shojaeefard, "Numerical simulation of flow over an airfoil with a cross flow fan as a lift generating member in a new aircraft model," *Aircraft Engineering and Aerospace Technology*, vol. 81, no. 1, pp. 59–64, 2009.

[11] S. Askari, M. H. Shojaeefard, and K. Goudarzi, "Experimental study of stall in an airfoil with forced airflow provided by an integrated cross-flow fan," *Journal of Aerospace Engineering*, vol. 225, no. 1, pp. 97–104, 2010.

[12] S. Askari and M. H. Shojaeefard, "Experimental and numerical study of an airfoil in combination with a cross flow fan," *Journal of Aerospace Engineering*, vol. 227, no. 7, pp. 1173–1187, 2012.

[13] S. Askari and M. H. Shojaeefard, "Shape optimization of the airfoil comprising a cross flow fan," *Aircraft Engineering and Aerospace Technology*, vol. 81, no. 5, pp. 407–415, 2009.

[14] N. Thouault, C. Breitsamter, and N. A. Adams, "Numerical and experimental analysis of a generic fan-in-wing configuration," *Journal of Aircraft*, vol. 46, no. 2, pp. 656–666, 2009.

[15] Z. F. Tang, M. Tang, and H. D. Wu, "The analysis of the influence of rotor anti-torque system's structure parameters on its aerodynamic influence," in *Proceeding of the 2nd Asian/Australian Rotorcraft Forum and the 4th International Basic Research Conference on Rotorcraft Technology*, pp. 1–9, Tianjin, China, 2013.

[16] Y. Lian, T. Broering, K. Hord, and R. Prater, "The characterization of tandem and corrugated wings," *Progress in Aerospace Sciences*, vol. 65, pp. 41–69, 2014.

[17] Z. Tong and M. Sun, "Flow analysis of twin-rotor configurations by Navier-Stokes simulation," *Journal of the American Helicopter Society*, vol. 45, no. 2, pp. 97–105, 2000.

[18] D. Siliang, T. Zhengfei, X. Pei, and J. Mengjiang, "Study on helicopter antitorque device based on cross-flow fan technology," *International Journal of Aerospace Engineering*, vol. 2016, Article ID 5396876, 12 pages, 2016.

[19] R. P. Tang, *Aerodynamic Experimental Research on Fan-Wing, M.S. Thesis*, Aeronautics and Astronautics Dept., Nanjing University of Aeronautics and Astronautics, 2014.

2

Research on the Fatigue Performance of TC6 Compressor Blade under the CCF Effect

Zhang Yakui[1] **and Guo Shuxiang**[2]

[1]*Aeronautics and Astronautics Engineering College, Air Force Engineering University, Xi'an 710038, China*
[2]*Department of Mathematics and Physics College, Air Force Engineering University, Xi'an 710038, China*

Correspondence should be addressed to Zhang Yakui; 379610749@qq.com

Academic Editor: Giovanni Delibra

This paper studied the influence of high and low combined fatigue (CCF) on compressor blade fatigue performance. We investigated the coupling between low cycle fatigue (LCF) loading from centrifugal force with high cycle fatigue (HCF) loading from vibration and determined the blade disc vibration frequency using static analysis at maximum rotational speed. We designed and constructed a combined fatigue test rig, and CCF tests were performed on a TC6 compressor blade to analyze fatigue life characteristics. Results showed that CCF could significantly shorten blade life compared with pure LCF and that larger HCF caused more significant fatigue life reduction. Fatigue source characteristics and CCF fracture appearance were observed and analyzed using a scanning electron microscope (SEM).

1. Introduction

Aircraft engine blade fatigue failure is a typical failure mode that drastically affects engine reliability and flight performance [1–7]. During operation, aircraft engines always endure low-amplitude, high-frequency vibration combined with high-level, low-frequency stress causing a combined fatigue effect. Modern compressor blades are designed to withstand stresses generated at resonance, but it is important to ensure blade integrity and life not only under HCF but also when superimposed with other damage mechanisms, such as LCF, which is inevitable in the compressor engine environment [8]. Most previous fatigue behavior studies of structural materials considered either LCF or HCF loading alone, although engineering components experience varying load history [9]. Pure LCF and HCF fatigue life predictions have large deviations compared to observation and did not reflect actual operational damage processes.

Several CCF studies have been reported in the last few decades. Fuchs et al. [10] were among the first to study CCF fatigue problems. The US Air Force subsequently acknowledged the CCF importance and designed experiments to explore interactions between LCF and HCF, showing that HCF fatigue limit increased under CCF [11]. Moshier et al. [12] also obtained similar conclusions and showed that HCF fatigue limit was higher under CCF than pure HCF conditions. Various foreign studies have also shown that LCF life is significantly shortened under CCF effect. Dungey and Bowen [13] confirmed this conclusion and tested titanium alloys under CCF and showed that the HCF life negligibly effected when the cycles exceeded 10^5. Wang et al. [14] conducted the CCF test to study the extrapolation of turbine disc. Schweizer et al. [15] studied the mechanism and modeled fatigue crack growth under CCF. Both studies found that centrifugal and thermal stresses caused LCF turbine disc failures, whereas aerodynamically induced vibration stresses caused HCF failures. Powell et al. [16] systematically studied the influence of the number of HCF cycles per LCF cycle on fatigue crack growth. Gelmedin and Lang [17] studied IN 713C fatigue behaviors under LCF, HCF, and CCF conditions and found that, compared with the pure LCF loading, the superimposed HCF could dramatically shorten blade lifetime if its amplitude was high enough. Many studies have considered the same problems of the combined fatigue effects, researching that

Table 1: Material properties of the studied material.

Property 20°C	σ_b (MPa)	$\sigma_{0.2}$ (MPa)	ψ	E (GPa)	Poisson ratio	Density ρ/kg/m^3
	980	1005	0.25	125	0.3	4450

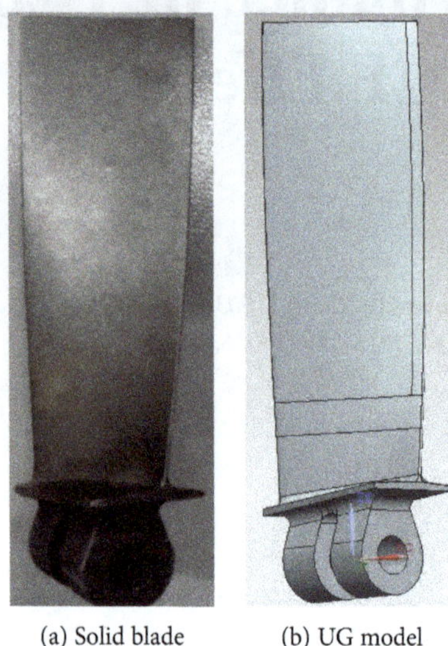

(a) Solid blade (b) UG model

Figure 1: Blade model.

combined fatigue life is more important and complex, and most studies considered turbine blades with few considering compressor blades [18–22].

In this paper, we first established a finite element model for the compressor blade and experimentally verified model accuracy by measuring the natural frequency of the root-retaining blade using impulse the hammering method. Static and vibration analyses of the blade disc system were then performed to obtain CCF test conditions and identify LCF load and HCF vibration frequency that produced maximum stress in the blade hazardous location. We formulated a CCF load spectrum based on those studies and realized a combined fatigue test rig. Fracture characteristics under CCF were analyzed to verify test design rationality. Last, the blades were also tested under pure LCF as well as CCF showing that the superimposition of vibration stress can lead to a significant reduction of LCF life.

2. Experimental Design

2.1. Finite Element Analysis

2.1.1. TC6 Characteristics. The most common blade material for high-pressure compressors employed in engines is TC6, strengthened martensitic thermally stable stainless steel with excellent overall mechanical properties, high thermal strength, and good stress corrosion resistance, hence very suitable for aeroengine blades. Table 1 shows the material properties [23, 24].

2.1.2. Blade Model Building. Figure 1 shows the three-dimensional blade model built in Unigraphics NX software following its design drawings, and Table 2 shows the geometric parameters.

Table 2: Blade geometric description.

Blade	Geometric size (mm)
High	184.5
Wide	44.5
Inner diameter	11.5
Outer diameter	24.3
Earring thickness	12.5
Inner surface spacing	6.7

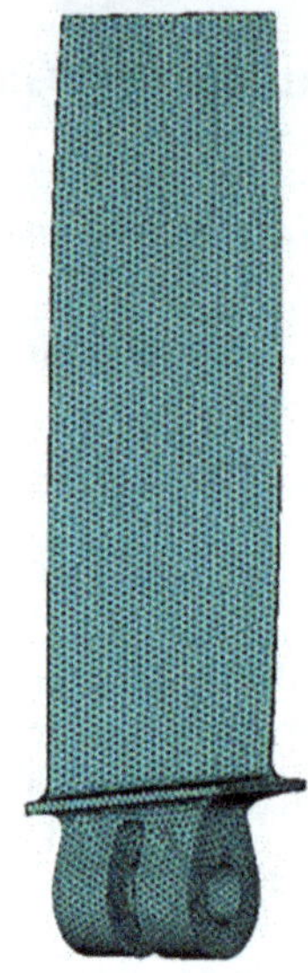

Figure 2: Blade mesh model.

Table 3: Blade model simulation results.

Order	Natural frequency (Hz)	Vibration type
The first order	251.68	First bend
The second order	873.58	First scroll
The third order	1400.8	Second bend
The fourth order	1967.9	Second bend torsion

We meshed the blade model using Solid187 10-node tetrahedron solid units (Solid187 is a higher order 3D 10-node solid structural element). The mesh size was set to 1 mm, as shown in Figure 2.

After establishing the blade finite element analysis (FEA) model, we calculated the blade model and defined boundary conditions such that all nodes on the inner ring of the two pinholes were fixed and constrained. We used Block Lanczos, an eigenvalue solver based on the Lanczos algorithm that is widely used in structural mode calculation, to simulate the first-four natural vibration frequency orders. The results were observed using the general post-processor POST1, as shown in Table 3.

The second-order natural blade vibration frequency (873.58 Hz) was much larger than the first-order frequency (251.68 Hz). The considered engine had 11,320 rpm design

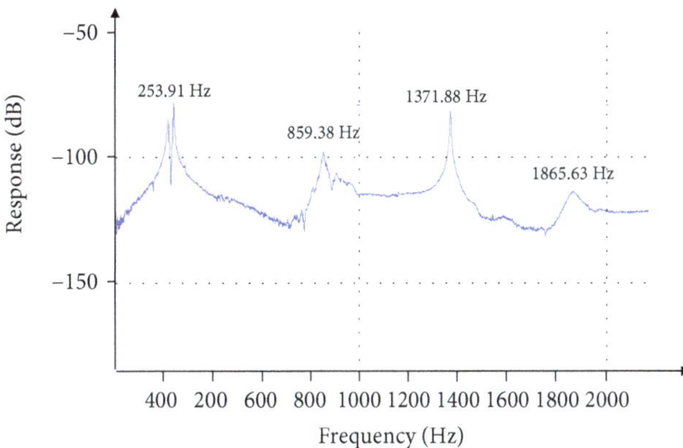

Figure 3: Frequency response of a1.

Table 4: Natural vibration frequencies of the blades.

Blade number	Natural frequency f/Hz The first order	The second order	The third order	The fourth order
a1	253.91	859.38	1371.88	1865.63
a2	243.81	840.62	1312.51	1824.32
a3	247.72	861.69	1347.71	1851.24
a4	247.14	853.13	1342.24	1849.08
a5	249.21	852.27	1353.89	1847.85
a6	253.88	868.83	1400.77	1868.21

speeds, hence periodic excitation generated by the engine rotor on the blade during operation would not exceed 251.68 Hz. Thus, first-order bending resonance had the highest occurrence probability and would be the most harmful to the blade. Therefore, this study only considered fatigue damage caused by first-order blade bending resonance.

2.1.3. Blade Model Experimental Verification. We measured the natural frequency of the root-retaining blade using the impulse hammering method and obtained the first-four vibration order shapes using laser scanning.

Six blade specimens were randomly selected, coded as a1–a6. Frequency response for a1 was obtained by holistically analyzing five pulse simulations following the experimental parameters and settings, as shown in Figure 3, where response peaks noted blade natural frequencies. Moreover, frequency response curves of a2–a6 are basically the same as that of a1.

According to the frequency response curve of each blade, the first-four order natural frequencies were collected, as shown in Table 4.

Table 5 compares simulation and experimental results and mean error probability for the natural frequencies for all blades = 2.8% on average, with maximum < 6%.

Having defined the natural frequencies, multiple measurement points were fixed on the leaf body and the exciter frequency adjusted to the natural rates to realize resonance at each stage of the blade. We used a Polytec laser vibrometer to detect displacement response and identify the vibration

Table 5: Comparative analysis of natural frequencies.

Order	The first order	The second order	The third order	The fourth order
Simulation value (Hz)	251.68	873.58	1400.8	1967.9
Experimental average value (Hz)	249.28	855.99	1354.83	1851.06
Relative error	0.95%	2.01%	3.28%	5.94%

Table 6: Blade disc geometric description.

Blade disc	Geometric size
Inner diameter (mm)	220
Outer diameter (mm)	354
Thickness (mm)	30

type. Experimental and model results were consistent, verifying that the blade modeling was accurate and reliable.

2.2. Combined Fatigue Test. The blade disc system, i.e., the compressor rotor system, connects the blades to the disc through pins, with the blades distributed around the edge of the disc. The considered blade disc system included 29 blades with other geometry as shown in Table 6.

During operation, the blade disc system is affected by unbalanced excitation causing the blade to vibrate. Static and vibration analyses of the blade disc system were performed to determine an appropriate test load spectrum, where static analyses locate hazardous points of the blade and vibrational analyses identify dangerous resonance of the system and ideal HCF test frequency.

2.2.1. Blade Disc System Static Analysis. In the case of cumulative fatigue damage, blade cracks usually occur in the region with largest equivalent stress under LCF conditions, which are considered the potential hazardous positions of the blade. Stress distribution patterns will be affected by engine design speed (3500 and 11,320 rpm minimum and maximum, respectively) and velocity amplitude. Boundary conditions were defined around the center axis of the disc, fixing and constraining displacements of all nodes on the inner ring, i.e., displacement = 0. We applied surface-surface contact laws between the disc and the blade, where the disc was the target and the blade the contact body. Rotational speeds were set at 3500, 4500, 5500, 6500, 7500, 8500, 9500, 10,500, and 11,320 rpm. Maximum stress position and stress distribution were derived using nonlinear static analysis, and the effect of increasing speed loading on stress distribution was investigated.

The results showed similar stress distribution under the different centrifugal speed loads. The hazardous position with maximum stress was preliminarily located in the central plate lug joints and earring inner, as shown in Figure 4. Maximum principal stress = 592 MPa when speed was maximum (11,320 rpm).

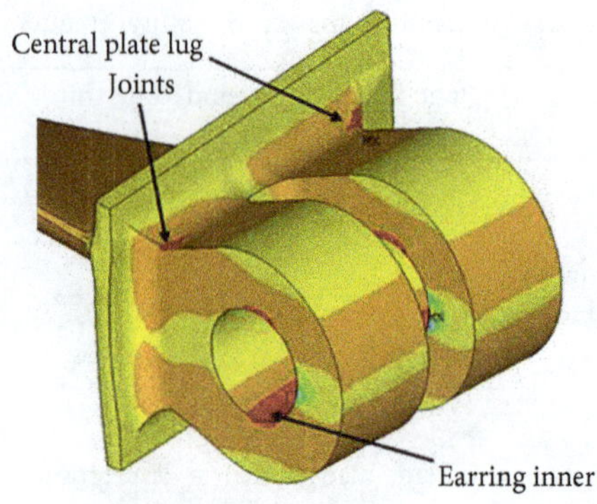

FIGURE 4: Local maximum stress of the blade.

2.2.2. Blade Disc System Vibration Analysis. The blade disc system undergoes centrifugation as it rotates, which increases blade bending stiffness and changes resonance frequencies from those in the stationary state. Therefore, vibrational frequencies change with rotational speed. The dynamic blade curve can be obtained by fitting the relationship between rotational speed and vibration frequencies.

The dangerous blade disc resonance speed was identified as 8425.51 rpm, and the first order bending vibration state (567.29 Hz) from static analysis is shown in Figure 5.

Single-blade vibration characteristics are meaningless and should not be used as the basis for HCF vibration conditions because practical situations almost invariably incorporate multiple coupled blade disc systems. Therefore, we considered coupled vibration characteristics of the system and its resonance frequency to derive the HCF conditions of the blade. If the first-order bending resonance of the system is likely to occur, we can determine the first-order f requency = 251.68 Hz of the blade equals to the high cycle vibration frequency, that is, the condition of the resonance failure frequency of the HCF in the CCF test.

2.2.3. Load Spectrum Derivation. We considered the relationship between stress (S) and total fatigue life (N) to investigate the load spectrum, covering pure LCF, pure HCF, and CCF conditions. A CCF test load spectrum was designed based on the typical service loading for an aircraft engine, as shown in Figure 6. The major low-frequency cycle had a trapezoidal stress wave with a 14.59 s period and included two short blocks and one long block, corresponding to climbing, cruising, and landing, i.e., typical aircraft flight. High-frequency vibration overlaid the low frequency loading, simulating resonance.

Vibration analysis of the blade disc system showed it was prone to first-order resonance and the sinusoidal excitation signal frequency $f_H = 251.68$ Hz. Similarly, the determined loading frequency ratio was consistent with the design requirements of low $T_L = 8250$ cycles and high $T_H = 3 \times 10^7$ blade cycle life. By calculation, the test frequency ratio is $R_f = 3636$ and low cycle loading frequency is $f_L = 0.067$ Hz.

Because maximum speed, $n_{max} = 11{,}320$ rpm, was constant, LCF load, σ_L, was also constant, as shown in Table 7. Compared with the resonant failure of the blade, forced vibration induced by aerodynamic force is another cause of blade failure. The aerodynamic force mainly manifests as flow pressure on the exhaust side of the blade, defined as approximately 10% of the maximum centrifugal force [25]. However, the aerodynamic effect should not be considered when CCF of the blade is studied independently. Therefore, we modified the low cycle load to 28,255 N, which was also the LCF test load.

2.3. Combined Fatigue Test Rig

2.3.1. Test Rig Design. Traditional HCF test rigs do not accurately represent CCF effects generated in compressor blades. Traditional HCF tests also only considered uniaxial machines, applying both steady and alternating loads along the major specimen axis. The purpose of the CCF test system design core is to optimally simulate actual working conditions using the test rig, and efficient reproduction of the specific blade damage location and failure mechanisms. To simulate LCF and HCF, the designed test rig was divided into LCF and HCF loading devices, as shown in Figure 7.

This design could also be used for pure LCF test. Figure 8 shows the three-dimensional test rig.

The actual test rig is shown in Figure 9. Selecting an appropriate monitoring point and accurately controlling the stress is critical. Because the blade hazardous position is closely connected to the device design, test monitoring points are usually chosen at the blade tip. However, due to severe vibration at the blade tip, the measurement location time domain displacement curve changes exceptionally irregularly, making it difficult to measure amplitude accurately. Therefore, we must select a different monitoring point. Vibration stress was transmitted through the exciting rod to the exciting point and only acts indirectly on the blade hazardous position. The amplitude at the exciting point is difficult to measure. Hence, we measured the amplitude at the hazardous position using the laser vibrometer and assumed these amplitudes were equal. So we finally set the monitoring point at the center of the internal fixture surface, as shown in Figure 10, because the time domain displacement curve for this location was sinusoidal with relatively small amplitude.

2.3.2. The Function of the Test Rig Realization. Hydraulic loading equipment can simulate centrifugal forces and balance the stress on the assessed blade part to the static stress for the selected engine rotating speed, producing tension on the steel cable along the radial direction of the blade. Figure 9 shows the loading force was controlled by an electro-hydraulic servo valve to meet the LCF requirements. Low cycle load was also monitored by the sensor.

The HCF loading device mainly consisted of the shaker, signal generator, and exciting rod. Under vibration, the excitation signal passes through the exciting rod to the blade, forcing the corresponding position to vibrate in the vertical direction of the blade. Thus, the vibration amplitude of the blade hazardous position varies with the HCF amplitude required by the test, which realizes CCF loading.

2.3.3. Vibration Stress and Amplitude Relationship. Transient thermal responses under seven different amplitudes were

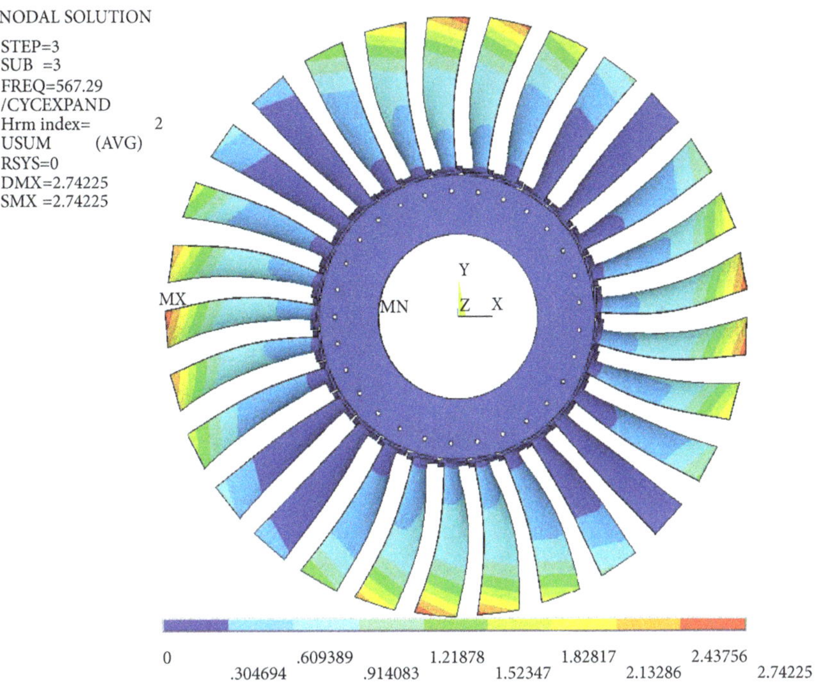

FIGURE 5: Static analysis of the blade disc system.

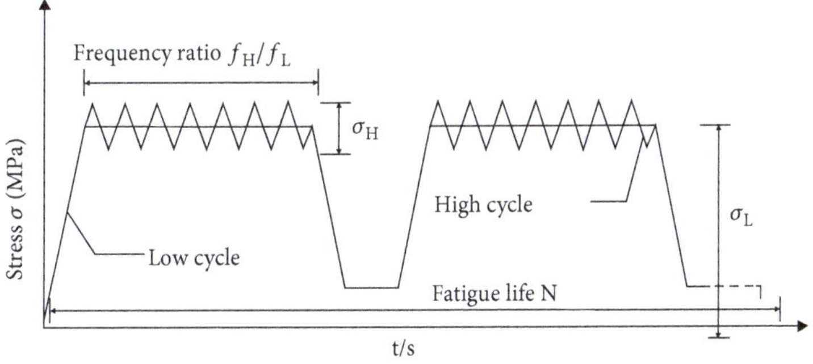

FIGURE 6: Load spectrum of the combined fatigue test.

TABLE 7: LCF load results.

Maximum rotating speed (rpm)	LCF load simulation value (N)	LCF load modified value (N)
11,320	25,686	28,255

calculated, and the average value was obtained. The statistical results are shown in Table 8.

The monitoring point amplitudes were continuously controlled by the shaker at 40, 50, 60, 70, 80, 90, and 100 μm, with corresponding measured amplitudes 47.49, 59.58, 71.66, 71.66, 83.74, 95.82, 107.91, and 119.99 μm, respectively, as shown in Table 8. Vibration amplitude and stress were fitted using the least squares method, as shown in Figure 11.

Vibration stress and amplitude are both linear. Therefore, the exponential can be expressed as

$$\sigma = 1.045 \times A + 70.04, \tag{1}$$

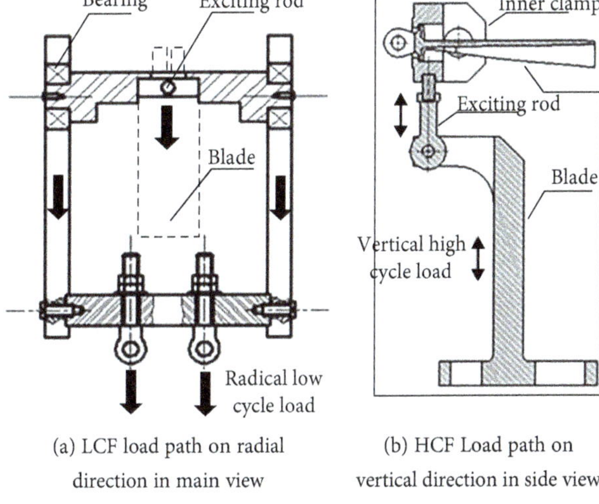

(a) LCF load path on radial direction in main view

(b) HCF Load path on vertical direction in side view

FIGURE 7: Load path of the CCF.

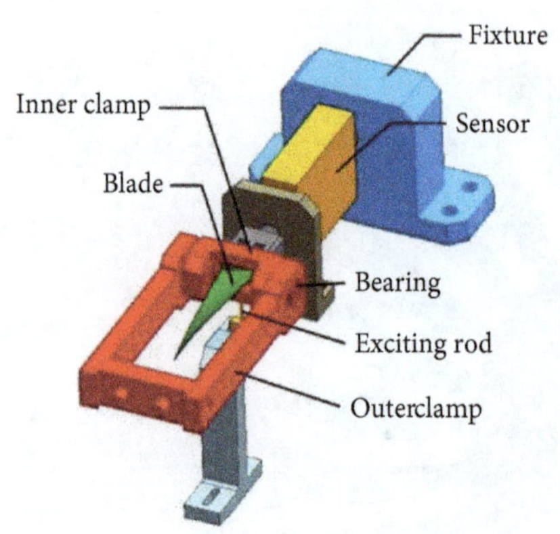

FIGURE 8: Combined fatigue test rig.

where σ is the vibration stress and A is the monitoring point amplitude. Therefore, vibration stress that cannot be measured directly needs to be converted to amplitude, because the vibration stress amplitude of the blade hazardous location at any amplitude can be determined.

According to the numerical results, we have explored the CCF test conditions, formulated the load spectrum, and realized the CCF test rig. Before studying the effects of different vibration stresses on LCF life under the low cycle load holding constant, we must first determine the correctness of the numerical results and the rationality of the combined test design based on fracture investigations of the test blade and study the relationship between vibration stress and amplitude. Taking into account these analyses, in the following studies, we conduct fracture tests and explore the effects of CCF on blade fatigue life.

3. Results and Discussion

3.1. Fractographic Investigations. Fracture characteristics changed under combined cycling with strong interaction among LCF, HCF, and crack initiation. Figures 12–17 show the major manifestations, which validate the CCF test system design.

A visual crack appeared at the root of the blade, corresponding to the static results of the hazardous location. Processed fracture patterns were applied to the specimens and scanned using SEM.

3.1.1. Fracture Behavior under CCF. Figures 13 and 14 show fracture modes indicating the crack initiation and propagation mechanism under CCF for different loading conditions. Figure 13 shows that region 1 is the crack source, region 2 is the high loading fracture zone, region 3 is the low loading fracture zone, and region 4 is the fracture morphology formed during the process. The fracture surface of the high cycle zone is smooth, while in the lower region it is rough.

Figure 14 shows grain boundary facets with cracks originating at grain boundary junctions alongside transgranular fatigue cracks, indicating build-up of time-dependent creep damage. The crack surface has an uneven oxide film, probably due to the prolonged exposure of the fracture in the air, which also affects the SEM analysis. Crack initiation was mostly transgranular, indicating fatigue was the predominant contributor to the overall damage. A similar trend was also evident for crack propagation modes under LCF, particularly when the blades were subjected to lower stabilized cyclic stress.

3.1.2. Fracture Analysis under LCF [26]. Figures 15 and 16 show the low expansion region. There is a multisource or linear-source fracture in the low cycle fatigue extended region, with occasional small dimples found nearby the fatigue source area. The fatigue band spacing of the extended region is larger, and separation distance generally exceeded micrometers.

There was obvious transgranular phenomenon and cavity nucleation. Fractures were predominantly along the grain boundaries, indicating they were caused by typical tensile forces. Intergranular cracks are developed through cavity formation at the grain boundaries due to high local stress concentration caused by grain boundary sliding. With increasing scanning time, transgranular initiated cracks were characterized by striations with secondary cracks present along near-surface regions, whereas stress-dependent tensile ratcheting became more prominent in the form of dimples (ductile rupture) towards the interior (Figure 15).

Thus, the predominance of tensile ratcheting in the crack propagation phase highlighted two failure modes: time-dependent intergranular fracture and stress-dependent ductile rupture. This process was also facilitated by impingement of slip bands at the grain boundary. However, at very high stresses, particles within a grain became decohered at the particle-matrix interface due to tensile ratcheting, resulting in dimple formation that eventually leads to ductile failure. Hence, these two processes are competitive and governed by both the applied stress and the dwell time. The predominance of compressive ratcheting due to the lower stabilized cyclic stress prevented fast damage accumulation as occurred when stress-dependent mechanisms were highly active, providing sufficient time for cavity nucleation at grain boundaries (Figure 16).

3.1.3. Fracture Analysis under HCF. Figure 17 shows selected portions of the fracture surface at higher magnification to assess HCF damage, including transverse fatigue striation and edge tearing in bright linear zones. Thus, fracturing was caused by transgranular expansion. The cross section between the fatigue source and extension area was bright and smooth, with many light fatigue arc curves in the fracture diagram. Furthermore, HCF striations were also directly ahead of grain boundary facets, indicating that minute HCF cracks are linked with intergranular cracks [26].

During the final failure, the uncracked ligaments between the arrested cracks presumably got torn off by shearing, forming extensive tear zones. Moreover, these striations tended to link with adjoining grain boundary facets, confirming that HCF damage acted as a channel to link LCF with crack source zones.

The cracks could not advance further by coalescing because a highly stabilized cyclic stress would lead to ruptures, as evidenced by extensive dimples. Thus, HCF damage

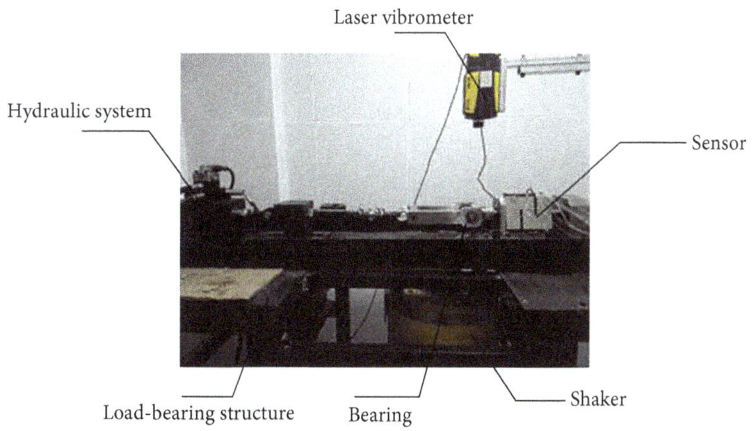

FIGURE 9: Assembly diagram of blade fixture system.

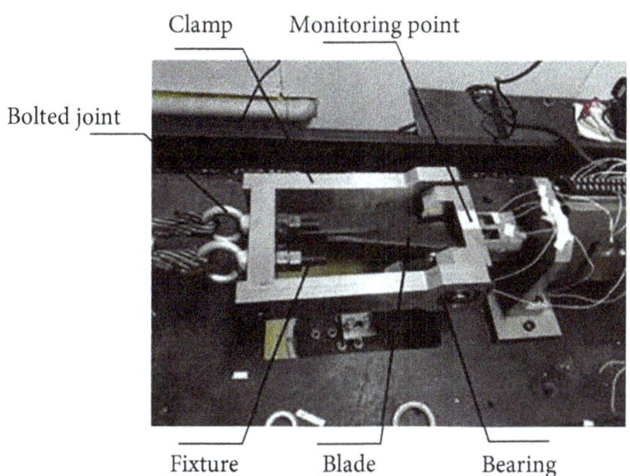

FIGURE 10: Vibration monitoring point diagram.

TABLE 8: Vibration stress and amplitude relationship of blade hazard position.

Amplitude of exciting point (μm)	Amplitude of monitoring point (μm)	Vibration stress (MPa)
47.49	40	112.03
59.58	50	122.27
71.66	60	133.64
83.74	70	141.59
95.82	80	152.23
107.91	90	166.28
119.99	100	173.9

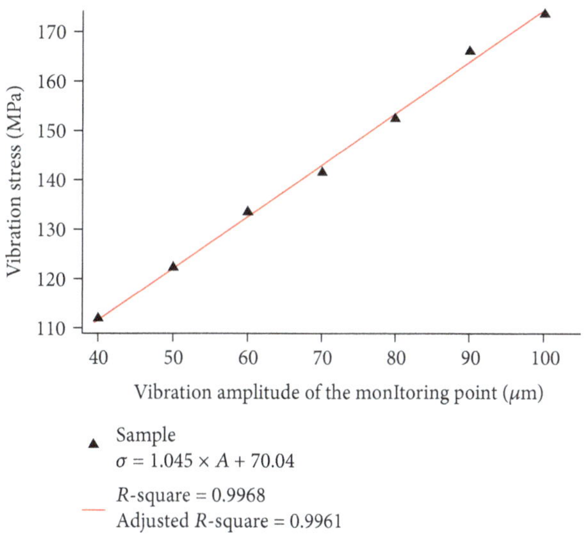

FIGURE 11: Relationship between vibration stress and amplitude.

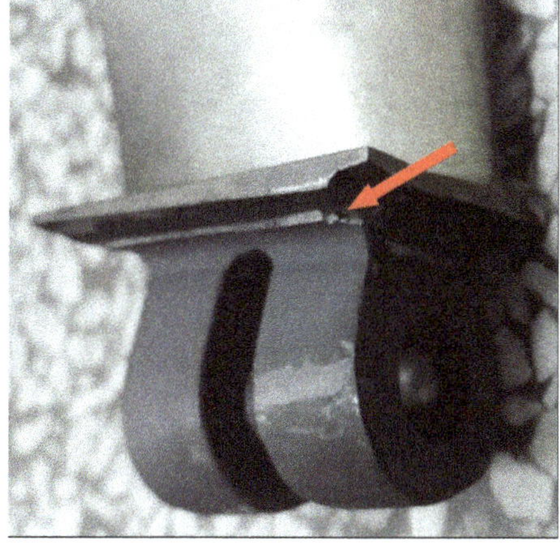

FIGURE 12: Fracture inspection place of the blade.

in such cases was negligible and might not have affected the final failure. When LCF damage is negligible and time-dependent damage is predominant in terms of the extensive intergranularity of fracture, HCF plays a similar role as revealed from the presence of striations within grain boundary facets which originated mostly from HCF cycling.

3.2. Fatigue Life Test. Through fracture results, we verified the rationality of the design of the experimental system, which can be used as a study of the combined fatigue effect of blades. Pure LCF and CCF tests were then performed to identify fatigue life characteristics.

3.2.1. Pure Low Cycle Fatigue Test and Result Analysis. Table 9 shows pure low cycle fatigue life test results for constant low cycle load (28,225 N) on seven blade specimens.

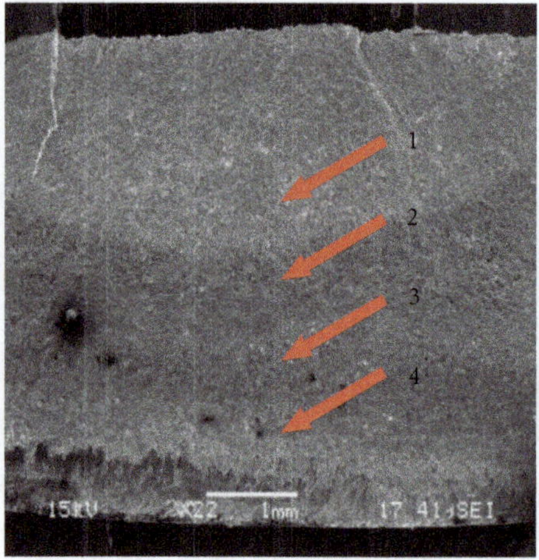

Figure 13: SEM fracture image.

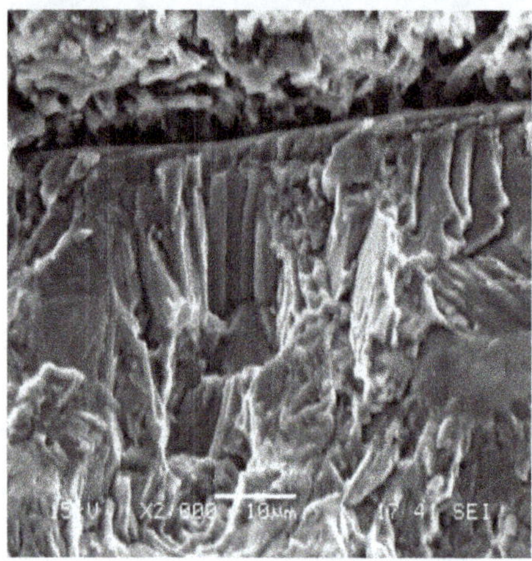

Figure 14: SEM fracture image, 2000x magnification.

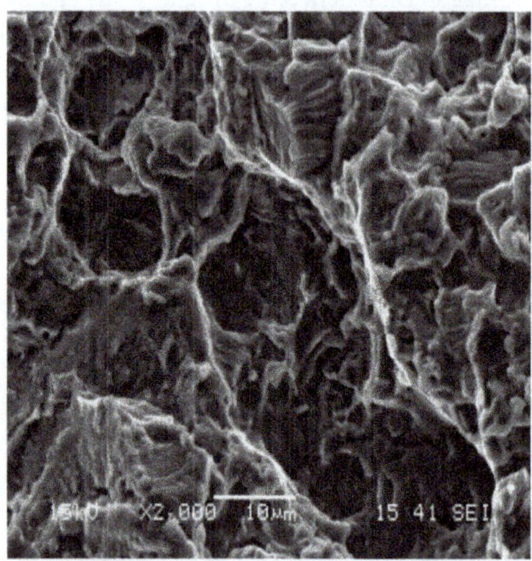

Figure 15: SEM LCF area, 2000x magnification.

3.2.2. Combined Fatigue Test and Result Analysis. Seven blades were selected for CCF tests under seven stress levels with constant low cycle load to compare with the pure LCF test and highlight HCF effects on fatigue life. Monitoring point amplitudes were controlled to 40, 50, 60, 70, 80, 90, and 100 μm, respectively, and the outcomes are shown in Table 10.

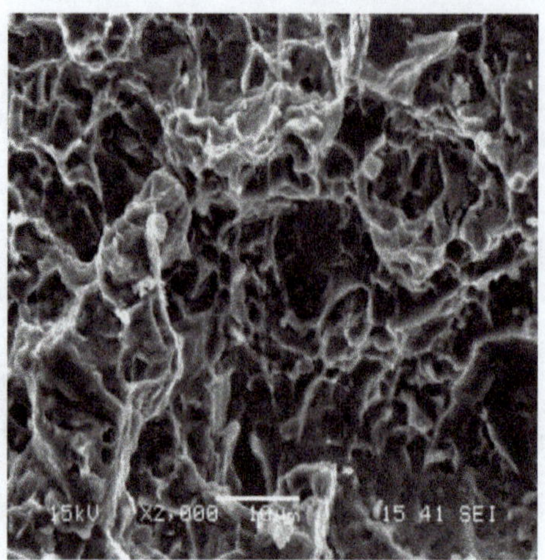

Figure 16: SEM cavity nucleation, 2000x magnification.

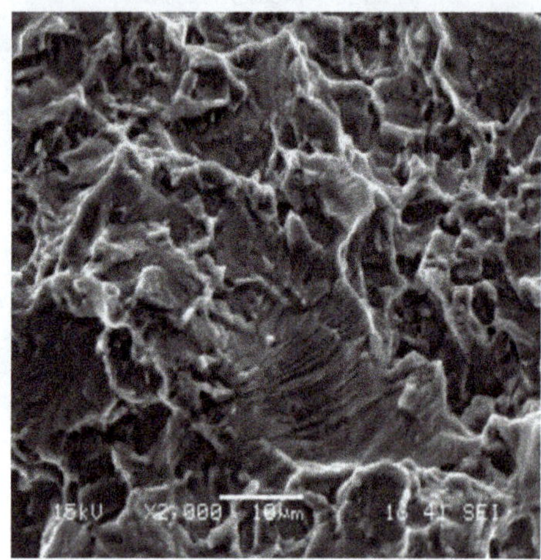

Figure 17: SEM HCF area, 2000x magnification.

Table 9: Pure low cycle fatigue test results.

Serial number	Low cycle load (N)	LCF life (cycle)
1	28,255	33,372
2	28,255	45,361
3	28,255	36,458
4	28,255	53,145
5	28,255	31,467
6	28,255	38,473
7	28,255	41,824

Table 9 indicates that pure LCF life can exceed 30,000 cycles whereas Table 10 shows that maximum LCF life under combined loading was only 11,241 cycles. The blade fatigue life under pure low cycle loading is much longer than that under the CCF, indicating that blade fatigue life can be significantly reduced by HCF loading and that the higher superimposed vibration stress level can lead to more significant reduction in the LCF fatigue life.

Table 10: HCF/LCF combined test records at different stress levels.

Serial number	Amplitude (μm)	LCF load (N)	HCF load (MPa)	LCF life (cycle)	HCF life (cycle)
1	40	28,255	112.03	11,241	40,872,276
2	50	28,255	122.27	10,120	36,796,320
3	60	28,255	133.64	9400	34,178,400
4	70	28,255	141.59	8675	31,542,300
5	80	28,255	152.53	8094	29,429,784
6	90	28,255	166.28	7642	27,786,312
7	100	28,255	173.9	7270	26,433,720

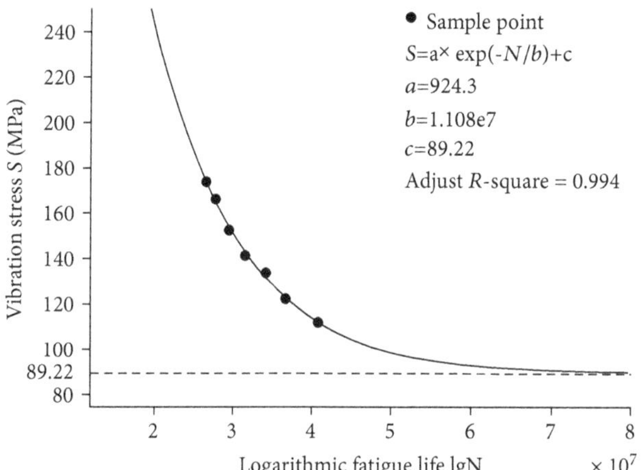

Figure 18: S-N curves in linear coordinates.

An overall reduction in CCF life compared with pure LCF was possibly due to HCF effects on fatigue crack propagation rate. The fraction of life spent in propagating a fatigue crack would be reduced but remain significant due to the lower stress range associated with minor cycles, hence significantly reducing total fatigue life. Accelerated crack growth could have been caused by the unloading associated with the LCF part of the CCF waveform, as demonstrated by Fleck and Smith [27] both in steel and aluminium alloy [20].

It is important to study HCF effects on LCF life in CCF testing and blade fatigue characteristics to predict blade life under any stress conditions. These results showed that HCF had a significant effect on LCF life, but the extent of this impact requires further study. So we need to analyze the combined test results in Table 10.

3.2.3. Median Cyclic Stress-Fatigue Life Curve. Fatigue characteristics are usually expressed using the relationship between maximum stress, S, and fatigue life, N, i.e., the S-N curve, which illustrates the endurance of specimens to changing stresses. Figure 18 shows the fitted S-N curve in linear coordinates.

Vibration stress and fatigue life are exponentially related, as

$$S = a \times \exp\left(-\frac{N}{b}\right) + c, \tag{2}$$

and the fitted result shows that $a = 924.3$ MPa, $b = 1.108 \times 10^7$ cycles, and $c = 89.22$ MPa, where c is the vibration stress threshold. Hence, the HCF vibration load had no significant effect on blade LCF life when vibration stress ≤ 89.22 MPa. Stress exceeded this threshold when we tested under maximum stress conditions, creating significant plasticity such that time spent initiating fatigue cracks was relatively short.

4. Conclusion

This study developed a CCF test system to investigate blade failure caused by centrifugal force and vibration combined. This combination more accurately represented practical conditions the blade would be subjected to in service. Hence, it becomes apparent that the previous and current design methods for blade material based upon HCF performance remain valid, because the original conservatism was retained. The fracture analyses have verified the rationality and feasibility of the designed combined fatigue test rig. Also, the life characteristics showed that the blade LCF life can be significantly reduced by the HCF loading. Increased vibration stress could lead to a more significant blade fatigue life reduction.

Conflicts of Interest

The authors declare that they have no conflicts of interest.

Acknowledgments

The author would like to thank the anonymous reviewers for their detailed and constructive comments and suggestions. The support from the National Natural Science Foundation of China (Grant No. 51175510) is also gratefully acknowledged.

References

[1] N. S. Xi, P. D. Zhong, H. Q. Huang, H. Yan, and C. H. Tao, "Failure investigation of blade and disk in first stage compressor," *Engineering Failure Analysis*, vol. 7, no. 6, pp. 385–392, 2000.

[2] N. Hou, Z. Wen, Q. Yu, and Z. Yue, "Application of a combined high and low cycle fatigue life model on life prediction of SC blade," *International Journal of Fatigue*, vol. 31, no. 4, pp. 616–619, 2009.

[3] Y. K. Zhang, R. X. Zhou, S. X. Guo, B. L. Shang, and W. T. Jia, "Design of compound fatigue test system of compressor blade and fatigue life analysis," *Journal of Aerospace Power*, vol. 32, no. 12, pp. 2880–2887, 2017.

[4] Y. Huang, Y. Liu, and W. Liu, "Analysis of structural-dynamic random reliability sensitivity based on the criteria of fatigue cumulative damage," *Aircraft Design*, vol. 30, no. 2, pp. 5–9, 2010.

[5] A. Fatemi and N. Shamsaei, "Multiaxial fatigue: an overview and some approximation models for life estimation," *International Journal of Fatigue*, vol. 33, no. 8, pp. 948–958, 2011.

[6] S. Liu, C. Liu, Y. Hu, S. Gao, Y. Wang, and H. Zhang, "Fatigue

life assessment of centrifugal compressor impeller based on FEA," *International Journal of Fatigue*, vol. 60, pp. 383–390, 2016.

[7] X. Y. Luo, R. G. Zhao, and W. He, "Analysis on low cycle fatigue properties and fractography of TC25 titanium alloy," *Chinese Journal of Solid Mechanics*, vol. 32, pp. 145–149, 2011.

[8] X. Yuan and S. Li, "Research and development of methods for the prediction of fatigue life," *Aeronautical Manufacturing Technology*, vol. 12, pp. 80–84, 2001.

[9] A. Sarkar, A. Nagesha, P. Parameswaran, R. Sandhya, K. Laha, and M. Okazaki, "Investigation of cumulative fatigue damage through sequential low cycle fatigue and high cycle fatigue cycling at high temperature for a type 316LN stainless steel: life-prediction techniques and associated mechanisms," *Metallurgical and Materials Transactions A*, vol. 48, no. 3, pp. 953–964, 2017.

[10] H. O. Fuchs, D. V. Nelson, and T. L. Toomay, *Fatigue Under Complex Loading*, Society of Automotive Engineers, Warrendale, PA, USA, 1977.

[11] T. Nicholas, "Critical issues in high cycle fatigue," *International Journal of Fatigue*, vol. 21, pp. 221–S231, 1999.

[12] M. A. Moshier, T. Nicholoas, and B. M. Hillberry, *High Cycle Fatigue Threshold in the Presence of Naturally Initiated Small Surface Cracks*, American Society for Testing and Materials, West Conshohochen, PA, USA, 2002.

[13] C. Dungey and P. Bowen, "The effect of combined cycle fatigue upon the fatigue performance of Ti-6Al-4V fan blade material," *Journal of Materials Processing Technology*, vol. 153-154, pp. 374–379, 2004.

[14] R. Wang, C. Cho, and J. Nie, "Combined fatigue life test and extrapolation of turbine disk mortise at elevated temperature," *Journal of Engineering for Gas Turbines and Power*, vol. 127, no. 4, pp. 863–868, 2005.

[15] C. Schweizer, T. Seifert, B. Nieweg, P. von Hartrott, and H. Riedel, "Mechanisms and modelling of fatigue crack growth under combined low and high cycle fatigue loading," *International Journal of Fatigue*, vol. 33, no. 2, pp. 194–202, 2011.

[16] B. Powell, T. Duggan, and R. Jeal, "The influence of minor cycles on low cycle fatigue crack propagation," *International Journal of Fatigue*, vol. 4, no. 1, pp. 4–14, 1982.

[17] D. Gelmedin and K.-H. Lang, "Fatigue behaviour of the superalloy IN 713C under LCF-, HCF- and superimposed LCF/HCF-loading," *Procedia Engineering*, vol. 2, no. 1, pp. 1343–1352, 2010.

[18] Z. Mazur, A. Hernández-Rossette A, and J. Porcayo-Calderón, "Fatigue investigation of the 69 MW gas turbine of a combined cycle unit," in *ASME Turbo Expo 2010: Power for Land, Sea and Air*, pp. 481–489, Glasgow, UK, 2010.

[19] S. A. Namjoshi and S. Mall, "Fretting behavior of Ti-6Al-4V under combined high cycle and low cycle fatigue loading," *International Journal of Fatigue*, vol. 23, pp. 455–461, 2001.

[20] M. Naeem, R. Singh, and D. Probert, "Implications of engine deterioration for a high-pressure turbine-blade's low-cycle fatigue (LCF) life-consumption," *International journal of Fatigue*, vol. 21, no. 8, pp. 831–847, 1999.

[21] J. Ding, R. F. Hall, J. Byrne, and J. Tong, "Fatigue crack growth from foreign object damage under combined low and high cycle loading. Part I: experimental studies," *International Journal of Fatigue*, vol. 29, no. 7, pp. 1339–1349, 2007.

[22] L. Mendia, F. J. Estensoro, C. Mary, and F. Vogel, "Effect of combined cycle fatigue on Ti6242 fatigue strength," in *The 11th International Conference on the Mechanical Behavior of Materials (ICM11)*, pp. 1809–1814, Como, Italy, 2011.

[23] Committee for Practical Handbook of Engineering Materials, "Magnesium Alloys and Titanium alloys," in *Practical Handbook of Engineering Materials: Aluminum alloys*, Standards Press of China, 1989.

[24] Beijing Aviation Materials research institute, *Material Data Sheet of Aircraft Engine Design*, China Aviation Engine Company, 1992.

[25] J. Hong, D. Y. Zhang, and L. L. Ghen, "Review on investigation of high cycle fatigue failures of the aero engine blade," *Journal of Aerospace Power*, vol. 24, no. 3, pp. 652–661, 2007.

[26] A. Sarkar, M. Okazaki, A. Nagesha, P. Parameswaran, R. Sandhya, and K. Laha, "Mechanisms of failure under low cycle fatigue, high cycle fatigue and creep interactions in combined cycling in a type 316LN stainless steel," *Materials Science and Engineering: A*, vol. 683, pp. 24–36, 2017.

[27] N. A. Fleck and R. A. Smith, "A discussion of mechanisms for acceleratedand retarded fatigue crack growth," in *Sixth International Conference on Fraction*, pp. 1832–1829, Delhi, India, 1984.

3

Aeroelastic Analysis of Wings in the Transonic Regime: Planform's Influence on the Dynamic Instability

Mario Rosario Chiarelli and Salvatore Bonomo

DICI, University of Pisa, Largo Lucio Lazzarino 1, 56122 Pisa, Italy

Correspondence should be addressed to Mario Rosario Chiarelli; chiarelli@ing.unipi.it

Academic Editor: Hikmat Asadov

This paper presents a study of transonic wings whose planform shape is curved. Using fluid structure interaction analyses, the dynamic instability conditions were investigated by including the effects of the transonic flow field around oscillating wings. To compare the dynamic aeroelastic characteristics of the curved wing configuration, numerical analyses were carried out on a conventional swept wing and on a curved planform wing. The results confirm that, for a curved planform wing, the dynamic instability condition occurs at higher flight speed if compared to a traditional swept wing with similar profiles, aspect ratio, angle of sweep at root, similar structural layout, and similar mass. A curved wing lifting system could thus improve the performances of future aircrafts.

1. Introduction

Modern technologies are aimed at increasing efficiency, in order to reduce operative costs and pollution and/or to increase the performances of aircrafts. For several years the Aerospace Engineering Unit of the Department of Civil and Industrial Engineering of the University of Pisa has been studying a novel geometry for wings with high aspect ratio. The wing has a curved planform: both the leading and trailing edges of the wing are described by curved lines. The in-plane curvature of the wing considerably reduces the aerodynamic drag especially in the transonic regime where the nonuniform distribution of the sweep angle of the curved wing leads to a reduction in the wave drag effects.

In the literature, various works focus on wing configurations with a curved leading edge or curved planform. The main topics discussed are a reduction in the induced drag or classical application of low aspect ratio wings for high supersonic configurations [1–5]. Only the authors in [2] discuss in detail the effects of a curved leading edge for a wing operating in the transonic regime; however, the trailing edge of the wing is assumed to be straight.

At the same time patents concerning the curved planform concept have been deposited, but only for a tip extension of wings of transport aircrafts [6, 7]. It is well known that the wing tip of the B787 aircraft has not only an out-off plane "C" shape but also a curved planform [8].

To the best of our knowledge, there are no studies on a fully curved planform shape with high aspect ratio wings (with both the leading edge and the trailing edge curved). From an engineering point of view, the research interest in such a wing configuration particularly concerns the strong reduction in drag and the important reduction in structural weight. Both of these synergic effects could lead to significant reduction in fuel consumption and pollution.

In transonic flight conditions, the flow field on the wings is strongly nonlinear, which is why the theoretical modelling of realistic aircraft configurations represents a challenge for researchers. Often, the technical literature regards the validation procedures of numerical techniques, adopted to describe the aerodynamic performances of wings or aircrafts operating in the transonic regime. Several works try to represent pressure and lift distributions. In these cases often the numerical results agree very well with the experimental data [9–11]. On the other hand, if the objectives include a realistic estimation of the drag for wing models or for complete aircraft configurations, the research also requires a large amount of human and computational resources [12–14].

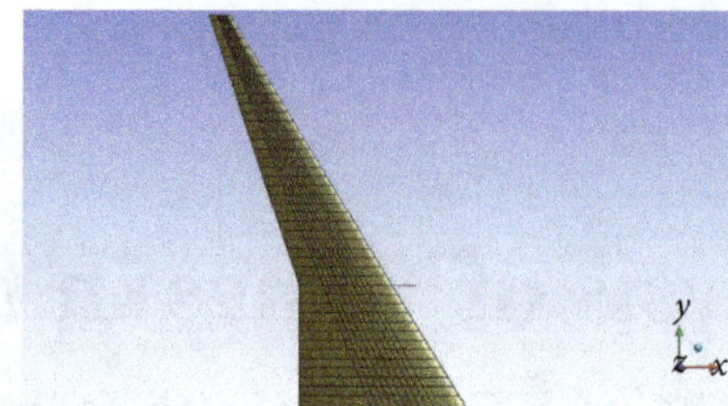

FIGURE 1: Swept wing model.

The present work compares the dynamic aeroelastic behaviour of wings, with different planform shapes, in a three-dimensional fully transonic flow field.

Today by means of a fluid structure interaction (FSI) technique, it is possible to represent the physics of transonic phenomena which develop around deformable lifting surfaces. There are several works that study the dynamic behaviour in the transonic regime of profiles mounted on elastic supports. However in these cases there are no three-dimensional effects of the flow around a real wing. On the other hand, the dynamic oscillations of a three-dimensional lifting surface make the problem very complex. The analysis becomes more complicated if dynamic interactions arise between the transonic flow field and the structural response of the wings. Thus numerical results obtained by comparative analyses carried out with a well-structured procedure for the construction of the aerodynamic grids with a similar topology, a similar number of cells, and a similar layout of cell dimensions for the problems analysed can guarantee a reliable technical comparison of the physical behaviours of the wing configurations under observation without having to use prohibitive computational resources. This is also true if the absolute values of the computed technical coefficients are affected by modelling errors. In fact, in the case of structured and similar fluid dynamic grids, these errors will have the same quantitative effects. Thus several numerical activities have been carried out at the University of Pisa to compare the aerodynamic behaviour of a curved wing with that of a conventional swept wing.

The results of the studies on the drag reduction obtained with a curved wing configuration can be found in [15–20]. In these works a preliminary analysis carried out with the NASTRAN® code also showed that such curved wings have a good dynamic behaviour; that is, flutter instabilities are not an issue for curved wings. To obtain more robust results for the dynamic case a new campaign of numerical analyses was carried out.

We applied the FSI technique by means of ANSYS Workbench® Rel. 15 commercial platform. The dynamic responses of a swept wing and a curved wing were studied and compared. The models of the two wings, as assumed in previous researches, were constructed with similar aerodynamic profiles, aspect ratios, sweep angles at the root section, and structural layouts. The geometry of the curved wing was obtained by shearing the swept wing in the longitudinal direction. Figures 1 and 2(a) show the planform shape of the two wings analysed. Table 1 summarizes the geometrical data of the models. Equation (1) describes the in plant shape of the leading edge of the curved wing (starting from a point at the leading edge position of the kink chord).

TABLE 1: Geometrical data of the two half-wing models.

	Swept wing	Curved wing
AR (aspect ratio)	9.5	9.5
Angle of sweep at root	32°	32°
Angle of sweep at tip	32°	53°
Half-wing span ($b/2$)	30 m	30 m
Reference surface area	379 m^2	379 m^2
Root chord	13.18 m	13.18 m
Tip chord	1.7 m	1.7 m
Kink section position ($b_K/2$)	9.3 m	9.3 m
Kink chord	7.373 m	7.373 m
Dihedral angle	5°	5°

In (1) t is a nondimensional parameter ($0 \le t \le 1$), b is the span of the wing, and b_K is the span at the kink chord (both expressed in meters; see Table 1).

The $s(t)$ coordinate (unit = m), which defines the geometry of the leading edge, is measured in the longitudinal direction (aerodynamic chord direction): $s(0) = 0$ corresponds to the leading edge of the kink section, while $s(1) = 19.359$ m corresponds to the leading edge of the tip section, as shown in Figure 2(b):

$$s(t) = 6.2487 \times 10^{-1} \cdot t \cdot \left(\frac{b - b_K}{2}\right) + 1.1054 \times 10^{-2} \cdot t^2 \cdot \left(\frac{b - b_K}{2}\right)^2 + 1.9022 \times 10^{-4} \cdot t^3 \cdot \left(\frac{b - b_K}{2}\right)^3. \tag{1}$$

For both wings an engine nacelle was modelled at the kink chord position. In the structural analyses for the two half wings, the root section was assumed clamped.

The combined fluid dynamic and structural analyses were carried out by taking into account the gravity effects and setting the proper geometrical angle of attack in order to get the same lift coefficient for both wings. Following previous experiences [18–20] the dynamic response analyses were also

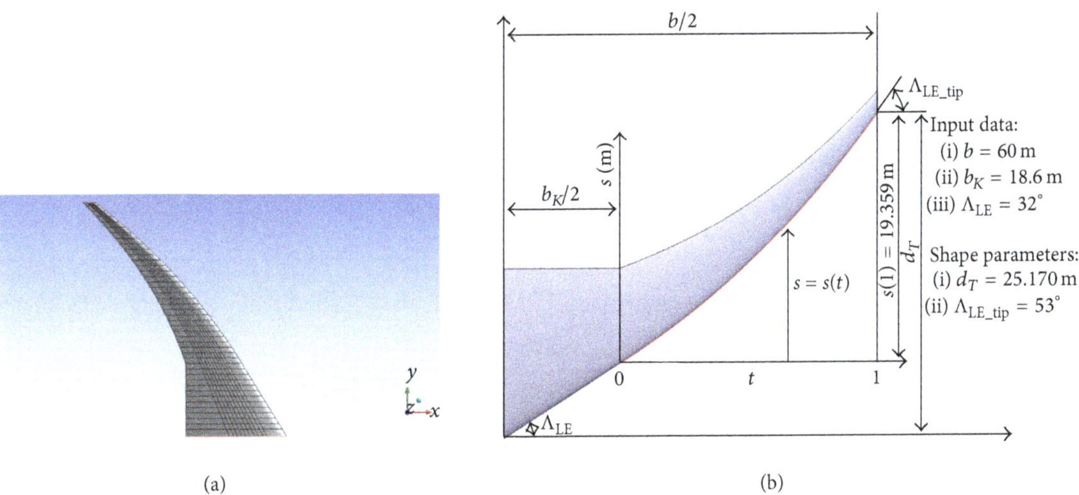

FIGURE 2: (a) Curved wing model. (b) Definition of leading edge geometry of the curved wing model.

executed by imposing different values for the asymptotic Mach number and the altitude of flight.

The elasticity and damping effects obviously influence the final response of the wings providing, for example, suitable displacement histories for fixed control points. The nodes of the finite element models positioned at the leading edge and at the trailing edge of the tip section of the two wings were assumed as control points. Thus, for each flight condition, the overall damping coefficient (which involves both structural and aerodynamic effects) was extracted by processing the displacement time history of the two wings.

The two wing models were constructed by adopting similar structural layouts (aluminium alloy material), similar thickness distribution for skins, and similar geometry for spars and stringers. The structural mass of the models is the same, also including the effect of fuel mass distributed along the span.

During a preliminary numerical campaign, carried out at sea level, neither of the wing models suffered from instability. In fact, the overall damping was always negative in the range values of the Mach number examined. This result depended on the scheme used to define the structural models. In fact in the present models, not only does the wing box affect the structural response but also the front and rear portions of the wing cross sections outside the wing box do. Thus in particular the estimated torsional natural frequencies of the wings were found to be unrealistic: the first torsional frequency was too high with respect to the first bending frequency. To overcome this problem, fictitious rotational inertia was added to the last three ribs at the tip region of the wings, thus keeping the total mass of the two wings unchanged.

This adaptation of the models provided the desired results: the reduction of the torsional frequency, the interaction of first bending and torsional modes, and the onset of dynamic instability for both wings.

Comparing the two wings, with similar aerodynamic profiles and structural layouts, reveals that for a curved geometry the dynamic instability conditions are reached with higher flight velocity values. This occurs at sea level for low subsonic flight conditions and at cruise altitude for high subsonic flight conditions (transonic flight).

The results obtained highlight the need for further research because, as demonstrated in previous studies [16, 18–20] the curved wing configuration itself leads to a reduction in the drag from the aerodynamic point of view (which means saving fuel). In addition, with a fixed flutter boundary, the curved configuration enables lighter structures to be used (which means further fuel savings). Alternatively, with an enlarged flutter boundary, the curved configuration enables faster machines to be designed without changing the aerodynamic efficiency.

2. Fluid Dynamic Models of the Wings

To carry out the fluid dynamic analyses the FLUENT® code was adopted. For the two wings structured meshes were constructed, maintaining a similar topology for both models. Thus numerical effects and/or numerical errors can be assumed as similar for the two models (this approach has also been adopted in previous works).

Firstly, a blocking procedure was used to define the control volumes around the wings models (e.g., Figures 3(a) and 3(b) show partial views of the block layout around the curved wing model). The volumes near the wings were constructed allowing for a good level of approximation in order to model the boundary layer. However, a lack of computational resources prevented a more detailed description of local phenomena (transition and/or boundary layer separation).

The section profile geometry that we adopted is similar to that used in previous research campaigns [16, 18–20], that is, the supercritical airfoil SC(2)-0410 [21].

The whole aerodynamic field analysed has the following dimensions: height 131 m, width 90 m, and length 278 m. To minimize the time needed for the analyses a whole grid of only 389 766 hexahedral cells and 400 544 nodes was used. Figure 4 shows the surface grid of the curved wing model.

The boundary conditions fixed for the lateral surfaces of the overall mesh volume are summarized in Table 2. In

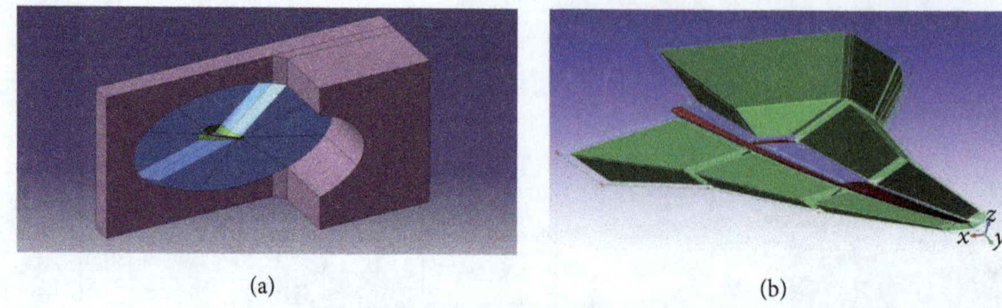

FIGURE 3: (a) Block layout around the curved wing model (partial view). (b) Block layout near the curved wing model (partial view).

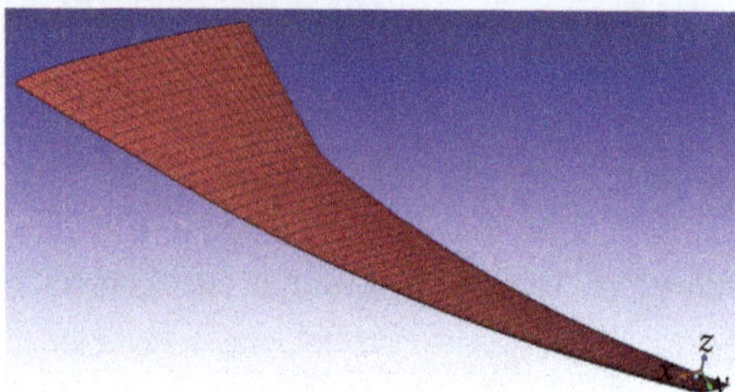

FIGURE 4: Surface grid of the curved wing model.

TABLE 2: CFD boundary conditions.

	Swept wing model	Curved wing model
Altitudes analysed	0 m–10000 m	0 m–10000 m
Pressure far field	Front side up down	Front side up down
Pressure outlet	Rear	Rear
Symmetry	Center-line plane	Center-line plane
Wall/no slip	Wing surface	Wing surface

order to take into account the viscosity effects, a viscous and turbulent flow was assumed during the CFD analyses. A standard K-ε model was used to describe the turbulence and an implicit unsteady analysis technique was adopted. The objective of our CFD analysis was to compare the dynamic behaviour of two different lifting systems and to draw damping-Mach curves for the two configurations. Starting from rigid CFD results, the dynamic analysis of the two wing models was based on similar values of the lifting coefficient equal to 0.36 for a Mach equal to 0.85 in hypothetical cruise flight conditions $h = 10\,000$ m.

3. Structural Models of the Wings

The models of the swept and curved wings were constructed by assigning the properties to the structural components (skin, stringers, ribs, and spars) in ANSYS R15.0. All the components of both structural wing models have the same dimensions. The structures were modelled with a metallic material (aluminium alloy). A three-spar configuration was assumed for the wing box layout. Upper and lower skin, ribs, and spar webs were modelled with shell elements, whereas the stringers and spar flanges were modelled with beam elements. Both the models consist of 8436 nodes and 4157 elements. Figures 5(a) and 5(b) show the finite element model of the curved wing.

The engine nacelle was modelled with beams with a very high stiffness and three-point masses describe the inertial effects of the engine. The structural mass and the fuel mass distributed along the wing are the same for both wings (swept and curved).

Fictitious inertia values were added on the tip region of the two wings to facilitate the dynamic instability and to overcome the effects of the boundary conditions at the root section of the wing models (the root section was assumed to be clamped for the wings). Two distinct moment inertia distributions were analysed: in the first case (Case 1 in Table 3), very low values for the instability Mach number were obtained. To describe a more realistic situation (in other words higher values for the flutter velocities), a second fictitious inertia distribution (Case 2 in Table 3) was assumed with a reduction of about 50% in the swept wing and 30% in the curved wing. Main characteristics of both finite element models are summarized in Table 3.

The structural damping factor ζ, as a function of the natural circular frequency ω_n, was introduced during the coupled CFD and transient response analyses according to the Rayleigh method (2).

The modal analyses enabled the values of α and β of (2) to be defined assuming a structural damping factor equal to 0.04 for two fixed natural frequencies; that is $f_1 = 1$ Hz and $f_2 = 6$ Hz. Consider

$$\zeta(\omega_n) = \frac{\alpha}{2 \cdot \omega_n} + \frac{\beta}{2} \cdot \omega_n; \qquad \alpha = \frac{2 \cdot \omega_1 \cdot \omega_2}{\omega_1 + \omega_2} \cdot \bar{\zeta}; \; \beta = \frac{2}{\omega_1 + \omega_2} \cdot \bar{\zeta} \tag{2}$$

TABLE 3: Structural analyses data.

	Swept wing	Curved wing
Material	Aluminium alloy	Aluminium alloy
Skin thickness	7 mm to 2.5 mm	7 mm to 2.5 mm
Rib thickness	7 mm to 1 mm	7 mm to 1 mm
Front spar thickness	12 mm to 5 mm	12 mm to 5 mm
Central spar thickness	12 mm to 8 mm	12 mm to 8 mm
Rear spar thickness	12 mm to 7.5 mm	12 mm to 7.5 mm
Total structural mass	15 372.2 kg	15 372.2 kg
Total fuel mass	20 000 kg	20 000 kg
Engine masses	4000 kg front, 4000 kg centre, and 2000 kg rear	4000 kg front, 4000 kg centre, and 2000 kg rear
Total model mass	45 394 kg	45 394 kg
Fictitious inertia (Case 1)	1300[a], 1000[a], 1000[a]	1300[a], 1000[a], 1000[a]
Fictitious inertia (Case 2)	750[a], 500[a], 500[a]	1050[a], 700[a], 700[a]

[a]Moment of inertia applied on ribs in the tip region (y direction): units = kg m^2.

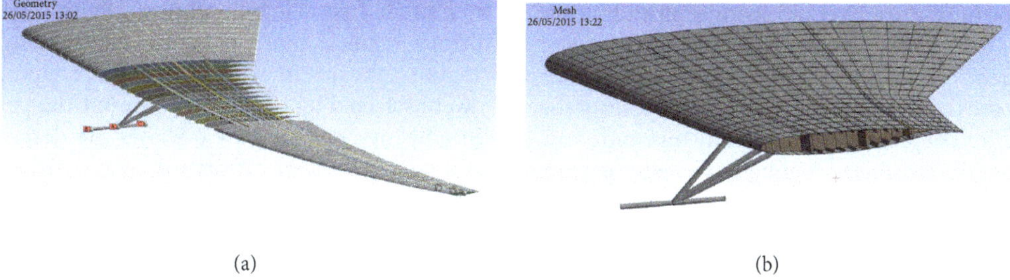

(a)

(b)

FIGURE 5: (a) The finite element model of the curved wing (upper and lower skins partially removed). (b) The finite element model of the curved wing (wing structure layout).

TABLE 4: Results of modal analysis: Case 0 (no fictitious moments of inertia were added to finite element models).

	Swept wing	Curved wing
First bending mode: Case 0	1.058 Hz Mode N. 1	0.916 Hz Mode N. 1
First torsion mode: Case 0	12.633 Hz Mode N. 8	12.733 Hz Mode N. 9

with $\overline{\zeta} = 0.04$ for $\omega_1 = 2 \cdot \pi \cdot f_1$ and $\omega_2 = 2 \cdot \pi \cdot f_2$; $\alpha = 0.431$ and $\beta = 1.82 \times 10^{-3}$.

3.1. Results of Modal Analysis of the Wings. Before activating the fluid structure interaction, modal analyses of the wing models were carried out to study the distribution of the natural frequencies and the shape of the associated normal modes. For the two models, Table 4 shows the bending and the torsional natural frequencies, whose coupling usually leads to dynamic aeroelastic instability for slender wings, without the effect of fictitious inertia positioned at the tip (Case 0).

Table 4 highlights that the first torsion mode has a frequency that is too high compared to the bending mode frequency and to typical engineering applications of similar wing structures. The classical flutter of a wing involves the interaction (inertial and aerodynamic) of bending and torsion modes. However if the frequencies of these modes are too far apart, their physical interactions are probably nil. Our results of the modal analyses are strongly affected by the discretization method of the wing structure. Unfortunately, because of geometrical matching requirements and in order to allow for a suitable data exchange between the mechanical and the fluid dynamic meshes, the portions of the wing models outside the wing box (see Figures 5(a) and 5(b)) deeply influence the mechanical response of the structure. As a consequence, for both wings, the frequency of the first torsion mode is too high with respect to the frequency of the first bending mode, as shown in Table 4. This situation does not correspond to the realistic dynamic behaviour of a transport aircraft wing; in fact, for these wing models, the preliminary FSI simulations did not reveal flutter. In order to reduce the first torsional frequency down to a more realistic value, we decided to add fictitious rotary inertia at the tip of the two wings without changing their total mass. These inertia values were added around the y-axis of the models (as shown in Figures 1 and 2(a)) without affecting the first bending frequencies (Case 1 in Table 3). This method considerably reduced the first torsional mode frequencies, as shown in Table 5; however these wing models reached the flutter conditions for lower Mach numbers than expected. In fact, some of the literature results related to large transport aircrafts [22] had given indications on typical values of Mach numbers along the flutter boundary. Thus we decided to

Table 5: Results of modal analysis: Case 1 and Case 2.

	Swept wing	Curved wing
First bending mode (N. 1) (Case 1; see Table 3)	1.057 Hz	0.913 Hz
First torsion mode (N. 4) (Case 1; see Table 3)	3.402 Hz	3.089 Hz
First bending mode (N. 1) (Case 2; see Table 3)	1.057 Hz	0.914 Hz
First torsion mode (N. 5) (Case 2; see Table 3)	4.417 Hz	3.729 Hz

reduce the fictitious rotary inertia in order to increase the frequency of the first torsion mode of both wings. On the other hand, as will be discussed in the next section, in Case 1 the swept wing showed a more critical behaviour with respect to the curved wing both at sea level and at cruise altitude. Thus a stronger reduction of the inertia was applied to the swept wing finite element model (Case 2 in Table 3). It was thus possible to compare the dynamic aeroelastic behaviour of the two wings with different planforms under fully transonic flow conditions. It was also possible to highlight the better performance of the curved wing also under more critical mechanical conditions: a smaller gap between the first bending and first torsion modes frequencies (as summarized in Table 5).

Finally, as expected the first bending frequencies remain unchanged in the three cases examined (Case 0: Table 4; Case 1 and Case 2: Table 5).

4. Results of the Fluid Structure Interaction Analyses

On the basis of the preliminary results of the modal analyses, the final fluid structure interaction analyses were carried out for two altitude values: sea level (0 m) and cruise condition (10000 m).

The time step for the fluid dynamic analyses was fixed at 0.01 s. The time step for the structural analyses was fixed at 0.0025 s. To minimize errors during the data exchange between the structural and the fluid dynamic modules five coupling iterations per time step were set. To obtain the overall damping factor ζ, from the time histories of the vertical displacement $V(t)$ of a node positioned at the leading edge of the wing tip sections, two pairs of relative maxima and minima were extracted as explained below.

Assuming that, close to the flutter condition, a damped harmonic oscillation occurs, $V(t)$ can be expressed as

$$V(t) = \widetilde{V} \cdot e^{-\zeta \cdot \omega_n \cdot t} \cdot \cos(\omega_d \cdot t + \phi)$$
$$\text{with } \omega_d = \sqrt{1 - \zeta^2} \cdot \omega_n, \tag{3}$$

where t is the time variable, ϕ is the phase, and ω_n is the undamped natural circular frequency. The damping factor ζ is estimated according to the following standard relationships:

$$\delta = \frac{1}{j} \cdot \ln\left(\frac{V_1^{\mathrm{MAX}} - V_1^{\mathrm{MIN}}}{V_{1+j}^{\mathrm{MAX}} - V_{1+j}^{\mathrm{MIN}}}\right);$$
$$\zeta = \frac{\delta}{\sqrt{(2 \cdot \pi)^2 + \delta^2}}, \tag{4}$$

where δ is the logarithmic decrement, j is the number of cycles between the first couple and the last couple of data extracted from the time history, and V_1^{MAX} and V_1^{MIN} are the maximum and minimum values of the first useful cycle of the generic time history data, while V_{1+j}^{MAX} and V_{1+j}^{MIN} are the corresponding values after j cycles. To estimate the damping factor, (4) was assumed to be also valid far from the pure flutter condition where the oscillation is dominated by a single frequency: the flutter frequency. When the time history seemed more complex than a single component response, the maxima, minima, and the appropriate value for j were selected by taking into account the periodicity of the time history itself. Figure 6 shows the described procedure considering the time history for Case 1 relevant to the curved wing at the sea level.

From a practical point of view a dynamic instability condition exists if the parameter ζ reaches a negative value corresponding to a monotonic growth of the amplitude of the wing structure oscillations. Obviously in the present case there is an interaction between the aerodynamic loads and the (dynamic) deformed shape of the structure. The aerodynamic loads generally introduce dissipative effects, which increase the overall damping of the system. However for each altitude of flight, there is a limit value for the airflow velocity (or Mach) corresponding to a limit for the dynamic stability of the system. As it is well known a pure harmonic motion exists for $\zeta = 0$, and from (3) $\omega_d = \omega_n$ represents the flutter natural circular frequency, and the corresponding velocity (or Mach) is said to be the flutter speed (or Mach).

For both wings, Case 0 was found to be stable for all Mach numbers (at sea level and at the cruise altitude). This depends on the very high values of the torsion natural frequency of both wings (Table 4).

For Case 1 (see Tables 3 and 5), the sea level flight condition was studied first. The highest value of the moment of inertia was applied on the rib positioned exactly at the wing tip. Figures 7 and 8 show the vertical displacement histories of the control node positioned at the leading edge of the tip section for the swept wing and for the curved wing, respectively.

At the sea level ($h = 0$ m) both wings are unstable, but for the swept wing, the instability condition corresponds to a lower Mach number ranging from 0.4 to 0.45 (see also Figure 15).

The time histories for the cruise flight conditions ($h = 10000$ m) are represented in Figures 9 and 10 for the swept wing and for the curved wing, respectively. Also in this case the swept wing reaches the instability condition for lower Mach number (between 0.7 and 0.8 as shown in Figure 15).

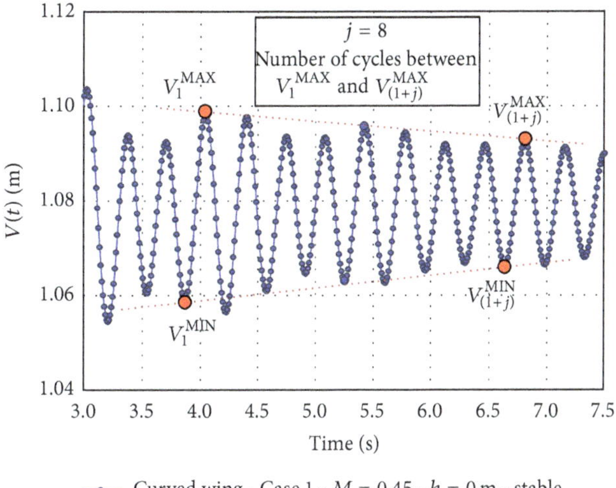

FIGURE 6: Time history of LE vertical displacement of the curved wing (Case 1: sea level, see Figure 8).

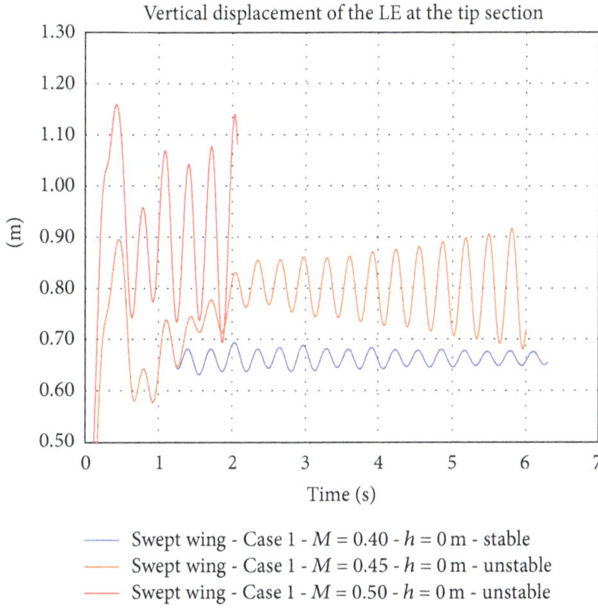

FIGURE 7: Results of the FSI analyses of the swept wing (Case 1, sea level).

On the basis of this first set of analyses, related to the distribution of fictitious moments of inertia corresponding to Case 1, both the wings clearly reach the instability conditions at sea level and at the cruise flight altitude. However the flutter conditions observed happen for subsonic airflows; in other words around the two wings, the flow is subsonic at each point of the aerodynamic field. This is an interesting result, because, for fully subsonic flight conditions, with all other design parameters fixed, the planform shape of the curved wing leads to an important increase in wing flutter speed. In Case 1 the flutter instabilities involve the interaction of the first bending mode with the first torsion mode of the two wings.

To confirm this last result also for a transonic flight condition, typical of modern transport aircrafts of a medium and/or long operative range, a second distribution of fictitious moments of inertia was adopted. In Case 2 (see Tables 3 and 5) the moments of inertia were halved for the swept wing, while as a precaution, for the curved wing a weaker reduction was assumed. As expected, the new distributions of moments of inertia provide different natural frequency values for first torsion modes (Table 5).

In this second case, for both wings, the instability condition corresponds to higher Mach numbers and the supersonic zones around the two wings occupy large zones of the aerodynamic field near the surfaces of wings (see Figures 11(a), 11(b), 12(a) and 12(b)).

The Reynolds numbers related to the represented supersonic zones are $Re = 5.835 \times 10^7$ for $M = 0.875$ and $Re = 6.001 \times 10^7$ for $M = 0.90$.

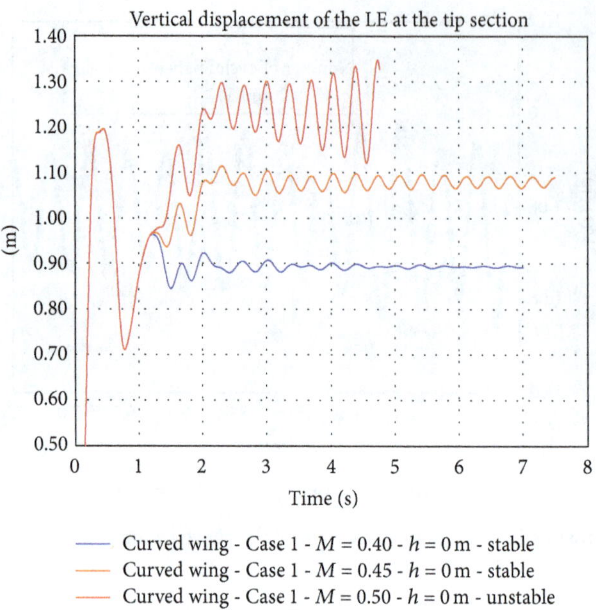

FIGURE 8: Results of the FSI analyses of the curved wing (Case 1, sea level).

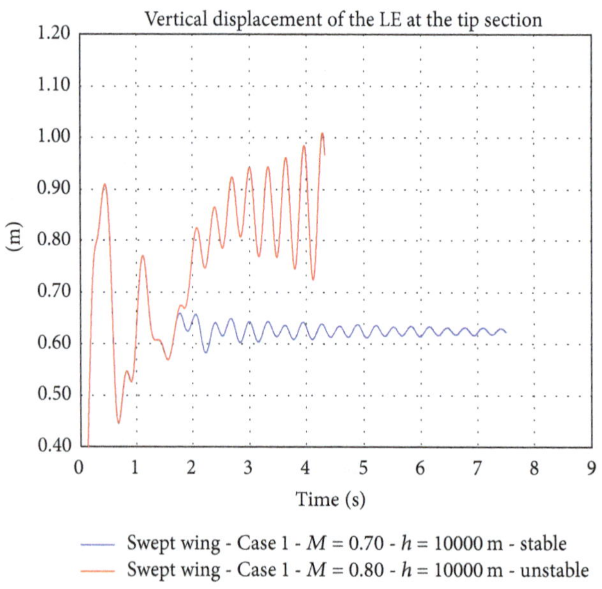

FIGURE 9: Results of the FSI analyses of the swept wing (Case 1, cruise altitude).

The aerodynamic fields are now fully transonic and shock waves develop on both the wings during the dynamic oscillations. As it is well known, this physical phenomenon represents a source of a strong nonlinearity from a mathematical point of view.

Nevertheless, in the present work, the well-structured aerodynamic grids were able to describe the complex aeroelastic behaviour of the two wing models very well, even though the adopted level of grid refinement was not very high due to the available computational resources. In a previous work it was also demonstrated that adopting well-structured grids (similar to fluid dynamic grids used in the present work), also with a low level of refinement, the numerical distributions of the pressure coefficient agree very well with the available experimental data [17]. For this reason, the present results describe in a reliable way the pressure field and the physical phenomena that develop on the two wings.

During the oscillating motion of the wings (flexural and torsional) the shock waves move in a chordwise direction due to the continuous change in the angle of attack along the wing span. The motion of the shock wave increases the complexity of the aeroelastic phenomena and the computational difficulties.

For Case 2, Figures 13 and 14 show the displacement time histories at the control node of the two wings calculated only for the cruise flight condition ($h = 10000$ m). It is well

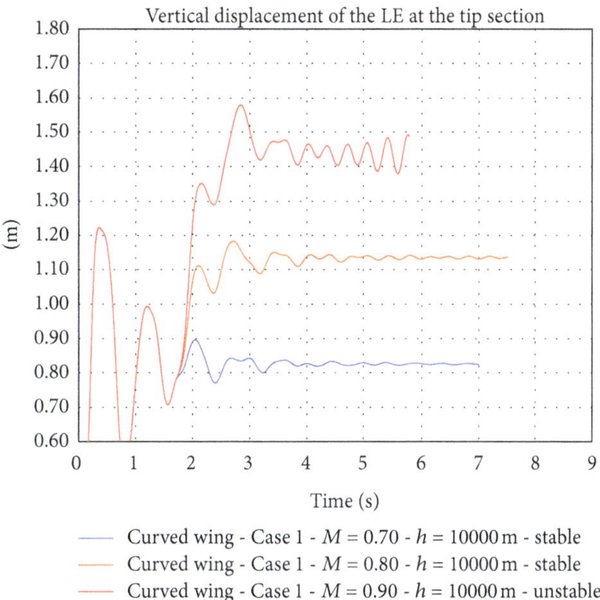

FIGURE 10: Results of the FSI analyses of the curved wing (Case 1, cruise altitude).

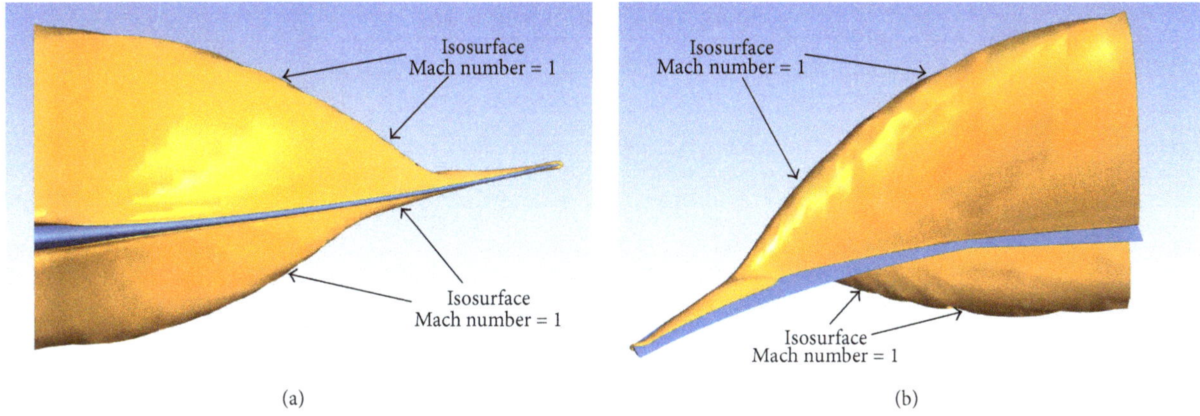

FIGURE 11: (a) Front view of the supersonic zone around the curved wing (Case 2, Mach = 0.90, time = 6.5 s). (b) Rear view of the supersonic zone around the curved wing (Case 2, Mach = 0.90, time = 6.5 s).

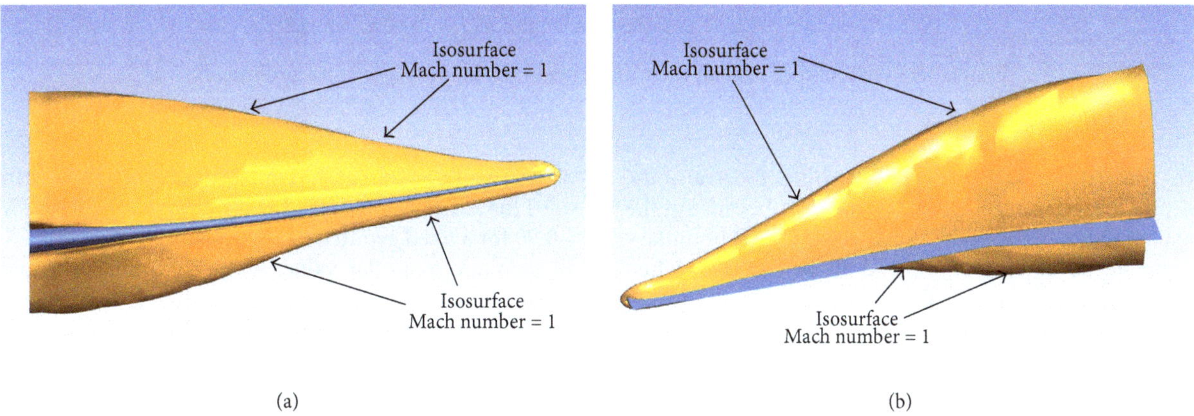

FIGURE 12: (a) Front view of the supersonic zone around the swept wing (Case 2, Mach = 0.875, time = 7 s). (b) Rear view of the supersonic zone around the swept wing (Case 2, Mach = 0.875, time = 7 s).

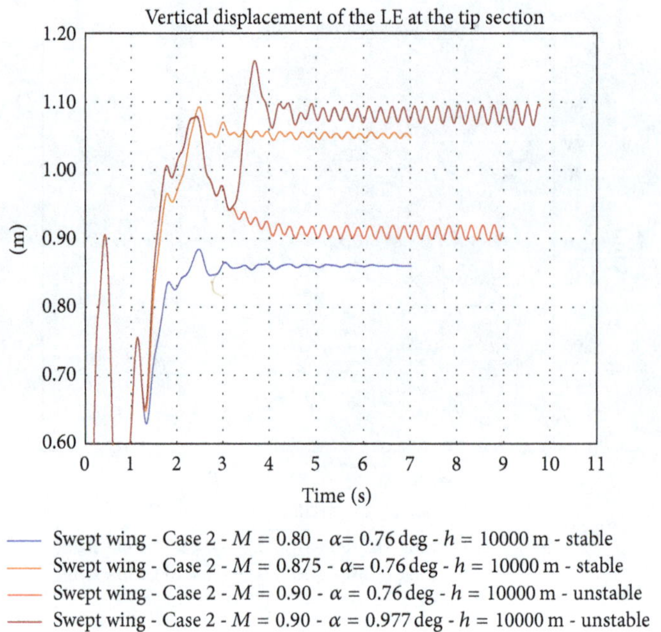

FIGURE 13: Results of the FSI analyses of the swept wing model (Case 2, cruise altitude).

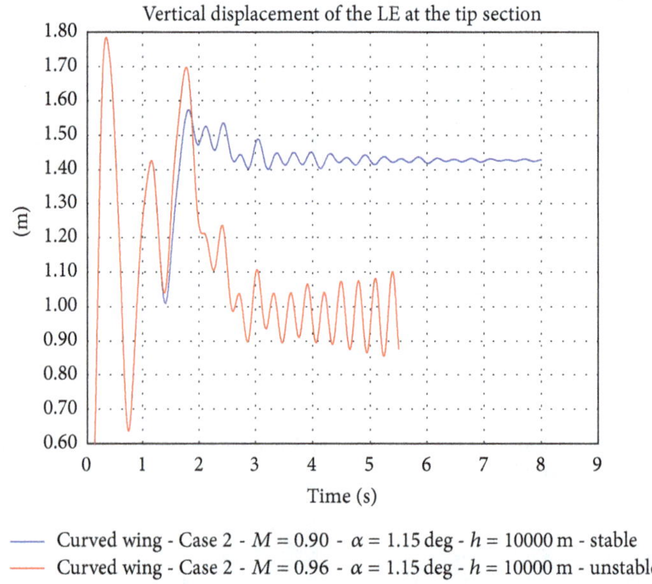

FIGURE 14: Results of the FSI analyses of the curved wing model (Case 2, cruise altitude).

known (see [11] or [23]) that approaching the transonic regime, for a fixed value of the angle of attack, because of the strong interaction between the shock wave and the boundary layer, the lift of rigid profiles tend to be reduced when the Mach number increases. This, in addition to the three-dimensional effects generated by the sweep angle, reduces the lift coefficient of the wing (if the angle of attack at the root section is fixed, as in the present analyses). In addition, taking the elasticity effects into account, a reduction in the displacement at the wing tip can be observed.

This phenomenon related to the swept wing is clearly evident in Figure 13. The asymptotic mean value of the vertical displacement of the control node reaches a maximum value for Mach = 0.875 and subsequently falls down for Mach = 0.90 for a fixed geometric angle of attack equal to 0.76 deg. To maintain a similar value of the rigid lift coefficient for Mach = 0.90, the geometric angle of attack of the swept wing model was increased to 0.977 deg, thereby obtaining the upper curve in the graph of Figure 13.

Likewise, for the curved wing, as shown in Figure 14, the asymptotic mean value of vertical displacement falls down for Mach = 0.96.

As the present analyses are aeroelastic, both conditions considered, for which the lift and the deformed shape fall

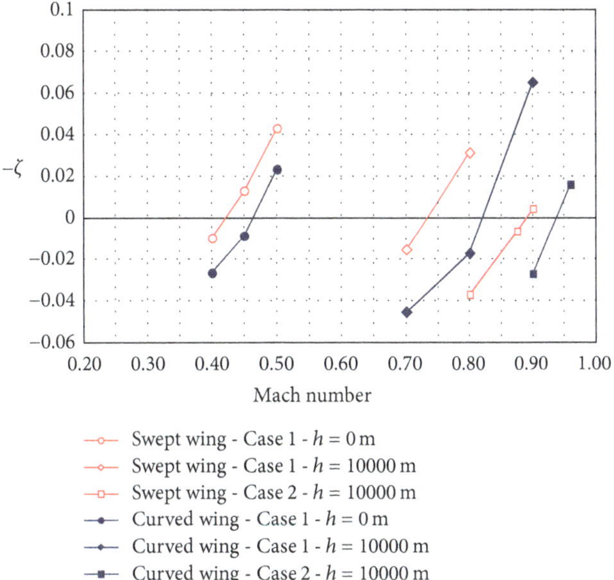

FIGURE 15: Damping ratio versus Mach (the damping data are represented with the opposite sign).

down, represent unstable motions of the wings for Case 2. For the swept wing the unstable small amplitude bending oscillations probably depend directly on the development of buffet phenomenon: a dynamic interaction between the shock wave and the separation of the boundary layer which produces an unsteady variation in the pressure distribution along the chord of the wing. The buffet phenomenon is well known in literature and is studied, as an example, in [24–26] and in the experimental research discussed in [27]. Even if all of these cited works refer to three-dimensional configurations, these studies on transonic flow instabilities concern rigid models: in other words the elasticity effects of the wing box structure are not considered assuming that the buffet phenomenon does not depend on elastic deformations of the lifting surfaces. Thus, in the cited references, the instability conditions of the transonic flow field depend only on the interaction between shock waves and boundary layer separation. To the best of our knowledge, in literature are available some studies that include the elasticity effect for analyzing flutter-buffet interaction but only for two-dimensional configurations (see as an example [28]).

As said above, the instability condition of the swept wing model examined in the present work seems to be related to a flutter-buffet interaction. In Figure 13 it can be seen that the displacement amplitude of the two unstable histories increases slowly: this fact is directly related to periodic oscillations of shock waves, on both upper and lower skin, near the tip zone of the swept wing model. On the other hand for the curved wing, the unstable oscillations are typical of a bending-torsion flutter instability condition. This fact can be confirmed by observing the displacement amplitude of the unstable history in Figure 14: for the curved wing the amplitude of the oscillations increases rapidly, showing the effect of a classical bending-torsion flutter phenomenon.

Figures 13 and 14 highlight that, also in Case 2 (i.e., a fully transonic regime), the curved planform provides increases in flutter speed. For a typical cruise altitude, the swept wing reaches the instability for a Mach number in the range of 0.8–0.9, while the curved wing reaches the instability for a Mach number in the range of 0.9–0.96. In the transonic regime this depends on the different level of energy associated with the development of the shock waves. For a curved planform, the shock waves are weaker, especially toward the tip of the wing (compare Figures 11(a) and 12(a)), and the pressure variations in the chordwise direction are smoother (as discussed in a previous work [16]). Thus by adopting a curved planform wing, the effects of the boundary layer separation can be reduced, and buffeting can thus be delayed.

Figures 15 and 16 compare the analyses. The graphs of the overall damping ratio of the wings were reproduced as a function of the Mach number and True Air Speed, respectively. The damping parameter was estimated using (4) for all the conditions examined.

Case 2 represents a realistic situation by comparing the instability conditions found with the data summarized in [22]. Thus, on the basis of the numerical results obtained in the present research, an increase in the flutter speed greater than 5% can be estimated for the curved wing compared with a conventional swept wing.

An increase in the flutter speed enables aircrafts to be operated with higher commercial velocity values, thus increasing the productivity of a fleet. On the contrary, the better dynamic response of the curved planform wing enables lighter aircrafts to be designed with a similar flutter boundary of traditional swept wing configurations. In this second case, the saving in weight means a reduction in operative costs. The consequent reduction in fuel consumption also reduces the level of pollution.

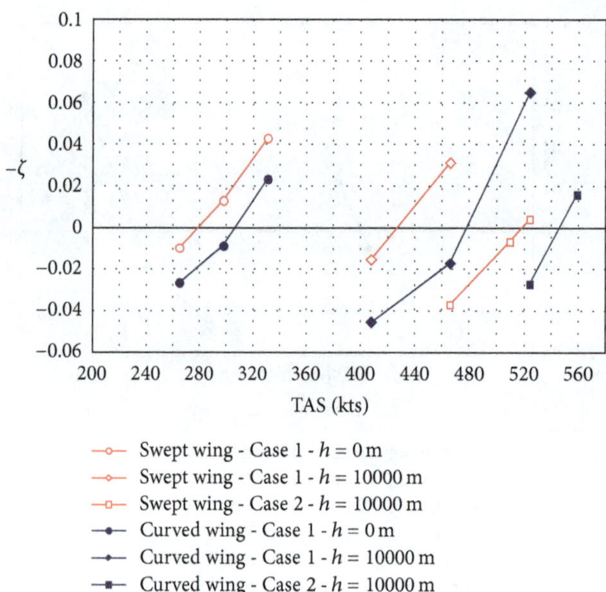

FIGURE 16: Damping ratio versus True Air Speed (the damping data are represented with the opposite sign).

5. Conclusions

A campaign of dynamic fluid structure interaction analyses was carried out to investigate the effects of the planform on the stability properties of high aspect ratio wings. As in previous studies, a comparison between a traditional swept wing and a curved planform wing was performed. The numerical models of the wings were constructed maintaining the same aerodynamic profiles, the same span, aspect ratio, value of the sweep angle at root section, structural layout, and total weight. To carry out the analyses commercial software was used (ANSYS Workbench Rel. 15). Using a blocking procedure (Figures 3(a) and 3(b)) structured grids were generated with the same level of refinement for the aerodynamic wing models. Also the structural meshes of the wings were constructed by adopting a similar layout of the elements. Validation of the modelling technique of the structured aerodynamic grids can be found in [17]: in this work, for a forward swept wing model, the computed distributions of the pressure coefficient were compared with experimental pressure measurements executed in a transonic wind tunnel. The comparison between the numerical and experimental data was excellent.

Previous studies have demonstrated that a curved planform wing reduces the drag coefficient. Present aeroelastic analyses show that this type of wing improves the dynamic performances of transport aircrafts. Adopting a fictitious distribution of moments of inertia applied at the tip of the two finite element models, aimed at reducing the first torsion natural frequency of clamped half wings, the fluid structure interaction analyses provided unstable conditions for both wing models. In addition from the numerical time histories of structural displacements, an estimation of the global damping ratio was obtained.

Our results highlight that (a) in the first case analysed both the swept wing and the curved wing reached the flutter condition (at the sea level and cruise altitude) for a subsonic flow field, and the bending-torsion flutter speeds of the curved wing were greater than swept wing; (b) in the second case (with lower values of the fictitious moments of inertia) at the cruise altitude, the flow fields analysed were fully transonic (as shown in Figures 11(a) and 12(a)), and the swept wing reached a flutter-buffet instability condition (in this case the instability phenomenon was related to the interaction between the shock wave movement and the second bending mode of the wing) earlier than the curved wing, for which a classical bending-torsion flutter developed at higher speed.

The results of the present study demonstrate that from a dynamic point of view (i) a curved planform wing shows an excellent performance also in fully subsonic flight conditions and (ii) in the transonic regime, for a curved planform wing, shock phenomena are less critical compared to those occurring on a conventional swept wing.

Figure 16 represents the calculated damping ratio as a function of the True Air Speed. For the second case examined, from this figure it can be estimated as an increment in the critical speed greater than 5%.

These results agree with the preliminary aeroelastic results obtained with the use of the NASTRAN code discussed in [16] and related to a half wing-body configuration. In other words the curved wing shows a good aeroelastic behaviour compared with a conventional swept wing (in [16] the flutter speed of curved wing was found to be higher especially at sea level). The aerodynamic forces in the NASTRAN code are computed according to a linear approach and do not take into account local phenomena due to boundary layer separation effects or to the dynamic interaction of the shock wave with the oscillations of a wing. In this sense, the previous results were obtained with some limitation on the unsteady aerodynamic field description. Instead, in the present work, by exploiting the hypotheses concerning the topology of the compared fluid dynamic grids used in nonlinear and unsteady FSI analyses, we have established that, also at cruise altitude and in a fully transonic regime reliably simulated, the flutter speed of a curved wing is higher with respect to a conventional swept wing configuration. Thus these results represent a further step in the technical validation of our proposed novel wing configuration.

Competing Interests

The authors declare that there are no competing interests regarding the publication of this paper.

References

[1] G. H. Holdaway and J. A. Mellenthin, "Evaluation of blended wing-body combinations with curved plan forms at mach numbers up to 3.50," NASA 19980227180, 1960.

[2] R. C. Lock, "Design of wing plan forms for transonic speeds," in *The Aeronautical Quarterly, Royal Aeronautical Society*, vol. 12, pp. 65–93, Royal Aeronautical Society, London, UK, 1961.

[3] M. E. Vaughn Jr. and J. E. Burkhalter, "Pressure loading on curved leading edge wings in supersonic flow," *Journal of Aircraft*, vol. 23, no. 7, pp. 574–581, 1986.

[4] J. Ashenberg and D. Weihsradius, "Minimum induced drag of wings with curved planform," *Journal of Aircraft*, vol. 21, no. 1, pp. 89–91, 1984.

[5] C. P. Van Dam, P. M. H. W. Vijgen, and B. J. Holmes, "Experimental investigation on the effect of crescent planform on lift and drag," *Journal of Aircraft*, vol. 28, no. 11, pp. 713–720, 1991.

[6] G. Heller et al., Wing Tip Extension for a Wing, Pub. No. US2002162917A1, November 2002, http://ep.espacenet.com/.

[7] L. B. Gratzer, "Curved wing tip," US Patent 20100181432 A1, 2010.

[8] http://www.boeing.com/commercial/787/.

[9] G. Lombardi, "Experimental study on the aerodynamic effects of a forward-sweep angle," *Journal of Aircraft*, vol. 30, no. 5, pp. 629–635, 1993.

[10] G. Lombardi, M. V. Salvetti, and M. Morelli, "Appraisal of numerical methods in predicting the aerodynamics of forward-swept wings," *Journal of Aircraft*, vol. 35, no. 4, pp. 561–568, 1998.

[11] A. Kuzmin, "Sensitivity analysis of transonic flow over J-78 wings," *International Journal of Aerospace Engineering*, vol. 2015, Article ID 579343, 6 pages, 2015.

[12] O. Brodersen and A. Stürmer, "Drag prediction of Engine-Airframe interference effects using unstructured Navier-Stokes calculations," in *Proceedings of the 19th AIAA Applied Aerodynamics Conference*, AIAA 2001-2014, Anaheim, Calif, USA, June 2001.

[13] J. D. Cole and N. D. Malmuth, "Wave drag due to lift for transonic airplanes," *Proceedings of the Royal Society of London—Series A: Mathematical, Physical and Engineering Sciences*, vol. 461, no. 2054, pp. 541–560, 2005.

[14] O. Gur, W. H. Mason, and J. A. Schetz, "Full-configuration drag estimation," *Journal of Aircraft*, vol. 47, no. 4, pp. 1356–1367, 2010.

[15] M. R. Chiarelli, M. Cagnoni, M. Ciabattari, M. De Biasio, and A. Massai, "Preliminary analysis of a high aspect ratio wing with curved planform," in *Proceedings of the Associazione Italiana di Aeronautica e Astronautica (AIDAA '09)*, pp. 978–988, Milan, Italy, July 2009.

[16] M. R. Chiarelli, M. Cagnoni, M. Ciabattari, M. De Biasio, and A. Massai, "High aspect ratio wing with curved planform: CFD and FE analyses," in *Proceedings of the 27th Congress of the International Council of the Aeronautical Sciences (ICAS '10)*, pp. 1524–1533, Nice, France, September 2010.

[17] M. R. Chiarelli, G. Lombardi, and A. Nibio, "A straight wing and a forward swept wing compared with a curved planform wing in the transonic regime," in *Proceedings of the International Conference of the European Aerospace Societies (CEAS '11)*, Venice, Italy, 2011.

[18] M. R. Chiarelli, M. Ciabattari, M. Cagnoni, and G. Lombardi, "The effects of the planform shape on drag polar curves of wings: fluid-structure interaction analyses results," in *STAR Global Conference 2013*, Orlando, Fla, USA, March 2013.

[19] M. R. Chiarelli, M. Ciabattari, M. Cagnoni, and G. Lombardi, "A comparison of the drag polar curves of wings using the fluid-structure interaction analyses," in *Proceedings of the Conference of the Italian Association of Aeronautics and Astronautics (AIDAA '13)*, Naples, Italy, 2013.

[20] M. R. Chiarelli, M. Ciabattari, M. Cagnoni, and G. Lombardi, "Fluid-structure interaction analyses of wings with curved planform: preliminary aeroelastic results," in *Proceedings of the Common European Asylum System (CEAS '13)*, Linköping, Sweden, June 2013.

[21] C. D. Harris, "NASA supercritical airfoils, a matrix of family-related airfoils," NASA Technical Paper 2969, Langley Research Center, Hampton, Va, USA, 1990.

[22] K. Koenig, "Pretension and reality of flutter-relevant tests," in *Advanced Aeroservoelastic Testing and Data Analysis*, AGARD-CP-566, North Atlantic Treaty Organization, Brussels, Belgium, 1995.

[23] D. G. Mabey, "Physical phenomena associated with unsteady transonic flows," in *Unsteady Transonic Aerodynamics—Progress in Astronautics and Aeronautics*, vol. 120, American Institute of Aeronautics and Astronautics, Washington, DC, USA, 1989.

[24] J. D. Crouch, A. Garbaruk, D. Magidov, and A. Travin, "Origin of transonic buffet on aerofoils," *Journal of Fluid Mechanics*, vol. 628, pp. 357–369, 2009.

[25] F. Sartor and S. Timme, "Reynolds-averaged Navier-Stokes simulations of shock buffet on half wing-body configuration," in *Proceedings of the 53rd AIAA Aerospace Sciences Meeting*, Kissimmee, Fla, USA, 2015.

[26] M. Iovnovich and D. E. Raveh, "Numerical study of shock buffet on three-dimensional wings," *AIAA Journal*, vol. 53, no. 2, pp. 449–463, 2015.

[27] J. Dandois, "Experimental study of transonic buffet phenomenon on a 3D swept wing," *Physics of Fluids*, vol. 28, no. 1, Article ID 016101, 2016.

[28] W. Zhang, C. Gao, Y. Liu, Z. Ye, and Y. Jiang, "The interaction between flutter and buffet in transonic flow," *Nonlinear Dynamics*, vol. 82, no. 4, pp. 1851–1865, 2015.

4

Numerical Simulation of Unsteady Conjugate Heat Transfer of Electrothermal Deicing Process

Zuodong Mu, Guiping Lin, Xiaobin Shen, Xueqin Bu, and Ying Zhou

Laboratory of Fundamental Science on Ergonomics and Environmental Control, Beihang University, Beijing 100191, China

Correspondence should be addressed to Xiaobin Shen; shenxiaobin@buaa.edu.cn

Academic Editor: Paul Williams

A novel 3-D unsteady model of in-flight electrothermal deicing process is presented in this paper to simulate the conjugate mass and heat transfer phenomena of water film runback, phase change, and solid heat conduction. Mathematical models of water film runback and phase change are established and solved by means of a loosely coupled method. At the current time step, solid heat conduction, water film runback, and phase change are iteratively solved until the heat boundary condition reaches convergence, then the temperature distribution and ice shape at the moment are obtained, and the calculation of the next time step begins subsequently. A deicing process is numerically simulated using the present model following an icing tunnel experiment, and the results match well with those in the literatures, which validate the present model. Then, an in-flight deicing process is numerically studied to analyze the effect of heating sequence.

1. Introduction

The water droplet in clouds may remain in a liquid state even if the temperature is below freezing point due to the surface tension and a lack of condensation nucleus. Supercooled water droplet would solidify when impinging on windward surfaces of aircraft, which would cause the deterioration in the aerodynamic performance due to the change of aerodynamic configuration [1]. Aircraft icing would impose serious adverse effects on flight safety, which has long been recognized as an important issue to prevent.

In view of the serious threats that ice accretion would impose on flight safety, ice protection methods must be applied to prevent or control the ice accretion. Hot air anti-icing method is widely applied in commercial jets. The bleed air from engine compressor impinges on the structure to heat the surface, so that the water evaporates or stays in a liquid state rather than freezes. Electrothermal ice protection method uses heating pads, which are incorporated in the multilayer structure, to heat the surfaces. Electrothermal pads are available in both anti-icing mode with a constant power density and deicing mode with a periodic power density. Besides, some other deicing methods have been developed, such as the thermomechanical expulsion deicing method [2], while they are not widely applied. A large amount of bleed air is needed in hot air anti-icing system, which would significantly affect the performance of an engine, especially during the takeoff or landing period. Meanwhile, the jet flow would cause excessive temperature at the impingement point, which may damage the composite materials that are widely used in a modern aircraft [3] or even the aluminum skin. The electrothermal ice protection method has a higher heating efficiency and a lower power density, which results in the advantages in structure safety and energy efficiency. As the concepts of more-electric-aircraft and all-electric-aircraft are increasingly employed in a modern aircraft design, the electrothermal ice protection method is drawing more attention. The ice protection system of B-787 [4], which is based on the electrothermal method, serves as an example.

Since the icing tunnel experiments and the flight experiments are complex, expensive, and not able to cover all environment conditions, the investigation of a deicing process by an experimental method is quite limited and seldom reported to date. The numerical simulation method, which is time-efficient and cost-effective, becomes an important tool for the design of the ice protection system. The in-flight deicing process is a conjugate mass and heat transfer phenomenon, which is very complex and consists of a variety of coupled processes such as the air-droplet flow, water film runback and phase transition, and multilayer solid heat conduction. The phase state varies with both spatial locations and time. Due to the periodic power density, spatial distribution of heaters, and sustained droplet impingement, phase transition phenomena, such as evaporation, solidification, liquefaction, and sublimation, may coexist on the surface of the protection area, which makes it complex and typically unsteady.

The early study of electrothermal deicing concentrated on the solution of multilayer structure heat conduction. Stallabrass [5] numerically analyzed the heat conduction during the deicing process, focusing on the effects of multilayer materials on the distribution of skin temperature, especially the effects of inner and outer insulating layers. Then, the latent heat was introduced to simulate the phase change during the deicing process by Baliga [6]. As the research went in depth, Marano [7] and Roelke et al. [8] developed the "enthalpy method," in which the unified energy equation was applied in the whole computational domain and the phase interface was determined by enthalpy distribution. Based on the enthalpy method, plenty of researches were conducted to simulate the phase transition process. Chao [9], Leffel [10], and Masiulaniec [11] applied the method to 2-D cases and analyzed the effects of surface curvature and heating pad spatial distribution. Chang [12, 13] numerically studied the 2-D deicing process by a finite volume method and analyzed the effects of the heating sequence and power density. Yaslik et al. [14] studied a 3-D deicing process, using a Douglas method to discretize the heat conduction equation. Xiao [15] applied a porous medium method to simulate the ice melting process. As is briefly reviewed above, researchers have done much work on the study of multilayer structure heat conduction and ice melting process based on enthalpy method, while the study of the in-flight deicing process, which involves sustained droplet impingement and water film runback, is quite rare.

Wright et al. [16] developed a 2-D deicing model, which was based on the classic Messinger mass and energy balance models [17]. The mass and energy balance equations took into account the input water caused by droplet impingement, but the water film runback mechanism was not taken into consideration. Habashi et al. [18–20] developed a conjugate model for the in-flight deicing process. The airflow and droplet impingement were calculated as initial solutions, and the solid conduction was solved, which was coupled with a water phase transition in a loosely coupled way. Finite element solvers were applied to the solution as the modules of the commercial software FENSAP-ICE, while the detailed solution method was not available in the open literature.

As electrothermal deicing method is increasingly employed in the new generation of aircraft, the study on the mechanism and simulation methods is urgently needed. While most current studies on the deicing process concentrated on the multilayer structure heat conduction and ice melting process, studies available on the in-flight deicing were quite rare.

This paper focuses on the conjugate heat transfer mechanism of the in-flight electrothermal deicing. A novel 3-D unsteady model is established based on the water film runback dynamic mechanism and phase transition thermodynamic model and solved by a loosely coupled method. Using the present model, the deicing process is numerically investigated, which contributes to a better understanding of electrothermal deicing so as to guide the design and optimization of an aircraft ice protection system.

2. Mathematical Model

The in-flight deicing process is a conjugate mass and heat transfer process, which involves air-droplet flow, solid heat conduction, water film runback, and the phase transition. The mass and energy balance, as is briefly shown in Figure 1, is determined by a variety of factors including convection, droplet impingement, evaporation, solidification, heat conduction, and film runback. Due to the periodic power density and spatial distribution of heaters, the phase transition process is typically unsteady.

2.1. Air-Droplet Flow. Since the volume fraction of droplet is very small, typically under 10^{-6} for icing conditions, the air-droplet flow can be solved by a one-way coupled method [21]. Reynolds-averaged Navier–Stokes equations (RANS) are applied to solve the air flow field. The mass and momentum equations are expressed as (1), and the energy equation is expressed as (2):

$$\frac{\partial \rho_a}{\partial t} + \frac{\partial(\rho_a u_i)}{x_i} = 0,$$
$$\frac{\partial}{\partial t}(\rho_a u_i) + \frac{\partial}{\partial x_j}(\rho_a u_i u_j) = -\frac{\partial p}{\partial x_i} + \frac{\partial}{\partial x_j}\left[\mu\left(\frac{\partial u_i}{\partial x_j} + \frac{\partial u_j}{\partial x_i} - \frac{2}{3}\delta_{ij}\frac{\partial u_l}{\partial x_l}\right)\right] + \frac{\partial}{\partial x_j}\left(-\rho_a \overline{u_i' u_j'}\right), \quad (1)$$

$$\frac{\partial}{\partial t}(\rho_a E) + \frac{\partial}{\partial x_i}[u_i(\rho_a E + p)] = \frac{\partial}{\partial x_j}\left[\left(\lambda + \frac{c_p \mu}{\Pr}\right)\frac{\partial T}{\partial x_j}\right] + \frac{\partial}{\partial x_j}\left[u_i(\mu + \mu_t)\left(\frac{\partial u_i}{\partial x_j} + \frac{\partial u_j}{\partial x_i} - \frac{2}{3}\delta_{i,j}\frac{\partial u_l}{\partial x_l}\right)\right], \quad (2)$$

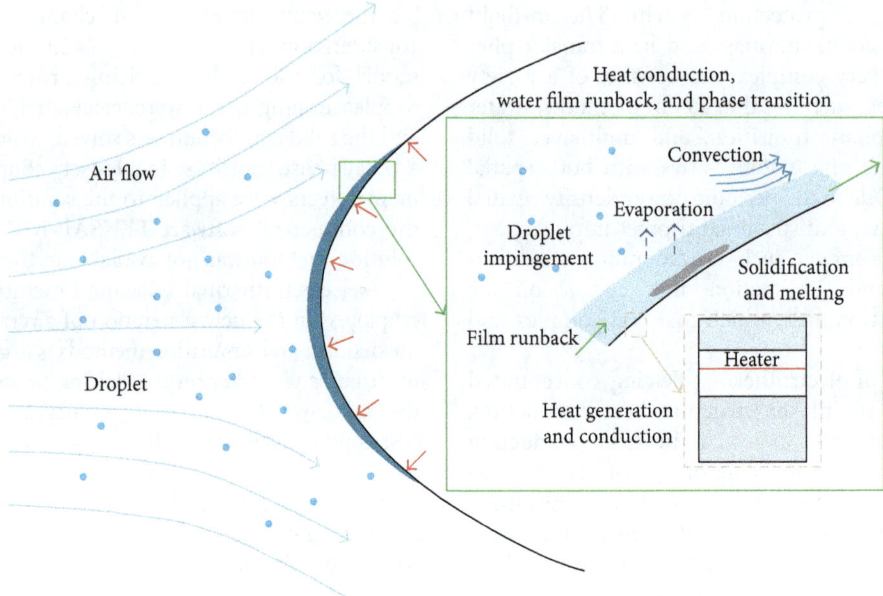

FIGURE 1: Mass and energy balance of deicing process.

where the Reynolds stress is defined by Boussinesq as

$$-\rho\overline{u_i'u_j'} = \mu_t\left(\frac{\partial u_i}{\partial x_j} + \frac{\partial u_j}{\partial x_i}\right) - \frac{2}{3}\left(\rho k + \mu_t\frac{\partial u_k}{\partial x_k}\right)\delta_{ij}, \quad (3)$$

where μ_t is the turbulent viscosity, u_i is the average velocity, and k is the turbulence kinetic energy.

The continuity and momentum equations of droplet flow are expressed as [22]

$$\frac{\partial(\rho\alpha)}{\partial t} + \nabla\cdot(\rho\alpha\mathbf{u}) = 0, \quad (4)$$

$$\frac{\partial(\rho\alpha\mathbf{u})}{\partial t} + \nabla\cdot(\rho\alpha\mathbf{u}\mathbf{u}) = \rho\alpha K(\mathbf{u}_a - \mathbf{u}) + \rho\alpha\mathrm{F}, \quad (5)$$

where ρ is the density of water, α is the droplet volume fraction, $\mathbf{u}$ is the velocity vector of droplet, $\mathbf{u}_a$ is the velocity vector of air, $\rho\alpha\mathrm{F}$ is the external body forces exerted on droplet, such as the gravity or inertial force, and K is the air-droplet momentum exchanger coefficient defined as

$$K = \frac{18\mu f_{\text{drag}}}{\rho d_p^2}, \quad (6)$$

where μ is the dynamic viscosity of air, d_p is the diameter of droplet, and f_{drag} is the drag function. The Schiller and Naumann model is adopted here.

$$f_{\text{drag}} = \frac{C_D \mathrm{Re}_r}{24}.$$

$$C_D = \begin{cases} \dfrac{24\left(1 + 0.15\mathrm{Re}_r^{0.687}\right)}{\mathrm{Re}_r}, & \mathrm{Re}_r \le 1000, \\ 0.44, & \mathrm{Re}_r > 1000. \end{cases} \quad (7)$$

Re_r is the relative Reynolds number and is given as

$$\mathrm{Re}_r = \frac{\rho_a|\mathbf{u}_a - \mathbf{u}|d_p}{\mu}. \quad (8)$$

2.2. Water Film Runback and Phase Transition. The differential form of continuity equation for runback water film on the surface of ice protection area is expressed as

$$\frac{\partial\rho}{\partial t} + \mathrm{div}(\rho v) = \dot{m}_{\text{imp}} - \dot{m}_{\text{ice}} - \dot{m}_{\text{evap}}, \quad (9)$$

where v is the velocity of water film, $\dot{m}_{\text{imp}}$ is the water mass of droplet impingement, $\dot{m}_{\text{ice}}$ is the icing rate, and $\dot{m}_{\text{evap}}$ is the evaporation rate. The water film is incompressible, and the development of water film is quick due to a quite small thickness. Therefore, the unsteady water mass term is neglected, and the continuity equation is derived by integrating the differential form in a control volume.

$$\sum\dot{m}_{\text{in},n} + \dot{m}_{\text{imp}} = \sum\dot{m}_{\text{out},n} + \dot{m}_{\text{ice}} + \dot{m}_{\text{evap}}, \quad (10)$$

where $\sum\dot{m}_{\text{in},n}$ is the total mass of water entering the current control volume from adjacent ones per unit time, $\sum\dot{m}_{\text{out},n}$ is the total mass of water flowing out of the control volume, and the mass flux through a face n can be expressed as

$$\dot{m}_{\text{out},n} = \rho l_n h(\mathbf{v}\cdot\mathbf{n}_n), \quad (11)$$

where h is the film thickness, l_n is the edge length, and $\mathbf{n}_n$ represents the unit vector normal to face n.

The mass flux of droplet impingement is determined by local collection efficiency β, as expressed as

$$\dot{m}_{\text{imp}} = u_\infty\cdot\mathrm{LWC}\cdot\beta\cdot A, \quad (12)$$

where u_∞ is the velocity of droplet at far field, LWC is the liquid water content, and A is the area of control volume.

The evaporation rate is determined by the Chilton-Colburn analogy theory where the convective mass transfer

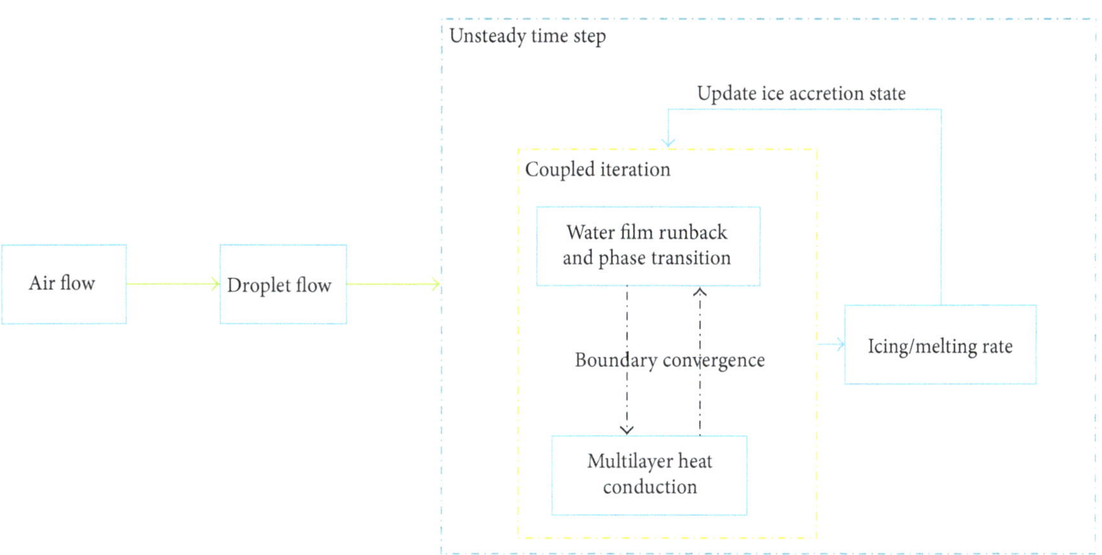

FIGURE 2: Diagram of coupling solution method.

coefficient k_c is obtained from the convective heat transfer coefficient h_c as shown below.

$$\frac{h_c}{c_p}\mathrm{Pr}^{2/3} = k_c \mathrm{Sc}^{2/3}, \tag{13}$$

where Sc is the Schmidt number. The convective heat transfer coefficient h_c is calculated according to the surface heat flux and the temperature difference.

$$h_c = \frac{q}{A \cdot \left[T - T_\infty\left(1 + f_{\mathrm{rec}}(\gamma - 1/2)Ma^2\right)\right]}, \tag{14}$$

where f_{rec} is the recovery coefficient; γ is the air specific heat ratio, of which the value is 1.4; and Ma is the Mach number of air flow. Then the evaporation rate is derived as

$$\dot{m}_{\mathrm{evap}} = \frac{h_c A}{c_{p,\mathrm{air}}} \cdot \frac{\mathrm{MW}_{\mathrm{water}}}{\mathrm{MW}_{\mathrm{air}}} \cdot \left(\frac{\mathrm{Pr}}{\mathrm{Sc}}\right)^{2/3} \cdot \left(\frac{p_{v,\mathrm{sat}}}{p_T} \cdot \frac{T_T}{T} \cdot \left(\frac{p_T}{p}\right)^{1/\gamma} - \frac{p_{v,\infty,\mathrm{sat}} r_h}{p_\infty}\right), \tag{15}$$

where MW is the air molecular weight, $c_{p,\mathrm{air}}$ is the air specific heat, $p_{v,\mathrm{sat}}$ is the vapor saturated pressure, p_T is the total pressure, T_T is the total temperature, and r_h is the relative humidity, of which the value on water film surface is 1.

The momentum equation of water film runback is given by incompressible Navier–Stokes equation as

$$\frac{\partial}{\partial t}(\rho \mathbf{v}) + \nabla \cdot (\rho \mathbf{v}\mathbf{v} + p\mathbf{I}) = \frac{\partial}{\partial y}\left(\mu \frac{\partial \mathbf{v}}{\partial y}\right) + \rho g, \tag{16}$$

where p is the pressure, $\mathbf{I}$ is the unit tensor, y is the coordinate normal to surface, and g is the gravitational acceleration.

The air-film boundary condition is defined as

$$\mu \frac{\partial \mathbf{v}}{\partial y} = \tau + \sigma\kappa, \quad y = h, \tag{17}$$

where τ is the shear stress, σ is the coefficient of tension, and κ is the water film surface curvature which is given by

$$\kappa = -\frac{\partial^2 h}{\partial x_j^2}\left[1 + \left(\frac{\partial h}{\partial x_j}\right)^2\right]^{-3/2}. \tag{18}$$

The solid-film boundary is depicted under the nonslip boundary condition.

$$\begin{gathered} \mathbf{v} = 0, \\ y = 0. \end{gathered} \tag{19}$$

Previous studies suggest that the effects of pressure gradient should only be considered for water film with a large thickness [23]. During the deicing process, the water film is very thin; therefore, the terms of gradient and gravity are negligible, of which the effect is slight compared with other terms such as shear stress. The tangential velocity gradient is also very small; therefore, the momentum diffusion term is negligible [24]. The value of surface curvature is generally very small for film which entirely wets the surface, and the shear stress is the dominant factor at the air-film interface. Under such condition, the momentum equation and boundary condition of water film are simplified as

$$\begin{gathered} \frac{\partial}{\partial y}\left(\mu \frac{\partial \mathbf{v}}{\partial y}\right) = 0, \\ \mu \frac{\partial \mathbf{v}}{\partial y} = \tau, \quad y = h, \\ \mathbf{v} = 0, \quad y = 0. \end{gathered} \tag{20}$$

The velocity distribution normal to the surface is derived as

$$\mathbf{v} = \frac{y}{\mu}\tau. \tag{21}$$

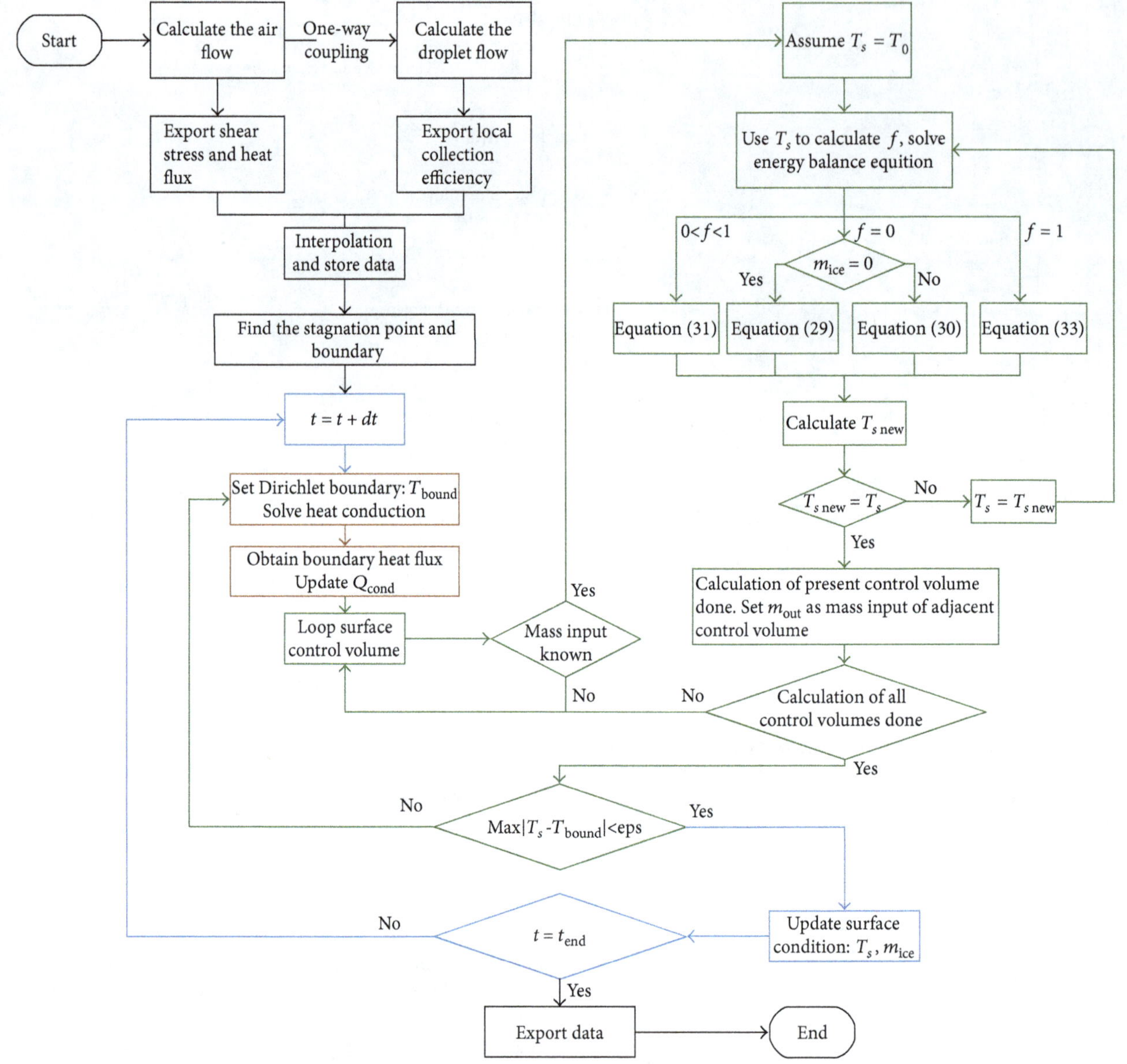

Figure 3: Flowchart of solution procedure.

The average velocity V is obtained by integrating the above equation along the film thickness direction, as expressed as

$$V = \frac{h}{2\mu}\tau. \tag{22}$$

The energy balance of water film is described by the following differential equation:

$$\frac{\partial(\rho c_w T)}{\partial t} + \operatorname{div}(\rho c_w T V) = H_{\text{imp}} + E_{\text{ice}} - Q_c - H_{\text{evap}} + Q_{\text{cond}}. \tag{23}$$

Integrating the above equation, the energy balance in a control volume is expressed as

Table 1: Environment conditions of NASA experiment.

Temperature (K)	Velocity (m/s)	LWC (g/m^3)	MVD (μm)	AoA
266.48	44.7	0.78	20	0

$$\frac{\rho A h c_w (T - T_{\text{pre}})}{\Delta t} + \sum_n H_{\text{out},n} - \sum_n H_{\text{in},n} = H_{\text{imp}} + E_{\text{ice}} - Q_c - H_{\text{evap}} + Q_{\text{cond}}, \tag{24}$$

where c_w is the specific heat of water, T and T_{pre} are the temperature at current time step and previous time step, Q_{cond} is the deicing heat flux from solid, obtained from the calculation of solid heat conduction during the iterations, and

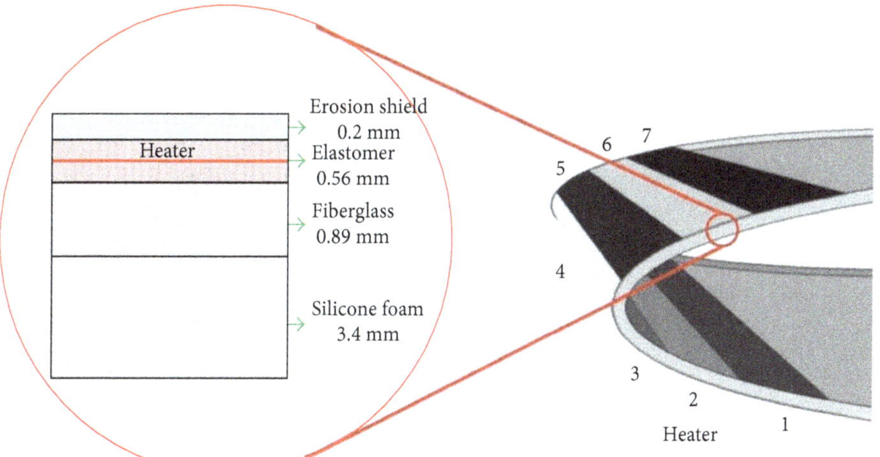

Figure 4: Material and structure of ice protection area.

$H_{\text{in},n}$ is the energy carried by runback water entering the current control volume through the face n, and it can be expressed as

$$H_{\text{in},n} = \dot{m}_{\text{in},n} c_w \left(T_{\text{in},n} - T_0\right). \quad (25)$$

H_{imp} is the energy of impinging water, which consists of two parts: the internal energy and the kinetic energy, as expressed below.

$$H_{\text{imp}} = \dot{m}_{\text{imp}} \left[c_w \left(T_\infty - T_0\right) + \frac{1}{2} u_\infty^2\right]. \quad (26)$$

Q_c is the convective heat flux obtained by the convective heat transfer coefficient h_c and the temperature difference between surface temperature T and recover temperature T_{rec}.

$$Q_c = h_c A \left(T - T_{\text{rec}}\right). \quad (27)$$

The latent heat of solidification E_{ice}, the out-flow energy $H_{\text{out},n}$, and the latent heat of evaporation H_{evap} are related to the phase condition of the control volume, which are discussed as follows by considering the freezing fraction f, namely the ratio of icing mass to the total entering water mass.

If $f = 0$ and the surface is clean, which means currently that there is no ice accretion, the water in the control volume remains at a liquid state, icing rate is zero, and the energy of runback water and evaporation can be expressed as

$$\begin{aligned} \dot{m}_{\text{ice}} &= 0, \\ E_{\text{ice}} &= 0, \\ H_{\text{out},n} &= \dot{m}_{\text{out},n} c_w \left(T_s - T_0\right) + \frac{1}{2} \dot{m}_{\text{out},n} V^2, \\ H_{\text{evap}} &= \dot{m}_{\text{evap}} \left[L_{\text{evap}} + c_w \left(T_s - T_0\right)\right], \end{aligned} \quad (28)$$

where L_{evap} is the latent heat of water evaporation.

If $f = 0$ and there is ice accretion in the current volume, the ice melts at a liquid-solid mixed state and the control volume remains at the phase transition temperature.

Table 2: Material properties of NASA experiment.

Material	Density (kg/m^3)	Heat conductivity (W/mK)	Heat capacity (J/kgK)
Erosion shield	8025.25	16.26	502.4
Elastomer	1383.96	0.2561	1256.0
Fiberglass	1794	0.294	1570.1
Silicone foam	648.75	0.121	1130.4

$$\begin{aligned} T_s &= T_0, \\ E_{\text{ice}} &= \dot{m}_{\text{melt}} L_s, \\ H_{\text{out},n} &= 1/2 \dot{m}_{\text{out},n} V^2, \\ H_{\text{evap}} &= \dot{m}_{\text{evap}} L_{\text{evap}}, \end{aligned} \quad (29)$$

where $\dot{m}_{\text{melt}}$ is the melting rate and L_s is the solidification latent heat.

If $0 < f < 1$, part of water freezes, and the control volume is at the phase transition temperature.

$$\begin{aligned} T_s &= T_0, \\ E_{\text{ice}} &= \dot{m}_{\text{ice}} L_s, \\ H_{\text{out},n} &= 1/2 \dot{m}_{\text{out},n} V^2, \\ H_{\text{evap}} &= \dot{m}_{\text{evap}} L_{\text{evap}}. \end{aligned} \quad (30)$$

If $f = 1$, all the water freezes and the energy terms can be expressed as

$$\begin{aligned} E_{\text{ice}} &= \dot{m}_{\text{ice}} \left[L_s + c_{\text{ice}} \left(T_0 - T_s\right)\right], \\ \dot{m}_{\text{out},n} &= 0, \\ H_{\text{out},n} &= 0, \\ H_{\text{evap}} &= \dot{m}_{\text{evap}} \left[L_{\text{sub}} + c_{\text{ice}} \left(T_s - T_0\right) - L_s\right], \end{aligned} \quad (31)$$

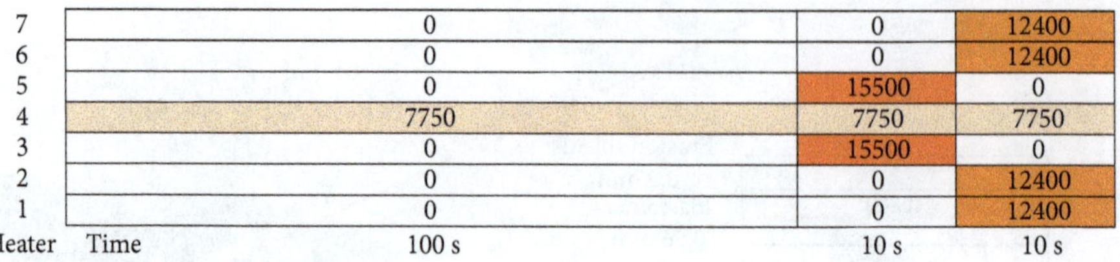

Heater	Time 100 s	10 s	10 s
7	0	0	12400
6	0	0	12400
5	0	15500	0
4	7750	7750	7750
3	0	15500	0
2	0	0	12400
1	0	0	12400

FIGURE 5: Heating sequence of NASA experiment (power in W/m2).

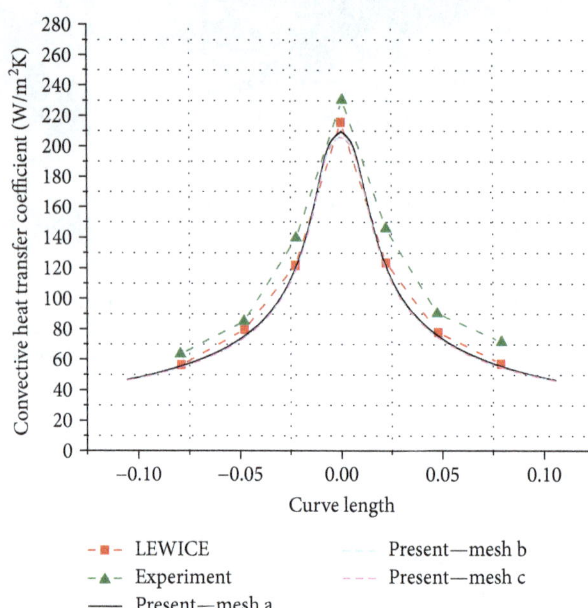

FIGURE 6: Comparison of convective heat transfer coefficient.

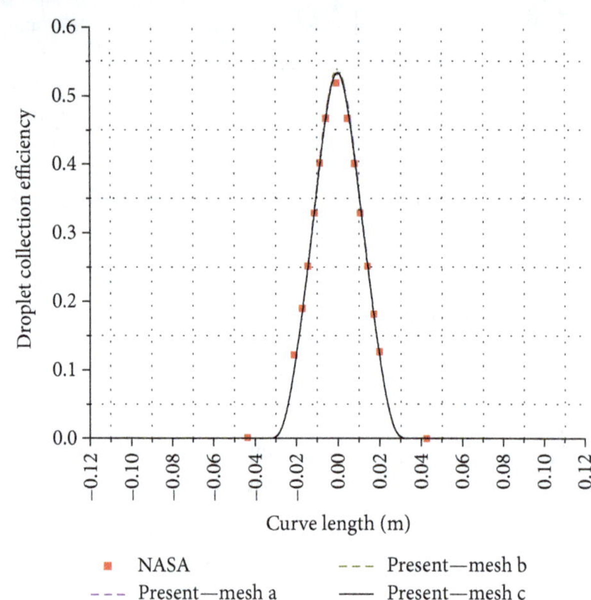

FIGURE 7: Comparison of droplet collection efficiency.

where c_{ice} is the specific heat of ice and L_{sub} is the sublimation latent heat.

2.3. Heat Conduction. The unsteady heat conduction through the multilayer materials is expressed as

$$\frac{\partial H}{\partial t} = \nabla \cdot (\lambda \nabla T) + S, \tag{32}$$

where H is the enthalpy of the material, λ is the thermal conductivity, and S is the heat source term. For electrothermal deicing process, the source term is determined by the spatial location of heaters and the power sequence.

The Dirichlet heat boundary condition of heat conduction is provided by the solution of water film runback and phase transition during the coupled solution. In return, the deicing heat flux is calculated and sent to the calculation of water film. The deicing heat flux is obtained at the boundary of a solid structure as shown in

$$Q_{\text{cond}} = -A\lambda \frac{\partial T}{\partial \mathbf{n}}, \tag{33}$$

where $\mathbf{n}$ is the normal vector of the boundary surface.

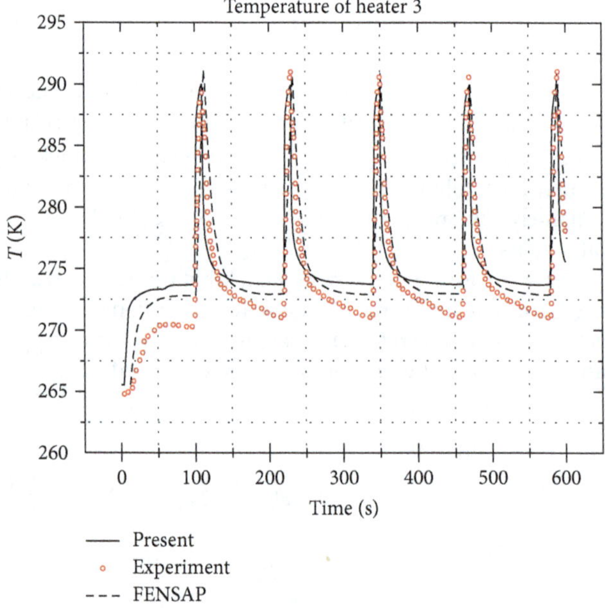

FIGURE 8: Comparison of temperature of heater 3.

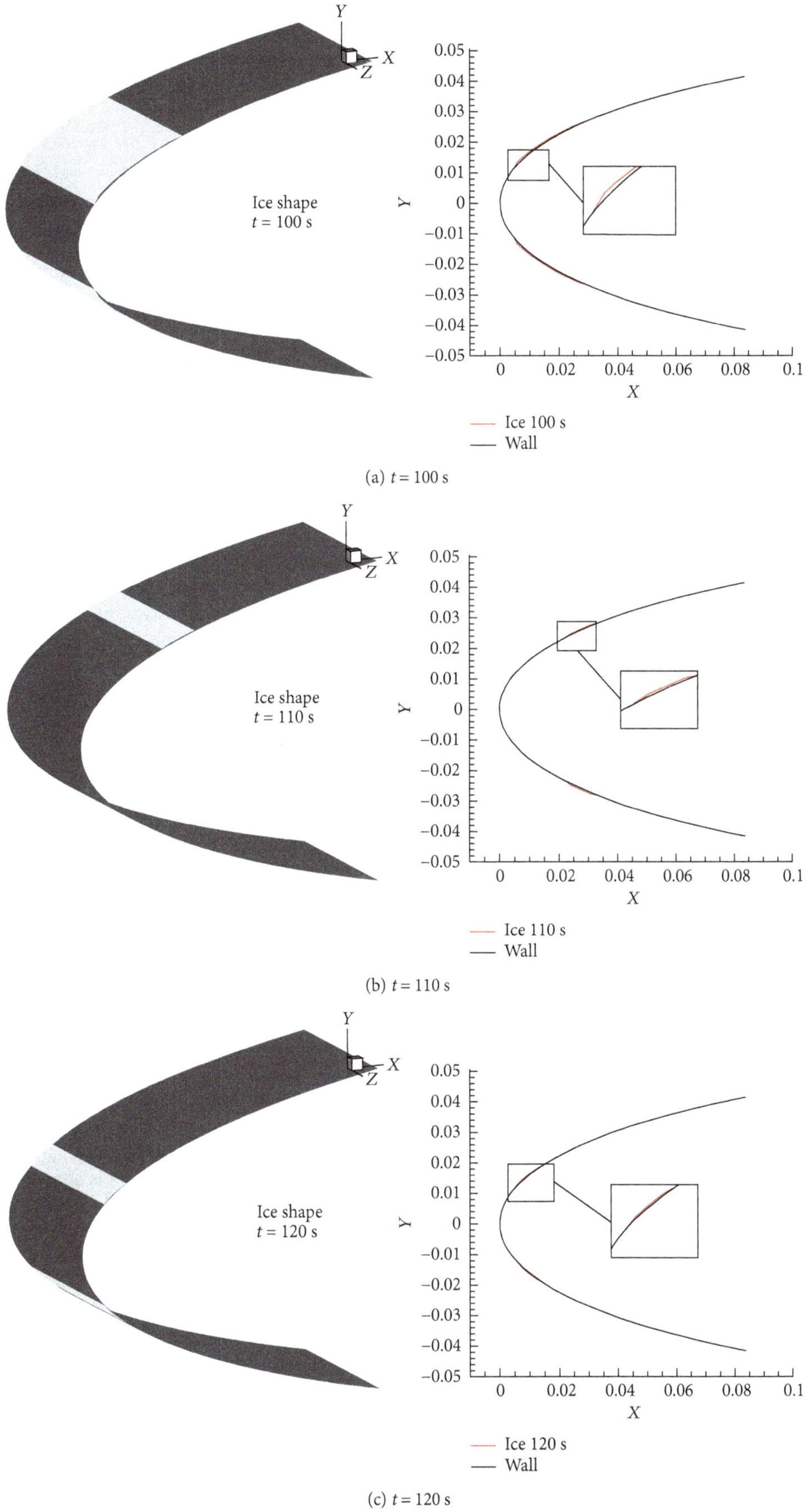

(a) $t = 100$ s

(b) $t = 110$ s

(c) $t = 120$ s

FIGURE 9: Simulated ice shape following NASA experiment.

3. Solution Procedure

During the solution of the above unsteady conjugate heat transfer model, the water film runback and phase transition are coupled with the solid heat conduction, due to the fact that the solution of the solid heat conduction provides the interface heat flux which is needed in the water film energy balance equation; on the other hand, the boundary condition of the solid heat conduction is provided by the solution of the film runback and phase transition. A loosely coupled method is applied to solve the present model, in which both the water film runback and the heat conduction are iteratively calculated until the heat boundary condition reaches convergence, and during each iteration, the surface temperature T and heat flux Q_{cond} are exchanged, as briefly shown in Figure 2.

Considering that the ice thickness during deicing process is typically controlled at a very small value, the effect of the ice shape on air-droplet flow is slight. As a result, the steady airflow solution is computed as an initial condition and is assumed unchanged during the simulation. Besides, it is accurate as long as the ice thickness does not exceed a certain limit due to a protection failure. The RANS equations of airflow are discretized, using a finite volume method in a second order upwind scheme. To simulate the turbulent flow, the transition SST model, which is shown to obtain good results for wall-bounded flows, such as the airfoil, is utilized. A CFD solver FLUENT is used to solve the governing equations. The governing equations of droplet flow field are solved by the finite volume method using the User-Defined Scalar (UDS) transport equation. The droplet volume fraction and velocity components are set as the UDS, and the convective terms, diffusion terms, and source terms are defined by codes which are programmed using the User-Defined Functions (UDF). The solution of air-droplet flow is the initial condition of water film runback and solid heat conduction, and the data exchange is achieved by interpolation due to the difference between flow field mesh and solid mesh.

A loosely coupled method is applied to solve the deicing model. At the current time step, both the solid heat conduction and the water film runback are iteratively calculated until convergence. During each iteration, the surface temperature and heat flux are exchanged, and the boundary conditions are updated. The ice shape is obtained by the icing rate or melting rate, and the calculation of the next time step then begins. The flowchart is shown in Figure 3, and some brief introduction is provided below for the flowchart.

(1) Solve the air-droplet flow and transfer the data by interpolation;

(2) Loop all the surface control volumes and check them. If the input water mass is already known, assume an initial temperature, solve the mass and energy balance equations, update the value of temperature and phase state, and provide input conditions for adjacent volumes. Keep the calculation of water film runback and phase transition until the calculations of all control volumes are done; the temperature distribution at boundary surface is obtained;

Table 3: Environment conditions of in-flight case.

Temperature (K)	Velocity (m/s)	LWC (g/m^3)	MVD (μm)	AoA
263	97.6	1	20	0

(3) Set Dirichlet boundary condition for solid heat conduction using the temperature distribution of step (2) and solve the heat conduction; calculate the deicing heat flux at boundary.

(4) Update the deicing heat flux of water film energy equation using the data of step (3).

(5) Repeat steps (2)–(4) until convergence, calculate ice accretion at the moment, and advance to the next time step.

4. Results and Discussion

Validations of the present model are conducted, and the results are compared with the experimental data. Then, a simulation of in-flight deicing process is conducted to optimize the heating sequence.

4.1. NASA Deicing Experiment in the Lewis Icing Research Tunnel. In order to perform the in-flight deicing simulation, a deicing experiment model is selected in the very rare records available in the open literature, which is conducted in the NASA Lewis Icing Research Tunnel (IRT) by Al-Khalil et al. [25].

The experiment model is a NACA 0012 airfoil with a chord of 0.914 m (36 in), and the environment conditions are listed in Table 1. The multilayer structure at the leading edge protection area is composed of four different layers, which are the erosion shield, the elastomer, the fiberglass, and the silicone foam, with a thickness of 0.2 mm, 0.56 mm, 0.89 mm, and 3.4 mm, respectively. The structure is shown in Figure 4, and the material properties are listed in Table 2. Seven heating pads are arranged in the elastomer layer, of which the heating sequence and power density are controlled independently. The length is 1.905 cm for heater 4; 2.54 cm for heaters 2, 3, 5, and 6; and 3.81 cm for heaters 1 and 7. The heating sequence is shown in Figure 5. Heater 4 acts as the heat blade, which keeps activated during the whole cycle to avoid ice accretion on the leading edge. The rear heating pads are activated alternately after 100 s.

The steady air-droplet flow was solved to obtain parameters such as the convective heat flux, the shear stress, and the local droplet collection efficiency. Then, the data were transferred to solid mesh by interpolation, and the unsteady deicing process was coupled solved. Structured grids were generated for the solution of air-droplet flow. To verify the mesh independence of the solution, three mesh files were applied, and the normal distance of the first layer is 0.01 mm (mesh a), 0.0075 mm (mesh b), and 0.005 mm (mesh c), respectively. The y^+ at the ice protection area of all three mesh files was controlled around or lower than 1.

Heater				
7		0	0	15000
6		0	0	15000
5		0	15000	0
4		10000	10000	10000
3		0	15000	0
2		0	0	15000
1		0	0	15000
Heater	Time	75 s	25 s	25 s

FIGURE 10: Initial heating sequence of in-flight case (power in W/m2).

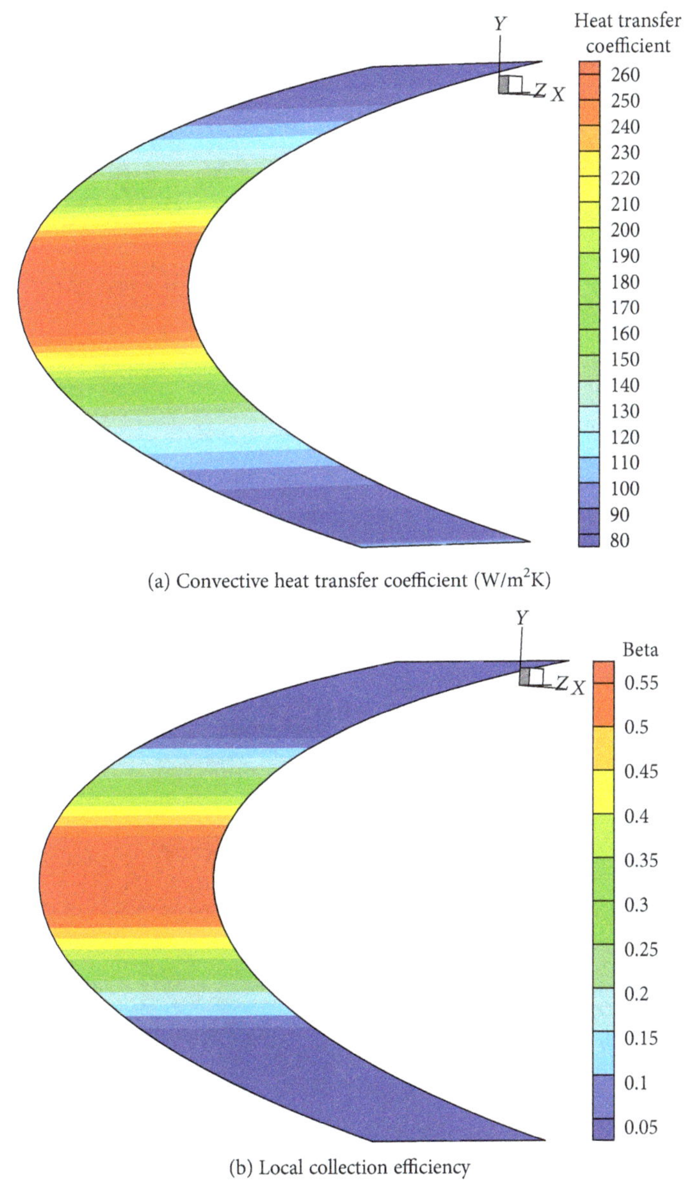

(a) Convective heat transfer coefficient (W/m^2K)

(b) Local collection efficiency

FIGURE 11: Contour of convective heat transfer and droplet impingement.

The simulated convective heat transfer coefficient curves are shown in Figure 6, which show good agreement with the experimental data and LEWICE solution. The results of three mesh files match well, indicating that the solutions are mesh independent. The convective heat transfer coefficient reaches the peak at the stagnation point and drops rapidly as it moves backwards due to the development of boundary layer. The experimental value is slightly larger than that of the simulated results, and the turbulence intensity of the icing tunnel test might be a possible explanation. The droplet collection efficiency was not recorded during the experiment, while it was simulated by NASA LEWICE code in the literature.

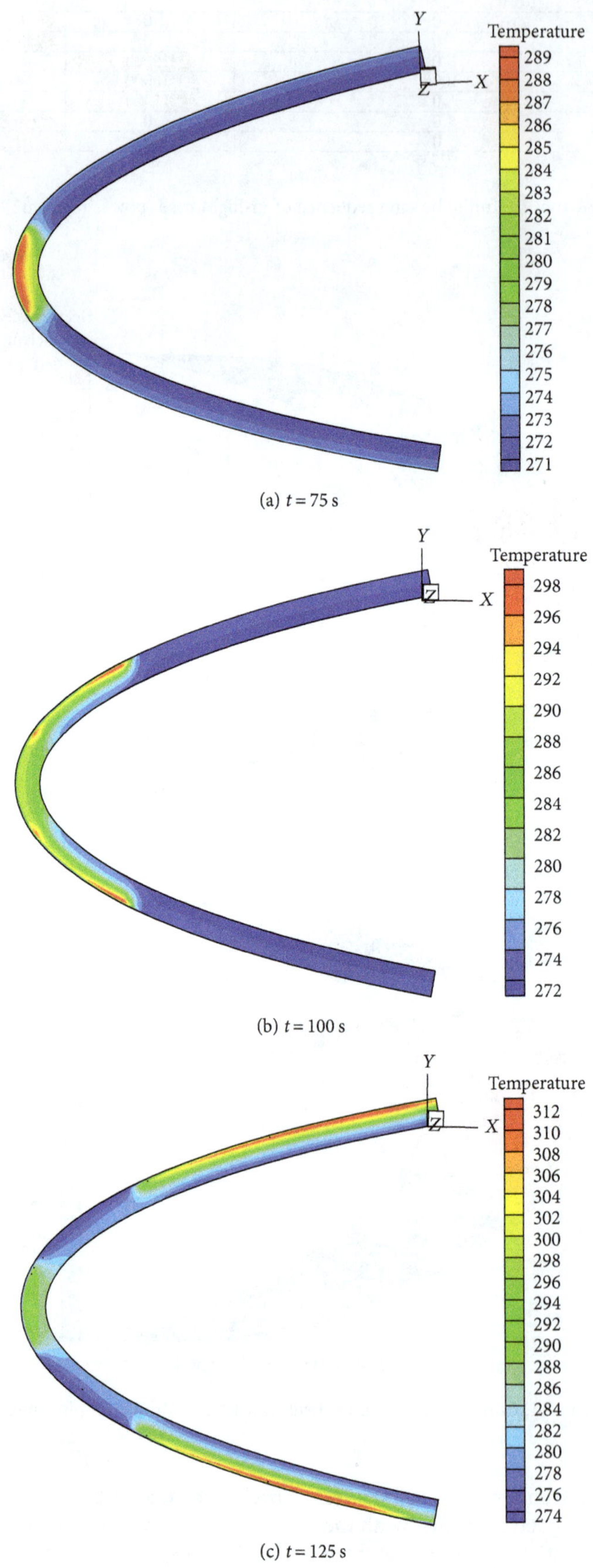

FIGURE 12: Temperature distribution at cross section (K).

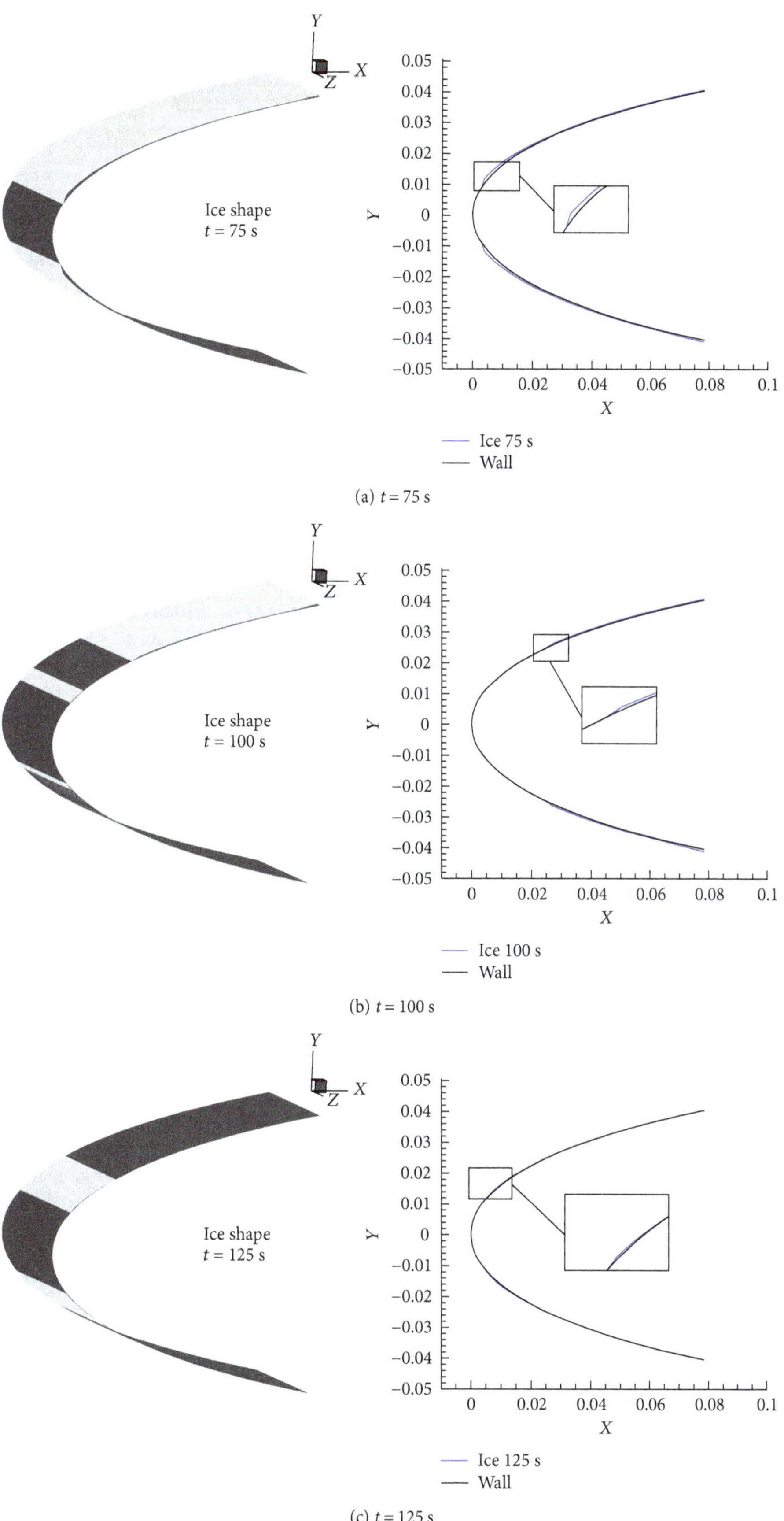

(a) $t = 75$ s

(b) $t = 100$ s

(c) $t = 125$ s

FIGURE 13: Simulated ice shape of in-flight case.

Heater			
7	0	0	15000
6	0	0	15000
5	0	15000	0
4	10000	10000	10000
3	0	15000	0
2	0	0	15000
1	0	0	15000
Time	75 s	30 s	20 s

FIGURE 14: Altered heating sequence.

The collection efficiency of the present model and LEWICE is shown in Figure 7, and the results match well. The value reaches its maximum at the stagnation point, and then the value decreases as it moves backwards. There is a droplet shadowed zone when the wrap distance exceeds 0.03 m, where the local collection efficiency is zero.

The simulated temperature of heater 3 is shown in Figure 8 with the experimental result and FENSAP simulated result provided in [18]. The temperature variation curve of each cycle is similar except for the beginning period when the solid structure starts to warm up. The result of the present model shows good agreement with the experimental data and FENSAP. The simulated ice shapes at 100 s, 110 s, and 120 s (the moment when heaters are turned on or off) are shown in Figure 9.

At the first stage (0–100 s), only the heat blade is activated. Due to the solid conduction and the runback liquid water, the temperature of heater 3 rises. Figure 9(a) shows that water remains liquid on the surface over the heat blade, and ice forms downstream where the heater is turned off. The runback ice covers part of the protection area. At the second stage (100–110 s), heaters 3 and 5 are activated and the temperature of heater 3 rapidly rises to 290 K. From Figure 9(b), it is observed that the ice over heater 3 melts and the runback ice area moves backward, due to the fact that the heating enlarges the water runback area. At the third stage (110–120 s), heater 3 is turned off and the temperature drops. Figure 9(c) shows that the runback ice, which forms during the second stage, melts and ice forms over heater 3 as it is turned off.

The simulated temperature of heater 3 is slightly higher than that of the experimental data. When heater 3 is off, the experimental temperature is around 270 K, and it is about 3 K lower than the simulated results. At this period, the surface is under a runback icing condition, which means a water-ice mixed state, and the temperature of such condition is set at 273.15 K (the freezing point) in the thermodynamic model during simulation. The possible reasons for the temperature difference between experiment and simulation are as follows: (1) The supercooled water film runback phenomenon, which is observed in the experiments [26–28], has not been considered in the present model. In simulation, the temperature of the runback water film would not be lower than the freezing point temperature (273.15 K). (2) The runback water might form beads or rivulet flow and does not completely wet the surface. Part of the surface is exposed to air convection, and the temperature is lower than that when the surface is completely wetted. (3) The temperature measuring point in the experiment is beneath the heating pad, and the precise location is not mentioned in the report; while in simulation, the average value of the heater area is calculated as the result. The average value might be larger when the heater is turned off, because the temperature at the margin is higher due to the heating of the adjacent heating pads.

4.2. Deicing Simulation under Different Heating Sequences. The above simulations validate the present model and the solution method. In this section, another deicing simulation is conducted under in-flight environment conditions to analyze the deicing performance of different heating sequences. The environment conditions are listed in Table 3. Compared with NASA icing tunnel experiment, the temperature is lower, and the air velocity and liquid water content are larger, which would lead to a severer ice protection condition.

The heating sequence is shown in Figure 10, with a cycle of 125 s. The heat blade (heating pad 4, at the leading edge) is kept activated during the whole cycle; heaters 3 and 5 are activated during 75–100 s, and other heaters are activated during 100–125 s. The ice protection structure model and the materials are the same with those in Section 4.2.

The convective heat transfer coefficient distribution is shown in Figure 11(a), and the contour of droplet collection efficiency is shown in Figure 11(b). The convective heat transfer intensity reaches its peak at the stagnation point and drops rapidly as it moves backwards due to the development of the boundary layer. Similarly, the droplet collection efficiency reaches its maximum at the stagnation point, and the value decreases as it moves backwards. There is no droplet impingement at the rear part.

Figures 12 and 13 show the solid temperature distribution of the cross section and the ice shapes at $t = 75$ s, 100 s, and 125 s, which are the end time of the three stages. During the first stage (0–75 s), the structure temperature at the leading edge increases due to the activated heat blade, and the temperature of the heat blade reaches 289 K as is shown in Figure 12(a). The temperature of the adjacent area also increases due to heat conduction, while the value is lower than that of the heat blade and the heated area is limited, because the thermal conductivity of the structure materials is low. The water remains at a liquid state over the heat blade, as shown in Figure 13(a), and ice forms downstream. Compared with the result of Section 4.2, the ice accretion area and the ice thickness are larger in this case. All the ice protection area is covered by ice, because the air velocity and liquid water content are larger and the temperature is lower in this case. Later in the second stage (75–100 s), heaters 3 and 5 are turned on, ice starts to melt, and solid temperature increases. At 100 s, ice over the rear part of heaters 3 and 5 melts completely, and the surface temperature exceeds 273.15 K

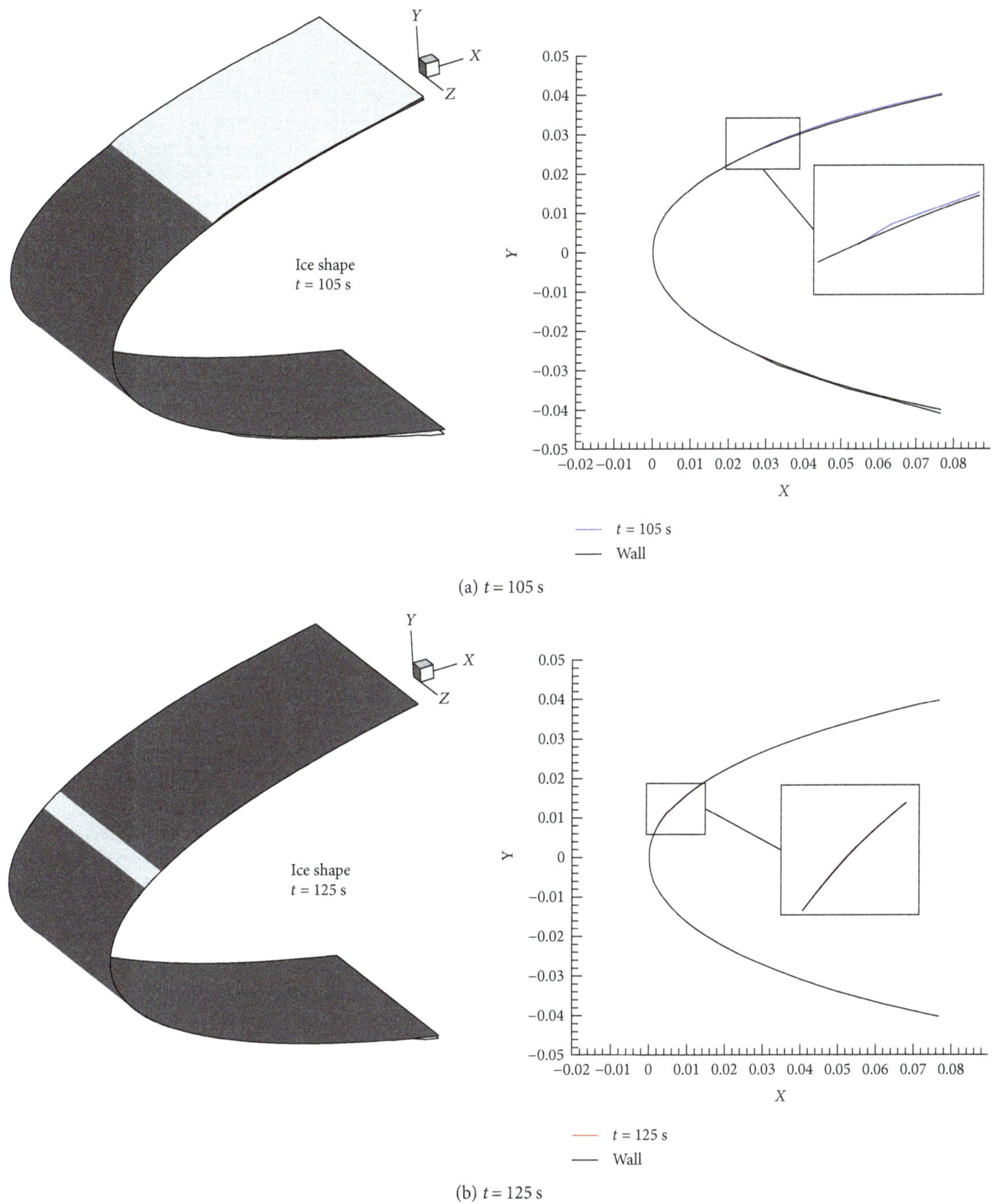

(a) $t = 105$ s

(b) $t = 125$ s

FIGURE 15: Simulated ice shape under altered heating sequence.

as shown in Figures 12(b) and 13(b). While there is ice left in the front of heaters 3 and 5, the accrete ice thickness is large there and the power is insufficient for all the accrete ice to melt. The ice continues to form on the nonheating surface area. In the third stage (100–125 s), the temperature at heaters 3 and 5 drops rapidly when the heaters are turned off, and runback water freezes over heaters 3 and 5, as shown in Figures 12(c) and 13(c). Ice over heaters 1, 2, 6, and 7 melts, and the temperature there increases to a high level since the convective heat dissipation there is relatively weak.

At the end of the second heating stage ($t = 100$ s), there is ice left unmelted in the front of the surface over heaters 3 and 5, which would lead to ice accretion on that area. According the discussion above, we know that the deicing power needed in the front is larger than that in the rear part. Therefore, the heating sequence is altered by extending the second stage to 30 seconds and shorten the third stage to 20 s, as shown in Figure 14. The total time of a cycle remains the same, while the power needed for a cycle is reduced, because the area of heaters 1, 2, 6, and 7 is larger. The simulated ice shape with the altered heating sequence is shown in Figure 15. Figure 15(a) shows that after the second stage the surface over heaters 3 and 5 is clean with all accrete ice melted. Figure 15(b) shows that after the whole deicing cycle, most part of the surface is clean, and only very slight ice accretion occurs at the location near the heat blade, which would be removed in the next cycle.

5. Conclusions

Based on the water film runback dynamics and energy balance theory, an unsteady conjugate heat transfer model for electrothermal deicing is established, and a loosely coupled solution method is developed. The model is applied in the simulation of deicing process, and the conclusions are as follows:

(1) In-flight deicing process is very complex due to factors such as droplet impingement and water film runback. The present model is capable of simulating the in-flight deicing process. Simulation following an icing tunnel experiment has been conducted to validate the present model, and the results show good agreement.

(2) The environment conditions would strongly affect the solid temperature distribution, the water film runback, and phase transition. A larger velocity or liquid water content would correspond to a larger runback icing range, and a higher power of longer heating duration is needed to perform the deicing process. Water remains at a liquid state over the heat blade, and ice forms on the surface where the heating power is insufficient.

(3) Heating sequence is a key factor for the deicing performance. A proper heating sequence not only leads to a better deicing performance, but also saves energy. The optimization of heat sequence can be conducted by means of numerical simulation.

(4) However, there are several factors not yet considered, including the ice shedding mechanism, the contact thermal resistance between multilayer materials, and the anisotropy of material properties.

Conflicts of Interest

The authors declare that they have no conflicts of interest.

Acknowledgments

The work was supported by the National Natural Science Foundation of China (Grant no. 51206008).

References

[1] G. Lin, X. Bu, and X. Shen, *Aircraft Icing and Anti-Icing Method*, Beihang University, Beijing, China, 2016.

[2] K. Al-khalil, "Thermo-mechanical expulsion deicing system - TMEDS," in *45th AIAA Aerospace Sciences Meeting and Exhibit*, Reno, NV, USA, 2007American Institute of Aeronautics and Astronautics.

[3] M. Sinnett, "787 no-bleed systems: saving fuel and enhancing operational efficiencies," *Aero Quarterly*, vol. 18, pp. 6–11, 2007.

[4] "Electro-thermal ice protection system for the B-787," *Aircraft Engineering and Aerospace Technology*, vol. 79, no. 6, 2007.

[5] J. R. Stallabrass, "Thermal aspects of deicer design," in *International Helicopter Icing Conference*, Ottawa, Canada, May 1972.

[6] G. Baliga, *Numerical Simulation of One-Dimensional Heat Transfer in Composite Bodies with Phase Change [Ph.D. thesis]*, the University of Toledo, Toledo, OH, USA, 1980.

[7] J. J. Marano, *Numerical Simulation of an Electrothermal De-Icer Pad [Ph.D. thesis]*, the University of Toledo, Toledo, OH, USA, 1982.

[8] R. J. Roelke, T. G. Keith, K. J. DE Witt, and W. B. Wright, "Efficient numerical simulation of a one-dimensional electrothermal deicer pad," *Journal of Aircraft*, vol. 25, no. 12, pp. 1097–1105, 1988.

[9] D. F. K. Chao, *Numerical Simulation of Two-Dimensional Heat Transfer in Composite Bodies with Application to De-Icing of Aircraft Components [Ph.D. thesis]*, the University of Toledo, Toledo, OH, USA, 1983.

[10] K. L. Leffel, *A Numerical and Experimental Investigation of Electrothermal Aircraft Deicing [Ph.D. thesis]*, the University of Toledo, Toledo, OH, USA, 1986.

[11] K. C. Masiulaniec, *A Numerical Simulation of the Full Two-Dimensional Electrothermal De-Icer Pad [Ph.D. thesis]*, the University of Toledo, Toledo, OH, USA, 1987.

[12] S. Chang, S. Ai, X. Huo, and X. Yuan, "Improved simulation of electrothermal de-icing system," *Journal of Aerospace Power*, vol. 23, no. 10, pp. 1753–1758, 2008.

[13] J. Fu, W. Zhuang, B. Yang, and S. Chang, "Simulation of heating control law of electrothermal deicing of helicopter rotor blade," *Journal of Beijing University of Aeronautics and Astronautics*, vol. 40, no. 9, pp. 1200–1207, 2014.

[14] A. D. Yaslik, K. J. De Witt, T. G. Keith, and W. Boronow, "Three-dimensional simulation of electrothermal deicing systems," *Journal of Aircraft*, vol. 29, no. 6, pp. 1035–1042, 1992.

[15] C. Xiao, *Study on Heat Transfer Characteristics and Effects of Electrothermal Aircraft Deicing [Ph.D. thesis]*, China Aerodynamics Research and Development Center, Sichuan, China, 2010.

[16] W. B. Wright, K. J. WittDe, and T. Keith, "Numerical simulation of icing, deicing, and shedding," in *29th Aerospace Sciences Meeting*, Reno, NV, USA, January 1991, American Institute of Aeronautics and Astronautics.

[17] T. G. Myers, "Extension to the Messinger model for aircraft icing," *AIAA Journal*, vol. 39, no. 2, pp. 211–218, 2001.

[18] T. Reid, G. S. Baruzzi, and W. G. Habashi, "FENSAP-ICE: unsteady conjugate heat transfer simulation of electrothermal de-icing," *Journal of Aircraft*, vol. 49, no. 4, pp. 1101–1109, 2012.

[19] M. Pourbagian and W. Habashi, "CFD-based optimization of electro-thermal wing ice protection systems in de-icing mode," in *51st AIAA Aerospace Sciences Meeting including the New Horizons Forum and Aerospace Exposition*, Grapevine, TX, USA, January 2013, American Institute of Aeronautics and Astronautics.

[20] M. Pourbagian and W. G. Habashi, "Aero-thermal optimization of in-flight electro-thermal ice protection systems in transient de-icing mode," *International Journal of Heat and Fluid Flow*, vol. 54, pp. 167–182, 2015.

[21] C. T. Crowe, "Review—numerical models for dilute gas-particle flows," *Journal of Fluids Engineering*, vol. 104, no. 3, p. 297, 1982.

[22] S. Yang, G. Lin, and X. Shen, "An Eulerian method for water droplet impingement prediction and its implementations," in *Proceedings of the 1st International Symposium on Aircraft Airworthiness*, pp. 72–81, Beijing, China, 2009.

[23] A. P. Rothmayer and J. C. Tsao, "Water film runback on an airfoil surface," in *38th Aerospace Sciences Meeting and Exhibit*, Reno, NV, USA, January 2000, American Institute of Aeronautics and Astronautics.

[24] Newmerical Technologies Int, *FENSAP-ICE, Software Package, Ver. 2011R1.0c*, Newmerical Technologies International, Canada, 2011.

[25] K. Al-Khalil, C. Horvath, D. Miller, and W. Wright, "Validation of NASA thermal ice protection computer codes. III - the validation of ANTICE," in *35th Aerospace Sciences Meeting and Exhibit*, Reno, NV, USA, 1997, American Institute of Aeronautics and Astronautics.

[26] R. List, F. Garcia-Garcia, R. Kuhn, and B. Greenan, "The supercooling of surface water skins of spherical and spheroidal hailstones," *Atmospheric Research*, vol. 24, no. 1–4, pp. 83–87, 1989.

[27] B. J. W. Greenan and R. List, "Experimental closure of the heat and mass transfer theory of spheroidal hailstones," *Journal of the Atmospheric Sciences*, vol. 52, no. 21, pp. 3797–3815, 1995.

[28] A. R. Karev, M. Farzaneh, and L. E. Kollár, "Measuring temperature of the ice surface during its formation by using infrared instrumentation," *International Journal of Heat and Mass Transfer*, vol. 50, no. 3-4, pp. 566–579, 2007.

5

A Multiconstrained Ascent Guidance Method for Solid Rocket-Powered Launch Vehicles

Si-Yuan Chen and Qun-Li Xia

School of Aerospace Engineering, Beijing Institute of Technology, Beijing 100081, China

Correspondence should be addressed to Si-Yuan Chen; 0035@bit.edu.cn

Academic Editor: Kenneth M. Sobel

This study proposes a multiconstrained ascent guidance method for a solid rocket-powered launch vehicle, which uses a hypersonic glide vehicle (HGV) as payload and shuts off by fuel exhaustion. First, pseudospectral method is used to analyze the two-stage launch vehicle ascent trajectory with different rocket ignition modes. Then, constraints, such as terminal height, velocity, flight path angle, and angle of attack, are converted into the constraints within height-time profile according to the second-stage rocket flight characteristics. The closed-loop guidance method is inferred by different spline curves given the different terminal constraints. Afterwards, a thrust bias energy management strategy is proposed to waste the excess energy of the solid rocket. Finally, the proposed method is verified through nominal and dispersion simulations. The simulation results show excellent applicability and robustness of this method, which can provide a valuable reference for the ascent guidance of solid rocket-powered launch vehicles.

1. Introduction

Hypersonic glide vehicles (HGV) are among the significant means of ultra-long range and fast attacks in the future. A solid rocket-powered launch vehicle is characterized by storage stability, short launch preparation time, and quick launch ability; it is also widely applied as an HGV carrier [1]. A boost-glide vehicle is equipped with an unconventional trajectory and strong maneuverability compared with the traditional ballistic missile. "Glide-insertion" launch trajectory has recently elicited widespread concern [2]. The solid rocket shuts off at the verge of the dense atmosphere, and the HGV directly proceeds with long-distance glide by aerodynamic force after separation. The launch vehicles fly along a low trajectory in the atmosphere and have posed a new challenge to the ascent guidance of launch vehicles.

Researchers have performed many studies on the method for the ascent guidance of launch vehicles, which mainly include reference trajectory design and tracking and atmospheric and exoatmospheric guidance. The reference trajectory design mainly adopts the direct method of trajectory optimization [3]. The famous launch vehicle trajectory optimization software programs POST [4] and OTIS [5] have found a solution that relies on direct shooting and point collocation methods, respectively. The pseudospectral method [6–8] has recently attracted widespread attention and relevant research achievements have formed special optimization software, such as the renowned GPOPS [9]. Despite good convergence and high computing efficiency, the direct method faces difficulties in online real-time application, particularly in offline trajectory design. In the trajectory tracking guidance, Lu [10] developed a nonlinear trajectory tracking guidance algorithm that tracks the precomputed trajectory and guarantees the satisfaction of the angle of attack and normal force path constraints. Seywald and Cliff [11] used a neighboring optimal control-based feedback law to guide a two-stage launch vehicle.

The indirect method of trajectory optimization is used by a majority of researchers to investigate the closed-loop ascent guidance [3]. Early research focused on the exoatmosphere, and the most representative methods are iterative guidance mode [12], which is used for the Saturn V rocket, and powered explicit guidance [13], which is intended for space shuttles. The principle is to infer a semianalytical solution under certain approximate assumptions and to avoid extensive numerical computation by determining a quasi-optimal value. Scholars have also developed the endoatmospheric ascent guidance method. Brown and Johnson [14] used a linearized aerodynamics model to obtain optimal control. A conventional shooting method and a homotopy procedure were used. However, reliable convergence was not always attained. Dukeman [15–17] developed a closed-loop ascent guidance algorithm, which adopts a multiple-shooting method to solve a two-point boundary-value problem and to obtain the optimal ascent thrust direction. Lu [18, 19] used the finite difference method to deal with endoatmospheric optimal guidance and presented a connection mode for the two algorithms of endoatmosphere and exoatmosphere. Problems in practical engineering application (i.e., large computing amount and inability to ensure convergence all the time) occur despite a considerable number of studies on endoatmospheric optimal closed-loop guidance.

In addition to using the closed-loop guidance method derived from optimal control theory, the exoatmosphere guidance of ballistic missiles tends to use Lambert guidance [20–22], which manipulates thrust direction by elliptic orbit equation in accordance with rocket and target positions. A solid rocket is not equipped with a thrust termination system to simplify the structure. The fuel must be completely consumed. Thus, an energy management method other than Lambert guidance must be employed to waste some of the rocket's excess energy. Patha and McGehee [23] developed an alternate attitude control energy management method, which is an open-loop guidance method and is only suitable for vacuum environment. Zarchan [24] proposed a general energy management method, which is a closed-loop guidance method. The vacuum flight assumption has minimal effect on guidance precision. However, HGV launch is free from the constraint of elliptic orbit. Traditional ballistic missile guidance and energy management methods are no longer applicable.

In terms of HGV launch issues, a majority of scholars consider that the rocket engine can be shut off, and exoatmospheric optimal guidance can be used for launch vehicle to ship it to an appropriate altitude with a desired velocity and flight path angle [2, 25]. Current literature has failed to develop unified and effective solutions to a HGV launch with solid rocket shut off by fuel exhaustion. Xu and Chen [26] developed a spline energy management guidance method but failed to limit the terminal altitude. To address this technical challenge, we have considered various methods, such as designing several groups of nominal trajectories offline and selecting an appropriate trajectory online to conduct the whole-phase tracking guidance based on the estimated terminal velocity and tracking the nominal trajectory in the first stage of launch vehicle and then tracking online generated trajectory after applying open-loop energy management in the second stage to achieve terminal constraint. However, the aforementioned methods are all limited by poor guidance accuracy and robustness. The final proposed method is featured by small calculated amount and good robustness; and it has a certain capacity which constrains the terminal angle of attack while satisfying the HGV separation requirements.

This paper is organized as follows: Section 2 presents the dynamics model and constraints for the ascent guidance problem. Section 3 optimizes and analyzes the ascent trajectory, which can be used for the tracking guidance of the first-stage rocket. Sections 4 and 5 introduce a spline guidance method and a bias energy management, respectively, which are both used for the second-stage rocket and are implemented together to achieve the HGV separation requirements. Following this, the proposed closed-loop guidance method is tested in Section 6. Finally, the conclusions are presented in Section 7.

2. Ascent Guidance Problem Formulation

2.1. Dynamics Model. Consider that the launch vehicle flies in a vertical plane formed between the launch and target position. Due to short flight time, the impact of earth rotation is negligible. The point-mass dynamics of the launch vehicle over a spherical earth is described by the following equations of motion [27]:

$$\dot{V} = \frac{T\cos\alpha - D}{m} - g\sin\gamma, \tag{1}$$

$$\dot{\gamma} = \frac{T\sin\alpha + L}{mV} - g\frac{\cos\gamma}{V} + V\frac{\cos\gamma}{h + R_0}, \tag{2}$$

$$\dot{h} = V\sin\gamma, \tag{3}$$

$$\dot{R} = V\cos\gamma\frac{R_0}{h + R_0}, \tag{4}$$

where V is the velocity, h is the height, R is the range, γ is the flight path angle, T is the thrust force, α is the angle of attack, R_0 is the average radius of the Earth, m is the mass of the launch vehicle, and $g = g_0(R_0/(h + R_0))^2$ is the gravity acceleration, where $g_0 = 9.81\,\text{m/s}^2$. The terms L and D are the aerodynamic lift force and drag force, that is, $L = qS_{\text{ref}}C_L$ and $D = qS_{\text{ref}}C_D$, where $q = 0.5\rho V^2$ is the dynamic pressure, ρ is the atmospheric density, S_{ref} is the reference area of the launch vehicle, and C_L and C_D are the lift and drag coefficients, respectively [28]. This study examines a two-stage solid rocket launch vehicle with relevant parameters that are reported in Table 1, and its payload adopts CAV-H [29].

2.2. Ascent Trajectory Constraints. The bending moment caused by aerodynamic force is a significant consideration in the ascent trajectory. This path constraint is typically in the form of the following inequality:

$$|q\alpha| \le b_{\max}, \tag{5}$$

where $b_{\max}$ is a nonnegative constant. Equation (5) ensures that the large angle of attack does not appear when the launch vehicle flies through high dynamic pressure areas.

Moreover, control variable α is restricted within a certain range of value, and its angular rate constraint is considered. Thus, the control constraint models are given by

$$\begin{gathered}\alpha_{\min} \le \alpha \le \alpha_{\max},\\ \dot{\alpha}_{\min} \le \dot{\alpha} \le \dot{\alpha}_{\max},\end{gathered} \tag{6}$$

Table 1: Launch vehicle properties.

	Stage 1	Stage 2
Total mass (kg)	16500	7093
Propellant mass (kg)	15173	5826
Mass flow rate (kg/s)	253	89.6
Burn time (s)	60	65
Specific impulse (s)	260	290
Thrust force (N)	645000	248500
Reference area (m^2)	3	3

where the subscripts "min" and "max" denote the minimum and maximum acceptable values, respectively.

The mission requirements indicate that a solid rocket should meet height h_f^*, velocity V_f^*, and flight path angle γ_f^* constraints at the time of fuel exhaustion:

$$h(t_f) = h_f^*,$$
$$V(t_f) = V_f^*, \quad (7)$$
$$\gamma(t_f) = \gamma_f^*.$$

Since the separation of the HGV at the verge of a dense atmosphere, some scholars argue that the rocket should also meet the terminal angle of attack constraint α_f^* [26]:

$$\alpha(t_f) = \alpha_f^*. \quad (8)$$

Finally, the ascent guidance problem can be described as follows: the two-stage launch vehicle is vertically launched from the ground and the thrust direction is a uniquely controlled variable. The path and the controlled variable constraints should be satisfied during the flight process; additionally, a specified HGV separation condition should be met when the rocket fuel is exhausted.

3. Ascent Trajectory Analysis

This section uses optimization theory to analyze the ascent trajectory characteristics. The optimization problem is solved by a pseudospectral method using GPOPS. The method and its applications are presented [9].

In this study, a "glide-insertion" ascent trajectory is considered. It is restricted by the constraints of a bending moment $b_{max} = 4500$ ps-deg, an angle of attack variation rate $|\dot{\alpha}| \leq 10°/s$, and a flight path angle variation rate $|\dot{\gamma}| \leq 2°/s$. The terminal height and flight path angle constraints are set as $h_f = 60$ km and $\gamma_f = 0°$, respectively. The cost function is maximum velocity at solid rocket shut-off point, $J = \max V(t_f)$. A comparison of different ignition modes is presented as follows.

Case 1. The solid rocket adopts "burn-burn" ignition mode.

Case 2. The solid rocket adopts "burn-coast-burn" ignition mode, and the angle of attack is limited to $\alpha = 0°$ at the first-stage rocket separation point to conduct the coasting flight of the second-stage rocket. The ignition time of the second-stage rocket is obtained by optimization.

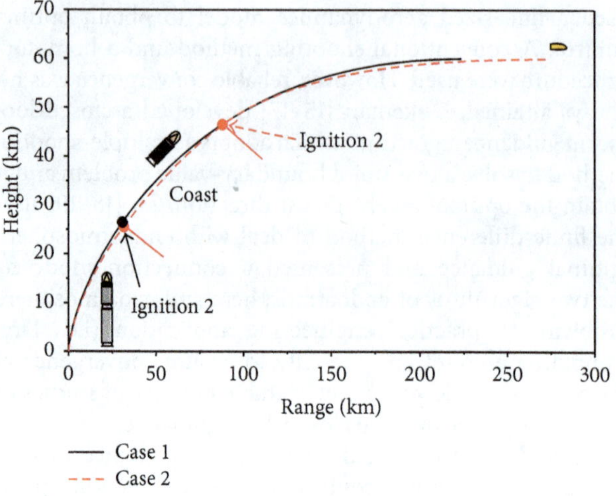

Figure 1: Ascent trajectories.

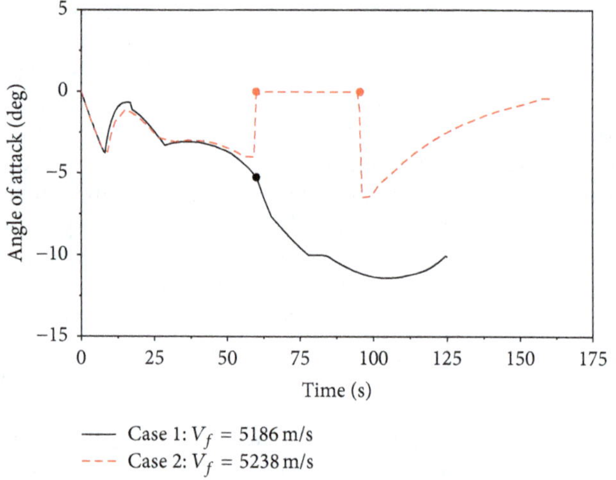

Figure 2: Angle of attack versus time.

Ascent trajectory and the corresponding angle of attack curves of the two cases are shown in Figures 1 and 2, respectively. The results show that Case 2 is superior to Case 1 regardless of range or terminal velocity. For the "glide-insertion" launch mission, the launch vehicle is required to turn agilely; the atmosphere at the altitude where the first-stage rocket shuts off is relatively dense. Therefore, the second-stage rocket ignited after a coasting flight to create a favorable flight environment for closed-loop guidance. As a result, the "burn-coast-burn" ignition mode of the solid rocket is employed for better performance of such launch mission.

This nominal trajectory can be used for tracking guidance of the first stage in order to guarantee a stable flight in the dense atmosphere. This paper does not provide a detailed introduction given that the tracking guidance method is relatively mature [27]. The second-stage rocket can use the closed-loop guidance method proposed in the following sections to achieve terminal multiconstraint requirements.

4. Spline Guidance Method

The second-stage rocket ignited after reaching an appropriate altitude. The atmospheric density is relatively thin and has minimal influence on the rocket. Thus, aerodynamic force could be ignored in the derivation of the closed-loop guidance method. First, the second derivative of height with respect to time is taken as

$$\ddot{h} = \dot{V}\sin\gamma + V\cos\gamma \cdot \dot{\gamma}. \tag{9}$$

Equation (9) is simplified by (1) and (2) as follows:

$$\ddot{h} = \frac{T}{m}\sin(\alpha+\gamma) - g + V^2\frac{(\cos\gamma)^2}{h+R_0}. \tag{10}$$

Equations (3) and (10) show that the altitude rate $\dot{h}$ and altitude acceleration $\ddot{h}$ are related to height, velocity, flight path angle, and angle of attack. Thrust force is constant, dry mass of launch vehicle is known, and gravity is related to height. Therefore, the terminal constraint variables $(t_f, h_f, V_f, \gamma_f, \alpha_f)$ can be converted to $(t_f, h_f, \dot{h}_f, \ddot{h}_f)$.

The preceding analysis shows that spline curve, planned in the height-time (H-T) profile, can be used to solve the two-point boundary problem. The mission requirements indicate that different spline curves are used to solve the problem. If the constraints of terminal height, velocity, and flight path angle are considered, initial conditions (t_0, h_0) and $(t_0, \dot{h}_0)$, as well as terminal conditions (t_f, h_f) and $(t_f, \dot{h}_f)$, can be selected to obtain the reference trajectory by solving (11a), which is a cubic spline curve equation:

$$h_{\text{ref}}(t) = a_3(t-t_0)^3 + a_2(t-t_0)^2 + a_1(t-t_0) + a_0. \tag{11a}$$

If a terminal angle of attack constraint is found, that is, the constraint $\ddot{h}_f$ is included, the terminal condition $(t_f, \ddot{h}_f)$ can be further introduced to utilize (11b) quartic spline curve equation to solve the problem

$$h_{\text{ref}}(t) = a_4(t-t_0)^4 + a_3(t-t_0)^3 + a_2(t-t_0)^2 + a_1(t-t_0) + a_0. \tag{11b}$$

The initial conditions $(t_0, \ddot{h}_0)$ can be further added to obtain the reference trajectory and ensure the continuity of control command by solving (11c), which is a quintic spline curve equation

$$h_{\text{ref}}(t) = a_5(t-t_0)^5 + a_4(t-t_0)^4 + a_3(t-t_0)^3 + a_2(t-t_0)^2 + a_1(t-t_0) + a_0. \tag{11c}$$

Thus, the curve h_{ref} is determined by parameters $\mathbf{A} = [a_0, a_1, a_2, a_3, a_4, a_5]$, which are obtained by solving the system of linear equations. The nominal curves $\dot{h}_{\text{ref}}$ and $\ddot{h}_{\text{ref}}$ can be obtained by taking the first and second derivatives of h_{ref} with respect to time, respectively. The comparison of different spline curves is given as in Figure 3.

Figure 3 shows the change of different spline curves. The curve $\ddot{h}_{\text{ref}}$ plays a decisive role. $\dot{h}_{\text{ref}}$ and h_{ref} in time history could be determined once $\ddot{h}_{\text{ref}}$ is confirmed. Given the strong controllability of the rocket, the control constraint is not introduced in the curve planning and is conducted during the actual flight. The guidance command α_{ref} could be obtained by taking the inverse of (10):

$$\alpha_{\text{ref}} = \arcsin\left(\frac{1}{\dot{W}}\left(\ddot{h}_{\text{ref}} + g - V^2\frac{(\cos\gamma)^2}{h+R_0}\right)\right) - \gamma, \tag{12}$$

where $\dot{W} = T/m$ is axial acceleration, which can be measured by the inertial navigation system. Equation (12) shows that $\ddot{h}_{\text{ref}}$ has the most critical influence on α_{ref} given that the changes of γ and g are small in a short cycle.

Launch vehicle flight in accordance with α_{ref} can meet the planned curve $(h_{\text{ref}}, \dot{h}_{\text{ref}}, \ddot{h}_{\text{ref}})$ under relevant assumptions. However, when the aerodynamic effect and control constraints are introduced, the launch vehicle cannot fly along the planned curve, and the guidance command needs to be conducted with closed-loop correction. Closed-loop command generation involves two methods. One is prediction guidance, which constrains the current and terminal state through spline curve in each guidance period. Therefore, command α_{ref} is needed in real-time computing. This method is suitable for the problem without initial control constraint. This method cannot be used by (11c).

The other method is using trajectory tracking guidance, which takes aerodynamic influence and system uncertainty as disturbance factors to track online generated spline curves. The tracking control law can be described as

$$\begin{aligned}\alpha_c &= \alpha_{\text{ref}} + \delta\alpha,\\ \delta\alpha &= -\mathbf{K}\cdot\delta\mathbf{x},\end{aligned} \tag{13}$$

where $\delta\mathbf{x} = (h\ \ \dot{h})^{\mathrm{T}} - (h_{\text{ref}}\ \ \dot{h}_{\text{ref}})^{\mathrm{T}}$. Feedback gains $\mathbf{K} = [k_1, k_2]$ can be calculated offline by linear quadratic regulator control law [25] and held constant. Constant gains cannot always ensure optimal performance. Nevertheless, they are suited for the situation in which the aerodynamic forces are negligible compared with the thrust force, and deviations from the nominal path are small.

Figure 4 shows the angle of attack curves with different spline guidance methods. The cubic and quartic spline prediction guidance do not introduce the initial angle of attack constraint. The beginning shows a convergence process that is caused by the deviation from guidance command and control constraint. Moreover, cubic spline prediction guidance does not introduce the terminal angle of attack constraint. The large command change occurs at the time close to the final time because the planned curve is updated in real time. The cubic spline tracking guidance first uses prediction guidance to eliminate initial deviation and then generates nominal trajectory to implement tracking guidance. Its terminal command is relatively smooth. The quintic spline tracking

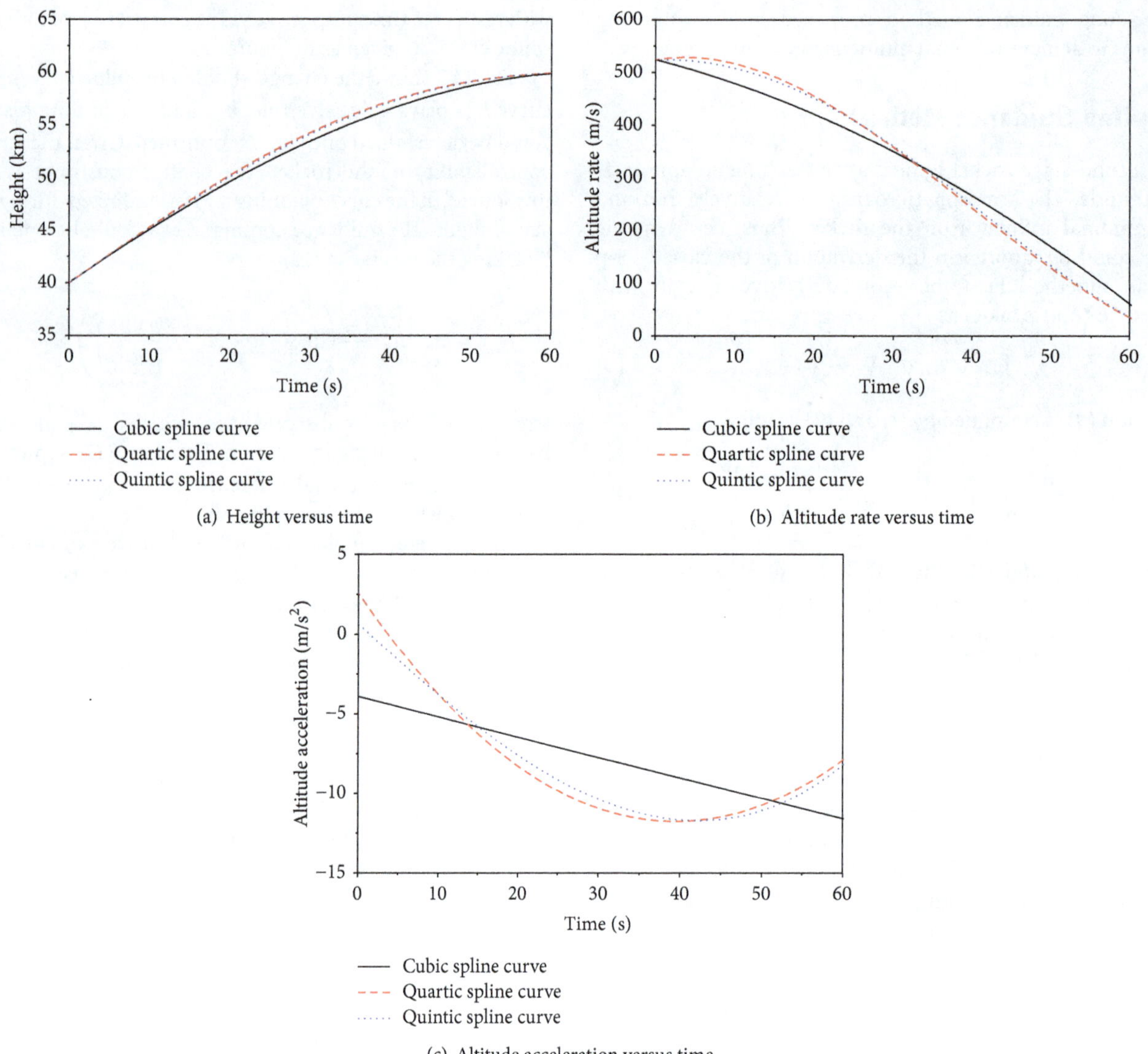

(a) Height versus time

(b) Altitude rate versus time

(c) Altitude acceleration versus time

FIGURE 3: Different spline curves.

guidance introduced the initial and terminal angle of attack constraints, and the guidance command changes slightly.

The closed-loop guidance based on spline curve is provided above. The design method adopts the equivalent state $(h_{ref}, \dot{h}_{ref}, \ddot{h}_{ref})$ rather than state (h, V, γ, α). If the terminal velocity $V(t_f)$ is not consistent with the expected speed V_f^*, the terminal constraints cannot be met even if the flight is conducted in accordance with the nominal curve. Therefore, a reasonable velocity control method should be conducted to achieve the desired state. An energy management strategy should be introduced for the solid rocket.

5. Bias Energy Management

The solid rocket is usually not equipped with a thrust termination system and shuts off by fuel exhaustion to reduce structure mass and improve reliability. The maximum velocity increment of the solid rocket can be expressed as [24]

$$W_f = \int_{\tau_0}^{\tau_f} \dot{W}\, dt = g_0 I_{sp} \ln \frac{m_0}{m_f}, \tag{14}$$

where τ_0 and τ_f represent the ignition and shut-off time of the rocket, respectively; m_0 and m_f are corresponding rocket mass; and I_{sp} is the engine specific impulse. The parameters on the right side of (14) can be confirmed in advance. The velocity produced by solid fuel is greater than the mission requirement. Therefore, energy management is needed to consume excess velocity.

The second-stage rocket guidance can be divided into the energy management and the closed-loop guidance phases. The energy management phase can be further divided into two phases. First, terminal velocity estimation method is

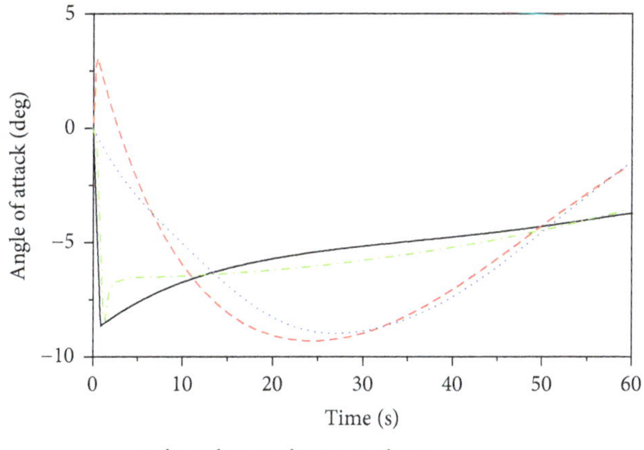

FIGURE 4: Different spline guidance methods.

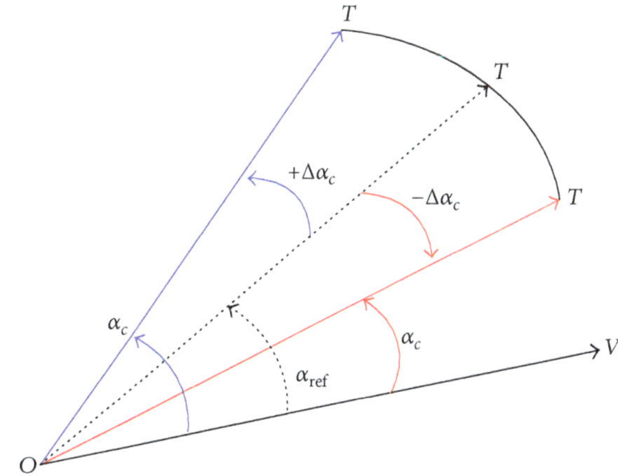

FIGURE 5: Schematic of thrust bias energy management.

used to calculate excess velocity. Second, energy management deceleration control is conducted based on excess velocity.

The analytical estimation method to estimate the terminal velocity is inaccurate under the effect of aerodynamic force. Therefore, the integral estimation method, which adopted closed-loop guidance, can be used. The closed-loop guidance command is also needed as a base command in the energy management process to ensure that the rocket is at a reasonable flight state. The analysis of the different spline guidance methods indicated that the cubic spline tracking guidance is suitable for estimating the terminal velocity without the need to update the curve in real time. Cubic spline prediction guidance can calculate the command in every guidance cycle, which is suitable as energy management base command.

The mass flow ratio $\dot{m}$ deviation is large. Although W_f is not affected, the deviation of τ_f occurs, which is a vital parameter for closed-loop guidance and terminal velocity estimation. Thus, τ_f and $\dot{m}$ estimation schemes should be introduced. Given that the axial velocity increment $W(t)$ is known at any time, the residual velocity increment can be expressed as

$$W_f - W(t) = g_0 I_{sp} \ln \frac{m_0 - \widehat{\dot{m}}(t-\tau_0)}{m_f}. \tag{15}$$

Based on the inverse of (15), the estimated mass flow ratio $\widehat{\dot{m}}$ is

$$\widehat{\dot{m}} = \frac{m_0 - m_f \exp\left(\left(W_f - W(t)\right)/g_0 I_{sp}\right)}{t-\tau_0}. \tag{16}$$

The estimated shut-off time $\widehat{\tau}_f$ can be expressed as

$$\widehat{\tau}_f = \frac{m_0 - m_f}{\widehat{\dot{m}}}. \tag{17}$$

The parameters $\widehat{\dot{m}}$ and $\widehat{\tau}_f$ can be computed in real time. Thus, excess velocity S is

$$S = V_{ref}\left(\widehat{\tau}_f\right) - V_f^*, \tag{18}$$

where $V_{ref}(t)$ is velocity variation generated by spline guidance command. Equation (18) shows that $V_{ref}(t)$ must be changed to reduce S. Moreover, the only controllable variable of the solid rocket is thrust direction α_c.

Figure 5 shows that the actual thrust direction bias $\Delta\alpha_c$ is relative to the guidance command α_{ref} and forms horizontal projection $T\cos(\Delta\alpha_c)$ and vertical projection $T\sin(\Delta\alpha_c)$ in the guidance command direction. Vertical projection changes the current flight status to make the rocket deviate from the desired trajectory and thus affects the next cycle guidance command. Horizontal projection can be considered as the main factor that causes the change of $V_{ref}(t)$ at the current time given that the aerodynamic effect of the angle of attack is small. Therefore, the excess velocity variation rate $\dot{S}$ caused by $\Delta\alpha_c$ can be expressed as

$$\dot{S} = \frac{T}{m}\left(\cos\Delta\alpha_c - 1\right). \tag{19}$$

The excess velocity S is usually within a reasonable and controllable range. Therefore, S could be converged to zero at the expected time as long as the reasonable $\dot{S}$ change law is designed. When the dynamic processes and control constraints are not considered, an exponential decay curve, which takes time as the independent variable, is used to describe S:

$$S_c = S_0\left(\frac{\tau_e - t}{\tau_e - \tau_0}\right)^k \quad (k > 1), \tag{20}$$

where τ_e is the end time of energy management, S_0 is the excess velocity at time τ_0, k is exponential decay coefficient, and $k > 1$ means that, with the change of time, the excess velocity is reduced as nonlinear. The first-order derivative of time to S_c is

$$\dot{S}_c = -\frac{kS_0}{(\tau_e - \tau_0)}\left(\frac{\tau_e - t}{\tau_e - \tau_0}\right)^{k-1}. \tag{21}$$

The actual $\dot{S}$ cannot be in accordance with $\dot{S}_c$ because of the aerodynamic effect and control constraints. Therefore, $\dot{S}_c$ must be conducted with correction in real time. That is, $\tau_0 = t$ and $S_0 = S$. $\dot{S}_c$ is further simplified as

$$\dot{S}_c = -\frac{kS}{(\tau_e - t)}. \tag{22}$$

Equation (22) is substituted to the left side of (19) and inverse operation is conducted to obtain the bias command $\Delta\alpha_c$:

$$|\Delta\alpha_c| = \arccos\left(\frac{\dot{S}_c}{\dot{W}} + 1\right). \tag{23}$$

Equation (23) shows that the sign of $\Delta\alpha_c$ can be either positive or negative, which means that the thrust deflection direction can be located on both sides of α_{ref}. In the energy management process, the rocket deviates from the desired path given that bias $\Delta\alpha_c$, which always exists between α_c and α_{ref}. α_{ref}, is increased or decreased constantly if unilateral bias is adopted during the entire process. The large α_{ref} is not conducive to consume S given the constraint of α_c. Therefore, the bias command reversal strategy should be introduced. Repeated reversals should be avoided given that energy management time usually does not last long, and control constraint exists. Therefore, a simple and efficient single reversal strategy is proposed that almost equally distributes the energy management time on both sides. Finally, energy management command and reversal strategy are shown in

$$\alpha_c = \alpha_{\text{ref}} + k_{\text{sgn}} \cdot |\Delta\alpha_c|, \quad (\alpha_{\max} \le \alpha_c \le \alpha_{\min}), \tag{24}$$

$$k_{\text{sgn}} = \begin{cases} \operatorname{sgn}(\alpha_{\text{ref}}(\tau_0)), & \tau_0 \le t < \tau_0 + 0.5\Delta\tau_E - \tau_{\text{Tra}}, \\ -\operatorname{sgn}(\alpha_{\text{ref}}(\tau_0)), & \tau_0 + 0.5\Delta\tau_E - \tau_{\text{Tra}} \le t \le \tau_E, \end{cases} \tag{25}$$

where k_{sgn} is the sign of thrust bias direction, $\Delta\tau_E = \tau_e - \tau_0$ is the energy management time, and τ_{Tra} is the reversal compensation time, which is used to conduct approximate compensation for the inversion time brought by the control variation rate constraint. Equation (25) shows that thrust bias direction is the same with closed-loop command direction at initial time. The thrust bias direction is reversed when the time exceeds the set reversal time. The energy management can be ended when (26) is met:

$$S \le \varepsilon_S, \tag{26}$$

where ε_S is a small constant value. Equation (26) shows that energy management can be ended and converted to the closed-loop guidance when S is reduced to a certain value. Energy management related change curves are given as in Figures 6 and 7.

Figure 6 shows that different S can be converged to zero in expected time within the range of rocket velocity waste capability. The larger the exponential coefficient k is, the faster the S decay becomes. Figure 7 shows α_c change curves under the control constraint and the thrust bias inversion strategy.

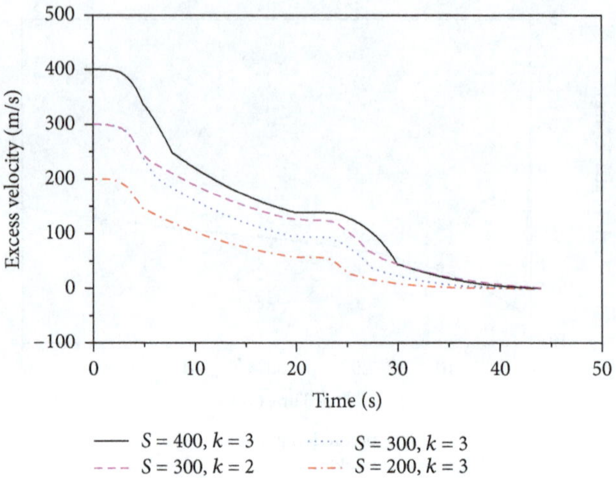

FIGURE 6: Excess velocity versus time.

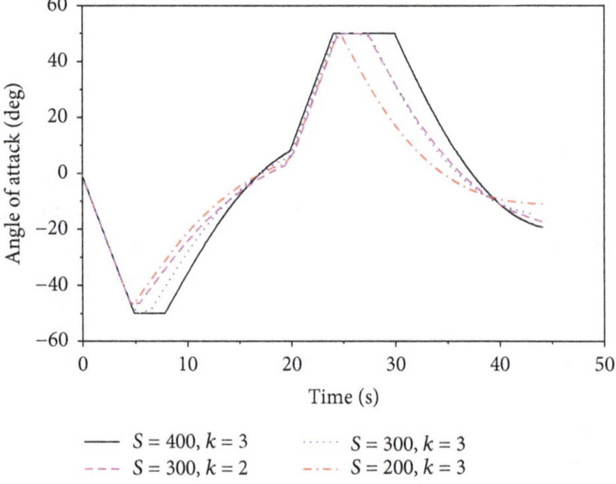

FIGURE 7: Angle of attack versus time.

The rocket is guaranteed to be at a reasonable position in space given that α_c contains the closed-loop guidance command.

The finished energy management can be converted to the closed-loop guidance. The different mission requirements indicate that corresponding spline guidance can be adopted. Cubic spline guidance can be used if no terminal angle of attack constraint exists. The quintic spline guidance can be utilized for its better continuity on control command than the quartic spline guidance when a terminal angle of attack constraint exists. The terminal velocity prediction is obtained by the cubic spline guidance. However, the influence of the control difference on terminal velocity is minimal under reasonable constraints. The multiconstrained closed-loop guidance flow chart is shown in Figure 8.

6. Simulation Results

This section conducts nominal and dispersion simulations to test the applicability and robustness of the multiconstrained

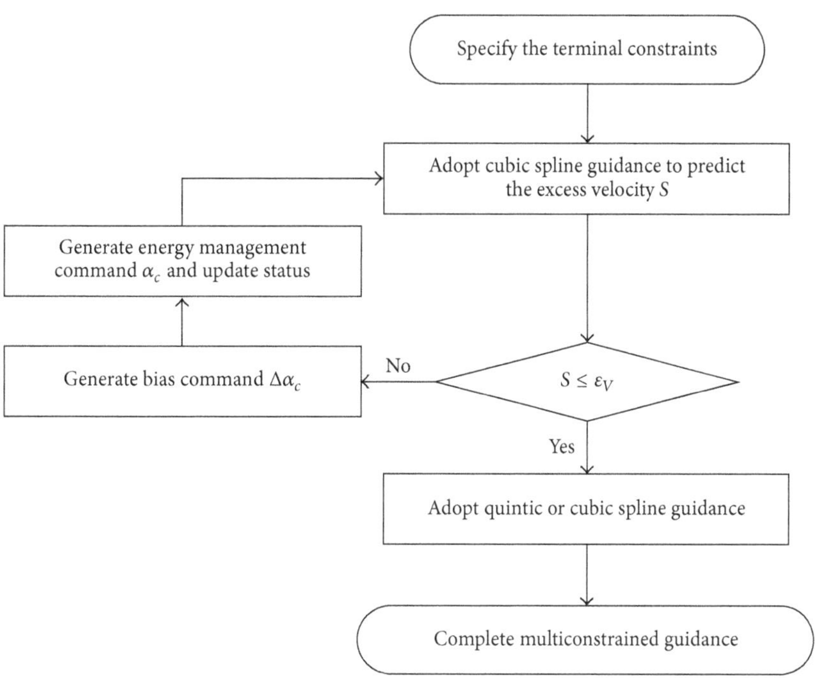

FIGURE 8: Flow chart of multiconstrained closed-loop guidance.

TABLE 2: Initial states.

h (km)	V (m/s)	γ (°)	α (°)
40	1671	18.3	0

TABLE 3: Terminal constraints.

	h_f^* (km)	α_f^* (°)	γ_f^* (°)	V_f^* (m/s)
Case 1	50	0	0	5000
Case 2	55	NAN	3	5100
Case 3	60	10	0	4900
Case 4	65	5	8	4800
Case 5	70	NAN	0	4950

closed-loop guidance method. The second-stage rocket is ignited when it reaches a suitable height. The initial states at the second-stage ignition time are shown in Table 2, and the five group terminal constraint conditions are presented in Table 3. The control constraints in the process of simulation are $|\alpha_c| \leq 50°$ and $|\dot{\alpha}_c| \leq 10°/\text{s}$. The energy management time is $\Delta\tau_E = 45$ s, and the exponential decay parameter is $k = 3$.

6.1. Nominal Cases. The cubic spline guidance in the closed-loop guidance phase is used for the cases (i.e., Cases 2 and 5) without a terminal angle of attack constraint. The others can adopt quintic spline guidance. Corresponding simulation results are indicated in Table 4 and Figures 9–12.

The simulation results show that this guidance algorithm can achieve the constraints of different terminal height, velocity, flight path angle, and angle of attack. Figure 9 shows the change of height under different terminal constraints. Figure 10 shows that the angle of attack change is relatively slight because its maximum angle and angular rate have been limited. The initial bias direction indicates that the initial closed-loop guidance command direction in Case

TABLE 4: Simulation results.

	h_f (m)	α_f (°)	γ_f (°)	V_f (m/s)
Case 1	50003	−0.14	−0.01	5004
Case 2	55001	4.89	2.99	5104
Case 3	60000	8.86	0.02	4894
Case 4	64994	4.72	7.99	4802
Case 5	69998	−11.87	0.00	4952

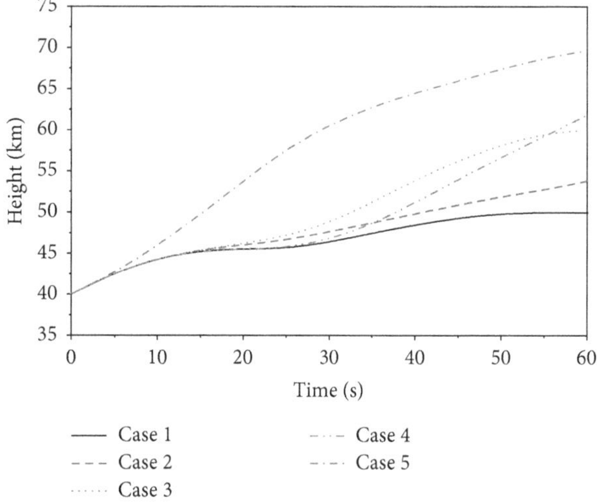

FIGURE 9: Height versus time.

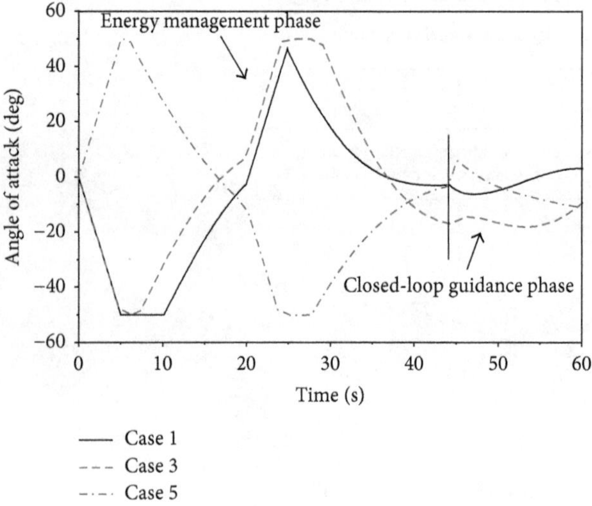

FIGURE 10: Angle of attack versus time.

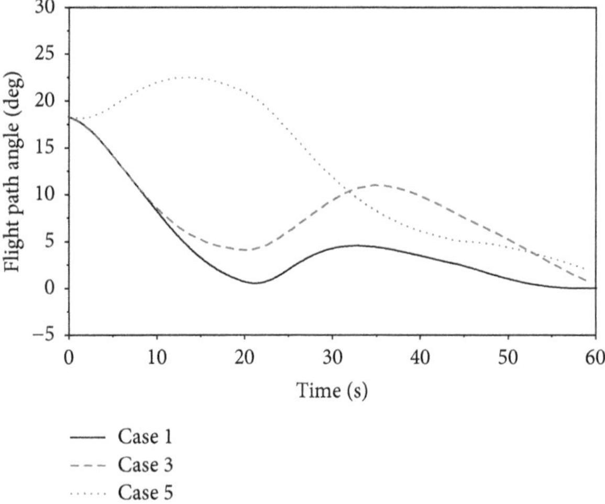

FIGURE 11: Angle of attack versus time.

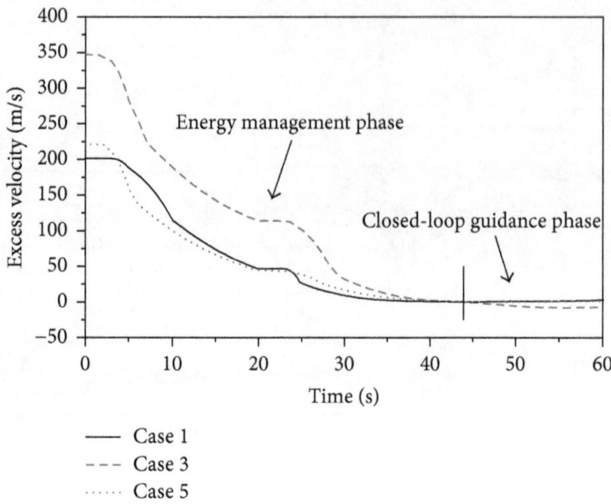

FIGURE 12: Excess velocity versus time.

TABLE 5: Dispersions used in Monte Carlo simulations.

Parameter	Mean	Value ($3-\sigma$)	Unit
Specific impulse	0	3	s
Mass flow ratio	0	5	%
Drag coefficient	0	20	%
Lift coefficient	0	20	%
Atmospheric density	0	20	%

TABLE 6: Monte Carlo simulations statistic results.

Terminal state deviation	Maximum/minimum	Mean	Standard deviation
$\Delta\gamma_f$ (°)	0.1127	0.0212	0.0268
Δh_f (m)	3.1650	0.4775	0.4546
ΔV_f (m/s)	−44.2075	−4.8080	12.7036
$\Delta\alpha_f$ (°)	−3.3150	−1.0896	0.7303
Δt_f (s)	−3.81	0.0698	1.0756

5 is opposite to that in Cases 1 and 3. Cubic or quintic spline guidance in terminal closed-loop guidance phase is selected according to whether an angle of attack constraint requirement exists or not. Figure 11 proves that the change of the flight path angle is consistent with the change of guidance command. Figure 12 shows that the rocket can complete the dissipation of excess velocity at the desired time through energy management.

6.2. Dispersion Testing. To further test the guidance performance and robustness under a wide array of common random dispersions in ascent flight, Monte Carlo simulations are implemented 1,000 times for Case 3. The specific impulse, mass flow ratio, aerodynamic coefficients, and atmospheric density are all dispersed. The dispersions obey the Gaussian distribution and are shown in Table 5. The simulations statistical results are presented in Figure 13 and Table 6.

The simulation results show that the dispersion of terminal state deviation is well-distributed and within the reasonable range. Therefore, this algorithm possesses favorable robustness and can meet the guidance requirements under the influence of uncertainties.

7. Conclusions

This paper proposes a multiconstrained ascent guidance method for a solid rocket-powered launch vehicle, which takes a hypersonic glide vehicle as the load and shuts off by fuel exhaustion. The pseudospectral method is used to analyze ascent trajectory characteristics. Moreover, determining the "glide-insertion" launch trajectory with rocket "burn-coast-burn" ignition mode is suitable for such flight missions. Based on the characteristics of the second-stage rocket flight phase, a multiconstrained closed-loop guidance method based on spline curve and the corresponding bias energy management strategy are proposed. Finally, the nominal and

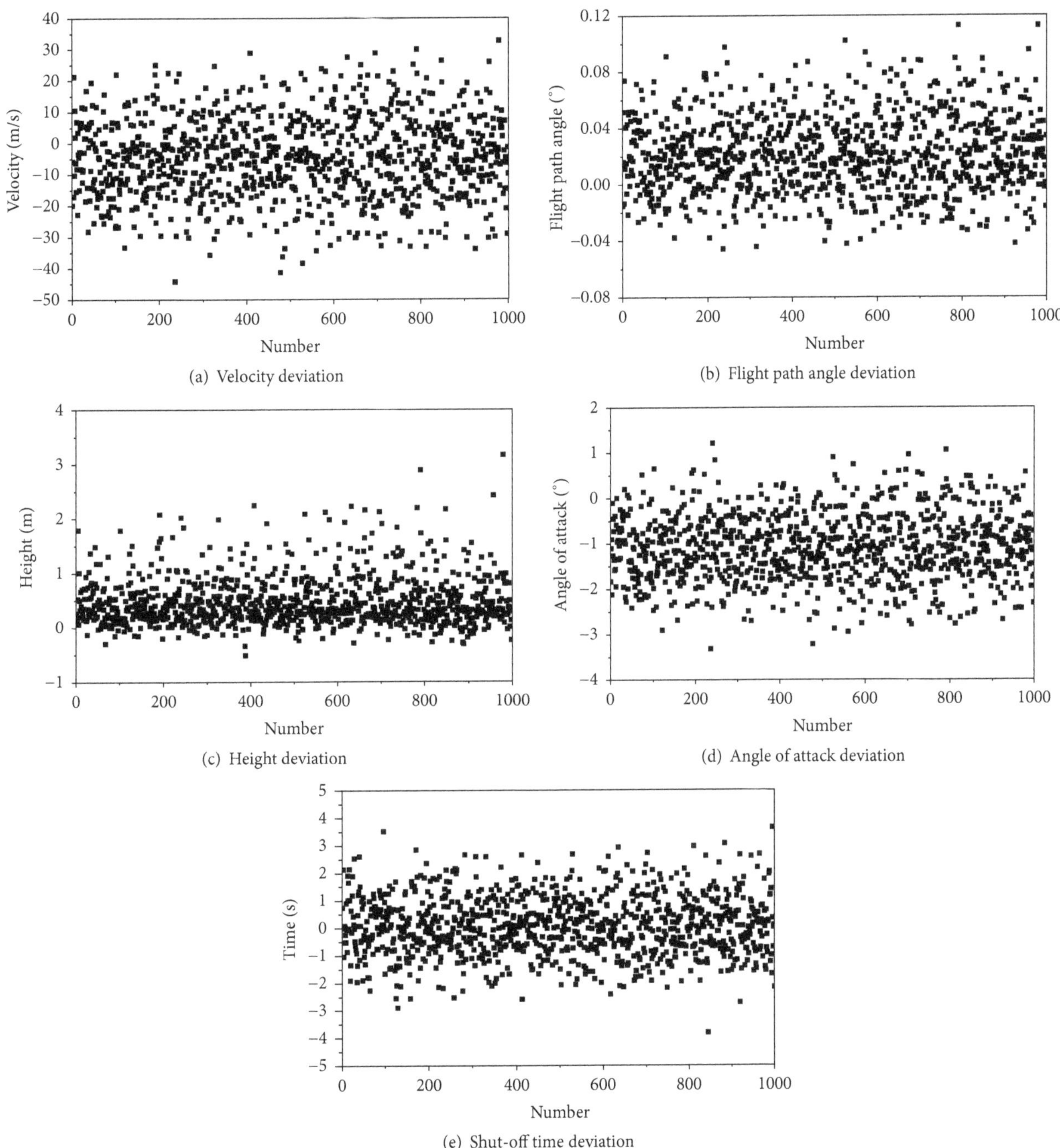

(a) Velocity deviation

(b) Flight path angle deviation

(c) Height deviation

(d) Angle of attack deviation

(e) Shut-off time deviation

FIGURE 13: Dispersion of terminal state deviation.

the dispersion simulations are used to verify the closed-loop guidance method. The simulation results prove the excellent applicability and robustness of this method, which can be used as a reference for the ascent trajectory design and guidance method of solid rocket-powered launch vehicles.

Competing Interests

The authors declare that there is no conflict of interests regarding the publication of this paper.

References

[1] S. H. Walker, J. Sherk, D. Shell, R. Schena, J. Bergmann, and J. Gladbach, "The DARPA/AF falcon program: the hypersonic technology vehicle #2 (HTV-2) flight demonstration phase," AIAA Paper 2008-2539, 2008.

[2] P. Lu, S. Forbes, and M. Baldwin, "Gliding guidance of high L/D hypersonic vehicles," AIAA Paper 2013-4648, 2013.

[3] J. T. Betts, "Survey of numerical methods for trajectory optimization," *Journal of Guidance, Control, and Dynamics*, vol. 21, no. 2, pp. 193–207, 1998.

[4] G. Brauer, D. Cornick, and R. Stevenson, *Capabilities and Applications of the Program to Optimize Simulated Trajectories (POST): Program Summary Document*, 1977.

[5] S. Paris and C. Hargraves, *OTIS 3.0 Manual*, vol. 24, Boeing Space and Defense Group, Seattle, Wash, USA, 1996.

[6] D. Benson, *A Gauss Pseudospectral Transcription for Optimal Control*, Massachusetts Institute of Technology, 2005.

[7] G. T. Huntington, D. Benson, and A. V. Rao, "A comparison of accuracy and computational efficiency of three pseudospectral methods," in *Proceedings of the AIAA Guidance, Navigation, and Control Conference*, vol. 6405 of *AIAA paper*, pp. 840–864, Hilton Head Island, SC, USA, August 2007.

[8] D. A. Benson, G. T. Huntington, T. P. Thorvaldsen, and A. V. Rao, "Direct trajectory optimization and costate estimation via an orthogonal collocation method," *Journal of Guidance, Control, and Dynamics*, vol. 29, no. 6, pp. 1435–1440, 2006.

[9] A. V. Rao, D. Benson, C. L. Darby et al., *User's Manual for GPOPS Version 5.0: A MATLAB Software for Solving Multiple-Phase Optimal Control Problems Using hp-Adaptive Pseudospectral Methods*, 2011.

[10] P. Lu, "Nonlinear trajectory tracking guidance with application to a launch vehicle," *Journal of Guidance, Control, and Dynamics*, vol. 19, no. 1, pp. 99–106, 1996.

[11] H. Seywald and E. M. Cliff, "Neighboring optimal control based feedback law for the advanced launch system," *Journal of Guidance, Control, and Dynamics*, vol. 17, no. 6, pp. 1154–1162, 1994.

[12] D. T. Martin, R. M. Obrien, A. F. Rice, and R. F. Sievers, "Saturn V guidance, navigation, and targeting," *Journal of Spacecraft and Rockets*, vol. 4, no. 7, pp. 891–898, 1967.

[13] R. L. McHenry, T. J. Brand, A. D. Long, B. F. Cockrell, and J. R. Thibodeau III, "Space Shuttle ascent guidance, navigation, and control," *Journal of the Astronautical Sciences*, vol. 27, no. 1, pp. 1–38, 1979.

[14] K. R. Brown and G. W. Johnson, "Real-time optimal guidance," *IEEE Transactions on Automatic Control*, vol. 12, no. 5, pp. 501–506, 1967.

[15] G. A. Dukeman, "Atmospheric ascent guidance for rocket-powered launch vehicles," in *Proceedings of the AIAA Guidance, Navigation, and Control Conference and Exhibit*, vol. 4559 of *AIAA Paper*, pp. 5–8, Monterey, Calif, USA, August 2002.

[16] G. Dukeman and A. J. Calise, "Enhancements to an atmospheric ascent guidance algorithm," AIAA Paper 2003-5638, 2003.

[17] G. A. Dukeman, *Closed-Loop Nominal and Abort Atmospheric Ascent Guidance for Rocket-Powered Launch Vehicles*, Georgia Institute of Technology, Atlanta, Ga, USA, 2005.

[18] P. Lu, H. Sun, and B. Tsai, "Closed-loop endoatmospheric ascent guidance," *Journal of Guidance, Control, and Dynamics*, vol. 26, no. 2, pp. 283–294, 2003.

[19] P. Lu and B. Pan, "Highly constrained optimal launch ascent guidance," *Journal of Guidance, Control, and Dynamics*, vol. 33, no. 2, pp. 404–414, 2010.

[20] R. H. Battin and R. M. Vaughan, "An elegant Lambert algorithm," *Journal of Guidance Control, and Dynamics*, vol. 1, no. 6, pp. 662–670, 1983.

[21] S. L. Nelson and P. Zarchan, "Alternative approach to the solution of Lambert's problem," *Journal of Guidance, Control, and Dynamics*, vol. 15, no. 4, pp. 1003–1009, 1992.

[22] G. Avanzini, "A simple lambert algorithm," *Journal of Guidance, Control, and Dynamics*, vol. 31, no. 6, pp. 1587–1594, 2008.

[23] J. T. Patha and R. K. McGehee, "Guidance, energy management, and control of a fixed-impulse solid-rocket vehicle during orbit transfer," in *Proceedings of the AIAA Guidance and Control Conference*, pp. 1–12, American Institute of Aeronautics and Astronautics, San Diego, Calif, USA, August 1976.

[24] P. Zarchan, *Tactical and Strategic Missile Guidance*, AIAA, Reston, Va, USA, 6th edition, 2012.

[25] P. H. Zipfel, "Orbital insertion control of a three-stage solid rocket booster modeled in six degrees-of-freedom," *Journal of Modeling, Simulation, Identification, and Control*, vol. 2, no. 1, pp. 31–44, 2014.

[26] H. Xu and W. Chen, "An energy management ascent guidance algorithm for solid rocket-powered launch vehicles," in *Proceedings of the 17th AIAA International Space Planes and Hypersonic Systems and Technologies Conference*, pp. 1–12, San Francisco, Calif, USA, April 2011.

[27] A. Tewari, *Advanced Control of Aircraft, Spacecraft and Rockets*, John Wiley & Sons, New York, NY, USA, 2011.

[28] M. J. Abrahamson, *Boost through Reentry Trajectory Planning for Maneuvering Reentry Vehicles*, Massachusetts Institute of Technology, 2007.

[29] T. H. Phillips, *A Common Aero Vehicle (CAV) Model, Description, and Employment Guide*, vol. 27, Schafer Corporation for AFRL and AFSPC, 2003.

Experimental Study of the Swirling Oxidizer Flow in HTPB/N_2O Hybrid Rocket Motor

Mohammad Mahdi Heydari and Nooredin Ghadiri Massoom

Malek Ashtar University of Technology, Babaei Highway, Lavizan, Tehran 15875-1774, Iran

Correspondence should be addressed to Mohammad Mahdi Heydari; mmheydary@gmail.com

Academic Editor: Franco Bernelli-Zazzera

Effects of swirling oxidizer flow on the performance of a HTPB/N_2O Hybrid rocket motor were studied. A hybrid propulsion laboratory has been developed, to characterize internal ballistics characteristics of swirl flow hybrid motors and to define the operating parameters, like fuel regression rate, specific impulse, and characteristics velocity and combustion efficiency. Primitive variables, like pressure, thrust, temperature, and the oxidizer mass flow rate, were logged. A modular motor with 70 mm outer diameter and variable chamber length is designed for experimental analysis. The injector module has four tangential injectors and one axial injector. Liquid nitrous oxide (N_2O) as an oxidizer is injected at the head of combustion chamber into the motor. The feed system uses pressurized air as the pressurant. Two sets of tests have been performed. Some tests with axial and tangential oxidizer injection and a test with axial oxidizer injection were done. The test results show that the fuel grain regression rate has been improved by applying tangential oxidizer injection at the head of the motor. Besides, it was seen that combustion efficiency of motors with the swirl flow was about 10 percent more than motors with axial flow.

1. Introduction

The hybrid rocket propulsion system is one kind of chemical propulsion systems which becomes attractive today. Safety, low investment, availability, and good performance are some of its features that make it a desirable propulsion system. These characteristics are obtained by a combination of liquid propulsion system and solid propulsion system features. By this alternative, it is possible to perform a vast range of missions from low orbit missions like sounding rocket [1, 2] and target drones [3] to space missions such as transportation vehicles [4, 5], thrusters [6], and upper-stage motors [7]. In this type of propulsion system, usually, the solid fuel is a cylinder with port(s) in it, placed between injector(s) and nozzle (shown in Figure 1). Injected liquid or gas oxidizer flows through the port(s).

One important characteristic of classic hybrid motors is the low regression rate of fuel, which is mentioned as a deficiency for this type of propulsion system. Another problem in using this system is nonuniform regression of fuel in longitudinal direction. Jones et al. [8] showed that the fuel regression rate can be increased by inducing a vortex in the combustion chamber. Their study indicated an increase of 16.7% in fuel regression rate for the vortex injection over the axial injection. They used High Density Polyethylene (HDPE) as the fuel and gaseous oxygen (GOX) as the oxidizer. PP and PMMA fuel grains were burned under the swirling and nonswirling oxygen flow conditions by Masugi et al. [9]. They used a small hybrid rocket engine with a large quartz glass window to directly observe the combustion chamber from the front. They stated that, for both PP and PMMA fuels, the swirling flames were found to develop closer to the grain surface than those without swirl, resulting in an increase in the regression rates. A study on vortex injection in hybrid rocket motors with nitrous oxide as the oxidizer and paraffin as the fuel has been performed by Bellomo et al. [10]. Measured performances showed an increase in regression rate up to 51%. Yuasa et al. pushed experiments further and developed a small sounding hybrid rocket with a swirling-oxidizer-type engine [11]. They studied the combustion mechanism of a GOX/PMMA small hybrid rocket of swirling oxidizer flow and found that applying

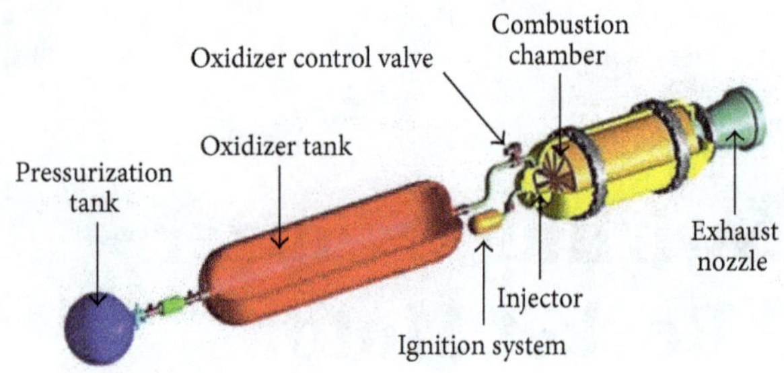

FIGURE 1: Classic hybrid motor.

swirl to an oxidizer flow increased fuel regression rates [12, 13]. Knuth et al. investigated the solid-fuel regression rate behavior and operating characteristics of vortex hybrid rocket engines [14]. They had tested motors with thrusts up to 960 N with gaseous oxygen and HTPB solid fuel. They measured average fuel regression rates up to seven times larger than classical hybrid motors. Kumar investigated the effect of swirl on fuel regression rate numerically. Parametric study of geometric characteristics showed that swirl is more effective for short grains ($L/D < 5$) and large diameter grains [15].

An important fuel and oxidizer composition which is used in the most famous application of hybrid motors, named Space Ship One, is HTPB/N_2O [16]. There are no studies examining oxidizer swirl flow effects on the regression rate and motor performance for this propellant composition. In the present study, this composition was used as propellant. Four tangential injectors were used to generate swirl flow in the motor. Tests have been conducted to study the effects of applying oxidizer swirl flow on this composition. The oxidizer mass flow rate was constant during tests as cavitating venturi was used.

2. Experimental Apparatus and Test Procedure

Figure 2 shows a schematic of the hybrid rocket motor laboratory. Nitrous oxide is used as oxidizer. It is pressurized by high pressure air. After test, high pressure gaseous nitrogen is used to purge the feed line and motor and to cool down motor as well. This helps to prevent undesired burning after test and provides a sharp thrust termination. Pressure transducers, thermocouples, flow meter, and load cell were used to log test data. Pressure transducers are TML PWF-PB series. A TML CLA-1KNA load cell is used to measure thrust of the motor, which is mounted on a rail and wagon mechanism. As it is shown in Figure 2, there is a T-junction in axial injection feed line, which one of its ends sits on the load cell. The load cell itself is mounted on a structure bolted to stand. Because the oxidizer inlets are perpendicular with respect to axial flow, it will not affect the measured thrust of the motor. The type K thermocouple is used to measure oxidizer temperature in the oxidizer run tank.

The oxidizer mass flow rate was measured in two ways. Both cavitating venturi and differential pressure flow meter were used in oxidizer feed line. The model of the smart differential pressure transmitter is SHHDP-9600-F12. By using a cavitating venturi in oxidizer feed line, the oxidizer mass flow rate was regulated during tests. Drawing of cavitating venturi is shown in Figure 3.

2.1. Test Motor. A motor has been designed which is flexible for different injector test, variable with length of prechamber and postchamber. A fuel grain with different length and port diameter can be tested in this motor configuration. Figure 4 shows a schematic of designed motor.

A prechamber is considered at the inlet of the combustion chamber, to make sure of complete evaporation of liquid oxidizer before reaching the fuel surface. Postchamber is considered to complete the combustion of unburned fuel vapor at the end of the fuel grain. The length of postchamber has a significant effect on combustion efficiency [17].

Injectors are used to atomize and vaporize oxidizer before reaching fuel grain. Injectors with 1-, 1.5-, and 2-millimeter orifice diameters were used for axial injection tests. For swirl flow tests, a new injector head flange was designed with four tangential injector slots and one axial injector slot as shown in Figure 5.

2.2. Test Procedure. Five tests were conducted based on test design. Four tangential injection tests and one axial injection test were conducted. The axial injection test which is called the benchmark test is representing a series of tests which have been conducted with the axial injection configuration. The oxidizer mass flux was changed in swirl flow tests in order to study the regression change with it.

To conduct a test, nitrous oxide is fed to the run tank from reservoir tanks. High pressure air is used to pressurize nitrous oxide in the run tank. By pushing the fire button, the oxidizer valve opens and oxidizer injects to the motor. After 0.5 seconds, igniter is triggered and combustion starts. After test time, which is usually set to 5 seconds, oxidizer valve will close and purge line will open. There is a bypass line to an environment which opens as well. This line helps to terminate thrust instantly and to have a sharp pressure and thrust profile at the end of the test, which leads to more accurate average results.

Cavitating venturi is used for maintaining the oxidizer mass flow rate. Cavitating venturi regulates the flow rate while downstream pressure fluctuates or changes. Experiments show that cavitating venturi can withstand downstream pressure changes up to 85 percent of upstream pressure and delivers almost constant mass flow rate [18]. Although there would be two-phase flow at cavitating venturi throat, oxidizer changes to liquid immediately after venturi and it can be ensured that oxidizer enters the motor in liquid form as there is a far distance between venturi and motor.

2.3. Data Reduction. In order to study the performance of motors, pressure before nozzle, igniter pressure, injection pressure, thrust, fuel mass variation, and the oxidizer flow rate were measured. Using these data, ballistic parameters of the motor, such as fuel regression rate, combustion efficiency, and specific impulse, were derived.

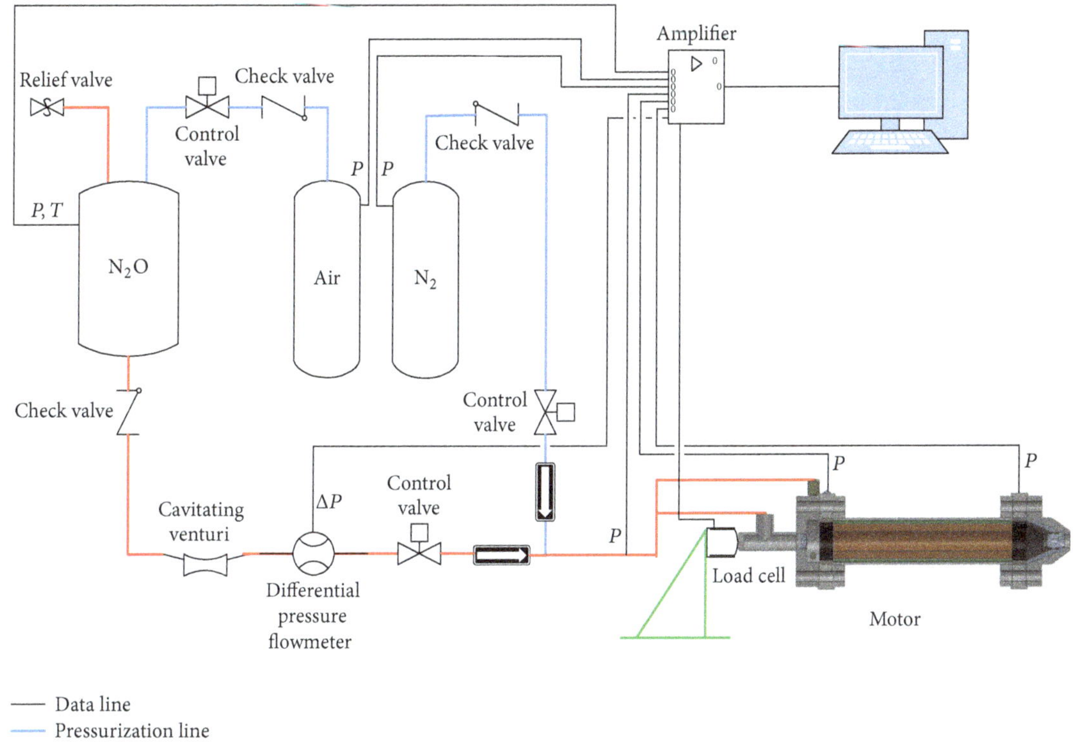

FIGURE 2: Schematic of the hybrid rocket motor system.

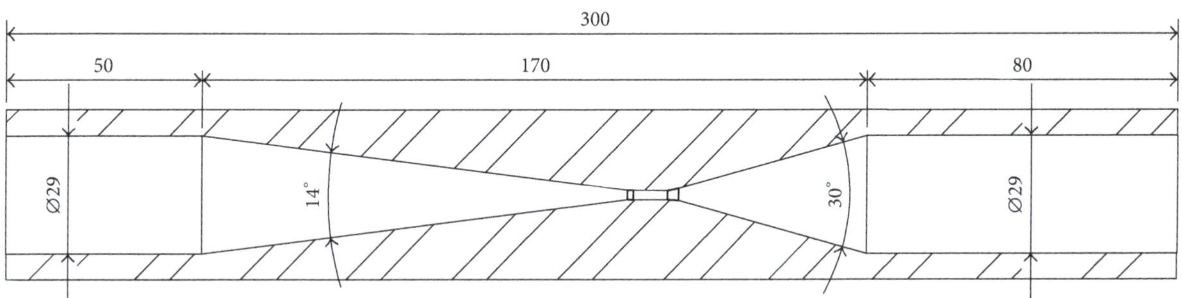

FIGURE 3: Cavitating venturi drawing.

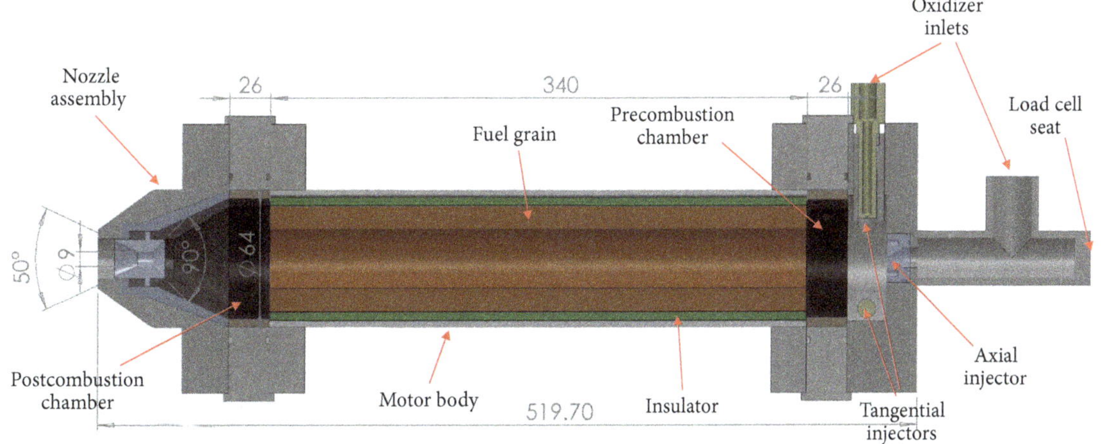

FIGURE 4: Subscale test hybrid motor.

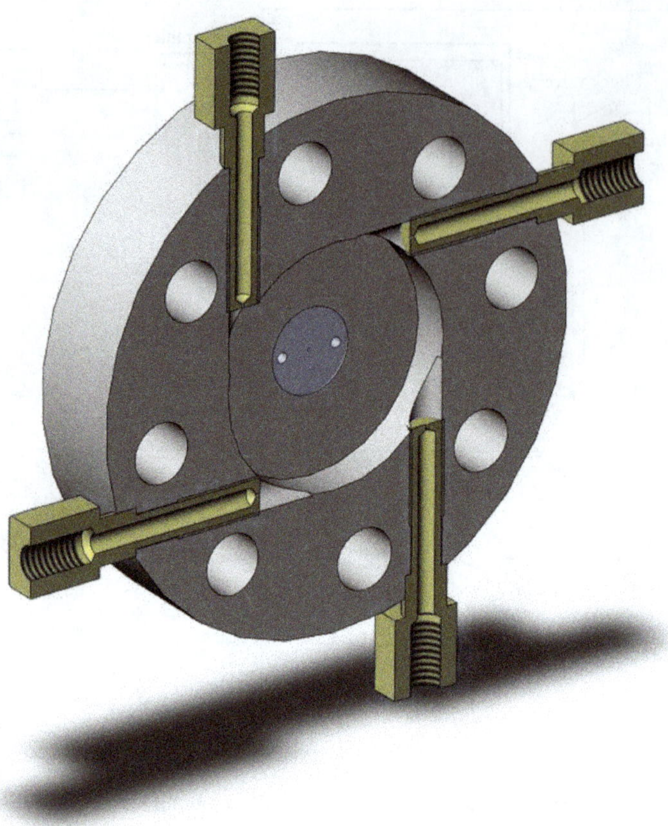

FIGURE 5: Section view of tangential injection cap.

Regression Rate. The average spatial and temporal regression rate can be evaluated by

$$\bar{\dot{r}} \cong \frac{\overline{D_f} - D_i}{2t_b}. \quad (1)$$

In (1), D_i is initial grain port diameter, which is measured before the test, t_b is burning time, and $\overline{D_f}$ is average final grain port diameter. Burning time is defined as the time between ignition and thrust termination. As the setup is designed in a way to have fast thrust termination, the calculated average regression rate would be accurate. The final port diameter is usually nonuniform along fuel grain; hence, the average final port diameter will be estimated with an equivalent average final diameter by

$$\overline{D_f} = \sqrt{\frac{4\left(m_{\mathrm{fuel}_i} - m_{\mathrm{fuel}_f}\right)}{\pi\rho_{\mathrm{fuel}}L} + D_i^2}. \quad (2)$$

In (2) m_{fuel_i} and m_{fuel_f} are initial and final fuel grain masses, respectively. The fuel density is defined by ρ_{fuel} which is $930\,\mathrm{kg/m^3}$ and L is the fuel grain length.

The calculated regression rate for the swirling oxidizer flow was compared with axial oxidizer flow regression rate. Classical relation to hybrid solid-fuel regression rate, introduced by Chiaverini [19], is

$$\dot{r} = aG_{\mathrm{ox}}^n. \quad (3)$$

In this relation a and n are empirical constants and G_{ox}^n is oxidizer mass flux passing through the fuel grain port which will be discussed later.

Oxidizer Mass Flow Rate and Mass Flux. The mass flow rate ($\dot{m}_{\mathrm{ox}}$) of a cavitating venturi is calculated by

$$\dot{m} = C_d A_{\mathrm{th}} \sqrt{2\rho\left(P_{\mathrm{up}} - P_{\mathrm{th}}\right)}. \quad (4)$$

In (4), C_d, A_{th}, and P_{th} are venturi discharge coefficient, throat area, and throat pressure, respectively, ρ being oxidizer density, and P_{up} is upstream pressure. Cavitating venturi has been designed such that throat pressure is less than the vapor pressure of fluid at working temperature. Therefore, P_{th} will be vapor pressure of oxidizer. For venturies with converging angle more than 10° the discharge coefficient is above 99 percent as shown by Reader-Harris et al. [20]. The converging angle of venturi in this study was 30° as shown in Figure 3. Therefore, the discharge coefficient in venturi formula has been neglected.

Oxidizer mass flux determines the regression rate of the fuel grain. As the regression rate is measured with time and space average during burn time, it is convenient to measure oxidizer mass flux the same way. Average oxidizer mass flux is calculated by

$$\overline{G}_{\mathrm{ox}} \cong \frac{\dot{m}_{\mathrm{ox}}}{\overline{A}_p}. \quad (5)$$

In (5), $\dot{m}_{ox}$ is oxidizer mass flow rate and $\overline{A}_p$ is average port area and it is estimated by following equation:

$$\overline{A}_p = \frac{\pi D_{avg}^2}{4}. \quad (6)$$

Here, D_{avg} is average port diameter that is average of initial and final port diameter. Substituting (6) and average port diameter in (5) yields (7), which can be measured by test instruments.

$$\overline{G}_{ox} = \frac{16\dot{m}_{ox}}{\pi\left(\overline{D_f} + D_i\right)^2}. \quad (7)$$

Combustion Efficiency. Combustion efficiency of the motor will be obtained by dividing the characteristics velocity of the test to theoretic ideal characteristics velocity.

$$\eta_{combustion} = \frac{C^*_{exp}}{C^*_{theory}}. \quad (8)$$

Characteristics velocity indicates the energy released by combustion of propellant composition. The theoretic ideal characteristics velocity is calculated by a chemical equilibrium code like CEA [21]. Characteristics velocity of tests has been calculated as below.

$$C^*_{exp} = \frac{\overline{P_c} A_{th}}{\dot{m}_{nozz}}. \quad (9)$$

In (9), P_c is the average combustion chamber pressure, A_{th} is nozzle throat area, and $\dot{m}_{nozz}$ is the average mass flow rate through nozzle which is approximated by oxidizer mass flow rate and average fuel mass flow rate.

2.4. Error Analysis. Some of the variables such as regression rate, thrust, and oxidizer mass flux were reduced from test data. In order to quantify the uncertainty of these variables, an error analysis has been performed for them. The first variable to be analyzed is the final port diameter, which is calculated by (2). The relative error of the final diameter consists of the relative errors in the measurement of change in fuel grain mass, fuel density, and fuel grain length and can be estimated by (10).

$$E_{D_f} = \sqrt{\left(\frac{D_i}{D_f}E_{D_i}\right)^2 + \left(\frac{1}{2}\frac{D_f - D_i}{D_f}\right)^2\left(E_{\Delta M}^2 + E_{\rho_{fuel}}^2 + E_L^2\right)}. \quad (10)$$

Next variable which is reduced from test data is the regression rate, which is calculated by (1). Based on this equation, the relative error of regression rate can be calculated using the following expression:

$$E_r = \sqrt{\left(\frac{D_f}{D_f - D_i}E_{D_f}\right)^2 + \left(\frac{1}{D_f - D_i}E_{D_i}\right)^2 + E_t^2}. \quad (11)$$

TABLE 1: Relative errors of test variables.

Variable	Relative error
D_i	0.004
ΔM	0.0001
ρ_{fuel}	0.0107
L	0.0004
t	0.002
D_t	0.01
T_{ox}	0.0071
ΔP	0.0001
D_f	0.0034
r	0.0208
$\dot{m}_{ox}$	0.0203
$\overline{G}_{ox}$	0.0209

Oxidizer mass flux is a function of oxidizer mass flow rate and initial and final diameter, as (7). Therefore, the relative error of this variable can be evaluated as follows:

$$E_{\overline{G}_{ox}} = \sqrt{\left(\frac{2D_f}{D_f + D_i}E_{D_f}\right)^2 + \left(\frac{2D_i}{D_f + D_i}E_{D_i}\right)^2 + E_{\dot{m}_{ox}}^2}. \quad (12)$$

The relative error of oxidizer mass flow rate itself consists of relative error of feed pressure, diameter of cavitating venturi nozzle, and oxidizer temperature, based on (4).

$$E_{\dot{m}_{ox}} = \sqrt{\left(2E_{D_t}\right)^2 + \left(\frac{1}{2}E_{T_{ox}}\right)^2 + \left(\frac{1}{2}E_{\Delta P}\right)^2}. \quad (13)$$

The relative errors for one of the tests have been calculated and are presented in Table 1.

3. Test Results and Discussion

Before swirl flow tests, a series of tests had been conducted using a single oxidizer axial injector. The fuel regression rate for axial oxidizer injection had been derived by using these tests. One of these tests has been reported as a benchmark for comparison between the swirl flow test results with axial flow test results. Ballistic characteristics of axial flow tests are presented in Table 2.

The experimental regression rate of axial flow tests with the same fuel and oxidizer combination as this study had been reported as

$$\dot{r} = 0.40G_{ox}^{0.37}. \quad (14)$$

Motor configurations for tests conducted in this study are presented in Table 3.

The two first vortex flow tests were performed with one axial injector's and four tangential injectors' configuration. Pressure and thrust profiles of these tests are shown in Figures 6 and 7.

The third test was done by using the four tangential injectors' and one axial injector's configuration. The fuel

TABLE 2: Ballistic characteristics of axial flow tests.

Test #	t_{burn} (sec)	$P_{chamber}$ (bar)	T_{motor} (kgf)	m_f (g/s)	D_{final} (mm)	O/F	C^*_{exp} (m/sec)	η_{c^*}
1-S0A1	7.60	39.2	43.2	53	54.81	3.23	1302.59	0.89
2-S0A1	9.00	37.5	41.3	51	55.11	3.54	1236.80	0.81
3-S0A1	7.43	36.7	40.72	48	53.24	3.54	1328.86	0.89
4-S0A1	7.54	34.47	38.53	41	41.82	4.26	1307.77	0.84
5-S0A1	7.72	24.29	38.55	41	38.36	4.42	1210.54	0.77

TABLE 3: Motor configurations for tests.

	Axial injector number	Tangential injector number	Precombustion chamber	Postcombustion chamber
0-bench	1	0	Yes	No
1-S4A1	1	4	Yes	No
2-S4A1	1	4	Yes	No
3-S4A1	1	4	No	Yes
4-S4A1	1	4	Yes	No

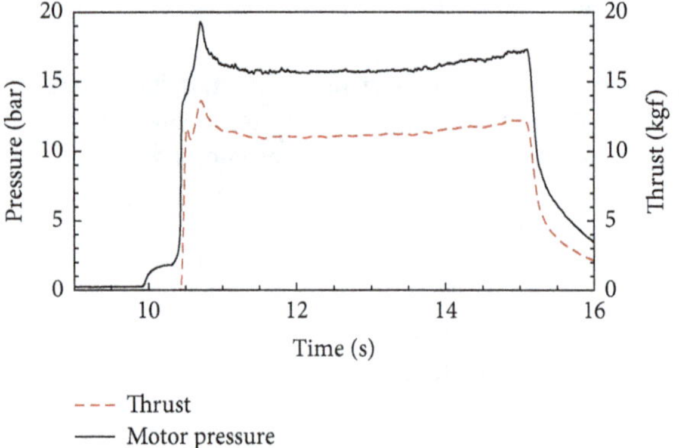

FIGURE 6: Pressure and thrust profile of test #1.

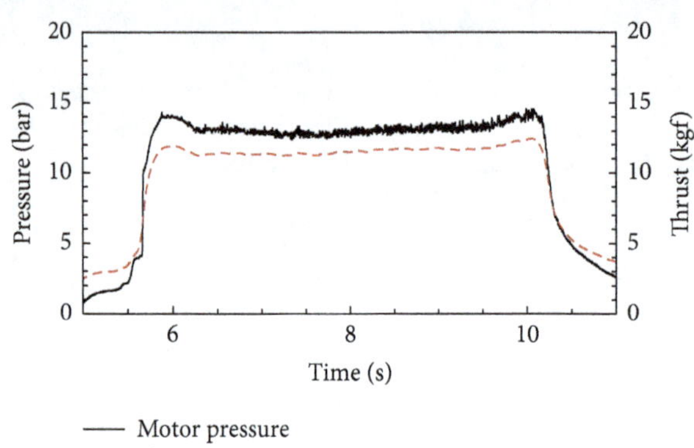

FIGURE 7: Pressure and thrust profile of test #2.

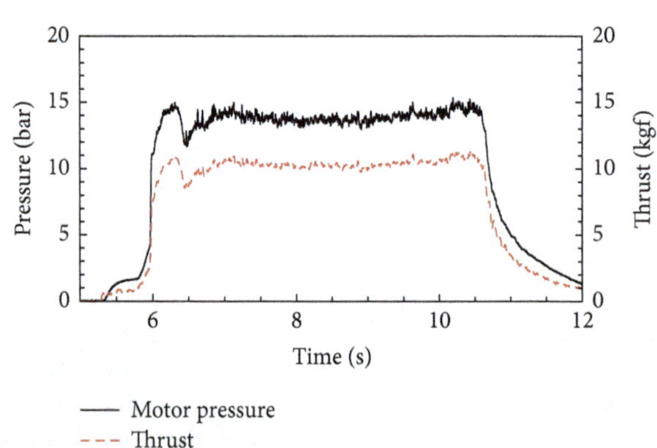

FIGURE 8: Pressure and thrust profile of test #3.

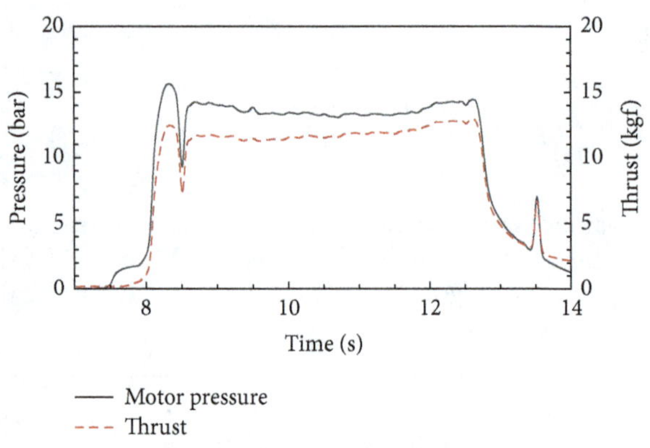

FIGURE 9: Pressure and thrust profile of test #4.

grain was pushed toward the nozzle entrance, according to Figure 4, to provide a larger precombustion chamber and to give more time for evaporating the oxidizer. It was expected to have more uniform fuel grain surface after test, compared to test with axial injector. Pressure and thrust versus time are shown in Figure 8.

The fourth test was done by the same injector configuration as test #3. Precombustion chamber configuration was changed to its initial state, that is, as tests #1 and #2. Postcombustion chamber was used in its original place at the aft end of the motor, before nozzle entrance. The pressure and thrust profiles of this test are shown in Figure 9.

Specifications of swirl flow tests and axial flow test are shown in Table 4.

The oxidizer mass flow rate either could be calculated by (4) and using test data or could be measured by a differential pressure flow meter used in oxidizer feed line. Results are shown in Table 5.

TABLE 4: Specification of tests.

Test #	Motor specs		Fuel specs (before test)		Injector specs		
	D_{th} (mm)	D_i (mm)	L_{grain} (mm)	m_{fuel_i} (g)	Axial injector number	Tangential injector number	$D_{injector}$ (mm)
0-bench	9.0	49.0	251	706.4	1	0	1.5
1-S4A1	9.1	42.0	249	856.8	1	4	1.0
2-S4A1	9.1	45.5	248	660.0	1	4	1.0
3-S4A1	9.1	48.8	240	623.8	1	4	1.0
4-S4A1	9.1	43.3	248	846.0	1	4	1.0

TABLE 5: Oxidizer mass flow rate.

Test #	T_{N_2O} (°C)	ρ_{N_2O} (kg/m^3)	P_{sat} (bar)	P_{up} (bar)	m_{ox} (g/s)
1-S4A1	25.0	743.9	56.6	69.7	60.0
2-S4A1	26.8	708.3	57.6	70.0	56.5
3-S4A1	25.8	721.7	56.3	69.2	59.7
4-S4A1	26.5	712.4	57.2	68.7	55.1

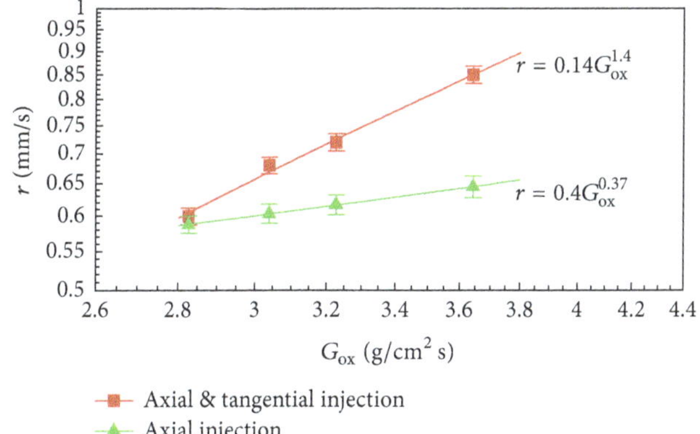

FIGURE 10: Fuel regression rate versus oxidizer mass flux.

Motor performance and ballistic analysis have been done for tests. The results of these calculations are provided in Table 6, as well.

Test results show that although oxidizer mass flux in benchmark test is higher than swirl flow tests, the fuel regression rate is lower compared to swirl flow tests. By comparing these values with results of bulk regression rate, which derived from swirl tests, the results show enhancement of regression rate is due to tangential injection in the combustion chamber. Figure 10 shows the fuel regression rate versus oxidizer mass flux for swirling oxidizer flow and axial oxidizer flow.

The fuel regression rate for a combination of axial and tangential oxidizer injection has been derived and the relation is

$$\dot{r} = 0.14G_{ox}^{1.40}. \quad (15)$$

The difference between the regression rate exponent of swirl flow and axial flow motors can be explained as follows. Classical solid-fuel regression rate relation in hybrid motors is derived by relating heat transfer at the fuel surface to the momentum of the flow. This is done by boundary layer analogies such as Reynolds analogy or Chilton-Colborn analogy. Axial momentum or mass flux is being used in classical hybrid relation, whereas, in a swirl flow motor, there is a tangential mass flux and momentum which is not taken into account.

There are two reasons for explanation of increase in the fuel regression rate. First, centrifugal force caused by swirl flow compresses reacting boundary layer which increases heat feedback to the fuel surface and increases the fuel pyrolysis rate. Second, the convection heat coefficient will be enhanced by swirling the flow, which increases heat flux to the fuel grain. It has been shown numerically that velocity magnitude in the flow field increases when flow is swirled [15]. This results in increase in heat transfer at the fuel surface which increases the regression rate as a consequence.

3.1. Effect of Swirl Flow. The equivalent regression rate is introduced in order to better compare fuel regression rate between two types of flows. The test regression rate is assumed to consist of two parts as shown in

$$\dot{r}_{test} = \dot{r}_{equivalent} + \dot{r}_{tang}. \quad (16)$$

Part one is the contribution of axial flow and part two is the contribution of swirl flow. By using (14), equivalent fuel regression rate of the swirl flow tests mass fluxes has been calculated. The results are shown in Table 7.

It could be seen from Table 7 that, by applying the tangential injection, the regression rate of fuel grain has been improved. It was shown in Figure 11 that the increasing tangential mass flux of oxidizer will increase the regression rate contribution due to swirl flow.

According to Table 6, oxidizer-to-fuel ratio of benchmark test was more than swirl flow tests. This results in higher characteristics velocity. Variation of characteristics velocity with respect to oxidizer-to-fuel ratio for tests pressures is shown in Figure 12. Experimental characteristics velocity has been derived and is shown in Figure 12, too.

The propellant used in this study has an optimum O/F of around 6.5. Both axial flow and swirl flow tests had O/F ratios far enough from this value. Therefore, low O/F effect applies to both cases, where it is easier to achieve higher efficiency comparing to optimal O/F.

TABLE 6: Ballistic characteristics of tests.

Test #	t_{burn} (sec)	$P_{chamber}$ (bar)	T_{motor} (kgf)	m_f (g/s)	D_{final} (mm)	r_f (mm/s)	O/F	G_{ox} (g/cm^2 s)	C^*_{exp} (m/sec)	η_{c^*}
0-bench	6.59	24.90	21.39	22	58.20	0.56	3.88	3.829	1288.32	0.88
1-S4A1	4.48	15.60	12.06	28	49.61	0.85	2.12	3.643	1213.57	0.99
2-S4A1	4.66	12.83	11.30	24	51.83	0.68	2.36	3.039	1108.53	0.91
3-S4A1	5.04	12.96	11.11	22	54.66	0.60	2.72	2.840	1102.85	0.87
4-S4A1	4.62	13.43	11.55	25	49.96	0.72	2.25	3.228	1168.69	0.97

TABLE 7: Regression rate changes due to swirl flow.

Test #	$r_{equivalent}$ (mm/s)	r_{test} (mm/s)	Δr (%)
1-S4A1	0.64	0.85	33.00
2-S4A1	0.60	0.67	13.56
3-S4A1	0.58	0.60	3.80
4-S4A1	0.61	0.72	18.33

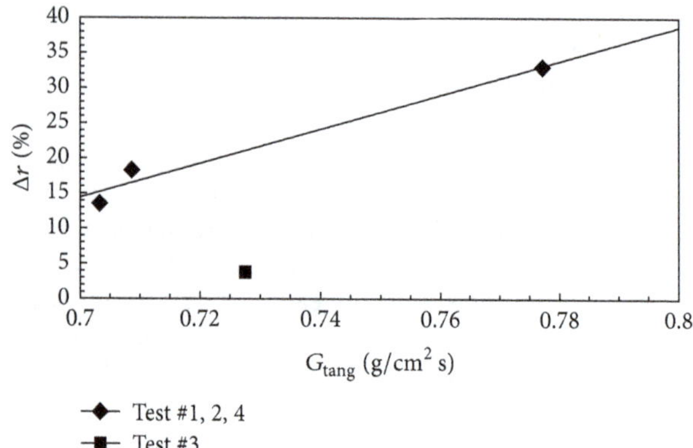

FIGURE 11: Effect of swirl flow on regression rate.

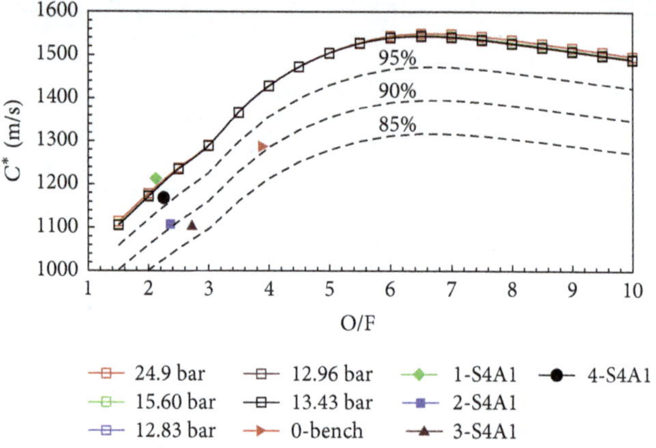

FIGURE 12: Effect of swirl flow on characteristics velocity.

It is clear that characteristics velocity is improved by swirl flow. Combustion efficiency or characteristics velocity efficiency has been calculated by dividing experimental characteristics velocity by theoretical characteristics velocity for each test with specific test conditions. Based on these results, it is clear that combustion efficiency of swirl tests was higher than benchmark test. Furthermore, it has been shown that, by removing postcombustion chamber in test #3, combustion efficiency was reduced compared to other swirl flow tests. Combustion efficiency of tests has been plotted in Figure 13 to show the effect of tangential oxidizer injection on the combustion performance of the motor. Axial injection efficiencies relate to tests which relation 14 has been derived from.

Figure 13 shows that, by applying tangential oxidizer injection, combustion efficiency increased by almost 10%. This improvement is mainly due to better mixing between the oxidizer and pyrolysed fuel. This will reduce the time needed for reactants to be consumed. In addition, reactants travel a longer path before they exit from the motor nozzle, when swirl flow is applied. Consequently, they have more time to react or to complete the reaction. Combustion efficiency will be improved in this way.

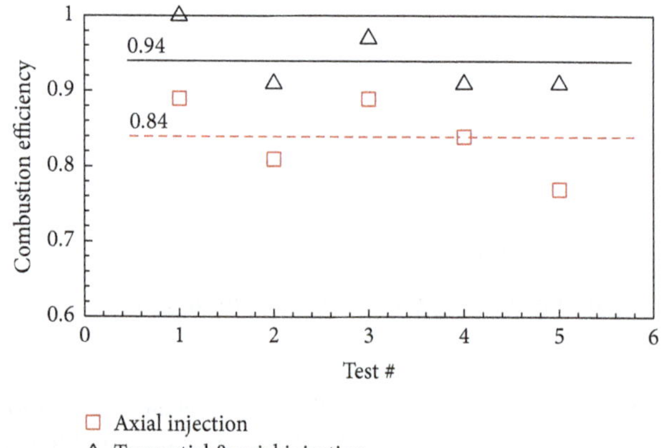

FIGURE 13: Comparison of combustion efficiency for oxidizer swirl flow and axial flow.

The discharge coefficient of the nozzle may be reduced due to swirl flow. This effect will increase the motor pressure. As characteristics velocity is only a function of propellant characteristics and combustion chamber properties, the nozzle behavior does not affect this variable directly. Also, checking the effect of chamber pressure on characteristics velocity shows it is negligible as shown in Figure 12. Therefore, swirl flow will not affect the combustion efficiency of motor by reducing discharge area of nozzle.

3.2. Effect of Pre- and Postcombustion Chamber. The square symbol point in Figure 11 is related to the regression rate of test #3, which had a longer precombustion chamber. Liquid N_2O atomizes and vaporizes as it enters the combustion chamber. This process needs a duration time or delay time to be completed. On the other hand, because the chemical reaction type of hybrid motor is gas phase combustion, the oxidizer should be vaporized before entering to the flame zone and reacting with fuel vapor. By applying tangential injection, oxidizer flow path will be extended before it reaches the fuel grain. This will help to have good mixing between the oxidizer and fuel from the lead of fuel grain and to have a uniform fuel grain surface after the test. Therefore, for comparison between swirl flow tests, the third test was done to see if giving more time to liquid oxidizer for vaporizing helps to improve motor performance and to have a more uniform fuel grain surface after the test.

Test results show that lengthening precombustion chamber not only did not improve the regression rate but also decreased the effect of swirl flow, although more study is needed to see if there is an optimum value for precombustion chamber length.

It was mentioned before that, for test #3, postcombustion chamber was shorter than other tests. It can be seen in Figures 12 and 13 that shorter postcombustion chamber results in lower combustion efficiency. The reason is shorter postcombustion chamber is equivalent to lower characteristics velocity (L^*) which is a well-known criterion for combustion efficiency.

3.3. Effect of Chamber Pressure. By looking at Table 6, one can conclude that pressure does not have a distinctive effect on regression rate or combustion efficiency. Test #3 cannot be compared with other tests as it has a different condition.

4. Conclusion

In this study, the swirling oxidizer flow effects on the fuel regression rate and motor performance of a HTPB/N_2O hybrid rocket motor have been investigated. A laboratory setup and a modular motor were developed and were fabricated. The motor and laboratory were designed to have a full control of parameters such as oxidizer mass flow rate, number and configuration of injectors, and pre- and postchambers, for further investigations. Swirl flow tests were conducted with one axial and four tangential injectors. Several conclusions are listed as follows:

(1) The fuel regression rate for swirl oxidizer flow tests was estimated by average spatial and temporal method and has been compared with axial oxidizer flow tests. Results show enhancement in the fuel regression rate.

(2) It was seen that fuel regression rate relation, derived from swirl flow tests, has different coefficient values compared to classical hybrid fuel regression rate relations. This indicates a different reacting boundary layer structure.

(3) It was shown that regression rate enhancement can be related to tangential oxidizer mass flux.

(4) Results of swirl flow tests show that, by applying swirling oxidizer flow, overall combustion efficiency increased almost by 10% and nearly complete combustion was achieved. This shows a better mixing between the oxidizer and fuel in the combustion chamber.

(5) Although combustion efficiency was increased by swirling the oxidizer flow, removing postcombustion chamber decreased the combustion efficiency. This confirms the need to use postcombustion chamber even in the swirl flow motor.

(6) It was shown that providing more length in the precombustion chamber does not improve the performance of the motor or the regression rate of the fuel grain. This indicates that liquid nitrous oxide had sufficient time to vaporize completely before reaching fuel grain in original motor design. Besides, larger length of the precombustion chamber leads to weaker swirl flow which is not desirable.

Conflicts of Interest

The authors declare that they have no conflicts of interest.

References

[1] J. P. de la Beaujardiere, M. Brooks, S. Chowdhury, B. Genevieve, and L. Roberts, "The phoenix hybrid sounding rocket program: a progress report," in *Proceedings of the 17th AIAA International Space Planes and Hypersonic Systems and Technologies Conference*, San Diego, California, CA, USA, 2011.

[2] L. Casalino and D. Pastrone, "Optimization of hybrid sounding rockets for hypersonic testing," *Journal of Propulsion and Power*, vol. 28, no. 2, pp. 405–411, 2012.

[3] J. Lieh, E. Spahr, A. Behbahani, and J. Hoying, "Design of hybrid propulsion systems for unmanned aerial vehicles," in *Proceedings of the 17th AIAA International Space Planes and Hypersonic Systems and Technologies Conference*, San Diego, California, CA, USA, 2011.

[4] A. Mack and J. Steelant, "FAST20XX: First progress on European future high-altitude high-speed transport," in *Proceedings of the 17th AIAA International Space Planes and Hypersonic Systems and Technologies Conference 2011*, April 2011.

[5] G. Cai, H. Zhu, D. Rao, and H. Tian, "Optimal design of hybrid rocket motor powered vehicle for suborbital flight," *Aerospace Science and Technology*, vol. 25, no. 1, pp. 114–124, 2013.

[6] M. Ambrogio, "The FP7 SPARTAN Program Status and Achievements," in *Space Propulsion*, Bordeaux, France, 2012.

[7] A. Karabeyoglu, J. Stevens, D. Geyzel, B. Cantwell, and D. Micheletti, "High performance hybrid upper stage motor," in *Proceedings of the 47th AIAA/ASME/SAE/ASEE Joint Propulsion Conference and Exhibit 2011*, San Diego, CA, USA, August 2011.

[8] C. C. Jones, D. D. Myre, and J. S. Cowart, "Performance and analysis of vortex oxidizer injection in a hybrid rocket motor," in *Proceedings of the 45th AIAA/ASME/SAE/ASEE Joint Propulsion Conference and Exhibit*, Denver, Colorado, CO, USA, August 2009.

[9] M. Masugi, T. Ide, S. Yuasa, T. Sakurai, N. Shraishi, and A. T. Shimad, "Visualization of Flames in Combustion Chamber of

Swirling-Oxidizer-Flow-Type Hybrid Rocket Engines," in *Proceedings of the in 46th AIAA/ASME/SAE/ASEE Joint Propulsion Conference Exhibit*, Nashville, Tennessee, TN, USA, 2010.

[10] N. Bellomo, F. Barato, M. Faenza, M. Lazzarin, A. Bettella, and D. Pavarin, "Numerical and experimental investigation on vortex injection in hybrid rocket motors," in *Proceedings of the 47th AIAA/ASME/SAE/ASEE Joint Propulsion Conference and Exhibit 2011*, San Diego, California, Ca, USA, August 2011.

[11] S. Yuasa, K. Yamamoto, H. Hachiya, K. Kitagawa, and Y. Oowada, "Development of a small sounding hybrid rocket with a swirling-oxidizer-type engine," in *Proceedings of the 37th Joint Propulsion Conference and Exhibit*, AIAA 2001-3537, AIAA, Salt Lake City, Utah, UT, USA, July 2001.

[12] S. Yuasa, O. Shimada, T. Imamura, T. Tamura, and K. Yamamoto, "A technjque for improving the performance of hybrid rocket engines," in *Proceedings of the in 35th AIAA/ASME/SAE/ASEE Joint Propulsion Conference Exhibit*, Los Angeles, California, CA, USA, 1999.

[13] T. Tamura, S. Yuasa, and K. Yamamoto, "Effects of Swirling Oxidizer Flow on Fuel Regression Rate of Hybrid Rockets," in *Proceedings of the in 35th AIAA/ASME/SAE/ASEE Joint Propulsion Conference Exhibit*, Los Angeles, 1999.

[14] W. H. Knuth, M. J. Chiaverini, J. A. Sauer, and D. J. Gramer, "Solid-fuel regression rate behavior of vortex hybrid rocket engines," *Journal of Propulsion and Power*, vol. 18, no. 3, pp. 600–609, 2002.

[15] C. P. Kumar and A. Kumar, "Effect of swirl on the regression rate in hybrid rocket motors," *Aerospace Science and Technology*, vol. 29, no. 1, pp. 92–99, 2013.

[16] M. Byko, "SpaceShipOne, the Ansari X Prize, and the materials of the civilian space race," *Feature Materials World*, vol. 56, no. 11, pp. 24–28, 2004.

[17] M. Chiaverini, "Fundamental of hybrid rocket combustion and propulsion," in *Proceedings of the Chemical Engineering Progress Symposium Series*, vol. 62, 1966.

[18] H. Ghassemi and H. F. Fasih, "Application of small size cavitating venturi as flow controller and flow meter," *Flow Measurement and Instrumentation*, vol. 22, no. 5, pp. 406–412, 2011.

[19] M. Chiaverini, "Review of solid-fuel regression rate behavior in classical and nonclassical hybrid rocket motors," in *Fundamentals of Hybrid Rocket Combustion and Propulsion*, pp. 37–126, American Institute of Aeronautics and Astronautics, Virginia, VA, USA, 2006.

[20] M. J. Reader-Harris, W. C. Brunton, J. J. Gibson, D. Hodges, and I. G. Nicholson, "Discharge coefficients of venturi tubes with standard and non-standard convergent angles," *Flow Measurement and Instrumentation*, vol. 12, no. 2, pp. 135–145, 2001.

[21] B. J. McBride and S. Gordon, *Computer Program for Calculation of Complex Chemical Equilibrium Compositions and Applications*, NASA Reference Publications, Ohio, OH, USA, 1994.

7

Effects of Freestream Turbulence on Cavity Tone and Sound Source

Hiroshi Yokoyama, Hiroshi Odawara, and Akiyoshi Iida

Department of Mechanical Engineering, Toyohashi University of Technology, 1-1 Hibarigaoka, Tempaku, Aichi 441-8580, Japan

Correspondence should be addressed to Hiroshi Yokoyama; h-yokoyama@me.tut.ac.jp

Academic Editor: William W. Liou

To clarify the effects of freestream turbulence on cavity tones, flow and acoustic fields were directly predicted for cavity flows with various intensities of freestream turbulence. The freestream Mach number was 0.09 and the Reynolds number based on the cavity length was 4.0×10^4. The depth-to-length ratio of the cavity, D/L, was 0.5 and 2.5, where the acoustic resonance of a depth-mode occurs for $D/L = 2.5$. The incoming boundary layer was laminar. The results for the intensity of freestream turbulence of Tu = 2.3% revealed that the reduced level of cavity tones in a cavity flow with acoustic resonance ($D/L = 2.5$) was greater than that without acoustic resonance ($D/L = 0.5$). To clarify the reason for this, the sound source based on Lighthill's acoustic analogy was computed, and the contributions of the intensity and spanwise coherence of the sound source to the reduction of the cavity tone were estimated. As a result, the effects of the reduction of spanwise coherence on the cavity tone were greater in the cavity flow with acoustic resonance than in that without resonance, while the effects of the intensity were comparable for both flows.

1. Introduction

Flows over open cavities such as the sunroofs of automobiles and landing gear configurations of airplanes often generate self-sustained oscillations and intense tonal sound. Therefore, to reduce the tonal sound from cavity is one of the most important issues for development of rapid transport vehicles such as airplanes or automobiles.

A mechanism for oscillations with fluid-acoustic interactions in compressible cavity flows was proposed by Rossiter [1]. The vortex impinging at the downstream edge of the cavity generates acoustic waves. These acoustic waves induce vortex shedding from the upstream edge of the cavity again. Sarohia [2] measured velocity fluctuations around cavities for laminar flows over shallow cavities (depth-to-length ratio of cavities $D/L < 0.35$ and freestream Mach number $M < 0.07$). He found that instability in the shear layer of the cavities was amplified by self-sustained oscillations. East [3] measured the acoustic pressure of deep cavities in the turbulent boundary layer ($1.0 < D/L < 8.5$ and $M < 0.2$). It was clarified that the self-sustained oscillations were amplified by coupling between instability in the shear layer and the acoustic mode in the direction of cavity depth. A cavity flow without acoustic resonance does not generate self-sustained oscillations or intense tonal sound in cavity flows at low Mach numbers in the turbulent boundary layer. Recently, direct numerical simulations were performed to investigate the three-dimensional stability of cavity flows [4].

The control of self-sustained oscillations in cavity flows and sound generated from cavities have also been investigated. Huang and Zhang [5] used plasma actuators implemented at the leading edge ahead of cavities in a recent study on the control of cavity flows and obtained significantly reduced levels of tonal sound. The flow induced by actuators traversed in the streamwise direction makes the shear layer thickness nonuniform in the spanwise direction.

Lusk et al. [6] investigated the attenuation effects of steady normal mass injection from the upstream of the leading edge of cavities on oscillations in supersonic cavity flows. The blowing flows from multisegmented slots spanning the cavity width broke up large-scale structures in cavity flows and reduced the level of pressure fluctuations in the cavities. These results indicate that disturbance in the incoming boundary layer is important to reduce the cavity tone.

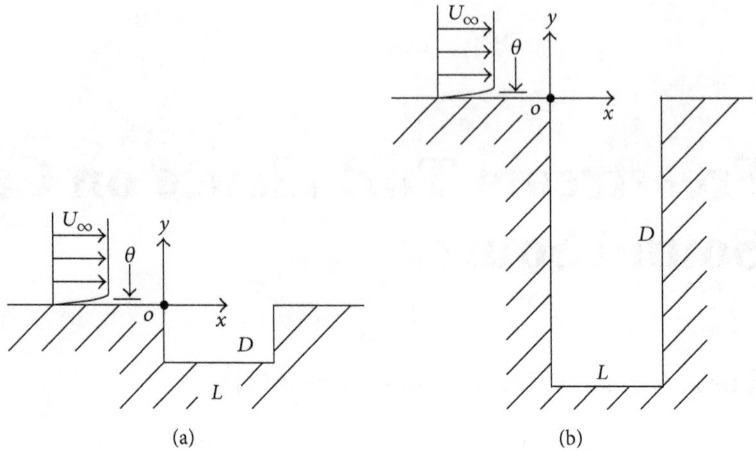

FIGURE 1: Configurations for flows around two-dimensional cavity. (a) $D/L = 0.5$. (b) $D/L = 2.5$.

TABLE 1: Computational and experimental conditions.

	D/L	Tu [%]	L_e/L	θ/L	M	Re_L	Re_D
Computation	0.5	0.0	—	0.0071	0.09	4.0×10^4	2.0×10^4
		0.9	0.095	0.0072			
		1.5	0.078	0.0074			
		1.8	0.090	0.0077			
		2.3	0.111	0.0078			
Computation	2.5	0.0	—	0.0071	0.09	4.0×10^4	9.9×10^4
		0.9	0.095	0.0072			
		1.5	0.078	0.0074			
		1.8	0.090	0.0077			
		2.3	0.111	0.0078			
Experiment	0.5 2.5	0.6	0.165	0.0072	0.09	4.0×10^4	2.0×10^4 9.9×10^4

The disturbances in the boundary layer are also possibly caused by freestream turbulence, freestream vortices, and acoustic waves, as reviewed by Saric et al. [7]. Brandt et al. [8] studied the effects of the integral scale of freestream turbulence on bypass transitions of the boundary layer by using direct numerical simulations. They found that transitions occurred earlier for larger-scale freestream turbulence. The condition of the incoming boundary layer affects the cavity tone, and this could cause the past measured scatter of sound pressure spectra of cavity tone particularly at a low Mach number [9]. However, the effects of freestream turbulence on cavity flows and generated sound have not yet been clarified.

Also, it has been considered to be important to clarify the structures of the sound source to understand the mechanism for reducing the cavity tone. Larsson et al. [10] directly simulated two-dimensional flows and acoustic fields and estimated the sound source based on the Curle's acoustic analogy [11] by using fluctuating surface pressure. They found that the sound source of the cavity tone was maximal at the downstream edge of the cavity. Ask and Davidson [12] discussed the cancellation of the sources at the cavity bottom and downstream wall. However, the effects of the intensity of the Lighthill's acoustic source [13] and spanwise coherence of the source on the sound have not been investigated.

The present study directly simulated flow and acoustic fields for cavity flows with various intensities of freestream turbulence. The sound source of the cavity tone based on Lighthill's acoustic analogy [13] was estimated to clarify the reduction mechanism for the cavity tone by freestream turbulence. The effects of the weakening of the intensity of the sound source and the lowering of spanwise coherence of the sound source on the cavity tone were separately investigated.

2. Flow Configurations

Figure 1 outlines the configurations for flows around a two-dimensional cavity. The origin of the coordinate system is located at the upstream edge of the cavity. The streamwise direction is the x-axis, the vertical direction is the y-axis, and the spanwise direction intersecting these two axes is the z-axis. The cavity length, L, is 20 mm. The freestream velocity, U_∞, is 30 m/s.

Table 1 summarizes the computational and experimental conditions. The depth-to-length ratio of the cavity is $D/L =$

0.5 and 2.5. In addition, acoustic resonance in the direction of cavity depth occurs in the flow over the cavity with $D/L = 2.5$, as explained in Section 5.2. The freestream Mach number, $M \equiv U_\infty/c$, is 0.09, where c denotes the speed of sound. The Reynolds number based on the cavity length, $\mathrm{Re}_L \equiv U_\infty L/\nu$, is 4.0×10^4, where ν denotes the kinetic viscosity and that based on the cavity depth, $\mathrm{Re}_D \equiv U_\infty D/\nu$, is 2.0×10^4 for $D/L = 0.5$ and 9.9×10^4 for $D/L = 2.5$.

The velocity profile in the flow over the flat plate without a cavity at the streamwise position corresponding to the location of the upstream edge of the cavity with and without freestream turbulence is in good agreement with that of the laminar Blasius boundary layer. The ratio of the momentum thickness in the upstream boundary layer to the cavity length was slightly changed by freestream turbulence (turbulent intensity of Tu = 0.0–2.3%) and varied in the range of $\theta/L = 0.0071$–0.0074 in the computation. This was approximately the same as that in the experiment ($\theta/L = 0.0072$), where the freestream turbulence is 0.6%.

The features of freestream turbulence in the present paper are described by the turbulent intensity, Tu, and the integral scale, L_e. The turbulent intensity, Tu, is calculated by using the rms values of the streamwise velocity fluctuations, $u_{\mathrm{rms}}/U_\infty$, in $y/L = 0.2$–0.5, where even the height of this lower limit is three times greater than the boundary layer thickness. The integral scale of the freestream turbulence, L_e, is calculated by using the autocorrelation function values of the streamwise velocity, u, in the same range of height. The intensity and integral scale are approximately constant in this range of height.

The computations were performed for turbulent intensity Tu = 0.0%, 0.9%, 1.5%, 1.8%, and 2.3%, where the integral scale was $L_e/L = 0.095$ for Tu = 0.9% and $L_e/L = 0.111$ for Tu = 2.3%. In addition, the ratio of the integral scale to the momentum thickness of the boundary layer, L_e/θ, was 13.3 for Tu = 0.9% and 15.5 for Tu = 2.3%. The turbulent intensity, Tu, was 0.6% in the experiment, and the integral scale, L_e/L, was 0.165.

It has been confirmed that no laminar-turbulent transitions in the boundary layer occurred with this freestream turbulence in both computations and experiments. Also, the preliminary computations for the flow over the cavity with $D/L = 0.5$ for freestream turbulence of $L_e/L = 0.078$ and 0.137 with turbulent intensity Tu = 1.5% indicated that the effects of difference of the integral scale on the reduced cavity tone were within 2 dB.

3. Experimental Methods

The experiments [14] were carried out using a suction-type, low-noise wind tunnel with a rectangular test section with a cross-section of 150 mm × 75 mm, as outlined in Figure 2. The intensity of freestream turbulence was less than 0.6% and nonuniformity of the freestream was less than ± 0.1% at the freestream velocity of 30 m/s. The background noise level was 58 dB (A) at the same velocity.

The test section was terminated in a spanwise manner by the end walls composed of acrylic plate and porous plate to visualize the flow within the cavity and suppress acoustic resonance in the spanwise direction. The velocity along the center of the cavity ($x/L = 0.5$) was measured using a hot-wire anemometer. The acoustic pressure in the far field ($x/L = 6.75$, $y/L = 23.5$) was measured with a nondirectional 1/2 inch microphone.

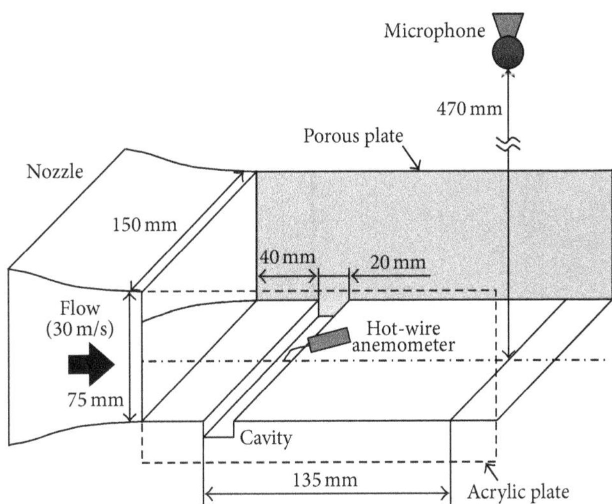

FIGURE 2: Schematics for experimental setup.

4. Numerical Methods

4.1. Governing Equations and Finite Difference Formulations. To clarify the fluid-acoustic interactions in the cavity flows, flow and acoustic fields were simulated simultaneously by directly solving the three-dimensional compressible Navier-Stokes equations in the conserved form.

The spatial derivatives were evaluated using a sixth-order-accurate compact finite difference scheme [15] (fourth-order-accurate scheme at the boundaries). The time integration was performed by the third-order-accurate Runge-Kutta method.

In order to reduce the computational cost, large-eddy simulations (LES) were performed in the present study. A tenth-order-accurate spatial filter dissipated the energy that should be transferred to subgrid scale vortices. This filter also removed numerical instabilities due to the compact finite difference scheme [16]. The details on these computational methods have been presented in Yokoyama and Kato [17].

4.2. Computational Grids. Figure 3 shows the computational grid for the flow over the cavity with $D/L = 2.5$. The spanwise length of the computational domain is $W/L = 1.0$ and the spanwise grid resolution is $\Delta z/L = 0.0125$. It has been preliminarily confirmed that the computational results with $W/L = 1.0$ approximately agree with those with a wider computational domain ($W/L = 2.0$). The computational grid for the flow over the cavity with $D/L = 0.5$ was the same as that for the range of $y/L \geq -0.5$ for the flow over the cavity with $D/L = 2.5$. The total grid points are 18 million.

Figure 4 outlines the computational domain that is divided into *vortical*, *acoustic*, and *buffer regions*. The vortical region is from the upstream edge to the downstream edge of

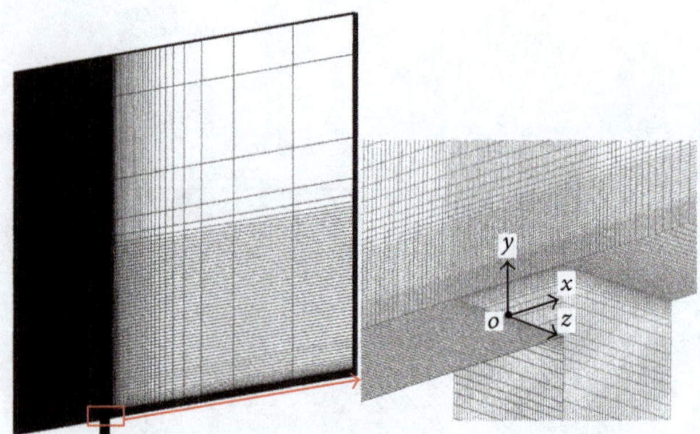

FIGURE 3: Computational grids for cavity flow with $D/L = 2.5$ (every second grid line is shown).

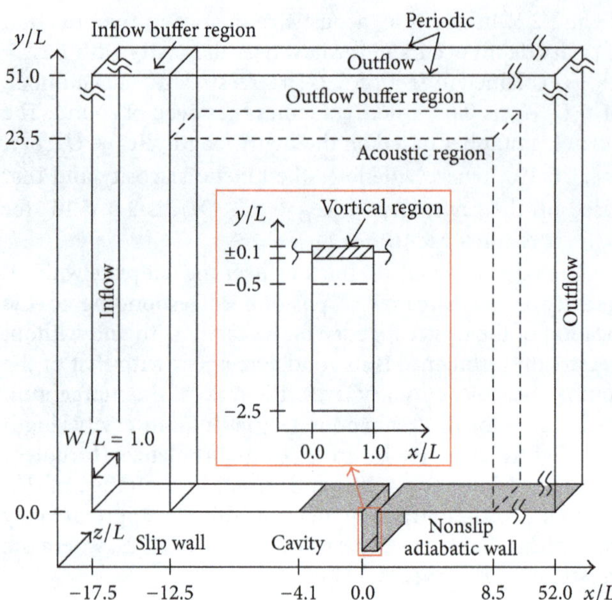

FIGURE 4: Computational domain and boundary conditions.

the cavity ($0.0 \leq x/L \leq 1.0$ and $-0.1 \leq y/L \leq 0.1$), where the shear layer is spread. The grid resolutions are $\Delta x/L = \Delta y/L \leq 0.01$ and sufficiently fine to capture the large-scale vortical structures and active fine scale vortices in the cavity flow. The predicted velocity spectra will be discussed later in Section 5.1.

The region surrounded by the dashed line ($-12.5 \leq x/L \leq 8.5$ and $y/L \leq 23.5$) in Figure 4 is the acoustic region. The acoustic region has an extent that is three and two times larger than the acoustic wavelength of the fundamental frequency of the cavity tone with $D/L = 0.5$ and 2.5, respectively. More than 10 points are used per acoustic wavelength that is two and four times greater than the fundamental frequency of the cavity tone with $D/L = 0.5$ and 2.5, respectively, in the grid resolutions in the acoustic region.

In the buffer region, the grid is sufficiently stretched to dissipate acoustic and voritcal disturbances near the artificial nonreflecting boundary conditions.

As discussed in Section 5, the comparison of the predicted flow and sound with those measured clarifies that it is possible to capture acoustic waves radiating from the cavity flow by using this computational grid and these numerical methods.

4.3. Boundary Conditions. Figure 4 also shows the boundary conditions. Nonreflecting boundaries [18–20] are used at the inflow and outflow boundaries. On the lower boundary in the upstream of the cavity, the boundary condition is changed from a slip wall to a nonslip and adiabatic wall. The position of this change was determined in order to set the boundary layer thickness to be that measured. Periodic boundary conditions are used in the spanwise direction.

A uniform steady flow is asymptotically imposed in the inflow buffer region ($x/L \leq -12.5$) for the cavity flow without freestream turbulence (Tu = 0.0%). In the case with freestream turbulence, a homogeneous turbulence field is imposed on the uniform steady flow in the inflow buffer region.

This homogeneous turbulence field was separately predicted as box turbulence, where the computation starts from initial flow fields based on the von Karman spectrum [21]. These methods were described in detail in the previous paper [17]. The turbulent intensity and scale were controlled by using the parameters of the von Karman spectrum. Figure 5 plots the energy spectrum of homogeneous box turbulence. The Taylor scale of homogeneous turbulence is almost 20 mm.

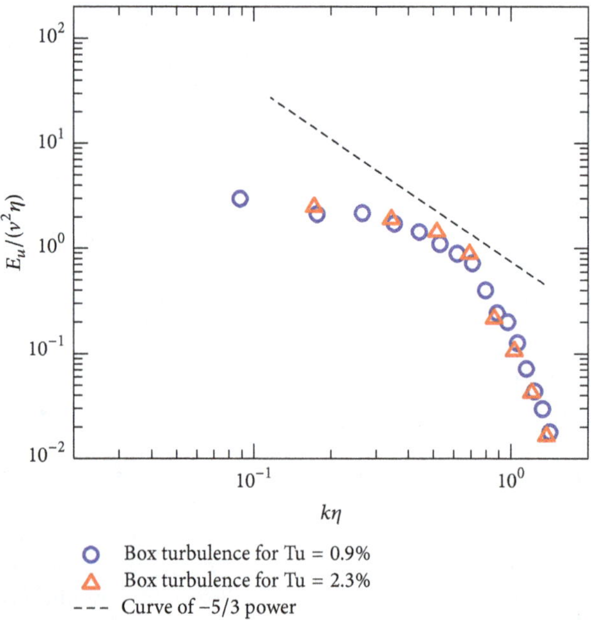

FIGURE 5: Energy spectra of homogeneous box turbulence.

4.4. Prediction of Acoustic Far Fields. The porous Ffowcs Williams and Hawkings (FW-H) method [22–24] was used to predict the acoustic pressure at the measurement point ($x/L = 6.75$ and $y/L = 23.5$) because predicting acoustic pressure in the far field with direct simulation greatly expends

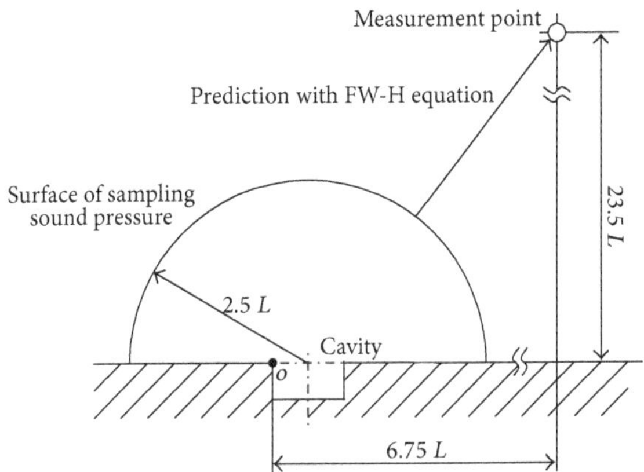

FIGURE 6: Schematics for prediction of acoustic far field by porous FW-H.

computational resources and time. The FW-H formula can be written as

$$L_{ij} = p\delta_{ij}$$
$$4\pi p'(\mathbf{x}_2, t) = \frac{1}{c}\frac{\partial}{\partial t}\int_s \left[L_{ij}(x_1, \tau)\frac{\widehat{\mathbf{r}}\widehat{\mathbf{n}}}{r}\right]_{\text{ret}} dS(\mathbf{x}_1), \tag{1}$$

where $\mathbf{r} = \mathbf{x}_2 - \mathbf{x}_1$, $r = |\mathbf{r}|$, $\widehat{\mathbf{r}} = \mathbf{r}/r$, and $\widehat{\mathbf{n}}$ represent the outward unit normal vector of surface S. The subscript "ret" denotes consideration of retarded time $\tau = t - r/c$. The pressure, p, was sampled at the semicylindrical surface of the radius, $r = 2.5L$, from the center of the cavity ($x/L = 0.5$ and $y = 0$), as outlined in Figure 6.

The spanwise computational domain, $W = L$, was smaller than that of the experiment, $W_e = 7.5L$, to reduce computational resources. To take the effects of this difference into consideration, the sound pressure levels, SPL(f), were estimated by using the equivalent coherent length, $L_c(f)$, following (2) [25–27]:

$$\text{SPL}(f) = \text{SPL}_s(f) + 10\log_{10}\left(\frac{W_e}{W}\right) \quad (L_c(f) \le W),$$
$$\text{SPL}(f) = \text{SPL}_s(f) + 20\log_{10}\left(\frac{L_c(f)}{W}\right) + 10\log_{10}\left(\frac{W_e}{L_c(f)}\right) \tag{2}$$
$$(W < L_c(f) \le W_e),$$
$$\text{SPL}(f) = \text{SPL}_s(f) + 20\log_{10}\left(\frac{W_e}{W}\right) \quad (W_e < L_c(f)),$$

where $\text{SPL}_s(f)$ is the sound pressure level predicted with the above porous FW-H method and the second and third terms on the right hand side are the correction terms. This length was determined by the coherence function values of the vertical velocity along the center of the cavity ($x/L = 0.5$, $y = 0$) in the spanwise direction, $\gamma^2(f, \Delta z) = 1/3$.

The coherence function was computed by the same way as that in the reference by Van Der Velden et al. [28] as indicated in (3):

$$\gamma^2(f, \Delta z) = \frac{|\Phi_{12}(f, z_1, z_2)|^2}{|\Phi_{11}(f, z_1, z_2)|\,|\Phi_{22}(f, z_1, z_2)|}, \tag{3}$$

where Φ_{11} and Φ_{22} denote autopower spectra and Φ_{12} denote the cross-power spectrum between two points along a given dimensional line $\Delta z = z_2 - z_1$. The equivalent coherent length, $L_c(f)$, also has been confirmed to approximately agree with the distance for the phase difference in Lighthill's stress tensor at the same position of $\phi = \pi/4$ (45 degrees).

When the coherence is larger than 1/3 in the entire computational region, the equivalent coherent length was computed by linear extrapolation using the values in the range of $\Delta z/L = 0.25$–0.5, where the gradient of the coherence is approximately constant.

5. Validation of Computational Accuracy

5.1. Flow Field. The predicted flow and acoustic fields for the cavity flow without freestream turbulence (Tu = 0.0%) were compared with those measured because the measured turbulent intensity in the boundary layer was negligibly small.

Figure 7 shows the predicted and measured profiles of mean velocity and turbulent intensity at the center of the cavity ($x/L = 0.5$) for the cavity flow with $D/L = 0.5$, where the contributions of vertical velocity v to the velocity measured by a hot-wire anemometer were considered as $u_h = (u^2 + (0.5v)^2)^{0.5}$ in the same way as our previous research [27]. It should be noted that the markers in Figure 7 in the predicted results represent only data to sample in the computational duration time, where the grid resolutions are finer. It is clarified that the predicted results agree well with those measured.

Figure 8 shows the predicted and measured power spectra of the velocity, u_h/U_∞, at the center of the cavity and the height for the maximal turbulent intensity ($x/L = 0.5$ and $y/L = 0.02$) for the cavity flow with $D/L = 0.5$. The frequency resolution of all spectral analyses is $\Delta\text{St} \equiv \Delta fL/U_0 = 0.104$. The predicted spectrum was averaged 20 times, where the samples were overlapped by 50% and the duration time of the samples for spectral analysis was $100L/U_0$. Also, the experimental spectrum was averaged 9400 times with respect to time.

As shown in Figure 8, there is the main peak of $\text{St}_f = 1.56$ (fundamental frequency) in both predicted and measured results, where the cavity tone radiates at this frequency. Moreover, the predicted power at the fundamental frequency is in good agreement with that measured.

To clarify the effects of the width of the computational domain, W, on the predicted flow fields, the coherence of the normal velocity at $x/L = 0.5$, $y = 0$ was computed for the flow fields predicted with different domains of $W/L = 1.0$ and 2.0. Figure 9 shows the predicted coherence at the fundamental frequency. It is presented that the distributions of the coherence predicted with the original width are in good agreement with those with the double width. This means

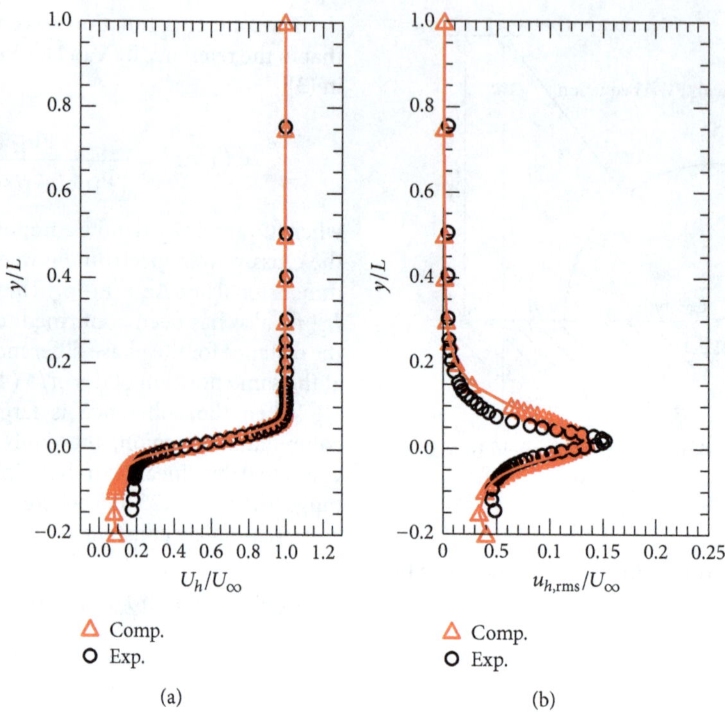

FIGURE 7: Predicted and measured profiles of mean velocity (a) and turbulent intensity (b) for the cavity flow with $D/L = 0.5$.

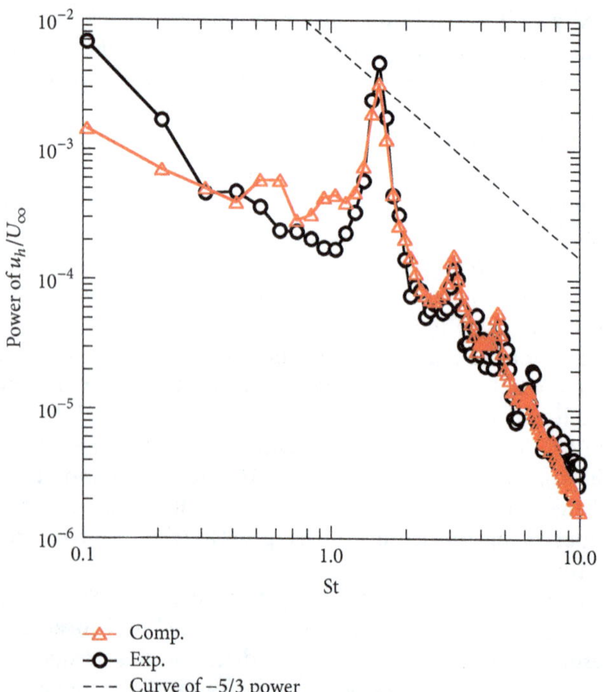

FIGURE 8: Predicted and measured power spectra of velocity fluctuations ($x/L = 0.5$ and $y/L = 0.02$) for cavity flow with $D/L = 0.5$.

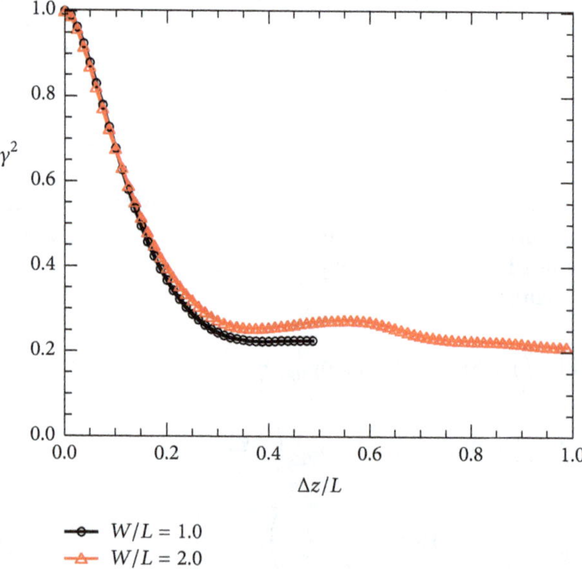

FIGURE 9: Predicted coherence of velocity, v/U_∞, at fundamental frequency in spanwise direction ($x/L = 0.5$ and $y = 0$) for cavity flow ($D/L = 0.5$) with original and wider computational domains of $W/L = 1.0$ and 2.0.

that the independency of the computational results on the width of the domain supports the adoption of the periodic conditions.

5.2. Sound Pressure Level. Figures 10(a) and 10(b) show the predicted and measured sound pressure spectra at the points of the far field ($x/L = 6.75$ and $y/L = 23.5$) for $D/L = 0.5$ and 2.5, respectively. The measured level was affected by the background noise of the wind tunnel in the range of St < 0.5. The bars in the figures present how the results with the same

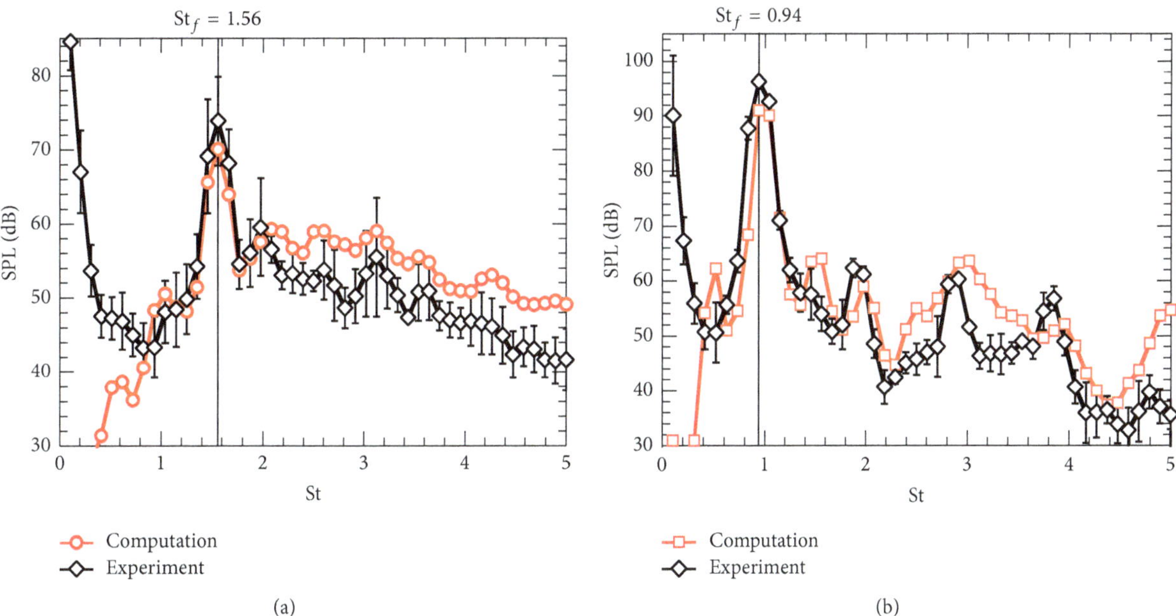

FIGURE 10: Comparison of predicted and measured sound pressure spectra ($x/L = 6.75$ and $y/L = 23.5$). (a) Cavity flow without acoustic resonance ($D/L = 0.5$). (b) Cavity flow with acoustic resonance ($D/L = 2.5$).

averaging number of 20 as that of simulation vary around the measured value with the long averaging number of 9400.

The predicted spectra in both cavity flows with $D/L = 0.5$ and 2.5 are in good agreement with those measured. Also, the fundamental frequency for the cavity flow with $D/L = 2.5$ and $St_f = 0.94$ is corresponding to the Rossiter mode of $n = 2$ [1], which is the streamwise number of vortices in the shear layer, while that for $D/L = 0.5$ and $St_f = 1.56$ is corresponding to $n = 3$. The present peak also agrees with the resonant frequency of one-quarter wavelength mode in the direction of cavity depth, $St_{res} = 0.84$ and 0.98, which are estimated by using the semiempirical formula by East [3] and by general open correction of a circular closed pipe of $8(D/2)/3\pi$ [29]. The slight difference with the resonant frequency estimated by East [3] is possibly due to the conditions of the incoming boundary layer, where the cavity was placed in a turbulent boundary layer in the experiments by East.

To confirm the occurrence of the acoustic resonance in the cavity flow with $D/L = 2.5$, the phase of the pressure fluctuations in the cavity was investigated. Figure 11 shows the phase differences in the pressure fluctuations, $C_p \equiv p/0.5\rho_\infty U_\infty^2$, at the fundamental frequency vertically along $x/L = 0.5$. The reference point for these phase differences is at the bottom of the cavity ($y/L = -2.5$). The phase of pressure fluctuations in the cavity is constant in the cavity ($y/L < -0.1$); that is, the standing waves due to acoustic resonance are generated in the cavity. These results indicate that acoustic resonance can be captured in the present computation.

To clarify the dependency of the predicted results on computational grid, the computations with two different meshes were also performed for cavity flow of $D/L = 0.5$ without freestream turbulence (Tu = 0.0%). One mesh consists of finer mesh of double grid points in the normal

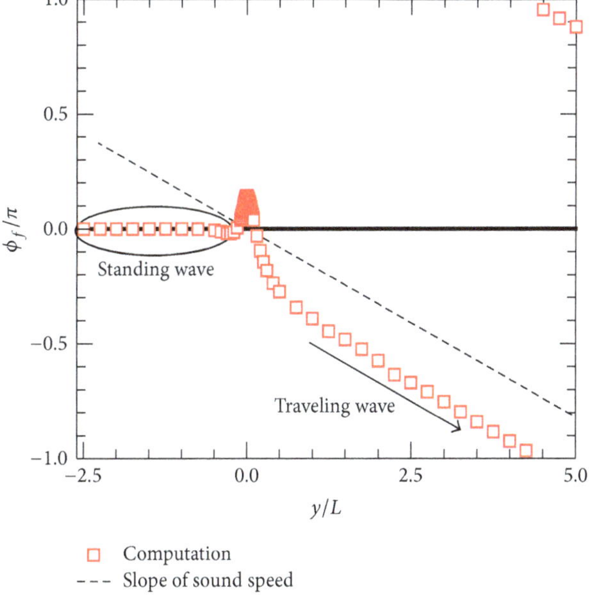

FIGURE 11: Predicted phase differences of pressure fluctuations, $C_p \equiv p/0.5\rho_\infty U_\infty^2$, at fundamental frequency vertically along $x/L = 0.5$ for cavity flow with acoustic resonance ($D/L = 2.5$).

direction in the cavity, where the normal stretch of grid resolution is lower than that of original mesh. The other one is twice as wide as the present mesh as mentioned in Section 5.1.

Figure 12 shows the predicted sound pressure spectra with these meshes. As shown in this figure, the tonal sound is predicted at the same fundamental frequency of $St_f = 1.56$ and

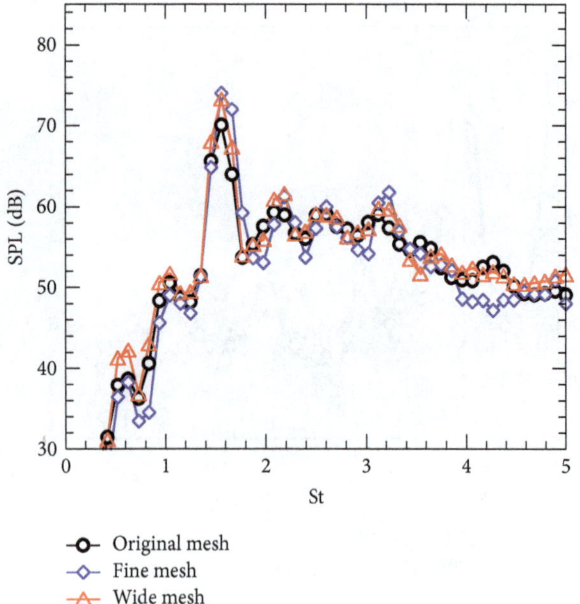

FIGURE 12: Comparison of predicted sound pressure spectra (x/L = 6.75 and y/L = 23.5) with original, finer, and wider meshes for cavity flow without freestream turbulence (D/L = 0.5).

the difference of level is within 5 dB. Therefore, it has been concluded that the dependency of the predicted results on the mesh is small.

It has been concluded along with the discussions of flow field in Section 5.1 that the present computations could adequately capture the flow and acoustic fields.

6. Results and Discussion

6.1. Shear Layer in Cavity Flow. Figure 13 shows the predicted mean velocity and turbulent intensity at the center of the cavity (x/L = 0.5) for the cavity flow with D/L = 0.5 with and without freestream turbulence. As shown in Figure 13(a), it is clarified that the shear layer is thickening with freestream turbulence.

The momentum thickness δ_m at $x/L = 0.5$ was computed as

$$\delta_m = \int_{y_{\min}}^{y_{\max}} \frac{U(y)}{U_\infty}\left(1 - \frac{U(y)}{U_\infty}\right) dy, \quad (4)$$

where $y_{\min}/L$ and $y_{\max}/L$ were set to be −0.2 and 0.2. This range is the same as that in reference [30]. The momentum thickness was clarified to become thicker from $\delta_m/L = 0.019$ (Tu = 0.0%) to 0.027 (Tu = 2.3%) by the freestream turbulence. Figure 13(b) also shows that the turbulent intensity becomes more intense by the freestream turbulence.

Figure 14 shows the predicted profiles for the cavity flow with D/L = 2.5. As shown in this figure, the shear layer becomes thicker, where the momentum thickness was changed from δ_m/L = 0.041 (Tu = 0.0%) to 0.049 (Tu = 2.3%) by the freestream turbulence. Figure 14 shows that the peak of the turbulent intensity becomes slightly weaker by the freestream turbulence. This is because the velocity fluctuations at the fundamental frequency become weaker. The effects of the freestream turbulence on velocity spectra are discussed in the next section. Also, as shown in Figure 14, the region for intense turbulent intensity is spread in the normal direction by the freestream turbulence of Tu = 2.3%. As a result, the mixing in the shear layer is enhanced and the shear layer becomes thicker.

6.2. Vortical Structures. Figure 15 shows instantaneous flow fields around the cavity for Tu = 0.0% and 2.3% for the cavity flow without acoustic resonance (D/L = 0.5). Figure 16 shows those for the cavity flow with acoustic resonance (D/L = 2.5).

Large-scale vortices are found for the cavity flow without freestream turbulence (Tu = 0.0%) for both cavity flows with and without acoustic resonance. These vortices are shed from the upstream edge of the cavity at the fundamental frequency and contribute to the periodic acoustic radiation from the cavity.

The large-scale vortical structures, which are coherent in the spanwise direction, are dominant for Tu = 0.0% particularly in the cavity flow with acoustic resonance. Also, streamwise rib structures and fine scale eddies in the shear layer can be observed for Tu = 0.0% for the cavity flow without acoustic resonance. As the turbulent intensity is increased, the fine scale vortices become active regardless of acoustic resonance.

Figures 17(a) and 17(b) show the power spectra of the vertical velocity, v/U_∞, at the center of the cavity (x/L = 0.5 and y = 0), where large-scale vortices are observed without freestream turbulence, for the cavity flow without acoustic resonance and that with acoustic resonance, respectively.

The power spectra for freestream turbulence Tu = 0.9% are almost the same as that for the cavity flow without freestream turbulence (Tu = 0.0%) in both cases with and without acoustic resonance. The level of velocity fluctuations at the fundamental frequency becomes significantly lower for Tu = 2.3%. This level corresponds to the power of velocity fluctuations due to the large-scale vortical structures, which became the sound source of the cavity tone.

These spectra are also compared with the curve for −5/3 power. All the slopes of the power spectra of velocity fluctuations with and without freestream turbulence in the range of St = 1–5 are along the curve of −5/3 power for the cavity flow without acoustic resonance. Also, the level in that range is approximately independent of freestream turbulence.

As shown in Figure 17(b), for the cavity flow with acoustic resonance, the slope of the spectrum becomes close to the curve for −5/3 power by adding freestream turbulence. Figures 18(a) and 18(b) show the power spectra near the downstream edge of the cavity (x/L = 0.9 and y = 0) for the cavity flow without acoustic resonance and that with acoustic resonance, respectively. Figure 18(b) shows that the slope of the power is clearly along the curve of −5/3 power in the range of St = 1–5 particularly for Tu = 2.3% for the cavity flow with acoustic resonance. Also, the power at high frequencies of St ≥ 3 becomes more intense with the freestream turbulence of Tu = 2.3%.

The above-mentioned results indicate that turbulent mixing in the shear layer is enhanced by freestream turbulence,

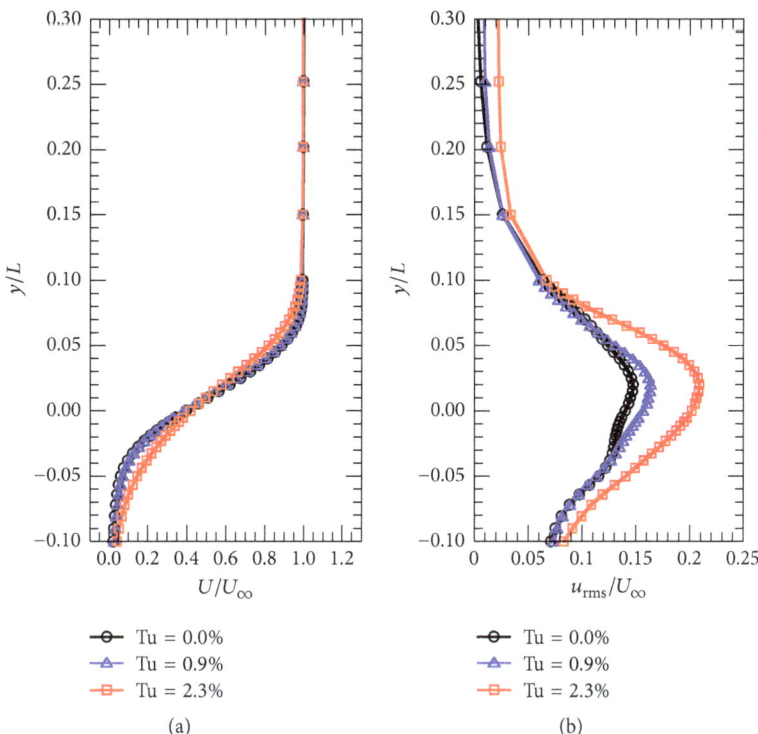

FIGURE 13: Predicted profiles of mean velocity (a) and turbulent intensity (b) for the cavity flow with $D/L = 0.5$.

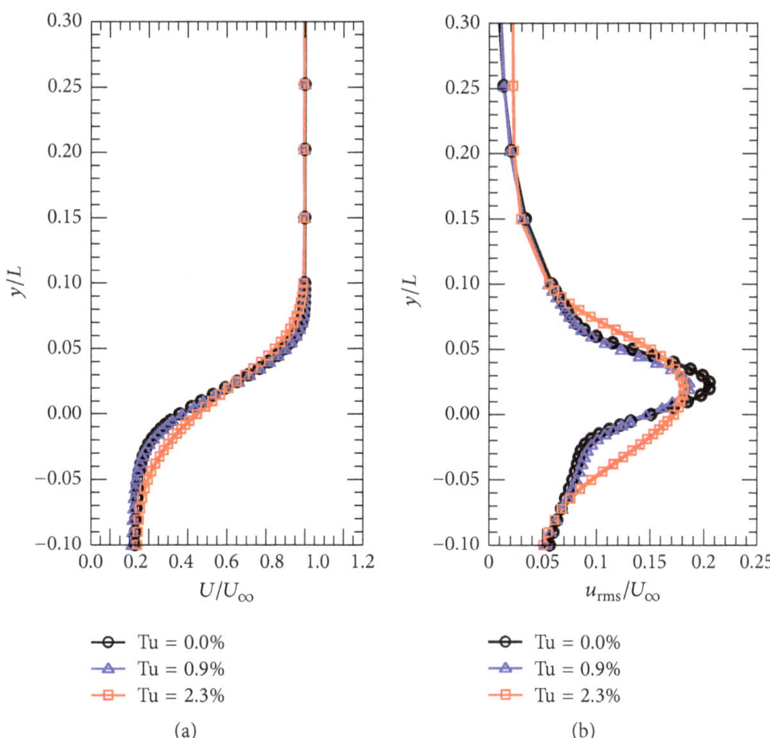

FIGURE 14: Predicted profiles of mean velocity (a) and turbulent intensity (b) for the cavity flow with $D/L = 2.5$.

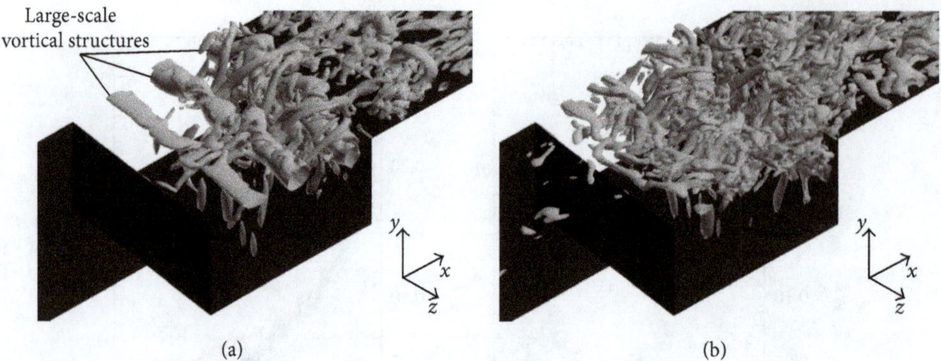

FIGURE 15: Instantaneous isosurfaces of second invariant of velocity tensor gradient ($q/(U_\infty/L)^2 = 20$) for cavity flows without acoustic resonance ($D/L = 0.5$). (a) Cavity flow without freestream turbulence (Tu = 0.0%). (b) Cavity flow with freestream turbulence, Tu = 2.3%.

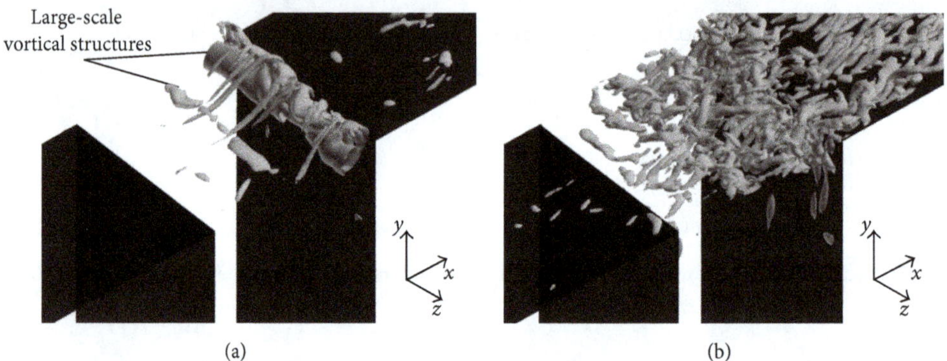

FIGURE 16: Instantaneous isosurfaces of second invariant of velocity tensor gradient ($q/(U_\infty/L)^2 = 20$) for cavity flows with acoustic resonance ($D/L = 2.5$). (a) Cavity flow without freestream turbulence (Tu = 0.0%). (b) Cavity flow with freestream turbulence, Tu = 2.3%.

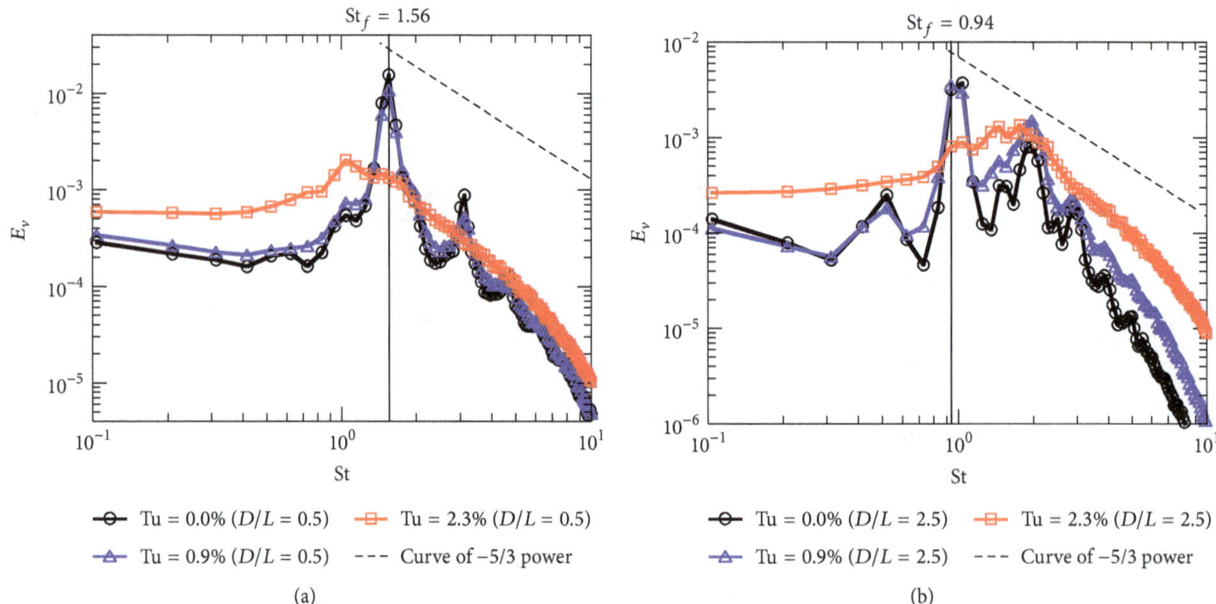

FIGURE 17: Power spectra of velocity v/U_∞ ($x/L = 0.5$ and $y = 0$). (a) Cavity flow without acoustic resonance ($D/L = 0.5$). (b) Cavity flow with acoustic resonance ($D/L = 2.5$).

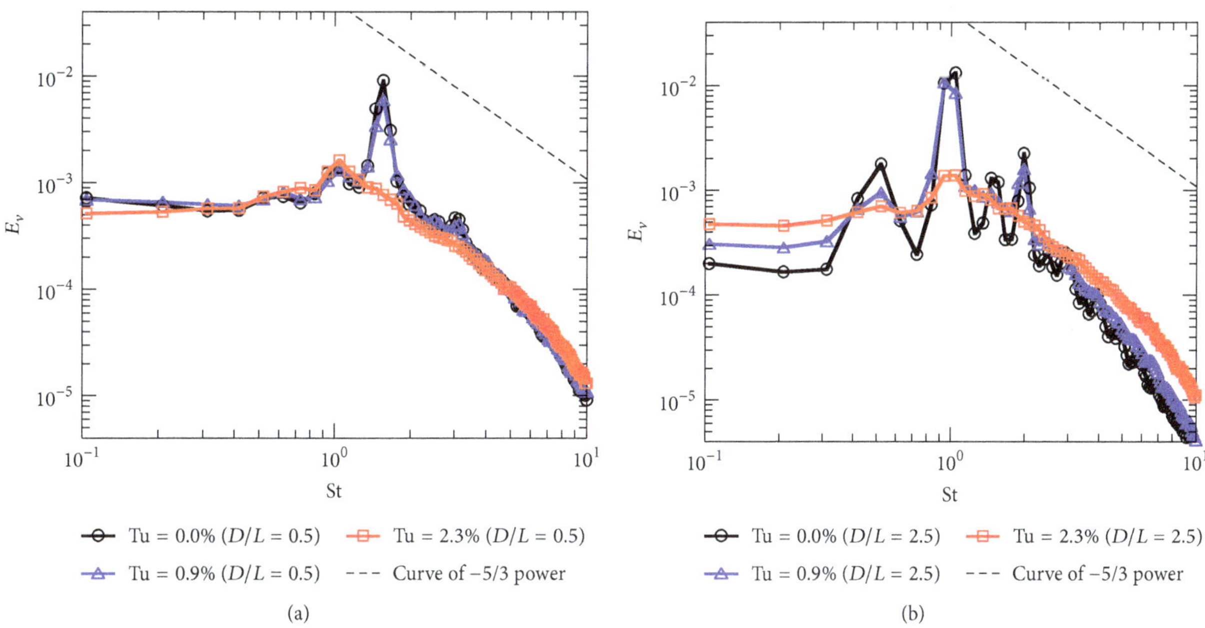

FIGURE 18: Power spectra of velocity v/U_∞ ($x/L = 0.9$ and $y = 0$). (a) Cavity flow without acoustic resonance ($D/L = 0.5$). (b) Cavity flow with acoustic resonance ($D/L = 2.5$).

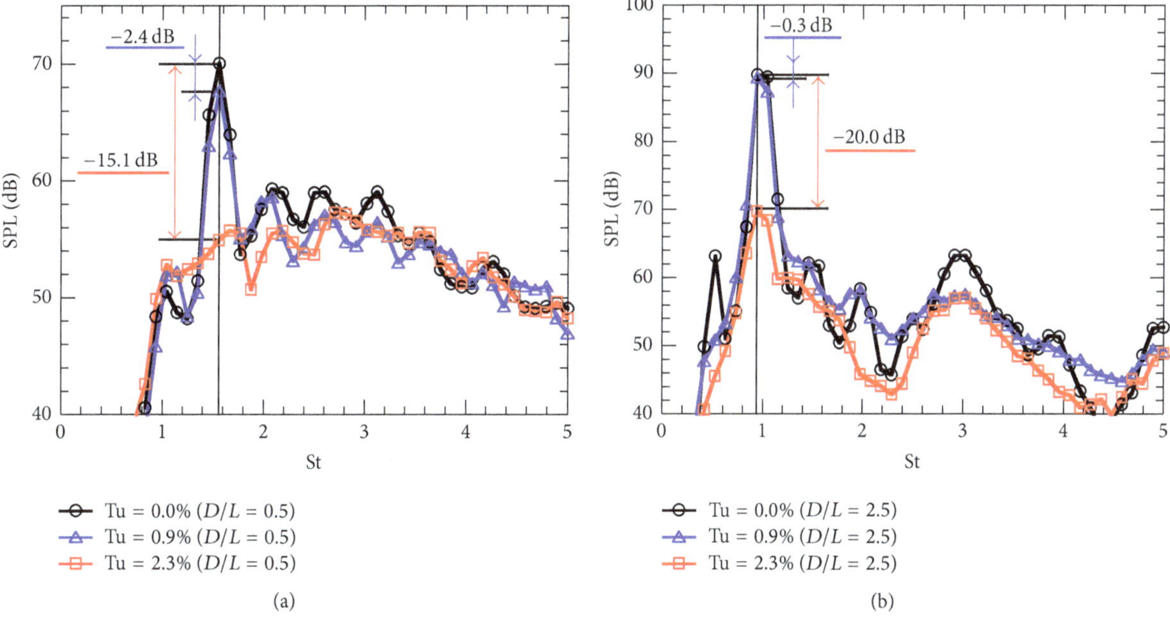

FIGURE 19: Sound pressure spectra ($x/L = 6.75$ and $y/L = 23.5$) and difference in tonal sound pressure levels due to effects of freestream turbulence. (a) Cavity flow without acoustic resonance ($D/L = 0.5$). (b) Cavity flow with acoustic resonance ($D/L = 2.5$).

and the turbulent transition in the shear layer occurs due to freestream turbulence.

6.3. Effects of Freestream Turbulence on Cavity Tone. Figure 19 shows sound pressure spectra in the acoustic far field ($x/L = 6.75$ and $y/L = 23.5$) for cavity flows with and without acoustic resonance. The differences in the tonal sound pressure levels at the fundamental frequency, ΔSPL, with reference to that for the cavity flow without freestream turbulence are also presented.

The ΔSPL is −2.4 dB and −15.1 dB for Tu = 0.9 and 2.3%, respectively, for the cavity flow without acoustic resonance. The ΔSPL is −0.3 dB and −20.0 dB for Tu = 0.9 and 2.3%, respectively, for the cavity flow with acoustic resonance.

The effects of freestream turbulence on the cavity tone are less than 2.5 dB for both cavity flows with and without acoustic resonance for freestream turbulence of Tu = 0.9%.

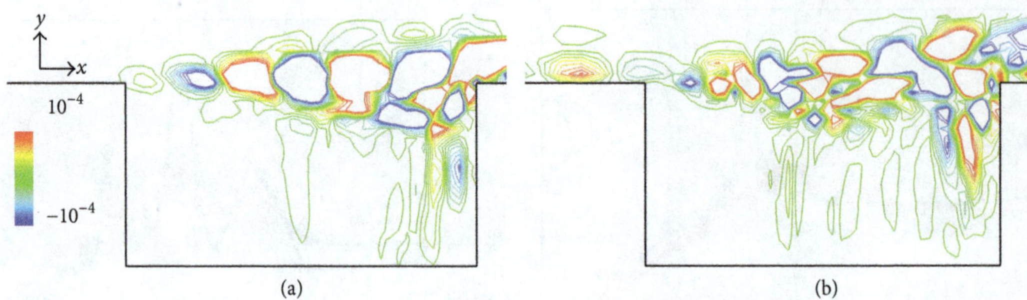

FIGURE 20: Contours of real part of power of Lighthill's stress tensor at fundamental frequency, $\partial^2 T_{ij}/\partial x_i \partial x_j/(\rho_\infty U_\infty{}^2/L^2)$, around cavity for cavity flows without acoustic resonance ($D/L = 0.5$). (a) Cavity flow without freestream turbulence (Tu = 0.0%). (b) Cavity flow with freestream turbulence, Tu = 2.3%.

FIGURE 21: Contours of real part of power of Lighthill's stress tensor at fundamental frequency, $\partial^2 T_{ij}/\partial x_i \partial x_j/(\rho_\infty U_\infty{}^2/L^2)$, around cavity for cavity flows with acoustic resonance ($D/L = 2.5$). (a) Cavity flow without freestream turbulence (Tu = 0.0%). (b) Cavity flow with freestream turbulence, Tu = 2.3%.

Tonal sound generated from the cavity flow with acoustic resonance is greatly affected by freestream turbulence for Tu = 2.3%.

The weakening of the intensity and lowering of the spanwise coherence of the sound source are considered to be responsible for the reduced cavity tone. The following sections discuss the effects of the freestream turbulence on the intensity and spanwise coherence of the sound source.

6.4. Intensity of Sound Source. Aerodynamic sound is related to the intensity of fluctuating pressures/velocities and their structures (coherent motion) according to acoustic analogies [11, 13, 22]. In addition, the velocity fluctuations near the downstream edge of the cavity contribute to the sound pressure level of the cavity tone since the cavity tone is generated by impinging large-scale vortical structures at the downstream edge of the cavity [17].

The Lighthill's equations [13] are

$$\frac{\partial^2 \rho}{\partial t^2} - c^2 \frac{\partial^2 \rho}{\partial x_i^2} = \frac{\partial^2 T_{ij}}{\partial x_i \partial x_j}, \tag{5}$$

$$T_{ij} \approx \rho u_i u_j, \tag{6}$$

where tensor T_{ij} is Lighthill's stress tensor. This tensor is approximately equal to the Reynolds stress term at the low Mach number [31] as shown in (6). In the present section, the intensity of the sound source was estimated by using the Reynolds stress term of Lighthill's stress tensor [13] at the fundamental frequency near the downstream edge ($0.75 \le x/L \le 1.0$). The range of $0.75 \le x/L \le 1.0$ is approximately equal to the scale of the large-scale vortical structures.

Figures 20 and 21 show the intensity of the above-mentioned sound source (the Reynolds stress term of Lighthill's stress tensor) at the fundamental frequency, around the cavity. Also, Figure 22 shows the power level of the sound source at the fundamental frequency, P_f, along $y = 0$ from the upstream edge to the downstream edge. As shown in the figure, the acoustic source is weakened by the freestream turbulence.

Moreover, the sound reduction level of the cavity tone due to the weakening of the intensity of the sound source, ΔSPL_i, was calculated by the integration of the power of source, P_f, in the range of $0.75 \le x/L \le 1.0$. The equation of ΔSPL_i is as follows:

$$\Delta\text{SPL}_i = 10 \log_{10} \left[\int_{0.75L}^{L} \left(\frac{P_f}{P_{f0}} \right) dx \right], \tag{7}$$

where the subscript "0" means the condition of the cavity flow without freestream turbulence (Tu = 0.0%).

Table 2 shows the values of ΔSPL_i in each condition. The value of ΔSPL_i is −8.5 dB and −8.2 dB for Tu = 2.3% with and without acoustic resonance, respectively, which represent greater effects than those for Tu = 0.9% (−0.3 dB and 0.8 dB). Also, this means that the cavity tone is greatly reduced by weakening of the sound source for Tu = 2.3%.

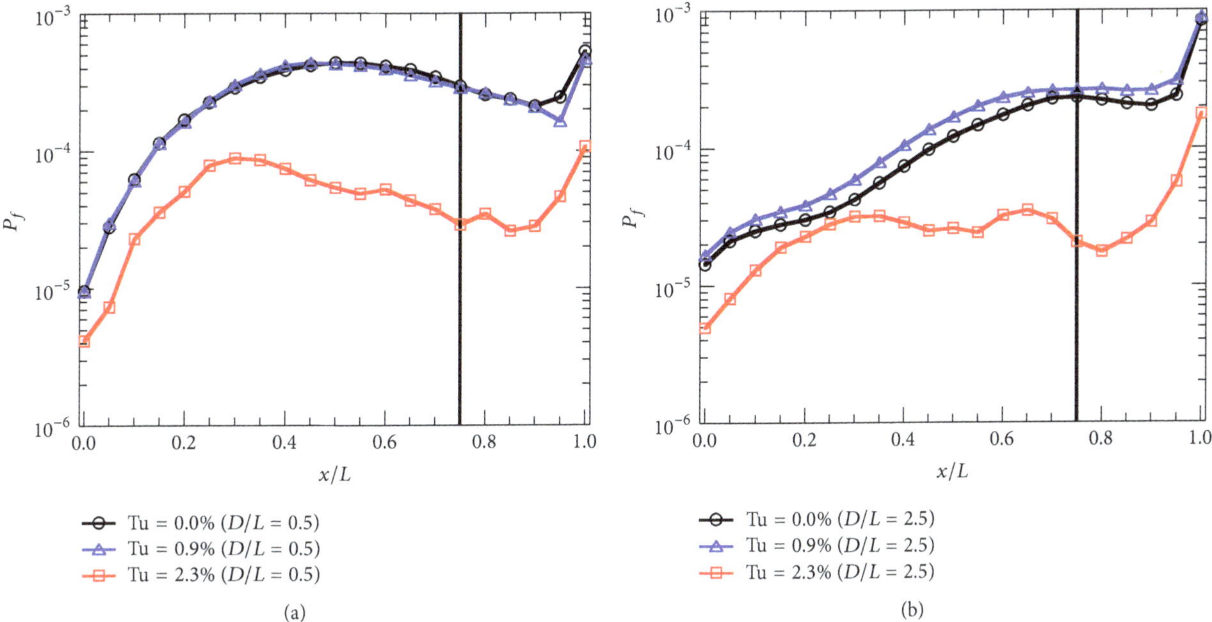

FIGURE 22: Power level of Lighthill's stress tensor at the fundamental frequency, P_f, along $y = 0$. (a) Cavity flow without acoustic resonance ($D/L = 0.5$). (b) Cavity flow with acoustic resonance ($D/L = 2.5$).

TABLE 2: Contributions of intensity and spanwise coherence of sound source to the sound pressure level.

D/L	Tu [%]	L_c/L	ΔSPL [dB]	ΔSPL_i [dB]	ΔSPL_c [dB]	ΔSPL_{total} [dB]
	0	0.22	—	—	—	—
0.5	0.9	0.12	−2.4	−0.3	−2.7	−3.0
	2.3	0.04	−15.1	−8.5	−4.2	−12.7
	0	6.40	—	—	—	—
2.5	0.9	5.76	−0.3	0.8	−0.5	0.3
	2.3	0.05	−20.0	−8.2	−17.1	−25.3

6.5. Spanwise Coherence of Sound Source. Two-dimensionality in the spanwise direction of large-scale vortical structures as shown in Figures 15 and 16 is quantitatively discussed. In the present section, the effects of freestream turbulence on the spanwise coherence of the sound source are discussed by computing the coherence of the vertical velocity, v/U_∞, at the fundamental frequency at the center of the cavity ($x/L = 0.5$ and $y = 0$), which is shown in Figure 23.

As shown in Figure 23, the coherence is decreased by the freestream turbulence for both cases with and without acoustic resonance. The equivalent coherent length, L_c/L, which was discussed in Section 4.4, is 0.22, 0.12, and 0.04 for the flow with freestream turbulence of Tu = 0.0, 0.9, and 2.3%, respectively, in the cavity flow without acoustic resonance, whereas it is 6.40, 5.76, and 0.05 for Tu = 0.0, 0.9, and 2.3% for the cavity flow with acoustic resonance.

Equivalent coherent length for the cavity flow with acoustic resonance without freestream turbulence, $L_c/L = 6.40$, is greater than that for the flow without acoustic resonance ($L_c/L = 0.22$). This means that the two-dimensionality of the large-scale vortices related with cavity tone becomes higher due to the acoustic resonance. As shown in Figures 15 and 16, three-dimensional fine scale vortices are less active for cavity flow with acoustic resonance.

In Section 4.4, the method for correction of SPL for the difference between the computational and experimental spanwise extent is discussed. With a similar method, the effects of lowering spanwise coherence of the sound source on the cavity tone, ΔSPL_c, are estimated as follows:

$$\begin{aligned} SPL_c &= 10\log_{10}\left(\frac{W_e}{\lambda_v}\right) \quad (L_c \le \lambda_v), \\ SPL_c &= 20\log_{10}\left(\frac{L_c}{\lambda_v}\right) + 10\log_{10}\left(\frac{W_e}{L_c}\right) \quad (\lambda_v < L_c \le W_e), \\ SPL_c &= 20\log_{10}\left(\frac{W_e}{\lambda_v}\right) \quad (W_e < L_c), \\ \Delta SPL_c &= SPL_c - SPL_{c0}, \end{aligned} \tag{8}$$

where λ_v is the streamwise scale of vortices related to cavity tone and estimated to be approximately $L/(4n)$ by the distance of the streamwise phase difference of velocity fluctuations of $\Delta\phi < \pi/2$ in the shear layer.

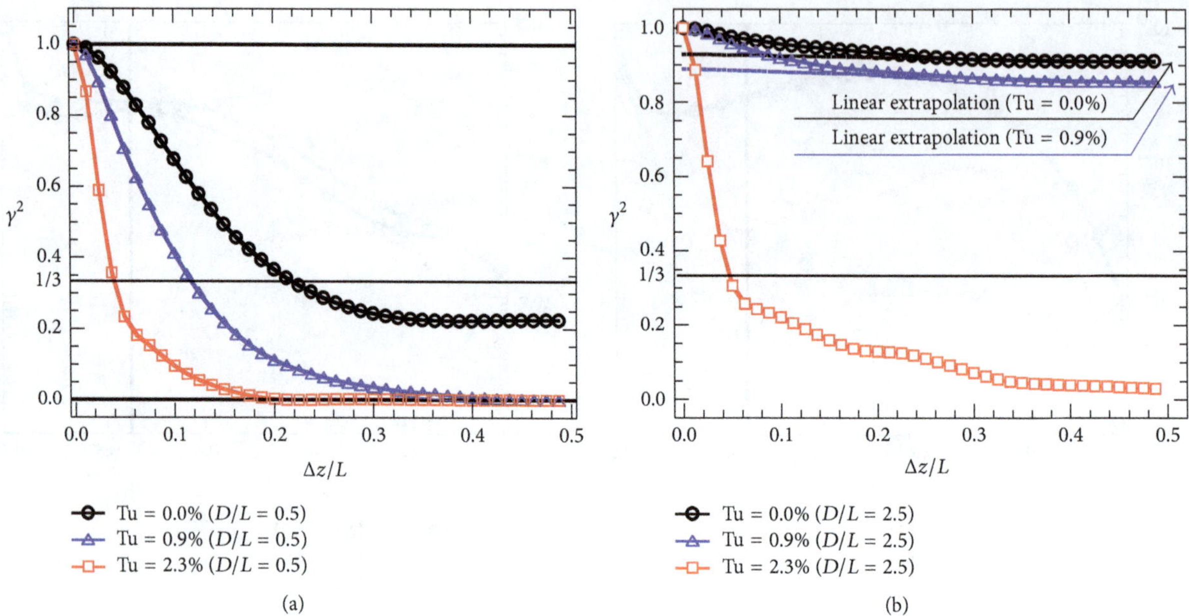

FIGURE 23: Coherence of velocity v/U_∞ at fundamental frequency in spanwise direction ($x/L = 0.5$ and $y = 0$). (a) Cavity flow without acoustic resonance ($D/L = 0.5$). (b) Cavity flow with acoustic resonance ($D/L = 2.5$).

Table 2 also shows the value of ΔSPL_c for each case. It is −0.6 dB and −4.2 dB for Tu = 0.9 and 2.3% in the cavity flow without acoustic resonance, while it is −0.5 dB and −17.1 dB for Tu = 0.9 and 2.3% for the cavity flow with acoustic resonance. These results present that the effects of the freestream turbulence in the cavity flow with acoustic resonance were greater than those in the flow without acoustic resonance. This is possibly because the turbulent transition occurs in the free shear layer of the flow with acoustic resonance while the shear layer is turbulent even without freestream turbulence in the flow without acoustic resonance.

6.6. Contributions of Intensity and Spanwise Coherence of Sound Source. Both the intensity and spanwise coherence of the sound source affect variation of the sound pressure level by freestream turbulence, as was previously mentioned. The contributions of these two factors are discussed in the present section. Table 2 and Figure 24 help to explain the contributions of various factors to the cavity tone. The total difference in the cavity tone due to contributions by these factors, $\Delta\text{SPL}_{\text{total}}$, is calculated as

$$\Delta\text{SPL}_{\text{total}} = \Delta\text{SPL}_i + \Delta\text{SPL}_c. \quad (9)$$

Figure 24 shows that the reduction level estimated with all factors, $\Delta\text{SPL}_{\text{total}}$, is comparable to the predicted one from the direct simulation, ΔSPL, for each case. Also, we found qualitative agreement such that the effects of freestream turbulence on the cavity tone were greater in the cavity flow with acoustic resonance than that without acoustic resonance. This is because the reduced sound pressure level due to the reduction of the coherence is greater for the cavity flow with acoustic resonance, where the turbulent transition occurs. Also, this agreement shows that the present method for estimating the effects of the intensity and coherence of the sound source on the cavity tone is reasonable.

7. Conclusions

Direct simulations of flow and acoustic field were carried out for cavity flows with and without acoustic resonance (D/L = 0.5 and 2.5) under various freestream turbulent conditions to clarify the effects of freestream turbulence on the cavity tone. The freestream Mach number was $M = 0.09$ and the Reynolds number based on the cavity length was $\text{Re}_L = 4.0 \times 10^4$. The incoming boundary layer was laminar with momentum thickness $\theta/L = 0.0071$–0.0074. The effects of turbulent intensities of Tu = 0.0–2.3% on flow and acoustic fields were investigated.

The sound source was estimated by using the flow field based on Lighthill's acoustic analogy. Moreover, the reduced cavity tone was decomposed into the effects of the intensity and spanwise coherence of the sound source to enable the reduction mechanism of the cavity tone to be understood. The total reduction level by summing these contributions by the intensity and coherence was clarified to be comparable to that predicted by direct simulations. Also, these analytical results can explain well the effects of the acoustic resonance and freestream turbulence on the reduced level, which were predicted from direct simulations. Thus, this indicates that the method for estimating the effects of the intensity and coherence of the sound source on the cavity tone in the present study is reasonable.

The effects of freestream turbulence on the cavity tone in a cavity flow with acoustic resonance, −20.0 dB, were greater than those in a cavity flow without acoustic resonance, −15.1 dB, for a cavity flow with freestream turbulence of Tu =

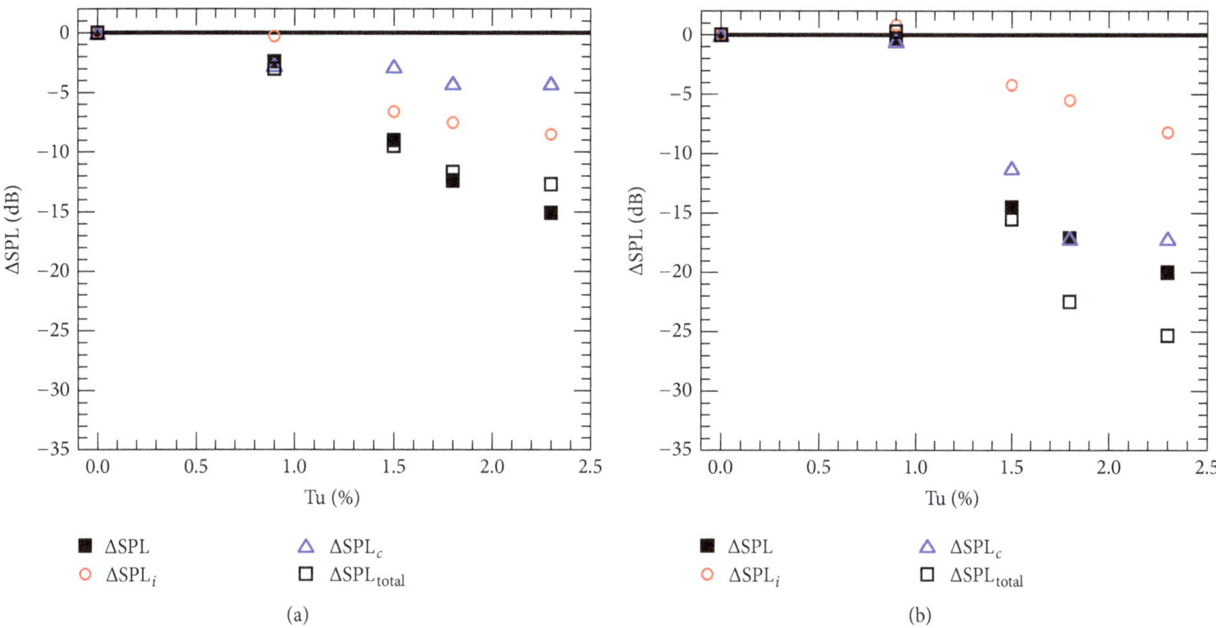

FIGURE 24: Effects of intensity and spanwise coherence of sound source on variation of cavity tone. (a) Cavity flow without acoustic resonance ($D/L = 0.5$). (b) Cavity flow with acoustic resonance ($D/L = 2.5$).

2.3%. This was caused by the difference in the flow field for the cavity flow without freestream turbulence Tu = 0.0%. That is, the large-scale vortical structures in the cavity flow with acoustic resonance were extremely coherent in the spanwise direction due to acoustic resonance, whereas they were less coherent even for Tu = 0.0% for the cavity flow without acoustic resonance. As a result of this difference in coherence, the cavity flow with acoustic resonance was affected more by freestream turbulence.

The analyses of the intensity and spanwise coherence of the sound source clarified that the effects of the intensity and spanwise coherence of the sound source, on the other hand, were almost the same as those without acoustic resonance. The effects of spanwise coherence of the sound source on the cavity tone were greater in the cavity flow with acoustic resonance than in that without resonance.

Competing Interests

The authors declare that there is no conflict of interests regarding the publication of this paper.

Acknowledgments

The present study was supported by JSPS KAKENHI Grant nos. 24760134 and 26820044 and through the Next-Generation Supercomputer Strategy Program made available by the Ministry of Education, Culture, Sports, Science, and Technology of Japan (MEXT).

References

[1] J. E. Rossiter, “Wind-tunnel experiments on the flow over rectangular cavities at subsonic and transonic speeds,” Aeronautical Research Council Report 3438, 1964.

[2] V. Sarohia, “Experimental investigation of oscillations in flows over shallow cavities,” *AIAA Journal*, vol. 15, no. 7, pp. 984–991, 1977.

[3] L. F. East, “Aerodynamically induced resonance in rectangular cavities,” *Journal of Sound and Vibration*, vol. 3, no. 3, pp. 277–287, 1966.

[4] G. A. Brés and T. Colonius, “Three-dimensional instabilities in compressible flow over open cavities,” *Journal of Fluid Mechanics*, vol. 599, pp. 309–339, 2008.

[5] X. Huang and X. Zhang, “Streamwise and spanwise plasma actuators for flow-induced cavity noise control,” *Physics of Fluids*, vol. 20, no. 3, Article ID 037101, 2008.

[6] T. Lusk, L. Cattafesta, and L. Ukeiley, “Leading edge slot blowing on an open cavity in supersonic flow,” *Experiments in Fluids*, vol. 53, no. 1, pp. 187–199, 2012.

[7] W. S. Saric, H. L. Reed, and E. J. Kerschen, “Boundary-layer receptivity to freestream disturbances,” *Annual Review of Fluid Mechanics*, vol. 34, no. 1, pp. 291–319, 2002.

[8] L. Brandt, P. Schlatter, and D. S. Henningson, “Transition in boundary layers subject to free-stream turbulence,” *Journal of Fluid Mechanics*, vol. 517, pp. 167–198, 2004.

[9] T. Colonius and S. K. Lele, “Computational aeroacoustics: progress on nonlinear problems of sound generation,” *Progress in Aerospace Sciences*, vol. 40, no. 6, pp. 345–416, 2004.

[10] J. Larsson, L. Davidson, M. Olsson, and L.-E. Eriksson, “Aeroacoustic investigation of an open cavity at low mach number,” *AIAA Journal*, vol. 42, no. 12, pp. 2462–2473, 2004.

[11] N. Curle, "The influence of solid boundaries upon aerodynamic sound," *Proceedings of the Royal Society. London. Series A. Mathematical, Physical and Engineering Sciences*, vol. 231, pp. 505–514, 1955.

[12] J. Ask and L. Davidson, "Sound generation and radiation of an open two-dimensional cavity," *AIAA Journal*, vol. 47, no. 6, pp. 1337–1349, 2009.

[13] M. J. Lighthill, "On sound generated aerodynamically. I. General theory," *Proceedings of the Royal Society. London. Series A. Mathematical, Physical and Engineering Sciences*, vol. 211, pp. 564–597, 1952.

[14] K. Terao, H. Yokoyama, Y. Ogoe, and A. Iida, "Proposition of new formula for frequency prediction based on generation mechanism of aerodynamic sound in cavity flows," *Transactions of the Japan Society of Mechanical Engineers B*, vol. 77, no. 779, pp. 1522–1532, 2011 (Japanese).

[15] S. K. Lele, "Compact finite difference schemes with spectral-like resolution," *Journal of Computational Physics*, vol. 103, no. 1, pp. 16–42, 1992.

[16] K. Matsuura and C. Kato, "Large-eddy simulation of compressible transitional flows in a low-pressure turbine cascade," *AIAA Journal*, vol. 45, no. 2, pp. 442–457, 2007.

[17] H. Yokoyama and C. Kato, "Fluid-acoustic interactions in self-sustained oscillations in turbulent cavity flows, I. Fluid-dynamic oscillations," *Physics of Fluids*, vol. 21, no. 10, Article ID 105103, 2009.

[18] K. W. Thompson, "Time dependent boundary conditions for hyperbolic systems," *Journal of Computational Physics*, vol. 68, no. 1, pp. 1–24, 1987.

[19] T. J. Poinsot and S. K. Lele, "Boundary conditions for direct simulations of compressible viscous flows," *Journal of Computational Physics*, vol. 101, no. 1, pp. 104–129, 1992.

[20] J. W. Kim and D. J. Lee, "Generalized characteristic boundary conditions for computational aeroacoustics," *AIAA Journal*, vol. 38, no. 11, pp. 2040–2049, 2000.

[21] J. O. Hinze, *Turbulence*, McGraw-Hill, New York, NY, USA, 1975.

[22] J. E. F. Williams and D. L. Hawkings, "Sound generation by turbulence and surfaces in arbitrary motion," *Philosophical Transactions of the Royal Society of London A: Mathematical, Physical and Engineering*, vol. 264, no. 1151, pp. 321–342, 1969.

[23] A. S. Lyrintzis, "Surface integral methods in computational aeroacoustics—from the (CFD) near-field to the (Acoustic) far-field," *International Journal of Aeroacoustics*, vol. 2, no. 2, pp. 95–128, 2003.

[24] M. Shur, P. Spalart, and M. Strelets, "Noise prediction for increasingly complex jets—part I: methods and tests," *International Journal of Aeroacoustics*, vol. 4, no. 3, pp. 213–246, 2005.

[25] C. Kato, A. Iida, Y. Takano, H. Fujita, and M. Ikegawa, "Numerical prediction of aerodynamic noise radiated from low Mach number turbulent wake," in *Proceedings of the 31st Aerospace Sciences Meeting and Exhibit*, Reno, Nev, USA, 1993.

[26] K. W. Chang, J. H. Seo, Y. J. Moon, and M. Roger, "Prediction of flat plate self-noise," in *Proceedings of the 12th AIAA/CEAS Aeroacoustics Conference*, pp. 1451–1464, May 2006.

[27] H. Yokoyama, K. Kitamiya, and A. Iida, "Flows around a cascade of flat plates with acoustic resonance," *Physics of Fluids*, vol. 25, no. 10, Article ID 106104, 2013.

[28] W. C. P. Van DerVelden, A. H. Van Zuijlen, A. T. De Jong, and H. Bijl, "On the estimation of Spanwise pressure coherence of a turbulent boundary layer over a flat plate," in *Proceedings of 11th World Congress on Computational Mechanics (WCCM '14)*, pp. 5710–5721, July 2014.

[29] J. W. S. Rayleigh, *The Theory of Sound*, Dover, Mineola, NY, USA, 1945.

[30] L. Larchevêque, P. Sagaut, I. Mary, O. Labbé, and P. Comte, "Large-eddy simulation of a compressible flow past a deep cavity," *Physics of Fluids*, vol. 15, no. 1, pp. 193–210, 2003.

[31] M. E. Goldstein, *Aeroacoustics*, McGraw-Hill, New York, NY, USA, 1976.

8

Unsteady Aerodynamic Modeling based on POD-ARX

Xiaopeng Wang,[1] Chen'an Zhang,[2] Wen Liu,[2] Famin Wang,[2] and Zhengyin Ye[1]

[1]*Northwestern Polytechnical University, Xi'an 710072, China*
[2]*State Key Laboratory of High-Temperature Gas Dynamic, Institute of Mechanics, Chinese Academy of Sciences, Beijing 100190, China*

Correspondence should be addressed to Chen'an Zhang; zhch_a@imech.ac.cn

Academic Editor: Zhiguang Song

The lack of stability is a problem encountered when applying the classical POD-Galerkin method to problems of unsteady compressible flows around a moving structure. To solve this problem, a hybrid reduced-order model named POD-ARX is constructed in this paper. The construction of this model involves two steps, including first extracting the fluid modes with the POD technique and then identifying the modal coefficients with the ARX model. The POD modes with the block of all modified primitive variables are extracted from the system response to the training signal. Once the POD modes are obtained, the snapshots are projected on these modes to determine the time history of modal coefficients and the resulting modal coefficients are used to identify the parameters of ARX model. Then, the ARX model is used to predict the modal coefficients of the system response to the validation signal. Sample two-dimensional aerodynamic force calculations are conducted to demonstrate this method. Results show that this method can produce a stable and accurate prediction to the aerodynamic response with significant improvement of computational efficiency for linear and even some nonlinear aerodynamic problems. In addition, this method also shows good wide-band characteristics by using the "3211" multistep signal as the training signal.

1. Introduction

In recent years, with the rapid development of the computer technology, computational fluid dynamics (CFD) has become a more and more common tool in the analysis of flow physics. The ability of performing high-fidelity unsteady flow simulation makes the CFD solver capture complex flow phenomenon accurately, such as separated flows and shock waves. However, due to the huge cost of time and computational resources, nowadays, the CFD technique is still not very suitable for analyzing fluid-structure interaction problems or other unsteady problem that needs repeated calculations. Therefore, it's very important to develop a surrogate model with high levels of both computational efficiency and accuracy.

In 1990s, a series of unsteady aerodynamic reduced-order models (ROM) were proposed. Compared to the direct CFD numerical simulation, these models can not only improve the computational efficiency significantly but also have the satisfying computational accuracy. Current researches on ROM can be generally divided into two groups. The first branch employs the system identification (SI) methods to build the relationship between the input and output data, including the autoregressive with exogenous input (ARX) model [1, 2], the Volterra series [3–5], and the neural networks [6–9]. The second branch is based on the eigenmodes of the flow field, including the proper orthogonal decomposition (POD), which projects the governing equations onto these eigenmodes to obtain a low-order dynamic model [10–12]. Due to the good performance in the accuracy and efficiency, the POD technique has become an active area of the ROM research and has been widely used in optimal design [13, 14], control design [15, 16], aeroelastic analysis [17–19], etc. However, due to the influence of inner product, simplification of the original governing equations, and the lack of dissipation in numerical schemes, the conventional POD ROMs are usually unstable and additional stabilization strategies have to be adopted to solve these problems [20–24].

In addition, some combinations between POD and system identification methods have also been proposed in recent years [25–27].

Lucia and Beran [26] proposed a hybrid approach by combining POD and Volterra theory. In this method, the POD technique is used to extract the fluid modes and the Volterra theory is employed to identify the modal coefficients of the flow field. This method has been demonstrated on a two-dimensional subsonic inviscid flow over a bump with forcing and been successfully used to predict the limit-cycle oscillation behavior over an elastic panel in supersonic flow [26, 27]. Although this method has shown its efficiency and accuracy in aerodynamic response prediction, it also presents some drawbacks. First, this method extracted the POD modes from the startup response of full-order model to validation signal itself and it showed in [27] that the POD basis derived from the impulse response data was not adequate for modeling the aeroelastic problems accurately, which means that these modes are difficult to represent the dominant feature of the flow generated from other input signals with different frequencies. Second, Lucia and Beran [26] and Lucia et al. [27] conducted the orthogonal decomposition for each variable separately. Although this method can provide the maximum POD power structure for each variable, it also makes the system relatively complicated and heavy. In addition, Placzek [28] points out that since the variables are correlated with each other, a part of physical problems may be ignored with this method. Therefore, it is preferable to form a set of eigenmodes from blocks containing all the fluid variables. But the traditional conservative variables are not appropriate because the Navier-Stokes (N-S)/Euler equations are not quadratic and the Galerkin projection will yield an inadequate implicit form of the modal coefficients.

Fortunately, some researchers, trying to apply the POD method in the compressible flow, provide some excellent ideas. Placzek [28] and Placzek et al. [29] introduced the modified primitive variables into the analysis of the nonlinear compressible flows around a rigid body with motion by POD. With the modified primitive variables, the N-S/Euler equations can be transformed to an explicit format about the modal coefficients by the Galerkin projection. However, the resulting reduced-order model lacks some dissipation and has to adopt the correction method to produce a stable response, which brings some new difficulties, such as the choice of the correction method for different flow conditions and the complex computation of the parameters.

In order to overcome the disadvantages of the above methods, a new POD-ARX model is constructed in this paper. The model is constructed by extracting the modes by POD technique and identifying the modal coefficients by the ARX model. As a difference model, the ARX model is very easy to implement with existing unsteady CFD codes and extremely efficient computationally. It is also well suitable for the multi-input/multioutput (MIMO) identification procedure and has been extensively used in aeroelastic problems [1, 2, 30–33]. Besides, recent researches have proved that the ARX model can be used in the nonlinear problems of an airfoil in pitching motion under large angle of attack in the near-instable flows [34–36]. The modified primitive variables are applied as a block to conduct the proper orthogonal decomposition, and POD modes are extracted from the snapshots obtained by full-order system response to the training signal with a wide range of frequencies. Finally, the performance of the method was validated by a subsonic inviscid flow sample around the moving NACA0012 airfoil.

2. Construction of the Reduced-Order Model

2.1. Overview of Proper Orthogonal Decomposition. The aim of POD is to find a set of optimal orthogonal basis $\mathbf{\Phi} = \{\phi_i, i = 1, 2, \dots, m\}$ to provide a best approximation to the behavior of the full-order system dynamics. Since the snapshots are usually centered, the problem can be transformed into finding the best basis to approximate the fluctuations of the snapshots around a mean state. Consequently, the problem in the discrete domain can be expressed as follows [26]:

$$q(t) = q_0 + \Delta q = q_0 + \sum_{i=1}^{m} a_i(t)\phi_i = q_0 + \mathbf{\Phi}\widehat{q}(t), \qquad (1)$$

where q_0 represents the full-order base solution, which can be a steady CFD solution or an average solution of the snapshots used to extract the POD modes; $\mathbf{\Phi}$ represents a linear transformation between the full-order solution $q(t)$ and the reduced-order solution $\widehat{q}(t)$; and the modal matrix $\mathbf{\Phi}$ is made up of fluid modes $\{\phi_i\}$, which is also called the POD modes. The modal coefficients $\{a_i\}$ make up the column of the matrix $\widehat{q}(t)$.The optimal basis functions can be yielded by solving the following eigenvalue problem [26]:

$$\mathbf{S}^T\mathbf{S}\mathbf{V} = \mathbf{V}\mathbf{\Lambda}, \qquad (2)$$

$$\mathbf{\Phi} = \mathbf{S}\mathbf{V}, \qquad (3)$$

where $\mathbf{S}$ is an $(N \times M)$ matrix, M is the number of snapshots and N is the number of data points in each snapshot. Snapshots, which are also called samples, represent the solutions of full-order system dynamics at different time. These solutions are generally collected to provide a good variety of flow field behaviors. $\mathbf{V}$ is the matrix of eigenvectors of $\mathbf{S}^T\mathbf{S}$, and the vectors make up the column of $\mathbf{V}$. $\mathbf{\Lambda}$ is the corresponding nonnegative diagonal matrix arranged in descending order that represents the eigenvalues of the system. Thus, once the snapshot matrix $\mathbf{S}$ is created, the eigenvalues and eigenvectors of the eigen-equation (2) can be solved. However, the matrix is so huge that it requires massive memory storage and is very time consuming to solve. In practice, the snapshot matrix $\mathbf{S}$ is usually redundant and can be eliminated by resizing the eigenvectors in $\mathbf{V}$ and eigenvalues in $\mathbf{\Lambda}$. Finally, the modal matrix $\mathbf{\Phi}$ can be solved using (3). The eigenvalue can be interpreted as the weight of contribution to each mode in the POD reconstruction and the "energy" captured by the retained modes relative to the whole set of eigenvalues of the correlation matrix $\mathbf{S}^T\mathbf{S}$ can be defined as follows:

$$\eta = \frac{\sum_{i=1}^{m}\lambda_i}{\sum_{j=1}^{N_s}\lambda_j}, \tag{4}$$

where η represents the "energy" captured by the retained POD modes. Since the eigenvalue sequence $\lambda_1, \lambda_2, \dots, \lambda_k$ descends rapidly, generally, the first few POD modes can capture more than ninety percent of the total "energy," which is usually enough to describe the system physics. More details about the method can be found in reference [26].

2.2. Model Reduction. Once the POD basis functions have been identified using the method of snapshots, the Euler equations must be recast to solve the modal coefficients in lieu of the full-system variables. For incompressible flows, this is generally accomplished using the Galerkin method. But for compressible flows, the reduced-order model of the Euler equations is generally difficult to solve due to the resulting implicit formulation. To solve the problem, Placzek et al. [29] introduces modified primitive variables $q = ((1/\rho), \mathbf{u}, p)^T$ into the unsteady N-S equations, which are then expressed as a polynomial form and adequate for the Galerkin projection.

The Euler equations in a moving frame with the modified primitive variables $q = ((1/\rho), \mathbf{u}, p)^T$ can be written as follows [29]:

$$\begin{cases} \dfrac{\partial \vartheta}{\partial t} + (\mathbf{u} - s) \cdot \nabla\vartheta = \vartheta \operatorname{div} \mathbf{u}, \\ \dfrac{\partial \mathbf{u}}{\partial t} + (\mathbf{u} - s) \cdot \nabla\mathbf{u} = -\vartheta\nabla p - \omega \cdot \mathbf{u}, \\ \dfrac{\partial p}{\partial t} + (\mathbf{u} - s) \cdot \nabla p = -\gamma p \operatorname{div} \mathbf{u}, \end{cases} \tag{5}$$

where $\vartheta = 1/\rho$ and ρ is the density, $\mathbf{u}$ is the velocity vector of the fluid, p is pressure, s is the velocity of the mesh and can be described by $s = s_0 + \omega \times \mathbf{r}$, where s_0 and ω represent the translation velocity and the angular velocity of the mesh, respectively. $\mathbf{r}$ is the position vector relating to the moving frame, γ is the heat capacity ratio.

Equation (5) can also be rewritten in the quadratic form:

$$\dot{q} = Q^C(q, q) + T(q, s_0, \omega), \tag{6}$$

where Q^C and T are defined by

$$Q^C(q, q) = \begin{bmatrix} -\mathbf{u} \cdot \nabla\vartheta + \vartheta \operatorname{div} \mathbf{u} \\ -\mathbf{u} \cdot \nabla\mathbf{u} - \vartheta\nabla p \\ -\mathbf{u} \cdot \nabla p - \gamma p \operatorname{div} \mathbf{u} \end{bmatrix},$$
$$T(q, s, \omega) = \begin{bmatrix} s \cdot \nabla\vartheta \\ s \cdot \nabla\mathbf{u} - \omega \cdot \mathbf{u} \\ s \cdot \nabla p \end{bmatrix}. \tag{7}$$

Introduce $q(t) = q_0 + \sum_{i=1}^{m} a_i(t)\phi_i$ into (6) and project the two sides of the system equations (6) onto each POD mode $\{\phi_i\}$; (6) becomes a quadratic ordinary differential equation:

$$\begin{aligned} \dot{a}_i(t) &= \langle Q^c(\bar{q}, \bar{q}), \phi_i\rangle + (\langle Q^c(\bar{q}, \Delta q), \phi_i\rangle + \langle Q^c(\Delta q, \bar{q}), \phi_i\rangle) \\ &\quad + \langle Q^c(\Delta q, \Delta q), \phi_i\rangle + \langle T(\bar{q}, s_0, \omega), \phi_i\rangle + \langle T(\Delta q, s_0, \omega), \phi_i\rangle \\ &= K_i + \sum_{i,j=1}^{m} L_{ij}a_j(t) + \sum_{j,k=1}^{m} R_{jik}a_j(t)a_k(t) + K_i^m u_m(t) \\ &\quad + \sum_{j=1}^{m} L_{ij}^m u_m(t)a_j(t), \end{aligned} \tag{8}$$

where $u_m = [s, \omega]^T$ and the parameters K_i, L_{ij}, R_{jik}, K_i^m, and L_{ij}^m are constants related to the basis functions, which can be computed from the analytical expressions and the details can be found in reference [28, 29].

2.3. Overview of ARX. The ARX model is a linear difference model, which describes the response of a system as a sum of scaled previous outputs and scaled values of inputs to the system. For a multi-input/multioutput system, the model can be written as follows:

$$\mathbf{y}_s(k+1) = \sum_{i=0}^{na} \mathbf{A}_i\mathbf{y}_s(k-i) + \sum_{i=0}^{nb-1} \mathbf{B}_i\mathbf{u}_s(k-i), \tag{9}$$

where $\mathbf{y}_s(k)$ is the vector of system output at the kth step, $\mathbf{u}_s(k)$ is the vector of system input at the kth step, matrices $\mathbf{A}_i$ and $\mathbf{B}_i$ are the constant coefficients to be identified, na and nb are called the ARX model orders. According to the equation, the system response at given time can be expressed as an algebraic series of multiplications and additions.

The accuracy of the model depends highly on the inputs used to obtain the training data. There must be as much information about the system's dynamics as possible to be packed into the training set of data for the identification procedure. By comparing the "3211" multistep input signal with other different input signals, Cowan [33] points out that the 3211 input signal is easy to implement in experiments and can excite the best frequency response. Once the system inputs and outputs are given, the matrices $\mathbf{A}_i$ and $\mathbf{B}_i$ can be calculated by the least-squared method and the construction of the ARX model is completed.

2.4. Construction of the POD-ARX Reduced-Order Model. It should be noted that the reduced-order model constructed in Section 2.2 is instable because the model lacks artificial dissipation in the process of construction. To reproduce the correct behavior of the original full-order model, Placzek et al. [29] contrasts several correction methods and recommends that the Tikhonov regularization method be a more robust method for the nonautonomous system. Using this method, the accuracy and stability of the response have been significantly improved for both the short- and the long-term

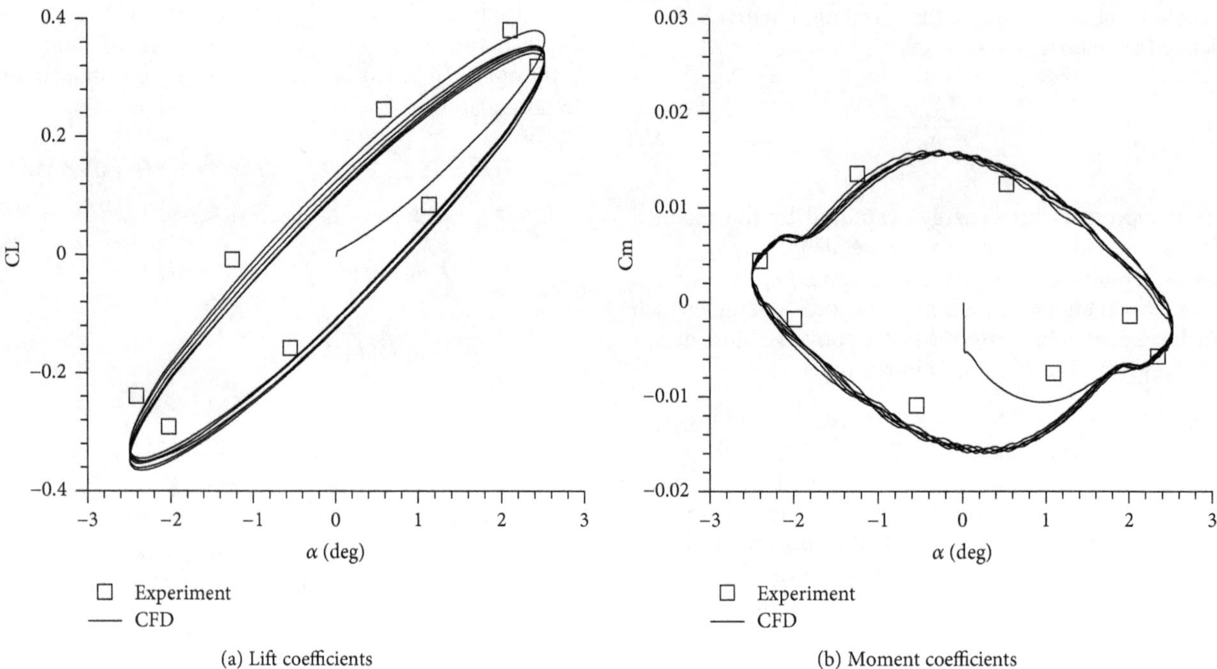

(a) Lift coefficients

(b) Moment coefficients

FIGURE 1: Comparison of aerodynamic coefficients between CFD codes and experiment.

behavior of the full-order model. Although this method is an identification method, it still needs the calculation of the parameters K_i, L_{ij}, R_{jik}, K_i^m, and L_{ij}^m, which is time consuming and complex. In addition, the choice of the parameters of the regularization method is also difficult.

The objective of this paper is to determine the modal coefficients of (8) using the system identification model with the motion signal being the system input and the modal coefficients being the system output. Furthermore, when $\Delta q \ll \bar{q}$, $Q^c(\Delta q, \Delta q)$, and $T(\Delta q, s_0, \omega)$ are small relative to $Q^c(\bar{q}, \bar{q})$, $Q^c(\bar{q}, \Delta q)$, $Q^c(\Delta q, \bar{q})$, and $T(\bar{q}, s_0, \omega)$, their projections $\sum_{j,k=1}^{m} R_{jik} a_j(t) a_k(t)$ and $\sum_{j=1}^{m} L_{ij}^{mfr}(t) a_j(t)$ are also small. That is to say, (8) can be treated as a linear or weak nonlinear system. Therefore, the ARX model is chosen here as the identification model. Then a hybrid ROM can be developed in this paper by combining the POD technique and the ARX model. In this method, the POD technique is used to extract the fluid modes from the snapshots and the ARX model is used to identify the modal coefficients of the flow field. The details of this method are described as follows:

(1) Observe and record the snapshots of the system for a predetermined input signal (training signal)

(2) Extract the POD basis functions from the snapshots gathered in step (1) by POD technique

(3) Project the snapshots onto the basis functions to obtain the time history of the modal coefficients

(4) Construct the ARX model using the SI technique, with the input signal in step (1) and the time history of the modal coefficients in step (3) being taken as the input and output data, respectively

(5) Predict the modal coefficients according to a new input signal by the ARX model obtained in step (4). Then construct the resulting flow field using the modal coefficients and the corresponding POD modes

3. CFD Code Validation

The full-order solver uses the cell-centered finite volume method to solve the Euler equation. The AUSM+ scheme is used to discretize the computation domain [37], while the implicit LU-SGS scheme is used for temporal integration [38]. For the unsteady calculation, the dual-time stepping method is adopted, with the fourth-order Runge-Kutta scheme for the subiteration.

The solver was validated with the pitching motion of NACA0012 in transonic flow. The airfoil was forced to pitch about its quarter chord and the motion is described as

$$\alpha(t) = \alpha_0 + d\alpha \sin(\omega t), \tag{10}$$

where α_0 is the mean angle of attack, α_m is the amplitude of the motion, and the reduced frequency k is defined by

$$k = \frac{\omega c}{2V_\infty}, \tag{11}$$

where V_∞ is the free-stream velocity of the flow and c is the chord of the airfoil. In this case, $\alpha_0 = 0.016\,\text{deg}$, $d\alpha = 2.51\,\text{deg}$, $Ma_\infty = 0.755$, and $k = 0.0814$. Figure 1 is the comparison of the aerodynamic coefficients obtained from CFD codes and experiment [39], which shows that the CFD codes are reliable for the calculation of unsteady aerodynamic forces.

4. Results

4.1. Wide-Band Characteristics of the ROM. Using the POD-ARX hybrid reduced-order model outlined in Section 2.4, the aerodynamic characteristics of a two-dimensional NACA0012 airfoil with pitching motion is investigated here. A C-shape structure mesh is used and the mesh consists of $N = 47944$ nodes, which means the DOFs of the full-order model are $Nv = 191766$. The full-order Euler solver is employed to provide snapshots for the POD modes, as well as the base solution for comparison with the ROMs results.

The model will be investigated in a subsonic flow with $Ma_\infty = 0.6$ and the motion of the airfoil will be limited to the pitching motion described as (11) at $\alpha_0 = 3$ deg.

Cowan [33] compared several different training signals and pointed out that the "3211" multistep input signal has a wide range of frequency. Therefore, the "3211" input signal, including a total of 310 time steps, is adopted here and shown in Figure 2. A set of $M = 310$ dimensionless flow field is extracted at each time from startup to provide snapshots for proper orthogonal decomposition. The frequency characteristic of the "3211" multistep signal is shown in Figure 3. It is found that the training signal has a good coverage to the frequency band from nearly 40 Hz to 90 Hz, with the dominant frequency being 65 Hz.

Equation (4) is adopted here to evaluate the "energy" captured by the POD modes. It is found that the first 8 POD modes contain more than 99% "energy" of the flow field. Consequently, the first 8 POD modes are extracted here to conduct the model reduction.

The snapshots are projected onto the basis functions to determine the time history of the modal coefficients. Then, the resulting data is used as the training data to identify the parameters A_i and B_i of the ARX model. A variety of ARX model orders (*na* and *nb*) are tried until the best fit for the training data is found. The model orders $na = 5$ and $nb = 5$ are ultimately chosen as the best fit for the training data. Then, the "3211" multistep signal is input to the newly constructed ARX model to test if it could accurately predict the modal coefficients of multistep response. Figure 4 shows the comparison of the multistep time history of modal coefficients obtained from the Euler solution and those from the ARX model. Note that the former were obtained by directly projecting the Euler solution onto the basis functions. We can see that the reduced-order model fits the training data extremely well. For a quantitative analysis, a relative error is defined to evaluate the performance of the ROM:

$$\sigma = \frac{\|y_{\mathrm{ROM}} - y_{\mathrm{CFD}}\|_F^2}{\|y_{\mathrm{CFD}}\|^2} \times 100\%, \tag{12}$$

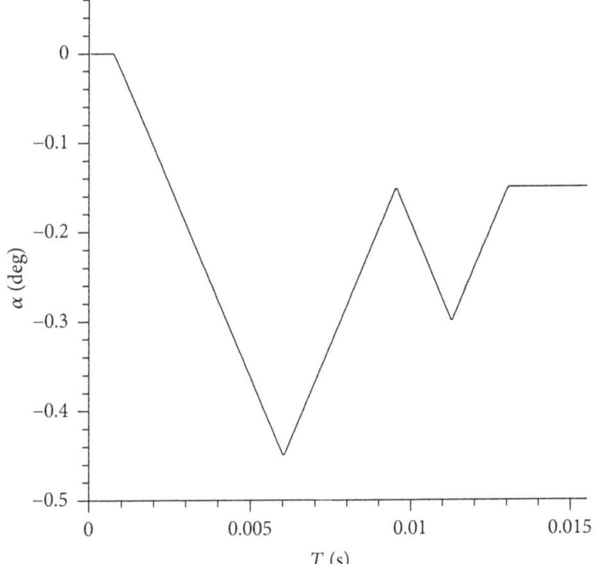

Figure 2: Training signal.

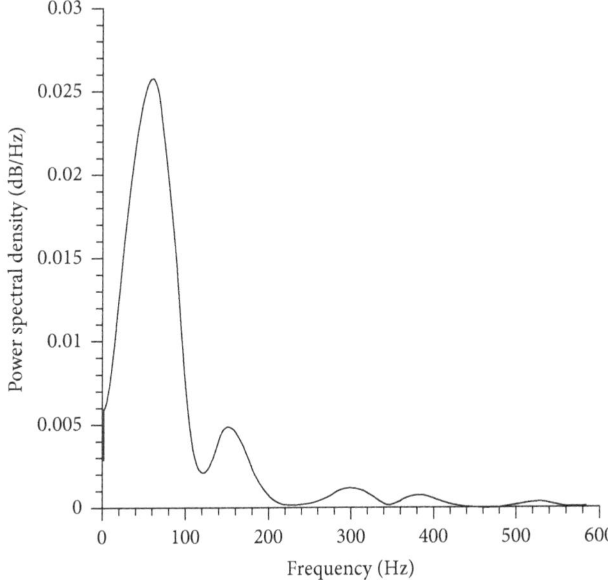

Figure 3: The power spectral density of the training signal.

where y_{ROM} and y_{CFD} represent the ROM output and the output of the Euler solver, respectively. According to the above equation, the relative errors of different modes are both less than 0.1%.

First, the validation signal described by (10) is set to be $f = 65$ Hz. The comparison of modal coefficients obtained from Euler and ROM solution to the oscillation signal is shown in Figure 5. Obviously, the results by the model match well with those from the Euler solution, with the relative error of each coefficient being 0.3%, 0.59%, 4.47%, 1.13%, 4.98%, 2.2%, 3.8%, and 2.9%, respectively. Besides, it can be found that the first 2 modes achieve better agreement than

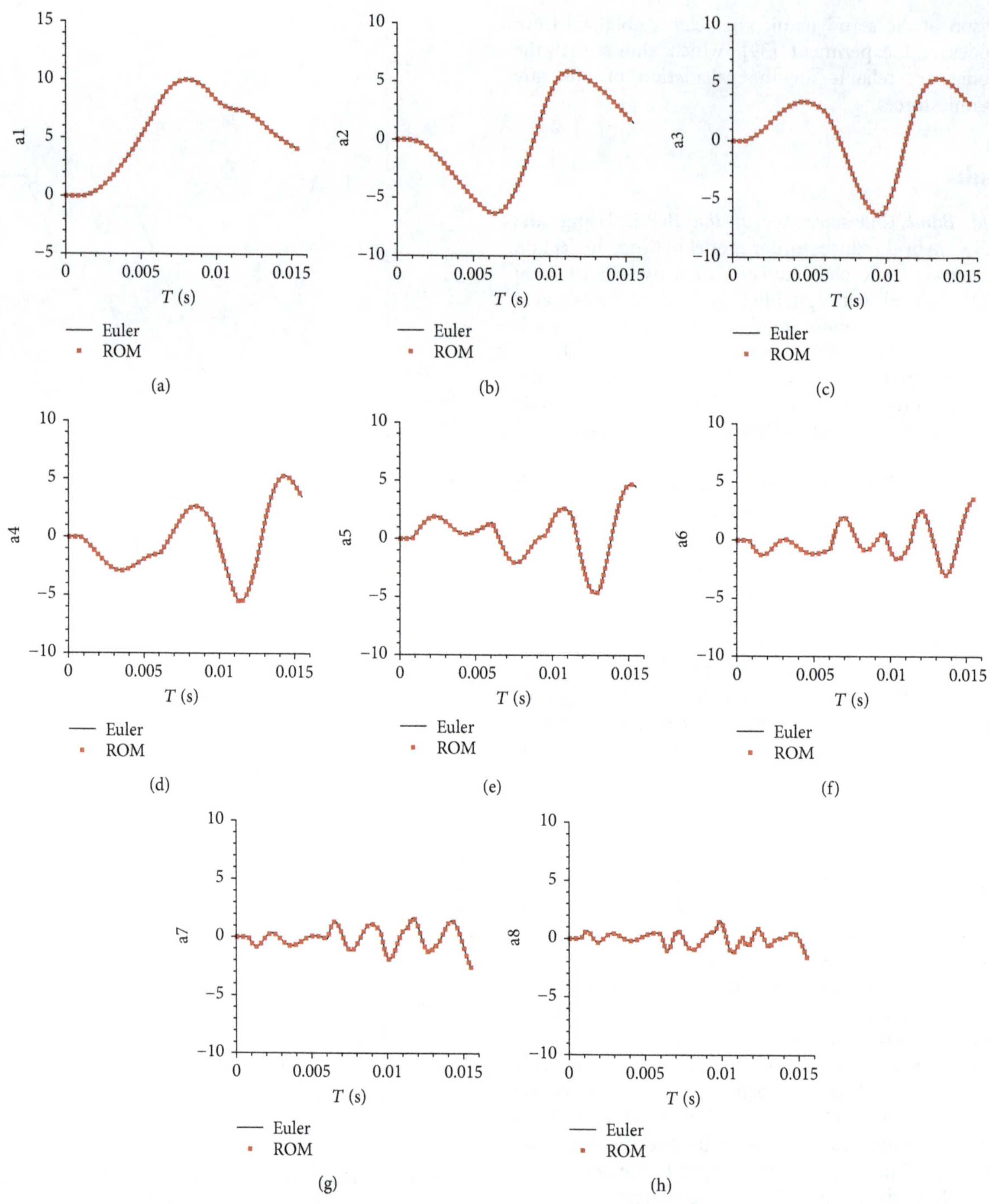

FIGURE 4: Comparison of modal coefficients obtained from Euler and ROM solution to "3211" signal.

the other modes. The reason for this phenomenon may lie in the fact that the first two modes can capture more linear characteristic of the system (the dominant characteristic of this linear system) while the last several modes maintain more nonlinear characteristic.

For a more intuitive evaluation to the result of the ROM, the flow field is reconstructed according to (1). The pressure distributions on the airfoil predicted by the CFD and ROM at four instants in time T/4, 2T/4, 3T/4, and T during the 5th period are shown in Figure 6. Apparently, the pressure distribution on the airfoil obtained from the ROM shows excellent agreement with the result from CFD.

In addition, the time histories of lift coefficients and moment coefficients are also compared between the ROM and CFD by the integral of the surface pressure on the airfoil, shown in Figure 7. The relative error of the lift coefficients and the moment coefficients is 1.1% and 4.2%, respectively. It is evident that the model has accurately predicted the aerodynamic response.

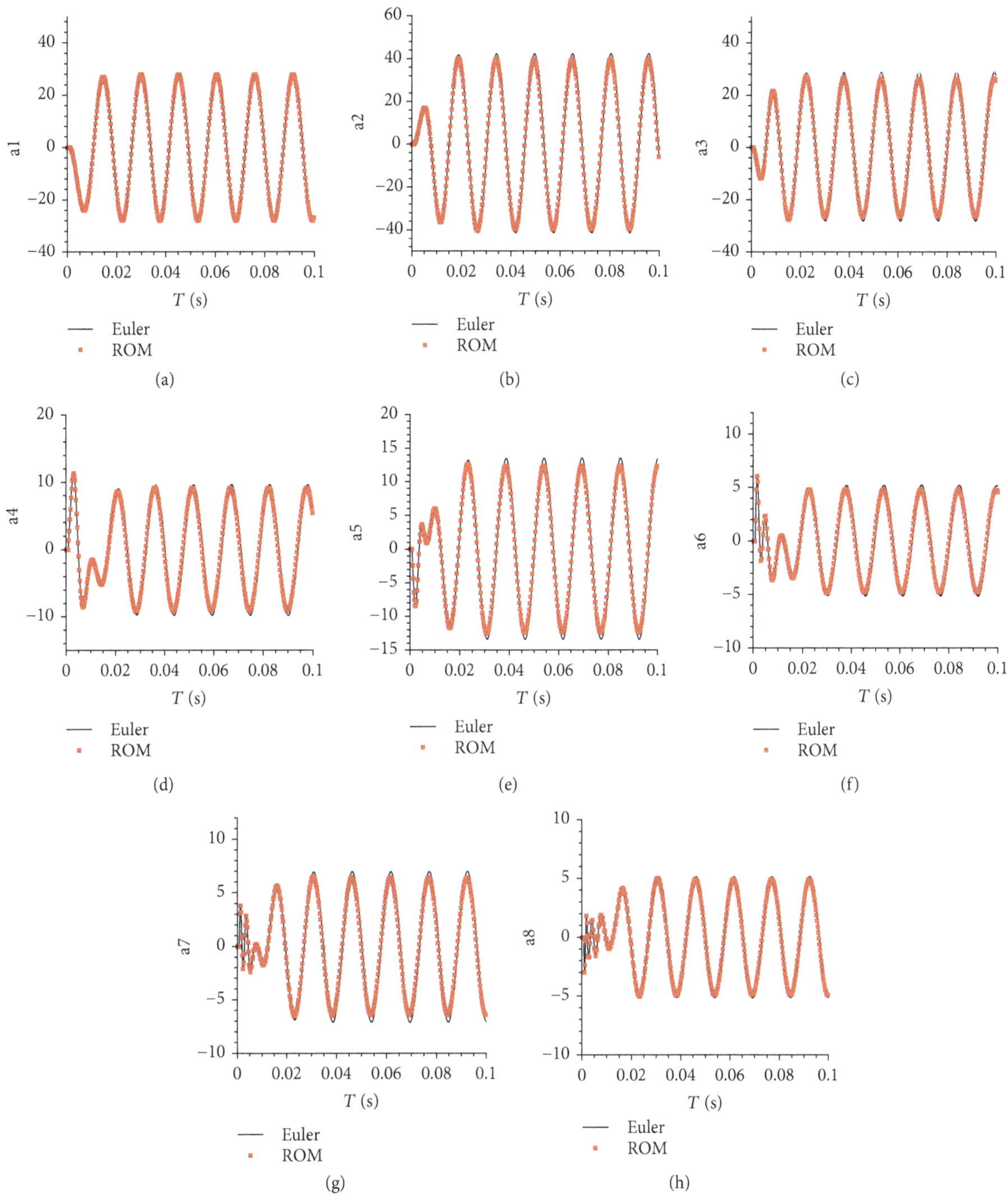

FIGURE 5: Comparison of modal coefficients obtained from Euler and ROM solution to oscillation signal with $f = 65$ Hz.

In order to further exhibit the characteristic of this model, two different validation signals with $f = 40$ Hz and $f = 90$ Hz are input to the ROM to predict the response. The aerodynamic coefficients and the pressure distributions on the airfoil at four instants in time T/4, 2T/4, 3T/4, and T during the 4th period predicted by CFD and ROM are shown in Figures 8–11. The results show that the pressure distributions and the aerodynamic coefficients from the ROM both agree well with those from the Euler solver at $f = 40$ Hz (the relative error of the lift coefficient and moment coefficient is 1.1% and 4.9%). By contrast, the relative error of the lift coefficients and moment coefficients is 1.5% and 6.8% at $f = 90$ Hz. Obviously, due to the frequency being far away from the dominant frequency of the "3211" signal, the errors are slightly larger at $f = 90$ Hz, but the accuracy is still acceptable in unsteady aerodynamic calculation. Therefore, we can see that the ROM developed in this paper performs well in a wide range of frequencies,

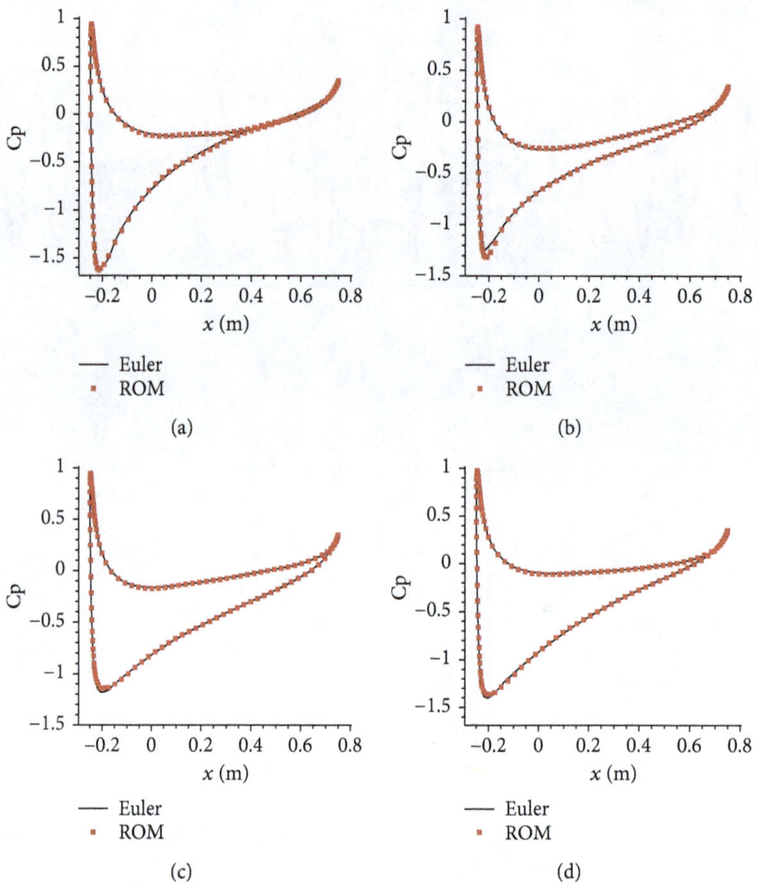

FIGURE 6: Pressure distribution predicted by the Euler and ROM at (a) T/4, (b) 2T/4, (c) 3T/4, and (d) T when $f = 65$ Hz.

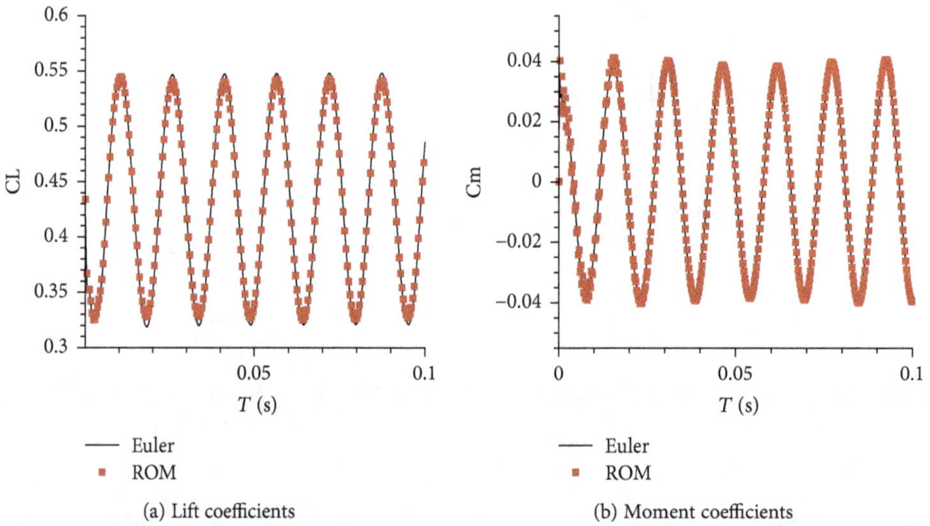

FIGURE 7: Comparison of the aerodynamic coefficients obtained from Euler and ROM when $f = 65$ Hz.

mainly resulting from the good combination of the model with the "3211" multistep signal.

4.2. Nonlinear Performance of the ROM. To evaluate the performance of the POD-ARX ROM in nonlinear problems, in this section, the ROM will be investigated under a larger mean angle of attack. Figure 12 is the lift coefficient curve of the airfoil when $Ma_\infty = 0.6$. It is obvious that the system begin to exhibit the nonlinear aerodynamic characteristics when $\alpha > 5$ deg.

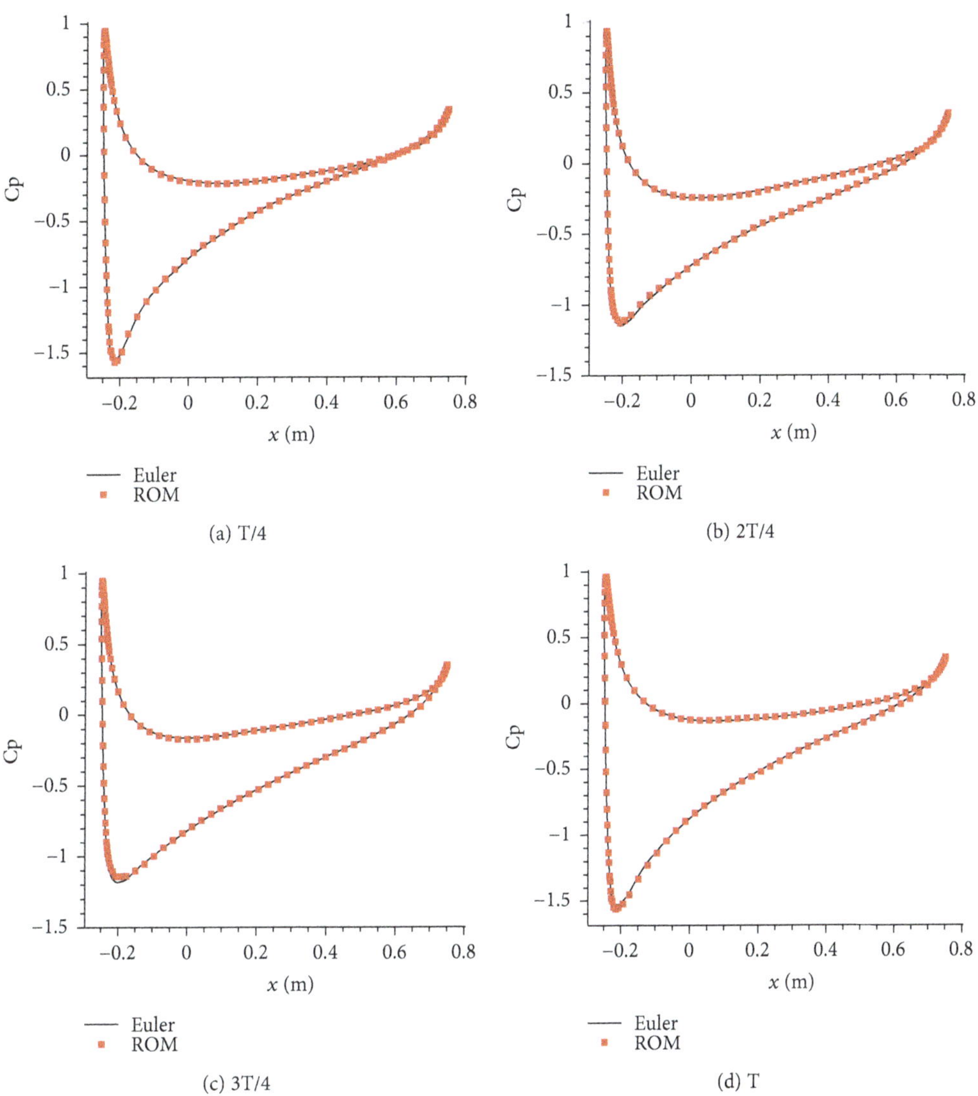

FIGURE 8: Pressure distribution predicted by the Euler and ROM at (a), (b), (c), and (d) when $f = 40$ Hz.

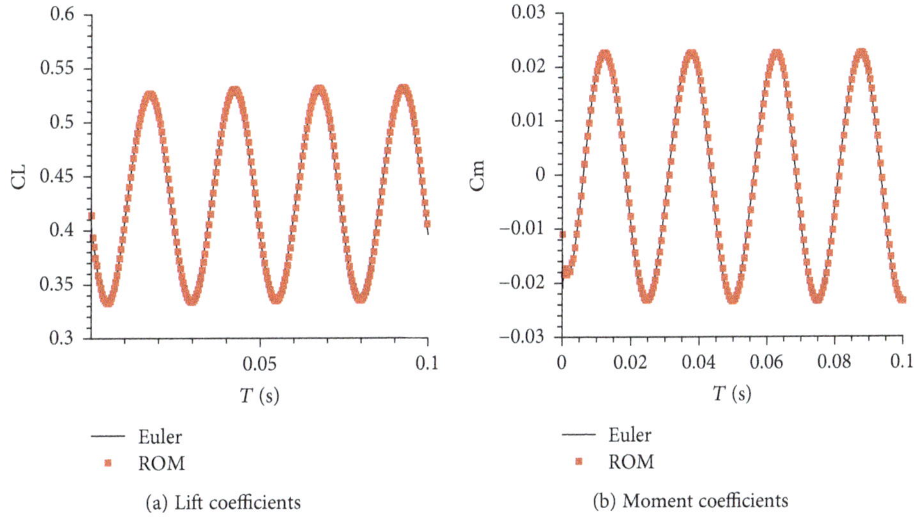

FIGURE 9: Comparison of the aerodynamic coefficients obtained from Euler and ROM when $f = 40$ Hz.

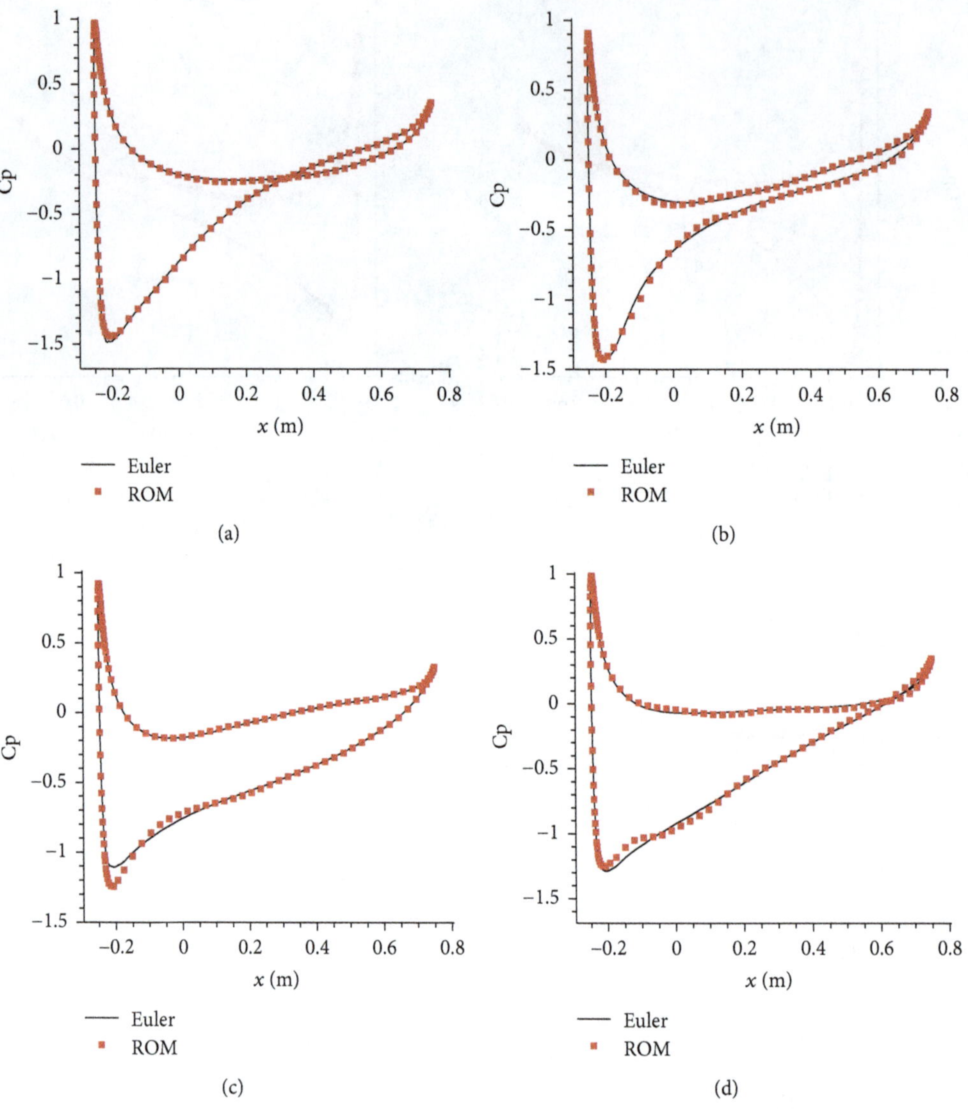

Figure 10: Pressure distribution predicted by the Euler and ROM at (a) T/4, (b) 2T/4, (c) 3T/4, and (d) T when $f = 90$ Hz.

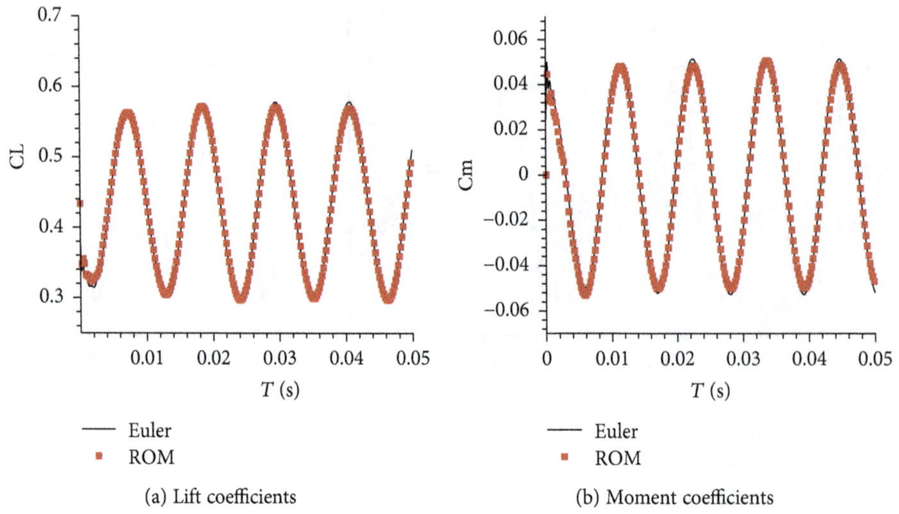

(a) Lift coefficients

(b) Moment coefficients

Figure 11: Comparison of the aerodynamic coefficients obtained from Euler and ROM when $f = 90$ Hz.

Figure 13 is the pressure distributions on the airfoil predicted by CFD and ROM at four instants in time T/4, 2T/4, 3T/4, and T during the 4th period when $\alpha_0 = 7$ deg and $f = 65$ Hz. The corresponding aerodynamic results are shown in Figure 14. The results show that the pressure distributions and the aerodynamic coefficients from the ROM both agree well with those from the Euler solver (the relative error of the lift coefficient and moment coefficient is 2.1% and 5.9%). However, it should be noted that due to the influence of the nonlinearity of the system, the "3211" multistep signal cannot fully excite the physical characteristics of the flow field which is very important for the extraction of the POD modes. Therefore, in this case, the first 500 steps (nearly one and a half period) of the validation signal are chosen as the training signal. In addition, more POD modes ($m = 16$) and more ARX model orders ($na = 10$ and $nb = 10$) are chosen to conduct the model reduction. In fact, in the case of large mean angle of attack, it is not only difficult to fully excite all the physical characteristics of the flow field by the training signal but also difficult to obtain a stable steady-state base flow. When we further increase the mean angle of attack α_0 to 8 deg, it is found that the steady-state flow becomes unstable and the ROM suffers from a failure.

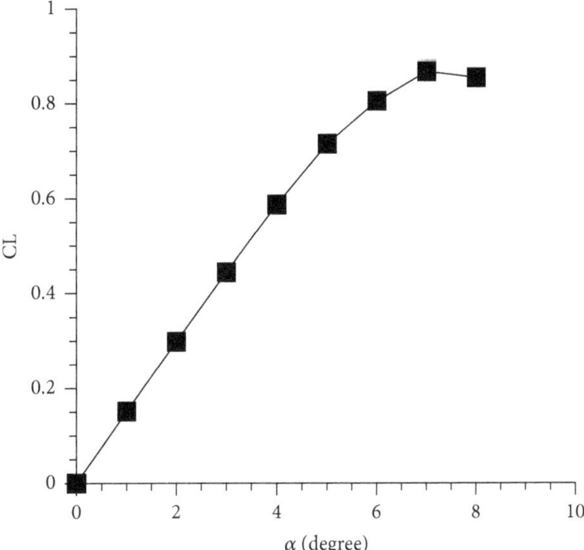

FIGURE 12: Lift coefficient curve of the airfoil when $Ma_\infty = 0.6$.

5. Computational Efficiency of the ROM

To evaluate the efficiency of the POD-ARX method in a more intuitive way, in this section, we will make a comparison about the computational time of the full-order solver and ROM. Results from the case in Section 4.1 are summarized in Table 1, where all the computations are run on a personal computer (CPU: 3.2 GHZ, Memory: 8.0 GB) and the calculation time is measured by the CPU time covering five periods of motion.

Apparently, the cost of identifying the ARX model is almost negligible. When the time for training the input signal by the full-order CFD solver (3056 seconds) and extracting the POD modes (5497 seconds) is taken into account, the whole model reduction process will cost 8553 seconds. Therefore, the computational efficiency can be improved by almost fifty percent. Once the ROM is built up, the calculation time is almost negligible when applying the model to the corresponding problems.

6. Conclusions

A new hybrid POD-ARX reduced-order model is developed in this paper by combining the POD technique and the ARX model. The method involves extracting the fluid basis functions with the POD technique and using the ARX model to identify the modal coefficients. The method is tested on a NACA0012 airfoil with pitching motion in subsonic inviscid flow. First, the motion at small angle of attack ($\alpha_0 = 3$ deg) is conducted. The "3211" multistep signal is used as the training signal to obtain the fluid snapshots and finally eight basis functions are extracted by the POD technique. Then, an eight-output ARX model is constructed. Three validation signals with different frequencies are input to the ARX model, respectively. The resulting time histories of the modal coefficients and the aerodynamic coefficients of the flow field both achieve good agreement with those obtained from the full-order CFD solver. When ignoring the time for obtaining the snapshots and extracting the POD modes, the computational time of the ROM can be reduced by nearly five orders of magnitude. Second, to evaluate the nonlinear performance of the method, a pitching motion at a larger angle of attack ($\alpha_0 = 7$ deg) is conducted. Since the nonlinear characteristics of the flow field, the "3211" multistep signal fails to excite the full characteristics of the flow field and hence the first 500 steps of the validation signal is used as the training signal. A total of 16 basis functions are extracted to construct the ROM. The results obtained from the ROM have a good agreement with those from the Euler solver.

The results have proven that the POD-ARX model can provide stable and accurate predictions for the unsteady aerodynamic response efficiently. Furthermore, compared with the POD-Volterra method, the POD-ARX model has good wide-band characteristics due to the easy combination with the "3211" multistep signals in the linear range, which is especially suitable to problems with uncertain frequencies or a wide range of frequencies. In addition, using all the primitive variables as a block also benefits to reduce the numbers of the identification terms. And compared with the correction method proposed in reference [29], the ARX model is very efficient and easy to implement without any additional calculations of the system parameters. However, it should be noted that for nonlinear problems, an appropriate training signal and a stable steady-state base flow is very important to the ROM. The initial investigation about this method is made just for two-dimensional aerodynamic problems in this paper. Future work will concern on the three-dimensional problems and the aeroelastic analysis.

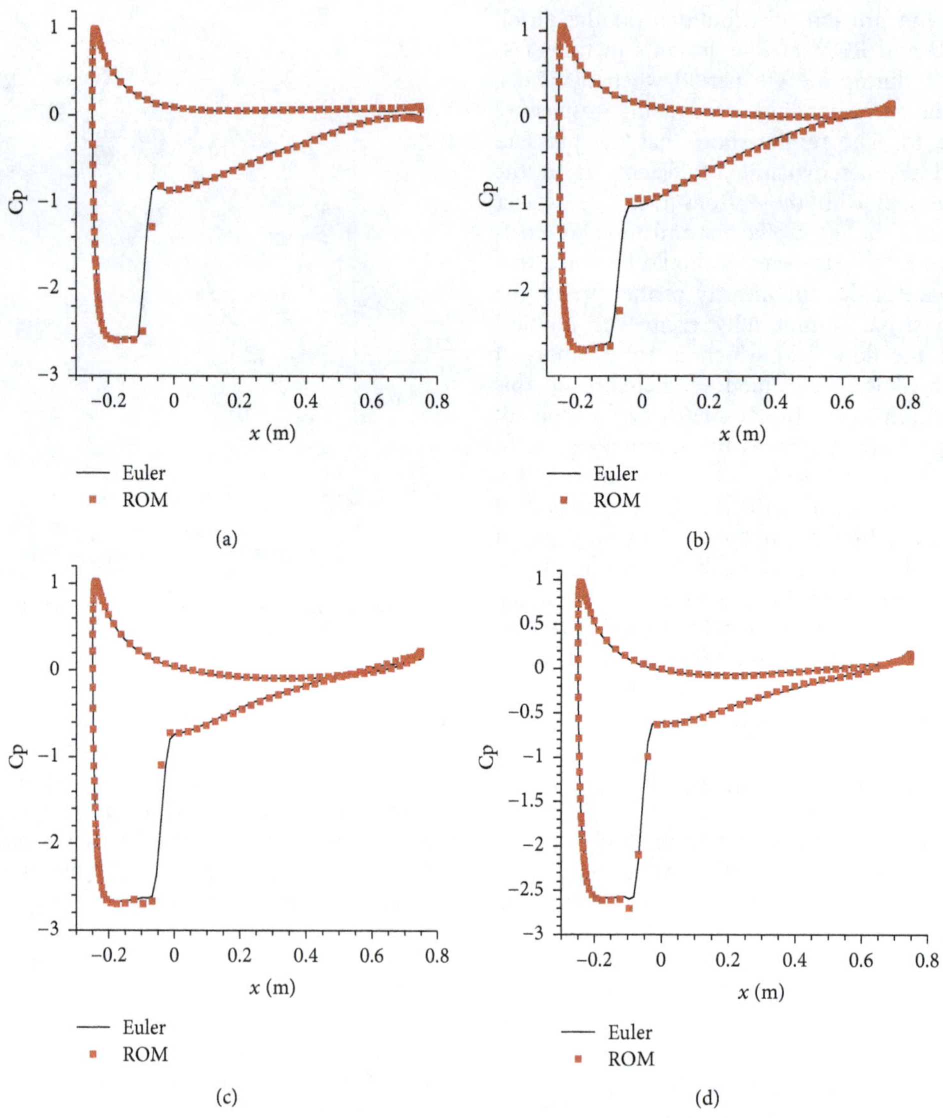

FIGURE 13: Pressure distribution predicted by the Euler and ROM at (a) T/4, (b) 2T/4, (c) 3T/4, and (d) T when $\alpha_0 = 7$ deg.

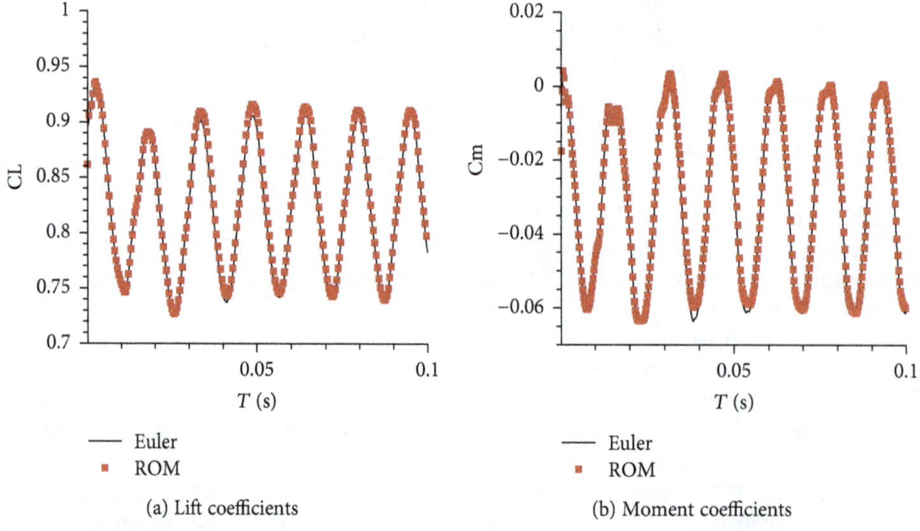

(a) Lift coefficients

(b) Moment coefficients

FIGURE 14: Comparison of the aerodynamic coefficients obtained from Euler and ROM when $\alpha_0 = 7$ deg.

TABLE 1: Computational efficiency.

Flow solver	Fluid DOFs	CPU time
Full order	191776	15165 sec
POD-ARX	8	0.2 sec

Conflicts of Interest

The authors declare that there is no conflict of interest regarding the publication of this paper.

Acknowledgments

This work was supported by the Strategic Priority Research Program of Chinese Academy of Sciences (Grant no. XDA17030100).

References

[1] T. J. Cowan, A. S. Arena, and K. K. Gupta, "Accelerating computational fluid dynamics based aeroelastic predictions using system identification," *Journal of Aircraft*, vol. 38, no. 1, pp. 81–87, 2001.

[2] J. Kou and W. Zhang, "Layered reduced-order models for nonlinear aerodynamics and aeroelasticity," *Journal of Fluids and Structures*, vol. 68, pp. 174–193, 2017.

[3] P. Marzocca, W. A. Silva, and L. Librescu, "Nonlinear open-/closed-loop aeroelastic analysis of airfoils via volterra series," *AIAA Journal*, vol. 42, no. 4, pp. 673–686, 2004.

[4] M. Balajewicz and E. Dowell, "Reduced-order modeling of flutter and limit-cycle oscillations using the sparse Volterra series," *Journal of Aircraft*, vol. 49, no. 6, pp. 1803–1812, 2012.

[5] T. Skujins and C. E. S. Cesnik, "Reduced-order modeling of unsteady aerodynamics across multiple mach regimes," *Journal of Aircraft*, vol. 51, no. 6, pp. 1681–1704, 2014.

[6] W. Zhang, B. Wang, Z. Ye, and J. Quan, "Efficient method for limit cycle flutter analysis based on nonlinear aerodynamic reduced-order models," *AIAA Journal*, vol. 50, no. 5, pp. 1019–1028, 2012.

[7] A. Mannarino and P. Mantegazza, "Nonlinear aeroelastic reduced order modeling by recurrent neural networks," *Journal of Fluids and Structures*, vol. 48, pp. 103–121, 2014.

[8] K. Lindhorst, M. C. Haupt, and P. Horst, "Nonlinear reduced order modeling for aeroelastic simulation with neural networks," in *Computational Flight Testing*, N. Kroll, R. Radespiel, J. Burg, and K. Sørensen, Eds., vol. 123 of Notes on Numerical Fluid Mechanics and Multidisciplinary Design, pp. 131–149, Springer, Berlin, Heidelberg, 2013.

[9] J. Kou and W. Zhang, "An approach to enhance the generalization capability of nonlinear aerodynamic reduced-order models," *Aerospace Science and Technology*, vol. 49, pp. 197–208, 2016.

[10] D. J. Lucia and P. S. Beran, "Projection methods for reduced order models of compressible flows," *Journal of Computational Physics*, vol. 188, no. 1, pp. 252–280, 2003.

[11] M. F. Barone and J. L. Payne, "Methods for simulation-based analysis of fluid-structure interaction," Tech. Rep. SAND2005-6573, Sandia National Laboratories, Albuquerque, NM, USA, 2005.

[12] B. A. Freno and P. G. A. Cizmas, "A proper orthogonal decomposition method for nonlinear flows with deforming meshes," *International Journal of Heat and Fluid Flow*, vol. 50, pp. 145–159, 2014.

[13] I. B. Oliveira and A. T. Patera, "Reduced-basis techniques for rapid reliable optimization of systems described by affinely parametrized coercive elliptic partial differential equations," *Optimization and Engineering*, vol. 8, no. 1, pp. 43–65, 2007.

[14] P. A. LeGresley and J. J. Alonso, "Airfoil design optimization using reduced order models based on proper orthogonal decomposition," in *Fluids 2000 Conference and Exhibit*, p. 2545, Denver, CO, USA, 2000.

[15] S. Lall, J. E. Marsden, and S. Glavaški, "A subspace approach to balanced truncation for model reduction of nonlinear control systems," *International Journal of Robust and Nonlinear Control*, vol. 12, no. 6, pp. 519–535, 2002.

[16] G. Chen, X. Wang, and Y. M. Li, "A reduced-order-model-based multiple-in multiple-out gust alleviation control law design method in transonic flow," *Science China Technological Sciences*, vol. 57, no. 2, pp. 368–378, 2014.

[17] D. Amsallem and C. Farhat, "Interpolation method for adapting reduced-order models and application to aeroelasticity," *AIAA Journal*, vol. 46, no. 7, pp. 1803–1813, 2008.

[18] Q. Zhou, G. Chen, A. Da Ronch, and Y. Li, "Reduced order unsteady aerodynamic model of a rigid aerofoil in gust encounters," *Aerospace Science and Technology*, vol. 63, pp. 203–213, 2017.

[19] W. Zhang, J. Kou, and Z. Wang, "Nonlinear aerodynamic reduced-order model for limit-cycle oscillation and flutter," *AIAA Journal*, vol. 54, no. 10, pp. 3304–3311, 2016.

[20] T. Lassila, A. Manzoni, A. Quarteroni, and G. Rozza, "Model order reduction in fluid dynamics: challenges and perspectives," in *Reduced Order Methods for Modeling and Computational Reduction*, pp. 235–273, Springer, 2014.

[21] A. Iollo, S. Lanteri, and J. A. Désidéri, "Stability properties of POD–Galerkin approximations for the compressible Navier–Stokes equations," *Theoretical and Computational Fluid Dynamics*, vol. 13, no. 6, pp. 377–396, 2000.

[22] K. Pathak and N. Yamaleev, "POD-based reduced-order model for arbitrary Mach number flows," in *6th AIAA Theoretical Fluid Mechanics Conference*, p. 3111, Honolulu, HI, USA, 2011.

[23] M. Bergmann, C. H. Bruneau, and A. Iollo, "Enablers for robust POD models," *Journal of Computational Physics*, vol. 228, no. 2, pp. 516–538, 2009.

[24] M. F. Barone, I. Kalashnikova, D. J. Segalman, and H. K. Thornquist, "Stable Galerkin reduced order models for linearized compressible flow," *Journal of Computational Physics*, vol. 228, no. 6, pp. 1932–1946, 2009.

[25] C. Yang, X. Y. Liu, and Z. G. Wu, "Unsteady aerodynamic modeling based on POD-observer method," *Science China Technological Sciences*, vol. 53, no. 8, pp. 2032–2037, 2010.

[26] D. Lucia and P. S. Beran, "Reduced-order model development using proper orthogonal decomposition and Volterra theory," *AIAA Journal*, vol. 42, no. 6, pp. 1181–1190, 2004.

[27] D. J. Lucia, P. S. Beran, and W. A. Silva, "Aeroelastic system development using proper orthogonal decomposition and Volterra theory," *Journal of Aircraft*, vol. 42, no. 2, pp. 509–518, 2005.

[28] A. Placzek, *Construction de modèles d'ordre réduit non-linéaires basés sur la décomposition orthogonale propre pour l'aéroélasticité, [Ph.D. thesis]*, Conservatoire National des Arts et Metiers, Paris, France, 2009.

[29] A. Placzek, D.-M. Tran, and R. Ohayon, "A nonlinear POD-Galerkin reduced-order model for compressible flows taking into account rigid body motions," *Computer Methods in Applied Mechanics and Engineering*, vol. 200, no. 49–52, pp. 3497–3514, 2011.

[30] G. Chen, J. Sun, and Y. M. Li, "Adaptive reduced-order-model-based control-law design for active flutter suppression," *Journal of Aircraft*, vol. 49, no. 4, pp. 973–980, 2012.

[31] D. J. Lucia, P. S. Beran, and W. A. Silva, "Reduced-order modeling: new approaches for computational physics," *Progress in Aerospace Sciences*, vol. 40, no. 1-2, pp. 51–117, 2004.

[32] W. Zhang and Z. Ye, "On unsteady aerodynamic modeling based on CFD technique and its applications on aeroelastic analysis," *Advances in Mechanics*, vol. 38, no. 1, pp. 77–86, 2008.

[33] T. J. Cowan, *Efficient Aeroelastic CFD Predictions Using System Identification, [M.S. thesis]*, Oklahoma State University, Stillwater, OK, USA, 1998.

[34] W. Zhang, X. Li, Z. Ye, and Y. Jiang, "Mechanism of frequency lock-in in vortex-induced vibrations at low Reynolds numbers," *Journal of Fluid Mechanics*, vol. 783, pp. 72–102, 2015.

[35] C. Gao, W. Zhang, J. Kou, Y. Liu, and Z. Ye, "Active control of transonic buffet flow," *Journal of Fluid Mechanics*, vol. 824, pp. 312–351, 2017.

[36] C. Gao, W. Zhang, X. Li et al., "Mechanism of frequency lock-in in transonic buffeting flow," *Journal of Fluid Mechanics*, vol. 818, pp. 528–561, 2017.

[37] M.-S. Liou, "Progress towards an improved CFD method - AUSM+," in *12th Computational Fluid Dynamics Conference*, p. 1701, San Diego, CA, USA, 1995, AIAA.

[38] R. F. Chen and Z. J. Wang, "Fast, block lower-upper symmetric Gauss-Seidel scheme for arbitrary grids," *AIAA Journal*, vol. 38, no. 12, pp. 2238–2245, 2000.

[39] I. J. Leontaritis and S. A. Billings, "Input-output parametric models for non-linear systems part I: deterministic non-linear systems," *International Journal of Control*, vol. 41, no. 2, pp. 303–328, 1985.

9

On the Topology Optimization of Elastic Supporting Structures under Thermomechanical Loads

Jie Hou, Ji-Hong Zhu, and Qing Li

Engineering Simulation and Aerospace Computing (ESAC), The Key Laboratory of Contemporary Design and Integrated Manufacturing Technology, Northwestern Polytechnical University, Xi'an 710072, China

Correspondence should be addressed to Ji-Hong Zhu; jh.zhu@nwpu.edu.cn

Academic Editor: Haibo Wang

This paper is to present a thermomechanical topology optimization formulation. By designing structures that support specific nondesignable domain, optimization is to suppress the stress level in the nondesignable domain and maintain global stiffness simultaneously. A global stress measure based on p-norm function is then utilized to reduce the number of stress constraints in topology optimization. Sensitivity analysis employs adjoint method to derive the global stress measure with respect to the topological pseudodensity variables. Some particular behaviors in thermomechanical topology optimization of elastic supports, such as the influence of different thermomechanical loads and the existence of intermediate material, are also analyzed numerically. Finally, examples of elastic supports on a cantilever beam and a nozzle flap under different thermomechanical loads are tested with reasonable optimized design obtained.

1. Introduction

Optimization design of thermoelastic structures is an important problem in aeronautics and aerospace products such as turbine engine components and thermal protection systems (TPS). As one of the most challenging topics in topology optimization, many difficulties are to be settled before effective engineering application.

One the one hand, literatures dealing with structural configurations design for different heat conduction purposes have been present using topology optimization, where the total potential energy was usually minimized as the design objective. For example, Li et al. [1] optimized the heat conduction path using the typical evolutionary structural topology optimization. The ineffective materials were gradually removed from the design domain iteratively. The final designs turned out to be some tree-like structures for volume-to-point heart conduction. Later, density based topology optimization methods were used to obtain similar optimized designs [2, 3]. Gao et al. [4] recognized the heat sources as design-dependent thermal loads; some more effective and clear optimized designs were finally obtained. Later, the design-dependent effect of heat conduction and convection in homogenization based topology optimization was further investigated [5]. Recently, Dirker and Meyer [6] extended the heat conduction topology optimization to the internal cooling system design.

One the other hand, topology optimization of coupled thermoelastic structures was generally considered as mean compliance minimization problems with different solutions in earlier works (e.g., [7–13]). Gao and Zhang [14] then extended such formulation to thermoelastic structures of multiphase materials. The concept of thermal stress coefficient (TSC) defined by Young's modulus and thermal expansion coefficients was proposed to characterize the dependence of thermal stress and thermal expansion coefficient upon element pseudodensities. Deaton and Grandhi [15] investigated the compliance design of the supporting structures in thermoelastic environment. Nonintuitiveness in design space and the design dependency that occurs with thermal loading were presented and different topology optimization formulations were used to solve the problem.

A recent literature survey by Deaton and Grandhi [16] has summarized the advances and applications of thermoelastic topology optimization.

Actually, eligible design of thermoelastic structures should meet both requirements of structural stiffness and stress to prevent structural failure. As a result, recent efforts have been continuously devoted to highlighting thermoelastic problems with the consideration of compliance, strain energy, and maximum stress. P. Pedersen and N. L. Pedersen [17, 18] proposed strength optimization minimizing the maximum Von-Mises stress for thermoelastic structures, where thermal stress was generated by thermal expansion of clamped structures. Zhang et al. [19] presented a topology optimization formulation of elastic supports for thermoelastic structures and investigated the strength design by directly minimizing the maximum Von-Mises stress in the nondesignable subregion. They further analyzed the differences of elastic strain energy and mean compliance of thermoelastic structures. Recently, Deaton and Grandhi [20] presented a topology optimization method with combined mechanical and thermal loads. A relaxation technique and a modified p-norm function were utilized to remove the singularity phenomenon and aggregate the large number of stress constraints.

In this paper, we continue to study the thermoelastic structures topology optimization suppressing stress level and maintaining global stiffness simultaneously. As shown in Figure 1, a typical thermoelastic continuum structure undergoes coupled thermal and mechanical loads. Subregion Ω is nondesignable and Ω_s is the elastic supporting structure assigned as the design domain. Large stress level appears in the nondesignable domain due to the coupled effect of thermal expansion and mechanical loads. Design objective here is to find proper structural configuration of the elastic supports that suppress the stress in the supported nondesignable domain and simultaneously maintain global stiffness evaluated by structural compliance. As the stress constraints are only applied on the nondesignable domain, the singularity problem in stress-based topology optimization is actually not involved here.

To have an in-depth understanding of the thermoelastic problems with elastic supports, we perform a detailed analysis of a three-bar truss analytical model to present some illustrative phenomena during optimization. The behaviors of structural compliance and stress are discussed, respectively. Numerical examples of elastic supports on a 2D cantilever beam and 3D nozzle flap are finally presented to verify the validity of the presented formulation.

2. Three-Bar Truss Analytical Model

The investigation on a truss-frame structure can reveal the underlying scheme of continuum topology optimization (e.g., [21, 22]). In this section, the analytical solutions of a three-bar truss system are derived under different thermomechanical loads with regard to the stress constraints. The structural compliance of thermoelastic problems is evaluated with further details discussed here.

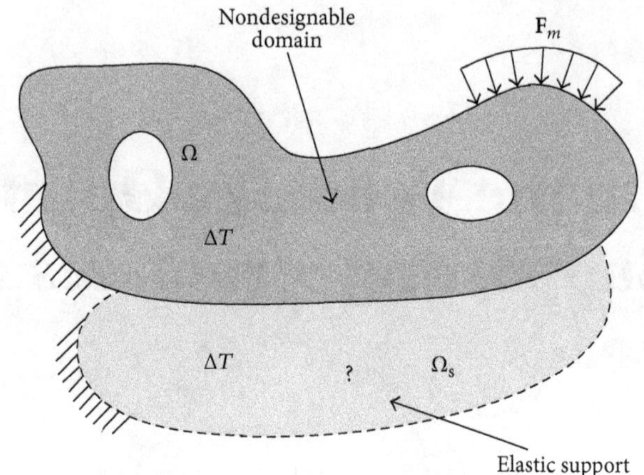

FIGURE 1: Topology optimization of thermoelastic structure with elastic support.

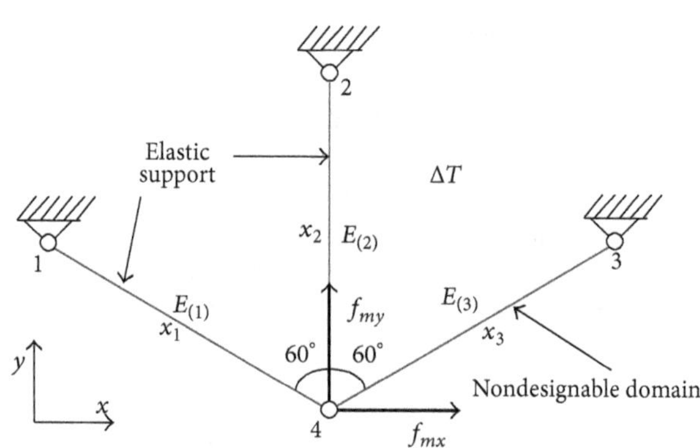

FIGURE 2: Three-bar truss system with elastic support.

As sketched in Figure 2, the truss system is comprised of three equal-length bars, that is, $E_{(1)}$, $E_{(2)}$, and $E_{(3)}$. The bars' cross-sectional areas are assigned as x_1, x_2, and x_3, respectively. One end of those bars is fixed and the other ends are hinged together to node 4. The mechanical loads applied on node 4 can be decomposed into two components f_{mx} and f_{my}. Meanwhile, all three bars undergo a temperature increase ΔT and the corresponding thermal expansion will cause thermal stresses. $E_{(1)}$ and $E_{(2)}$ constitute the elastic support and the cross-sectional areas, that is, x_1 and x_2, which are assigned as the design variables. $E_{(3)}$ is chosen as the nondesignable domain which means x_3 is fixed. In this example, the continuous design variables are used to evaluate the behaviors of the structural response.

According to the thermoelastic theory, the thermal load in each bar can be calculated as

$$F_{i\text{th}} = E\alpha\Delta T x_i, \tag{1}$$

where E is Young's modulus, α is the thermal expansion coefficient, ΔT is the temperature increase, and x_i is the cross-sectional area of the corresponding bar.

The strain and stress of the bar $E_{(i)}$ can be expressed as

$$\varepsilon_i = \frac{\overline{U_{ix}}}{L},$$
$$\sigma_i = E\varepsilon_{im} = E\left(\varepsilon_i - \varepsilon_{ith}\right) = E\left(\frac{\overline{U_{ix}}}{L} - \alpha\Delta T\right), \quad (2)$$

where $\overline{U_{ix}}$ is the displacement with respect to the elemental x-axis and ε_{im} and ε_{ith} are corresponding mechanical and thermal strain.

The elastic strain energy ϕ is defined as the potential mechanical energy in the elastic body, which is written as

$$\begin{aligned}\phi &= \frac{1}{2}\int \left(\boldsymbol{\varepsilon}_m\right)^T \mathbf{D}\boldsymbol{\varepsilon}_m dV \\ &= \frac{1}{2}\int \left(\boldsymbol{\varepsilon} - \boldsymbol{\varepsilon}_{\text{th}}\right)^T \mathbf{D}\left(\boldsymbol{\varepsilon} - \boldsymbol{\varepsilon}_{\text{th}}\right) dV \\ &= \frac{1}{2}\int \boldsymbol{\varepsilon}^T \mathbf{D}\boldsymbol{\varepsilon}\, dV - \int \boldsymbol{\varepsilon}^T \mathbf{D}\boldsymbol{\varepsilon}_{\text{th}} dV + \frac{1}{2}\int \boldsymbol{\varepsilon}_{\text{th}}^T \mathbf{D}\boldsymbol{\varepsilon}_{\text{th}} dV,\end{aligned} \quad (3)$$

where $\mathbf{D}$ is the elastic matrix and $\boldsymbol{\varepsilon}$ is the total strain vector consisting of mechanical and thermal items $\boldsymbol{\varepsilon}_m$ and $\boldsymbol{\varepsilon}_{\text{th}}$.

Then, we consider the mean compliance of structure

$$C = \frac{1}{2}\mathbf{F}^T\mathbf{U} = \frac{1}{2}\mathbf{F}^T\mathbf{K}^{-1}\mathbf{F}, \quad (4)$$

where $\mathbf{F}$ and $\mathbf{U}$ are nodal load and displacement vectors. $\mathbf{K}$ is global stiffness matrix.

According to the existing works, the strain energy minimization is more beneficial for the stress reduction while the mean compliance reflects the structural overall stiffness [19]. The purpose of this study is to find the optimized configuration of elastic supports preventing large deformations resulting from thermal and mechanical load. Naturally, the mean compliance minimization is selected as the design objective throughout this paper. Besides, two design constraints, that is, volume constraint and stress constraint, are involved in this study. The formulation of the optimization can be expressed as

$$\begin{aligned}&\text{Find:} \quad x_1, x_2 \\ &\text{Min:} \quad C \\ &\text{s.t.:} \quad \left(x_1 + x_2\right) L \leq \overline{V}; \\ &\qquad\quad \left|\sigma_3\right| \leq \overline{|\sigma|}.\end{aligned} \quad (5)$$

To demonstrate the relation between the design variables and the global compliance, parameters listed in Table 1 are used.

At the first place, the mechanical load is fixed with $f_{mx} = f_{my} = 7000\,\text{N}$. Structural compliance with different temperature increases is plotted by sets of contours in Figure 3. Thick solid line and dashed line denote stress and volume constraints, respectively. Along with the lower and upper limits of the design variables, these constraints define the feasible regions of optimization, which is indicated as the dark area in the figure. The black spot denotes the optimal point

Table 1: Constant parameters list.

Young's modulus E (Pa)	2×10^{11}
Thermal expansion coefficient α (°C^{-1})	1×10^{-5}
Length of bar L (m)	1
Cross-sectional area x_3 (m^2)	1×10^{-4}
Upper limit of stress in nondesignable domain $\overline{\lvert\sigma\rvert}$ (MPa)	80
Upper limit of design domain's volume $\overline{V}$ (m^3)	2.5×10^{-4}
Lower limit of design variables $\underline{x}$ (m^2)	5×10^{-5}
Upper limit of design variables $\overline{x}$ (m^2)	2×10^{-4}

where the compliance reaches a minimum. Those symbol signs are applied to Figures 3 and 4.

As illustrated in Figure 3, when the structure undergoes thermomechanical loads, a small temperature increase will lead to nonmonotone compliance with respect to the design variables. Consequently, the volume constraint may not be active as shown in Figures 3(b) to 3(d). In this case, using more material could weaken the structural mechanical performance due to the thermomechanical loading condition.

In another case, fixing $\Delta T = 20$°C, the mechanical loads are proportionally raised. The contour lines of compliance are shown in Figure 4. At the beginning, the stress is primarily caused by thermal load. Very compliant structures are used to offset thermal stress. Typically, when both mechanical and thermal loads are small as in Figure 4(a), the optimal point lies at the lower left corner of the feasible region. At this point, the stress constraint is inactive. As mechanical loads increase, more materials are required to strengthen the structure. The volume constraint is finally active in Figure 4. We can foresee that as the mechanical loads increase, there will be no feasible design for this problem.

Notably, there exist some critical optimal points for thermomechanical loads, for example, in Figures 3(a) and 4(d). The volume constraint is exactly active there. A slight increase of temperature or decrease of mechanical loads will lead to inactive volume constraint. These phenomena also happen in topology optimization of continuum structures with thermomechanical loads, where inactive volume constraints and elements with intermediate densities are always found (see [20]).

3. Formulation of Thermomechanical Topology Optimization

Based on the analyses in the previous section, we propose to use the topology optimization formulation as

$$\begin{aligned}&\text{Find:} \quad \mathbf{X} = \left(x_1, x_2, \ldots, x_n\right) \\ &\text{Min:} \quad C = \frac{1}{2}\mathbf{F}^T\mathbf{U} = \frac{1}{2}\left(\mathbf{F}_{\text{th}} + \mathbf{F}_m\right)^T \mathbf{U} \\ &\text{s.t.:} \quad V\left(\mathbf{X}\right) = \sum_{i=1}^{n} x_i v_i \leq \overline{V}; \\ &\qquad\quad \max_{j=1,2,\ldots,m} \left(\sigma_j^{\text{VM}}\right) \leq \overline{\sigma};\end{aligned}$$

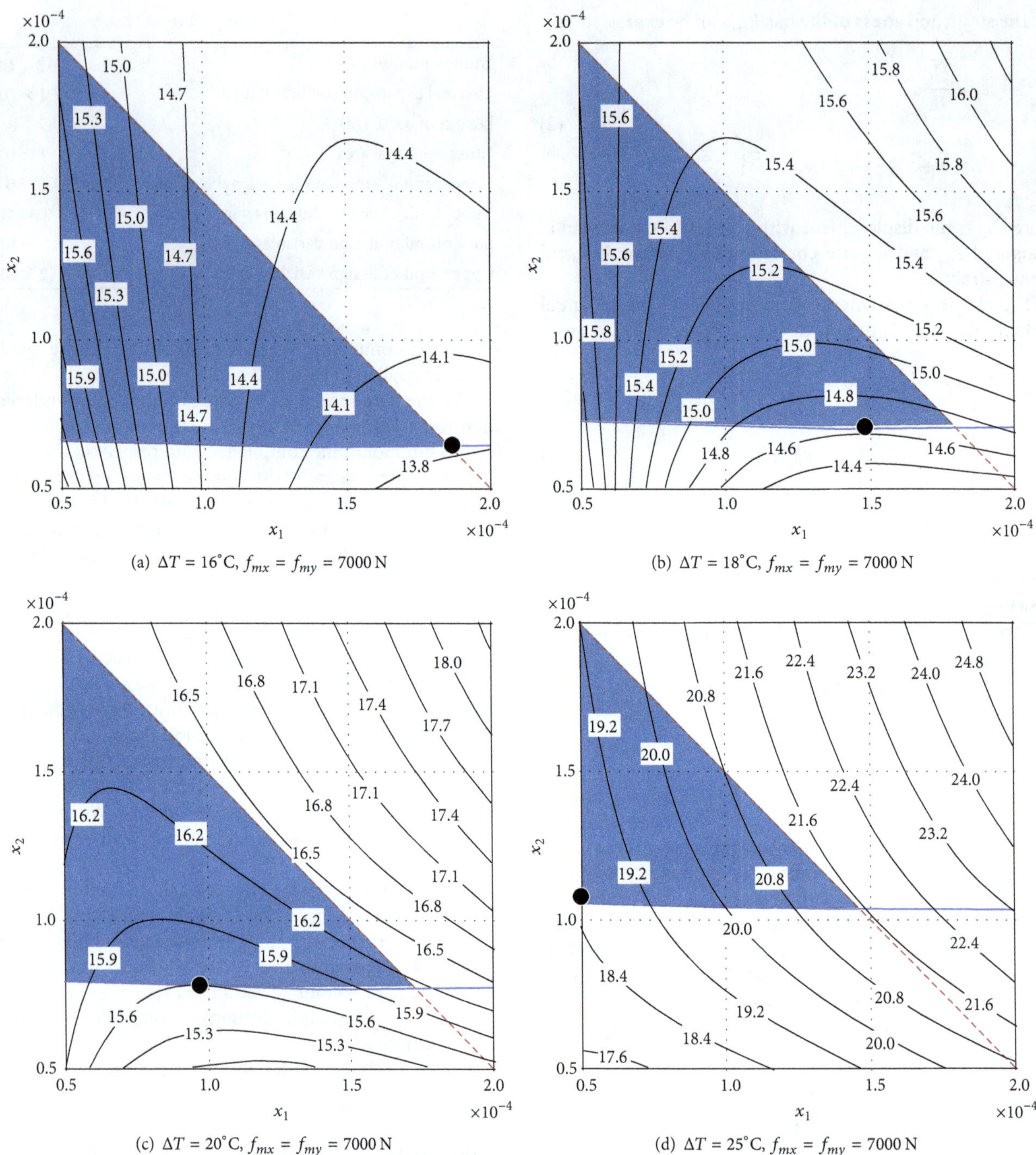

FIGURE 3: Optimizations under different temperature increases.

$$0 < \delta \leq x_i \leq 1,$$
$$i = 1, 2, \ldots, n, \tag{6}$$

where n and m are the numbers of elements in elastic support and nondesignable domain, respectively. $\mathbf{X}$ is the vector of design variables, that is, the pseudodensities which vary between 0 and 1 to describe the material distribution over the design domain. C is the global compliance of the structure. $\mathbf{F}_{th}$ and $\mathbf{F}_m$ are the thermal and mechanical load vectors composing the nodal load vector $\mathbf{F}$. σ_j^{VM} is the Von-Mises stress of the jth element in the nondesignable domain. $V(\mathbf{X})$ is the volume of the design domain and v_i is the volume of the ith element. δ is a small constant set as 0.001 in this paper to avoid singularity of stiffness matrix.

3.1. Design-Dependent Thermal Load. Unlike the static mechanical load, thermal load is a typical design-dependent load depending upon the material layout over the design

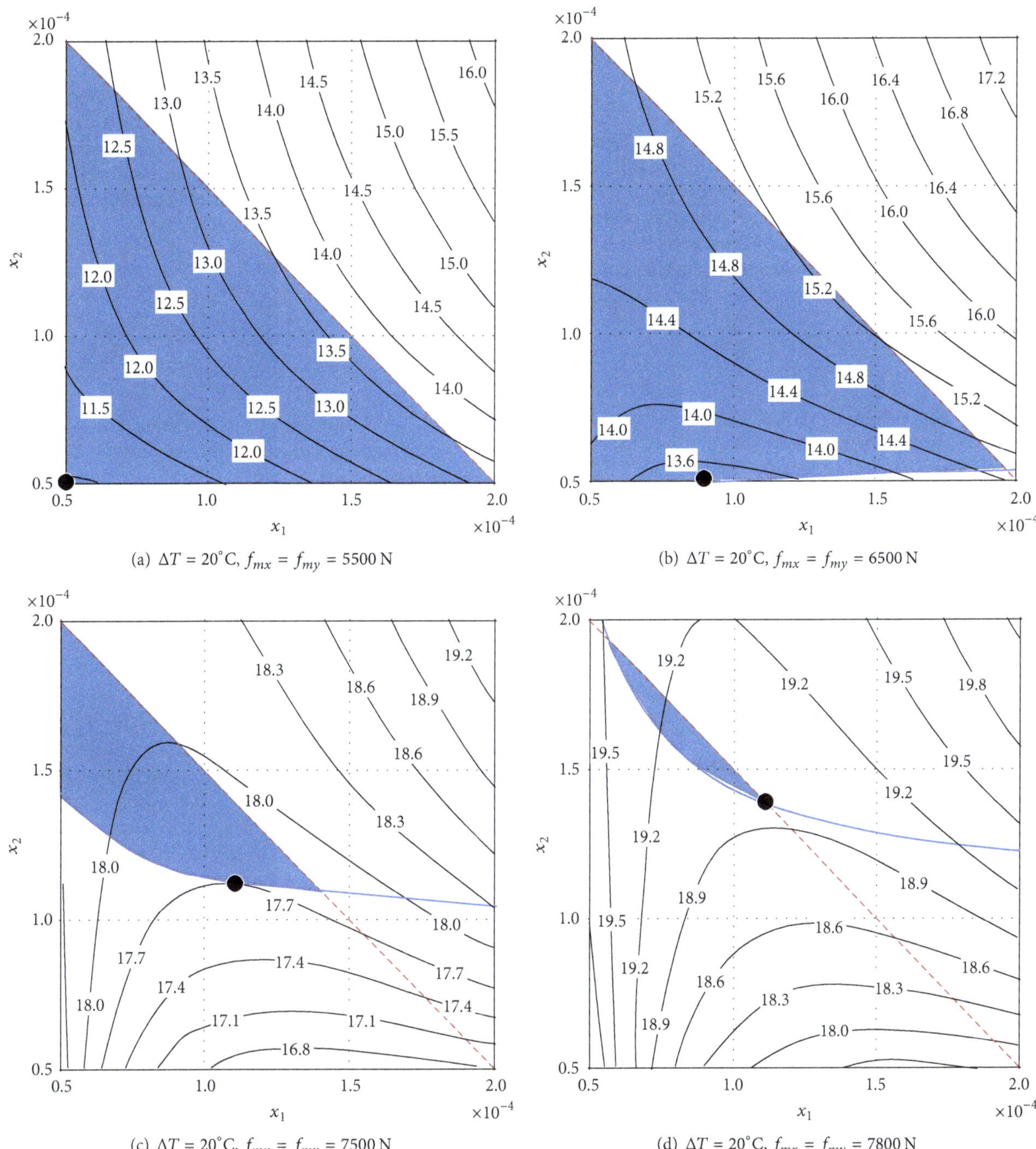

(a) $\Delta T = 20°\text{C}$, $f_{mx} = f_{my} = 5500$ N

(b) $\Delta T = 20°\text{C}$, $f_{mx} = f_{my} = 6500$ N

(c) $\Delta T = 20°\text{C}$, $f_{mx} = f_{my} = 7500$ N

(d) $\Delta T = 20°\text{C}$, $f_{mx} = f_{my} = 7800$ N

FIGURE 4: Optimizations under different mechanical loads.

domain. Thus, the matching relation between the thermal load and stiffness should be carefully handled.

According to the thermoelastic theory, the thermal load over the ith element is expressed as

$$\mathbf{F}_{i\text{th}} = \int_{\Omega_i} \mathbf{B}_i^T \mathbf{D}_i \boldsymbol{\varepsilon}_{i\text{th}} d\Omega, \quad (7)$$

where $\mathbf{B}_i$ and $\mathbf{D}_i$ are the element strain-displacement matrix and elasticity matrix, respectively. $\boldsymbol{\varepsilon}_{i\text{th}}$ is the thermal strain of the ith element. $\mathbf{D}_i$ is evaluated in terms of Young's modulus E_i. Here, a polynomial interpolation is used to link the pseudodensity variables to element elastic modulus [23]:

$$E_i = \left((1 - \beta) x_i^p + \beta x_i \right) E_{i0}, \quad (8)$$

where $p = 3$ and $\beta = 1/16$. This polynomial model works well in topology optimization with design-dependent loads, such as inertial loads, dynamic loads, and thermal loads. It maintains a positive gradient when the pseudodensity is zero.

Suppose the thermal expansion coefficient is temperature-independent and only the steady-state temperature field is taken into account. The thermal strain vector can be written as follows:

$$\boldsymbol{\varepsilon}_{i\text{th}} = \alpha_i \Delta T_i \boldsymbol{\alpha}^T. \tag{9}$$

Here, α_i is the thermal expansion coefficient. ΔT_i denotes the temperature increase. $\boldsymbol{\alpha}$ is a constant vector for the calculation of strain vector which is $[1, 1, 1, 0, 0, 0]$ in this paper [17].

Referring to the conception of thermal stress coefficient (TSC) [14], the element stiffness and the thermal stress load should be penalized independently in terms of element pseudodensity. A linear interpolation to the thermal expansion coefficient reads

$$\alpha_i = x_i \alpha_{i0}, \tag{10}$$

where α_{i0} is the original thermal expansion coefficient. Consequently, the elemental thermal expansion coefficient now varies with the pseudodensities in coordination with Young's modulus.

With the substitution of (8)–(10) into (7), the thermal load can be expressed as

$$\begin{aligned} \mathbf{F}_{i\text{th}} &= \gamma_i (x_i) \mathbf{F}_{i0\text{th}}, \\ \mathbf{F}_{i0\text{th}} &= \int_{\Omega_i} \mathbf{B}_i^T \mathbf{D}_{i0} \boldsymbol{\alpha} \, d\Omega, \\ \gamma_i &= E_i \alpha_i, \end{aligned} \tag{11}$$

where $\mathbf{F}_{i0\text{th}}$ is the thermal load when the element is solid with unit Young's modulus, unit thermal expansion coefficient, and unit temperature increase which is always constant in the optimization process. γ_i denotes the thermal stress coefficient (TSC) which can be treated as an inherent material property. With the introduction of TSC, the thermal load can be explicitly expressed as the function of design variables.

3.2. The Global Stress Measure. Difficulties of min-max problems are generally involved in stress-based topology optimization. Naturally, stress is evaluated and constrained in each element. In this way, the overall stress level is restricted below the prescribed limit. But the scale of the optimization problem and the computational cost increase dramatically. To remedy the drawback, aggregation methods combining all local stresses into one or several Kreisselmeier-Steinhauser (KS) or p-norm global functions [24, 25] are generally used. A tradeoff between the global and local stress approaches is that global stress measure reduces the computational cost while it introduces high nonlinearity in global measure functions and loses the control over local stresses. Different aggregation strategies have been brought in. For example, the active set approach [26] aggregates the active constraints into one set, which highly reduces the number of stress constraints. However, the changes of the active set can lead to poor convergence. Le et al. [27] introduced an adaptively updated p-norm aggregation method, which can precisely approximate the maximum local stress. A clustered method [28] was proposed as a compromise between global and local methods. The constraints were aggregated into several clusters. This method was proven to be efficient and robust by numerical examples but the results strongly depended upon the number of clusters.

Concerning the optimization of elastic support structure in this paper, the thermal stress constraints are imposed only in the nondesignable domain. Compared with the total elements number, the number of evaluated stresses is comparatively small. Without loss of generality, the standard p-norm measure is utilized which reads

$$\sigma_{PN} = \left(\sum_{j=1}^{m} \left(\sigma_j^{\text{VM}} \right)^P \right)^{1/P}, \tag{12}$$

where σ_{PN} is the p-norm of the elemental Von-Mises stress and P is the aggregation parameter. Apparently, the p-norm degenerates to the summation of the stress components when P is assigned as 1. Moreover, it can be proven mathematically that

$$\lim_{p \to \infty} \left(\sum_{j=1}^{m} \left(\sigma_j^{\text{VM}} \right)^P \right)^{1/P} = \max_{j=1,2,\ldots,m} \sigma_j^{\text{VM}}. \tag{13}$$

Obviously, with a sufficiently large value for P, the global measure can exactly match the maximum stress. However, the large value may lead to oscillation and ill-posed problems [29]. Consequently, to balance the effect of constraints' quality and convergence efficiency, the value of P should be properly chosen, which is assigned as 15 in this paper.

3.3. Adjoint Sensitivity Analysis. Firstly, the sensitivity of the global compliance with the design-dependent thermal load is derived and written as

$$\begin{aligned} \frac{\partial C}{\partial x_i} &= \frac{1}{2} \mathbf{U}^T \frac{\partial \mathbf{F}_{\text{th}}}{\partial x_i} + \frac{1}{2} \left(\mathbf{F}_m + \mathbf{F}_{\text{th}} \right)^T \cdot \frac{\partial \mathbf{U}}{\partial x_i} \\ &= \frac{1}{2} \mathbf{U}^T \frac{\partial \mathbf{F}_{\text{th}}}{\partial x_i} + \frac{1}{2} \left(\mathbf{F}_m + \mathbf{F}_{\text{th}} \right)^T \\ &\quad \cdot \mathbf{K}^{-1} \left(\frac{\partial \mathbf{F}_{\text{th}}}{\partial x_i} - \frac{\partial \mathbf{K}}{\partial x_i} \mathbf{U} \right) \\ &= \mathbf{U}^T \frac{\partial \mathbf{F}_{\text{th}}}{\partial x_i} - \frac{1}{2} \mathbf{U}^T \frac{\partial \mathbf{K}}{\partial x_i} \mathbf{U}. \end{aligned} \tag{14}$$

Sensitivities of the thermal load with respect to the design variables can be obtained by differentiating its definition formulated in (11). The derivative of the stiffness matrix can be easily evaluated according to the interpolation functions in (8).

Sensitivity analysis is then carried out to evaluate the global stress measure in the nondesignable domain with respect to the pseudodensity variables, which can be written as

$$\frac{\partial \sigma_{PN}}{\partial x_i} = \left(\sum_{j=1}^{m}\left(\sigma_j^{\mathrm{VM}}\right)^P\right)^{1/P-1} \sum_{j=1}^{m}\left(\left(\sigma_j^{\mathrm{VM}}\right)^{P-1} \cdot \frac{\partial \sigma_j^{\mathrm{VM}}}{\partial x_i}\right) = \left(\sigma_{PN}\right)^{1-P} \sum_{j=1}^{m}\left(\left(\sigma_j^{\mathrm{VM}}\right)^{P-1} \cdot \frac{\partial \sigma_j^{\mathrm{VM}}}{\partial x_i}\right). \quad (15)$$

Here, the Von-Mises equivalent stress is defined as

$$\sigma_j^{\mathrm{VM}} \sqrt{\boldsymbol{\sigma}_j^T \mathbf{V} \boldsymbol{\sigma}_j} = \sqrt{\left(\mathbf{B}_j\mathbf{U}_j - \boldsymbol{\varepsilon}_{j\mathrm{th}}\right)^T \mathbf{D}_j^T \mathbf{V} \mathbf{D}_j \left(\mathbf{B}_j\mathbf{U}_j - \boldsymbol{\varepsilon}_{j\mathrm{th}}\right)}, \quad (16)$$

where $\boldsymbol{\sigma}_j$ is the element stress vector. $\mathbf{U}_j$ and $\boldsymbol{\varepsilon}_{j\mathrm{th}}$ are the element nodal displacement vector and thermal strain vector. The matrix $\mathbf{V}$ is a constant matrix; that is,

$$\mathbf{V}^{2\mathrm{D}} = \begin{bmatrix} 1 & -\frac{1}{2} & 0 \\ -\frac{1}{2} & 1 & 0 \\ 0 & 0 & 3 \end{bmatrix},$$

$$\mathbf{V}^{3\mathrm{D}} = \begin{bmatrix} \begin{pmatrix} 2 & -1 & -1 \\ -1 & 2 & -1 \\ -1 & -1 & 2 \end{pmatrix} & 0_{3\times 3} \\ 0_{3\times 3} & \begin{pmatrix} 6 & & \\ & 6 & \\ & & 6 \end{pmatrix} \end{bmatrix}. \quad (17)$$

As only the Von-Mises stresses in the nondesignable domain are concerned, we have

$$\frac{\partial \sigma_j^{\mathrm{VM}}}{\partial x_i} = \frac{\left(\mathbf{B}_j\mathbf{U}_j - \boldsymbol{\varepsilon}_{j\mathrm{th}}\right)^T \mathbf{D}_j^T \mathbf{V} \mathbf{D}_j}{\sqrt{\left(\mathbf{B}_j\mathbf{U}_j - \boldsymbol{\varepsilon}_{j\mathrm{th}}\right)^T \mathbf{D}_j^T \mathbf{V} \mathbf{D}_j \left(\mathbf{B}_j\mathbf{U}_j - \boldsymbol{\varepsilon}_{j\mathrm{th}}\right)}} \mathbf{B}_j \frac{\partial \mathbf{U}_j}{\partial x_i} = \frac{1}{\sigma_j^{\mathrm{VM}}} \cdot \boldsymbol{\sigma}_j^T \mathbf{V} \mathbf{D}_j \mathbf{B}_j \frac{\partial \mathbf{U}_j}{\partial x_i}, \quad (18)$$

where $\partial \mathbf{D}_j / \partial x_i = \partial \boldsymbol{\varepsilon}_{j\mathrm{th}} / \partial x_i = 0$. In this case, the singularity phenomenon in stress-based topology optimization is actually not involved in this optimization problem.

We can also define

$$\mathbf{U}_j = \mathbf{A}_j \mathbf{U}, \qquad \frac{\partial \mathbf{U}_j}{\partial x_i} = \mathbf{A}_j \frac{\partial \mathbf{U}}{\partial x_i}. \quad (19)$$

Table 2: Material properties used in the optimization.

Properties	Elastic support	Nondesignable domain
Elastic modulus (MPa)	71000	3500
Coefficient of thermal expansion (10^{-6}/°C)	23	7
Poisson's ratio	0.33	0.4

The matrix $\mathbf{A}$ transforms the global nodal displacement vector to an element one.

Based on the differentiation of finite element equilibrium equation, we have the derivative of the nodal displacement vector

$$\frac{\partial \mathbf{U}}{\partial x_i} = \mathbf{K}^{-1}\left(\frac{\partial \mathbf{F}_{\mathrm{th}}}{\partial x_i} - \frac{\partial \mathbf{K}}{\partial x_i}\mathbf{U}\right). \quad (20)$$

The substitution of the above equation into the derivative of p-norm aggregation function produces

$$\frac{\partial \sigma_{PN}}{\partial x_i} = \left(\sigma_{PN}\right)^{1-P} \sum_{j=1}^{m}\left(\left(\sigma_j^{\mathrm{VM}}\right)^{P-2} \boldsymbol{\sigma}_j^T \mathbf{V} \mathbf{D} \mathbf{B}_j \mathbf{A}_j\right) \cdot \mathbf{K}^{-1}\left(\frac{\partial \mathbf{F}_{\mathrm{th}}}{\partial x_i} - \frac{\partial \mathbf{K}}{\partial x_i}\mathbf{U}\right). \quad (21)$$

Suppose

$$\mathbf{Q}^T = \left(\sigma_{PN}\right)^{1-P} \sum_{j=1}^{m}\left(\left(\sigma_j^{\mathrm{VM}}\right)^{P-2} \boldsymbol{\sigma}_j^T \mathbf{V} \mathbf{D} \mathbf{B}_j \mathbf{A}_j\right) \mathbf{K}^{-1} \quad (22)$$

is adjoint displacement; the derivative of the p-norm function can be obtained by an additional finite element calculation.

4. Numerical Study

In this section, a 2D cantilever beam and a 3D nozzle flap of a turbine engine are tested to study the topology optimization of elastic supports under thermomechanical loads.

4.1. 2D Cantilever Beam. As shown in Figure 5, a 2D cantilever beam consists of a nondesignable top surface and an elastic support assigned as design domain. The material properties are listed in Table 2. Since the thermal expansion coefficient of elastic support is higher than that of the nondesignable domain, significant thermal stresses will occur as a result of temperature increase.

In accordance with Section 3, the global compliance of the whole structure is minimized as the design objective. The volume fraction is constrained to 50% and the upper bound of the stress constraints in the nondesignable domain is 200 MPa.

The optimized results with various load combinations are listed and compared in Figure 6. The result in Figure 6(a) is set as a benchmark with clear configuration. Firstly, the mechanical load is fixed at 20 N/mm in Figures 6(a)–6(d).

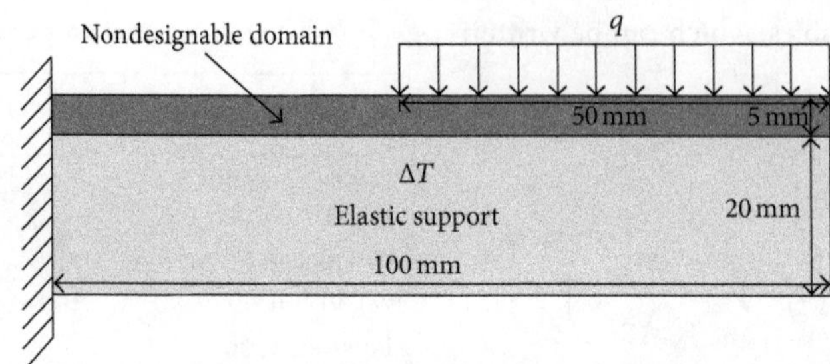

FIGURE 5: Definition of the 2D cantilever beam.

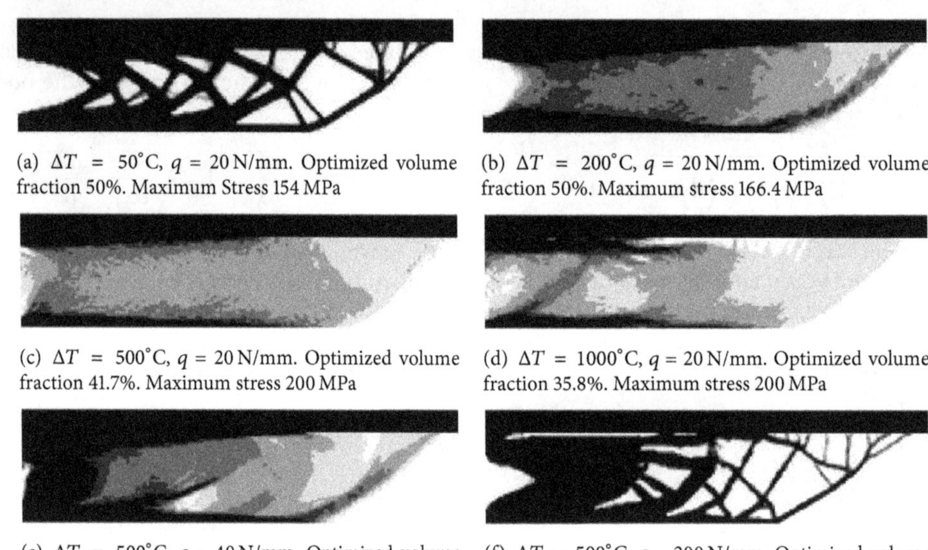

(a) $\Delta T = 50^\circ$C, $q = 20$ N/mm. Optimized volume fraction 50%. Maximum Stress 154 MPa

(b) $\Delta T = 200^\circ$C, $q = 20$ N/mm. Optimized volume fraction 50%. Maximum stress 166.4 MPa

(c) $\Delta T = 500^\circ$C, $q = 20$ N/mm. Optimized volume fraction 41.7%. Maximum stress 200 MPa

(d) $\Delta T = 1000^\circ$C, $q = 20$ N/mm. Optimized volume fraction 35.8%. Maximum stress 200 MPa

(e) $\Delta T = 500^\circ$C, $q = 40$ N/mm. Optimized volume fraction 50%. Maximum stress 200 MPa

(f) $\Delta T = 500^\circ$C, $q = 200$ N/mm. Optimized volume fraction 50%. Maximum stress 812.4 MPa

FIGURE 6: Optimized results of minimizing compliance under different loads.

As the temperature increases, the volume constraint starts to be inactive. More elements with intermediate densities arise. This is due to the nature of thermal stress constraint where the elements with compliant materials can offset the thermal stress better. Later, considering the changes in Figures 6(c), 6(e), and 6(f), clear structure patterns gradually appear when larger mechanical loads are applied. Thus, the elastic support is able to undergo the pressure with good stiffness. However, when the thermomechanical loads are too high, the optimization finds no feasible solution where the stress constraints are violated as shown in Figure 6(f).

The optimized results have shown that the relative magnitude of the thermal and mechanical load greatly influences the optimized results, which have good consistency with the analytical solution of the three-bar truss model in Section 3. Moreover, the existence of a large amount of intermediate density material is reasonable when the thermal load is dominant over the mechanical load. Eliminating the grey elements directly or using some numerical schemes may improve the global stiffness but will unfortunately lead to higher stress level.

4.2. 3D Nozzle Flap. As shown in Figure 7, a nozzle flap of a turbine engine is composed of titanium stiffeners and a ceramic plate. Material properties are listed in Table 3. Figure 7 also illustrates the thermoelastic loads applied on the model, including a uniform pressure of 1 MPa and global temperature increase of 500°C. Significant thermal stresses are generated in the plate due to the different thermal expansion coefficient. In practical design, the stiffeners are assigned as the elastic support design domain and the plate is nondesignable. The design objective is to minimize the global compliance with a 25 MPa stress constraint on the nondesignable plate and a 30% material volume constraint on the design domain.

TABLE 3: Material properties used in the optimization.

Material properties	Titanium	Ceramic
Elastic modulus (MPa)	11000	3500
Coefficient of thermal expansion (10^{-6}/°C)	10	7
Poisson's ratio	0.33	0.4

The optimized design of elastic support as shown in Figure 8 is obtained by topology optimization using the proposed formulation. The optimized design is presented by

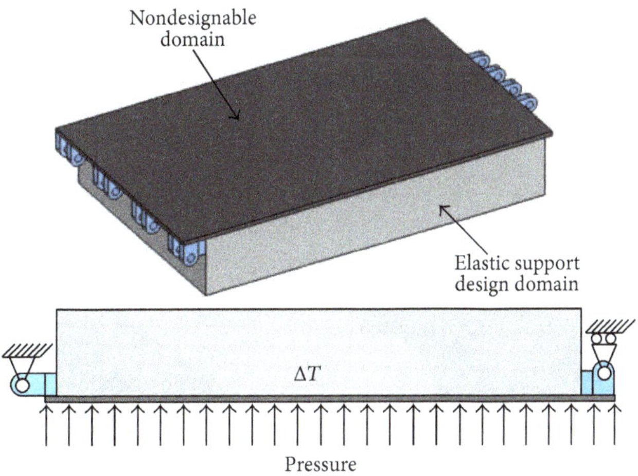

FIGURE 7: Nozzle flap model and its design domain, loads, and boundary conditions.

FIGURE 8: Topological optimized design with a temperature increase of 500°C.

hiding the elements with their pseudodensities under 0.5 to show a clear structural configuration. The two strongest stiffeners are composed of solid elements while the detailed structural branches are using intermediate material with pseudodensities between 0.5 and 0.9.

To further emphasize the effect of the thermal loads, two more designs are obtained by using different temperature increases, that is, 100°C and 1000°C, with identical mechanical loads. The optimized designs are shown in Figure 9. Compared with the structural topology in Figure 8, the optimized design in Figure 9(a) has shown a much clearer load carrying path with less intermediate material as the mechanical load is dominant. In Figure 9(b), an extremely high temperature increase of 1000°C is used. The optimized design is mostly composed of intermediate material as expected. No clear structural configuration is achieved.

CAD model as shown in Figure 10(a) is then rebuilt according to the optimized design in Figure 8. An existing design of the nozzle flap is shown in Figure 10(b) for comparison. To verify the effect of topology optimization, the two models are analyzed, respectively, with refined finite element mesh. Two models share identical boundary conditions and thermomechanical loads; the stress distribution in the bottom plate is shown in Figure 11. The overall comparison of the two designs is shown in Table 4.

Compared with the existing design, the optimized design reduces the maximum stress in the bottom plate significantly from 284.7 MPa to 25 MPa. The global compliance decreases from 271.9 KJ to 254.6 KJ. Meanwhile, material of $0.129 \times 10^7\ \text{mm}^3$ is saved, that is, 5.676 kg lighter than before.

5. Conclusion

In this paper, topology optimization of elastic supporting structures under thermomechanical loads is investigated. A three-bar truss model is firstly employed to reveal the particularity of thermoelastic problems, that is, nonmonotonous compliance, inactive volume constraint with high temperature increase, and so forth. Similar appearances also have been found in the topology optimization of a 2D cantilever beam structure presented in this paper. On account of stress-based topology optimization with large numbers of design constraints, global stress measure approach based on

TABLE 4: Comparison of the optimal design and the existing design.

	Topological optimized design	Rebuilt optimized design	Existing design
Maximum stress in the bottom plate (MPa)	25.00	25.00	284.7
Compliance of the whole structure (KJ)	302.2	254.6	271.9
Volume of the elastic support (mm^3)	2.482×10^7	2.078×10^7	2.207×10^7

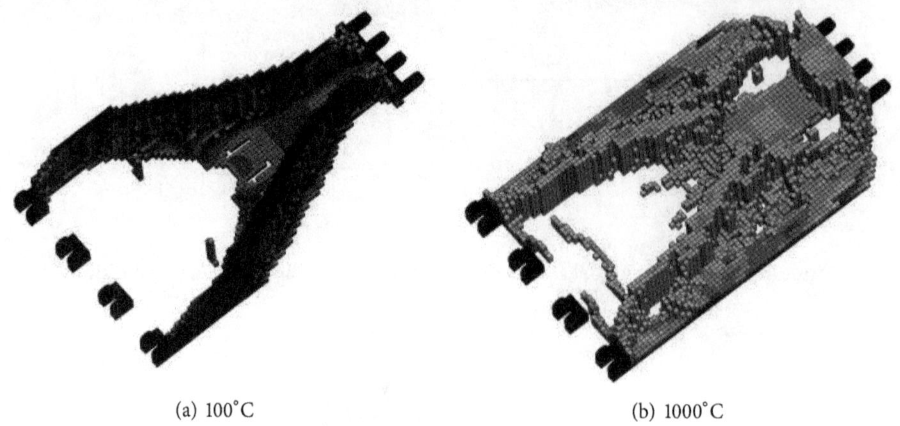

(a) 100°C (b) 1000°C

FIGURE 9: Topological optimized design with different temperature increases.

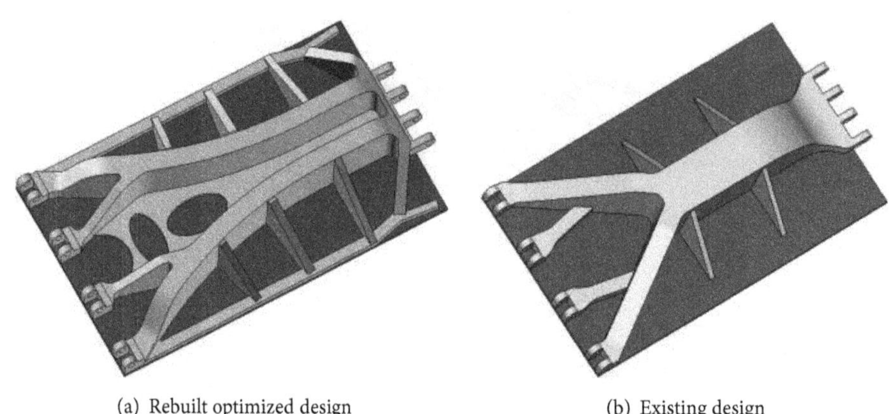

(a) Rebuilt optimized design (b) Existing design

FIGURE 10: Rebuilt optimized design and existing design to be compared.

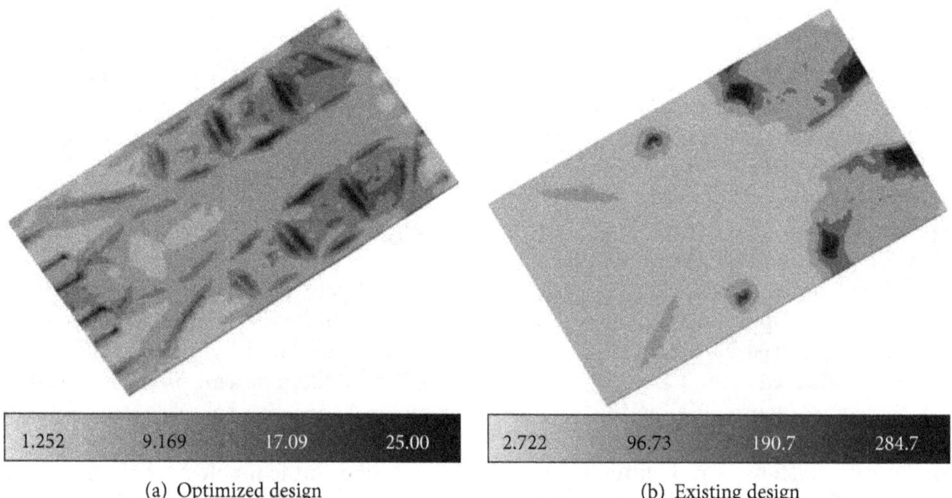

(a) Optimized design (b) Existing design

FIGURE 11: Stress distribution in the bottom plate (MPa).

p-norm function is used to aggregate the stress constraints involved in each iteration. Meanwhile, formulation of design-dependent thermal load is presented with different interpolation functions assigned for material elastic modulus and thermal expansion coefficients. Sensitivity analysis is then carried out to evaluate the global stress measure in nondesignable domain and the compliance with respect to the pseudodensity variables. In the Numerical Study, a 2D cantilever beam model and a 3D nozzle flap are optimized. Compared with the existing design, the optimized designs not only use fewer materials, but also are both stiffer and better in reducing the maximum stress in nondesignable domain.

Competing Interests

The authors declare that they have no competing interests.

Acknowledgments

This work is supported by the National Natural Science Foundation of China (11432011, 51521061), the 111 Project (B07050), and the Fundamental Research Funds for the Central Universities (3102014JC02020505).

References

[1] Q. Li, G. P. Steven, O. M. Querin, and Y. M. Xie, "Shape and topology design for heat conduction by evolutionary structural optimization," *International Journal of Heat and Mass Transfer*, vol. 42, no. 17, pp. 3361–3371, 1999.

[2] A. Gersborg-Hansen, M. P. Bendsøe, and O. Sigmund, "Topology optimization of heat conduction problems using the finite volume method," *Structural and Multidisciplinary Optimization*, vol. 31, no. 4, pp. 251–259, 2006.

[3] Y. Zhang and S. Liu, "Design of conducting paths based on topology optimization," *Heat and Mass Transfer*, vol. 44, no. 10, pp. 1217–1227, 2008.

[4] T. Gao, W. H. Zhang, J. H. Zhu, Y. J. Xu, and D. H. Bassir, "Topology optimization of heat conduction problem involving design-dependent heat load effect," *Finite Elements in Analysis and Design*, vol. 44, no. 14, pp. 805–813, 2008.

[5] A. Iga, S. Nishiwaki, K. Izui, and M. Yoshimura, "Topology optimization for thermal conductors considering design-dependent effects, including heat conduction and convection," *International Journal of Heat and Mass Transfer*, vol. 52, no. 11-12, pp. 2721–2732, 2009.

[6] J. Dirker and J. P. Meyer, "Topology optimization for an internal heat-conduction cooling scheme in a square domain for high heat flux applications," *Journal of Heat Transfer*, vol. 135, no. 11, Article ID 111010, 2013.

[7] H. Rodrigues and P. Fernandes, "A material based model for topology optimization of thermoelastic structures," *International Journal for Numerical Methods in Engineering*, vol. 38, no. 12, pp. 1951–1965, 1995.

[8] D. A. Tortorelli, G. Subramani, S. C. Y. Lu, and R. B. Haber, "Sensitivity analysis for coupled thermoelastic systems," *International Journal of Solids and Structures*, vol. 27, no. 12, pp. 1477–1497, 1991.

[9] Q. Li, G. P. Steven, and Y. M. Xie, "Displacement minimization of thermoelastic structures by evolutionary thickness design," *Computer Methods in Applied Mechanics and Engineering*, vol. 179, no. 3-4, pp. 361–378, 1999.

[10] S. Cho and J.-Y. Choi, "Efficient topology optimization of thermo-elasticity problems using coupled field adjoint sensitivity analysis method," *Finite Elements in Analysis and Design*, vol. 41, no. 15, pp. 1481–1495, 2005.

[11] Q. Xia and M. Y. Wang, "Topology optimization of thermoelastic structures using level set method," *Computational Mechanics*, vol. 42, no. 6, pp. 837–857, 2008.

[12] J. Yan, G. Cheng, and L. Liu, "A uniform optimum material based model for concurrent optimization of thermoelastic structures and materials," *International Journal for Simulation and Multidisciplinary Design Optimization*, vol. 2, no. 4, pp. 259–266, 2008.

[13] S. P. Sun and W. H. Zhang, "Topology optimal design of thermo-elastic structures," *Chinese Journal of Theoretical and Applied Mechanics*, vol. 41, no. 6, pp. 878–887, 2009.

[14] T. Gao and W. Zhang, "Topology optimization involving thermo-elastic stress loads," *Structural and Multidisciplinary Optimization*, vol. 42, no. 5, pp. 725–738, 2010.

[15] J. D. Deaton and R. V. Grandhi, "Stiffening of restrained thermal structures via topology optimization," *Structural and Multidisciplinary Optimization*, vol. 48, no. 4, pp. 731–745, 2013.

[16] J. D. Deaton and R. V. Grandhi, "A survey of structural and multidisciplinary continuum topology optimization: post 2000," *Structural & Multidisciplinary Optimization*, vol. 49, no. 1, pp. 1–38, 2014.

[17] P. Pedersen and N. L. Pedersen, "Strength optimized designs of thermoelastic structures," *Structural and Multidisciplinary Optimization*, vol. 42, no. 5, pp. 681–691, 2010.

[18] P. Pedersen and N. L. Pedersen, "Interpolation/penalization applied for strength design of 3D thermoelastic structures," *Structural and Multidisciplinary Optimization*, vol. 45, no. 6, pp. 773–786, 2012.

[19] W. H. Zhang, J. G. Yang, Y. J. Xu, and T. Gao, "Topology optimization of thermoelastic structures: mean compliance minimization or elastic strain energy minimization," *Structural and Multidisciplinary Optimization*, vol. 49, no. 3, pp. 417–429, 2014.

[20] J. D. Deaton and R. V. Grandhi, "Stress-based design of thermal structures via topology optimization," *Structural and Multidisciplinary Optimization*, vol. 53, no. 2, pp. 253–270, 2016.

[21] G. I. N. Rozvany, "Exact analytical solutions for some popular benchmark problems in topology optimization," *Structural Optimization*, vol. 15, no. 1, pp. 42–48, 1998.

[22] G. I. N. Rozvany, "Basic geometrical properties of exact optimal composite plates," *Computers & Structures*, vol. 76, no. 1, pp. 263–275, 2000.

[23] J. H. Zhu, W. H. Zhang, and P. Beckers, "Integrated layout design of multi-component system," *International Journal for Numerical Methods in Engineering*, vol. 78, no. 6, pp. 631–651, 2009.

[24] P. Duysinx and O. Sigmund, "New developments in handling stress constraints in optimal material distribution," in *Proceedings of the 7th AIAA/USAF/NASA/ISSMO Symposium on Multidisciplinary Analysis and Optimization, Multidisciplinary Analysis Optimization Conferences*, vol. 1, pp. 1501–1509, 1998.

[25] R. J. Yang and C. J. Chen, "Stress-based topology optimization," *Structural Optimization*, vol. 12, no. 2-3, pp. 98–105, 1996.

[26] P. Duysinx and M. P. Bendsøe, "Topology optimization of continuum structures with local stress constraints," *International*

Journal for Numerical Methods in Engineering, vol. 43, no. 8, pp. 1453–1478, 1998.

[27] C. Le, J. Norato, T. Bruns, C. Ha, and D. Tortorelli, "Stress-based topology optimization for continua," *Structural and Multidisciplinary Optimization*, vol. 41, no. 4, pp. 605–620, 2010.

[28] E. Holmberg, B. Torstenfelt, and A. Klarbring, "Stress constrained topology optimization," *Structural and Multidisciplinary Optimization*, vol. 48, no. 1, pp. 33–47, 2013.

[29] R. T. Haftka and Z. Gürdal, *Elements of Structural Optimization*, Kluwer Academic, Dordrecht, Netherlands, 1992.

10

Nonlinear Material Behavior Analysis under High Compression Pressure in Dynamic Conditions

Muhammad Zubair Zahid,[1] Shahid Ikramullah Butt,[1] Tauqeer Iqbal,[2] Syed Zohaib Ejaz,[3] and Zhang Faping[4]

[1] *School of Mechanical and Manufacturing Engineering, National University of Sciences and Technology, Islamabad, Pakistan*
[2] *Université Catholique de Louvain, Louvain-la-Neuve, Belgium*
[3] *Birmingham City University, Birmingham, UK*
[4] *School of Mechanical Engineering, Beijing Institute of Technology, Beijing, China*

Correspondence should be addressed to Shahid Ikramullah Butt; drshahid@smme.nust.edu.pk

Academic Editor: Paul Williams

Gun chamber pressure is an important parameter in proofing of ammunition to ensure safety and reliability. It can be measured using copper crushers or piezoelectric sensor. Pressure calculations in copper crusher method are based on linear plastic deformation of copper after firing. However, crusher pressure deformation at high pressures deviates from the corresponding values measured by piezoelectric pressure transducers due to strain rate dependence of copper. The nonlinear deformation rate of copper at high pressure measurements causes actual readings from copper crusher gauge to deviate from true pressure values. Comparative analysis of gun chamber pressure was conducted for 7.62 × 51 mm ammunition using Electronic Pressure, Velocity, and Action Time (EPVAT) system with piezoelectric pressure transducers and conventional crusher gauge. Ammunitions of two different brands were used to measure chamber pressure, namely, NATO standard ammunition and non-NATO standard ammunition. The deformation of copper crushers has also been simulated to compare its deformation with real time firing. The results indicate erratic behavior for chamber pressure by copper crusher as per standard deviation and relative spread and thus prove piezo sensor as more reliable and consistent mode of peak pressure measurement. The results from simulation, cost benefit analysis, and accuracy clearly provide piezo sensors with an edge over conventional, inaccurate, and costly method of copper crusher for ballistic measurements due to its nonlinear behavior.

1. Introduction

Proof is a destructive test in which small numbers of proof samples are selected as representative of a large group known as lot or batch of the ammunition by the same manufacturer and process. Proofing of ammunition is important to ensure the safety, reliability, and operational effectiveness of conventional ammunition. The importance of proofing is often poorly understood, leading to failure in ammunition safety and stability. Proof of ammunition is a systematic method of evaluating the properties, characteristics, and performance capabilities of ammunition throughout its life cycle. It is used to assess the reliability, safety, and operational effectiveness of stocks. Proof is the functional testing or firing of ammunition and explosives to ensure safety and stability in storage and intended use. In-service proof and the surveillance of ammunition are undertaken to ensure that the ammunition continues to meet the required quality standards throughout its life [1].

The acceptance or rejection criterion during test firing of guns and ammunitions depends upon many factors. Chamber pressure is defined as the force per unit area that explosive gas exerts on the walls of a gun chamber. Chamber pressure being the most important, researchers have been making unremitting efforts to improve the testing precision [2] as this must be tested and verified during research and development of product acceptance.

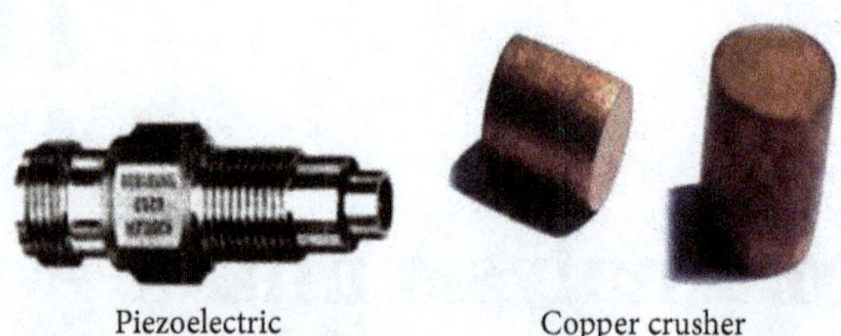

Figure 1: Piezoelectric sensor and copper crushers.

Crusher gauges were used to measure gas chamber pressure as commonly available standardized method till 1960s [3]. Usually a piston fitted to a piston hole into the chamber of the barrel compresses a copper or lead cylinder. The crusher gets permanently deformed due to the effect of the gas pressure generated by the burning of the explosive mixture on the base of the piston. After measuring the deformed crusher length, the peak pressure is estimated after comparing it to a calibrated conversion table provided by the manufacturer with each lot of crushers. Though convenient, it is incapable of recording the dynamic change of chamber pressure with time. Figure 1 shows piezo sensor and conventional cylindrical copper crusher.

With the development of charge amplifiers by W. P. Kistler in 1950s piezoelectric techniques were used in the area of interior ballistics [3] which led to the use of piezoelectric transducers over crusher gauges thus making it convenient to plot accurately pressure time curve in 1960s [4].

NATO [5–7], CIP [8], and SAAMI [9] are providing insight into research and development in the field of crusher and piezoelectric pressure measurement. Pressure measurement method used by these organizations is differentiated by the measurement point and the measuring techniques. Thus, NATO standard Kistler type 6215 and the conformal PCB type 117B104 piezoelectric pressure transducers were developed by various manufacturers [3].

Copper crushers and piezoelectric sensor methods both have their pros and cons for chamber pressure measurement. Copper crusher technique is simple enough to obtain a rapid estimation of the peak pressure in ammunition testing but it has drawbacks for being of limited accuracy and only gives the peak pressure [3]. Similarly, for higher pressures the deformations noted are lower in comparison to that obtained through piezoelectric sensors due to nonlinear variations which are observed in copper material. While piezoelectric pressure sensors offer the advantages of high sensitivity, good linearity, low hysteresis, and versatility, their main use is in higher cost and precision areas [10].

The crusher pressure readings may deviate quite considerably from the corresponding values measured by piezoelectric pressure transducers. The pressure readings obtained by crushers from different manufacturers are not consistent. It has become clear that the crusher and piezoelectric pressure transducer readings differ, depending on the pressure range and the used gun type, by as much as up to +20%, and that the deviation normally increases with pressure [11]. It is anticipated that the gun pressures in the future will be increasing, which means that the difference between pressure readings obtained by different measurement techniques is also going to increase.

Piezoelectric pressure transducers give the voltage at each stage of burning (starting from ignition by primer till complete burning) which is translated in the form of a graph; hence it helps in getting the voltage peak which in turn is converted into pressure through amplifier system. Since it gives exact voltage peak achieved during the combustion in chamber, hence it is useful in exactly knowing if the pressure is not crossing the allowed safety limits of gun chamber.

In our experimental study the chamber pressure was measured by the two different techniques (crusher gauges and piezoelectric sensors) for the "ammunition type 7.62 × 51 mm." This research will try to give an insight into the following knowledge gap areas for "ammunition type 7.62 × 51 mm" not explored previously:

(i) Nonlinear behavior of copper crusher gauge material parameters at high pressure

(ii) Comparison of chamber pressure values made for copper crusher gauge and piezoelectric sensors under the same controlled test conditions

(iii) Similar caliber ammunition type "7.62 × 51 mm" used to explore the possibilities of error in crusher gauges from different ordnance manufacturers, that is, NATO standard ammunition versus non-NATO ammunition

(iv) Another prime objective of cost benefit analysis for copper crusher gauges with piezoelectric sensors also not done by other researchers previously.

1.1. Copper Crusher Pressure and True Pressure. True pressure is that value of maximum gas pressure which actually exists and would be obtained from an ideal measuring system. Accuracy of a gauge is its ability to record or measure the "true" pressure without systematic error. True pressure is that value of maximum gas pressure which actually exists and would be obtained from an ideal measuring system [12]. Pressure measured by copper crusher gauge is always less than the true pressure due to the following reasons:

(i) Inertia inherent in the system

(ii) Consequent time required to compress the copper

(iii) Very transient nature of the peak pressure in a gun.

Previously, pressure measured by copper gauges was accepted as true pressure but, with the advent of inertialess gauges, it has been found that there is a reasonably constant relationship between pressure on copper and that measured by an inertialess gauge (true pressure) [13]. For present purposes modern transducer systems which record pressure as a time function are accepted as giving the best available estimate of "true pressure" [12].

2. Methodology

EPVAT testing is well-established NATO proofing system for proofing/inspecting of ammunition and ensures

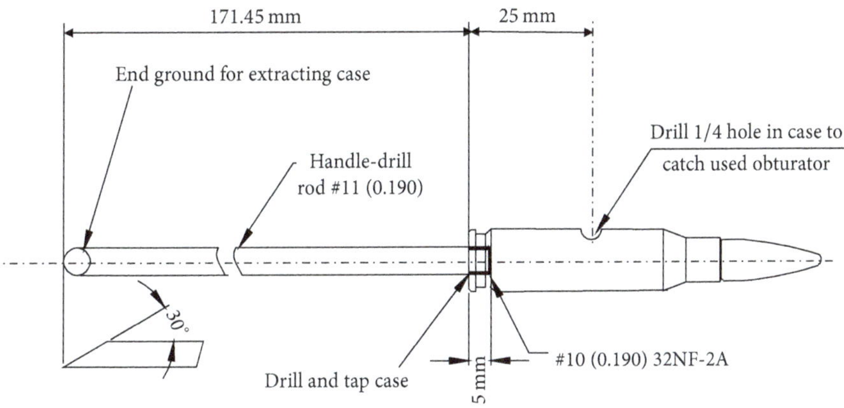

FIGURE 2: Drawing for preparing proof cartridge [05].

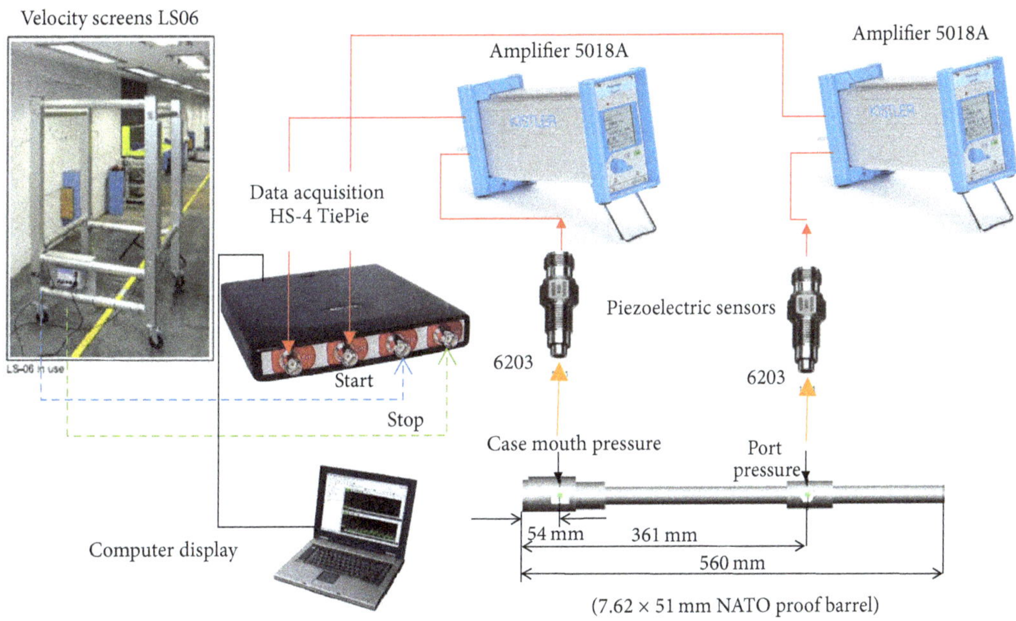

FIGURE 3: Schematic of Electronic Pressure, Velocity, and Action Time.

safety/quality. This procedure ensures the safety of the shooter and its 100% functionality at the target. It is a comprehensive procedure for testing ammunition using state-of-the-art instruments and computing devices. The procedure itself is described in NATO MOPI document AC/225 (Com. III/SC.1)D/200 [14].

The MOPI manual also provides guidelines to prepare cartridge cases for using copper crusher method. Figure 2 shows drawing layout for preparing cartridge.

In EPVAT system, case mouth pressure, port pressure, action time, and velocity of a bullet are measured simultaneously. This NATO system clearly defines piezoelectric sensor locations on the barrel, magnitude of pressure limits, bullet energy, action time, and velocity limits against a particular caliber, thus providing a unique opportunity to enhance quality of ammo. Action time is defined as the time that requires bullet to leave muzzle end from the ignition of bullet primer.

Internal ballistic parameters measured through EPVAT system are most important to study and analyze the effects on weapon operation, barrel length vis-à-vis velocity, recoil force, flash intensity, muzzle design, barrel redesign (if needed), and continuous propellant improvement. The schematic diagram of EPVAT system is shown in Figure 3.

NATO MOPI manual, as standard document for ballistic measurements, provides the following requirements for accurate pressure measurement through copper crusher/ piezoelectric sensors [15]:

(i) The mean peak chamber pressure of any type of ammunition shall not exceed 50,000 pounds per square inch (corrected) when the ammunition is conditioned at 21°C using the radial copper pressure cylinder as required by STANAG 2310 [16].

(ii) The corrected mean peak pressure of any type of ammunition, measured at case mouth position using

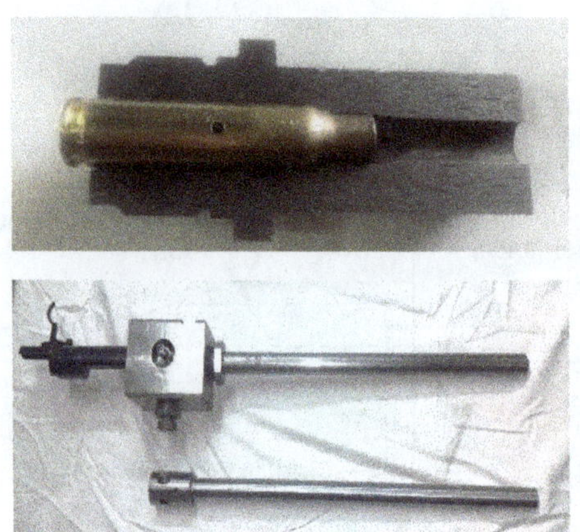

FIGURE 4: EPVAT system for case mouth pressure measurement.

piezoelectric transducer, shall not exceed 380 MPa (corrected) when the ammunition is conditioned at 21°C as required by STANAG 2310 [16].

(iii) The weight of all bullets to be used for proofing should be within limits 8.4 to 10 grams.

(iv) The minimum average energy for a true pressure through Electronic Pressure, Velocity, and Action Time (EPVAT) barrel shall be 2915 Joules.

(v) The action time should also be limited to 4 milliseconds.

Figure 4 indicates the EPVAT system for case mouth pressure measurement. The pressure to be measured during proof fire acts on the piezo sensor's diaphragm, which converts it to a proportional force. This force is transmitted on to the quartz. Quartz has an intrinsic property of producing an electrostatic charge when subjected to load. A spring carries this negative charge till connector. The charge is later on converted to positive voltage by an amplifier.

The piezo sensor system is well suited to measure quasi-static phenomenon of rapid dynamic nature. However, amongst its limitations are static measurements for an unlimited period of time. In order to measure chamber pressure using transducers, cartridge case should have a specific clearance of 0.2 mm from transducer face to allow dynamic measurement. Figure 5 shows an actual installation of piezo sensor to measure chamber pressure.

Piezoelectric transducers are also used to measure cartridge mouth pressure and are mainly used in ammunition acceptance and testing. The installation scheme is far simpler as compared to chamber pressure because no work has to be done on to the ammunition. No drilling is required for such measurement with an important parameter of peak measurement as outcome.

Case mouth measuring scheme is presented in Figure 6. Case mouth measuring method puts a heavy load on transducers and hence shortens its life due to the fact that there is huge pressure jump at case mouth which could potentially damage transducer's diaphragm [17].

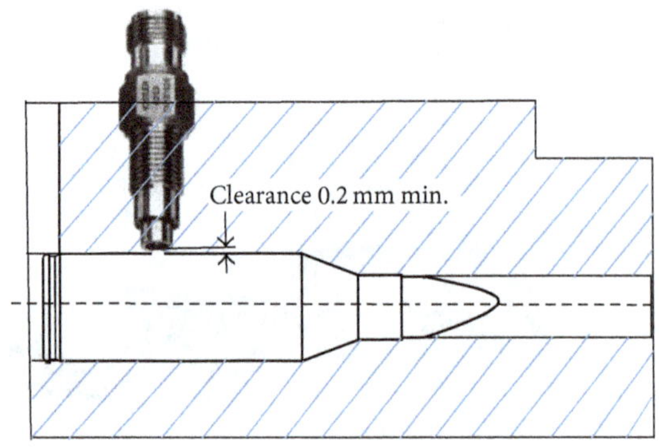

FIGURE 5: Piezo sensor for cartridge chamber pressure [17].

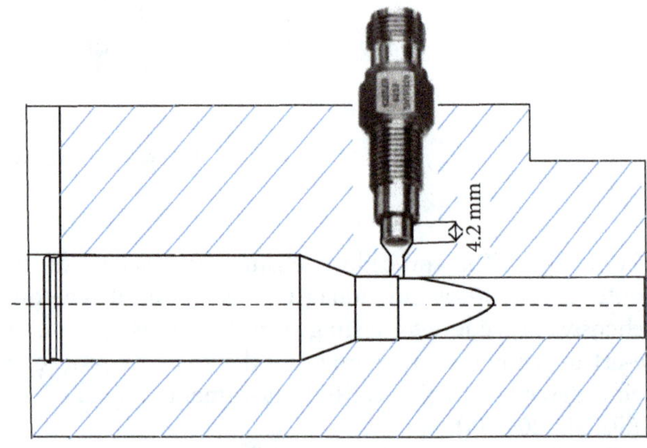

FIGURE 6: Piezo sensor for cartridge case mouth pressure.

3. The System Design

In normal pressure proof barrels, barrel has one drilled hole for chamber pressure measurement (either for copper crusher or for piezo transducer), whereas, in this study, a comparison of pressures measured by copper crusher as well as piezo is done at the same location using same barrel keeping all other parameters constant. Therefore, a dedicated mechanism is

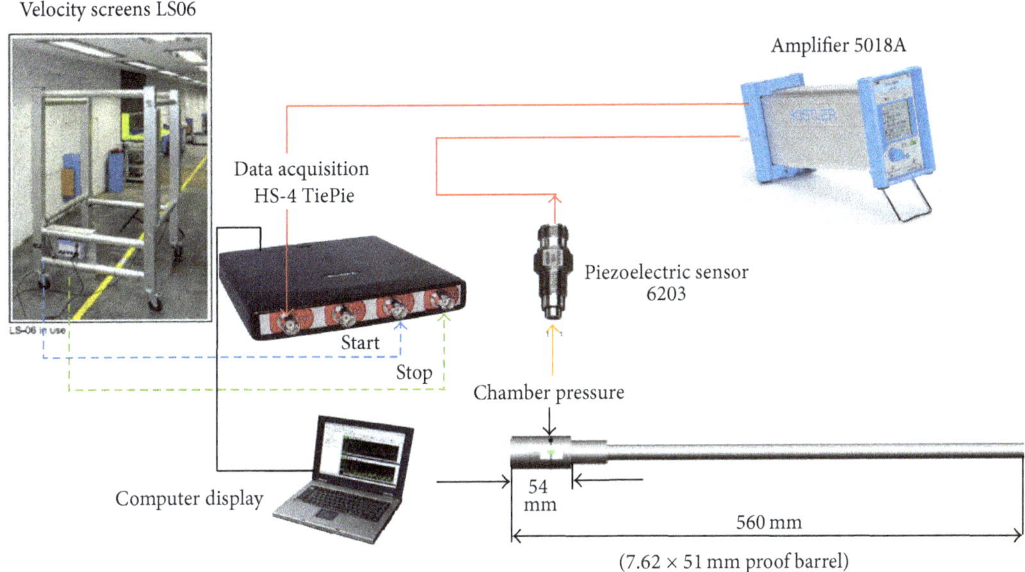

FIGURE 7: Schematic of chamber pressure measurement mechanism.

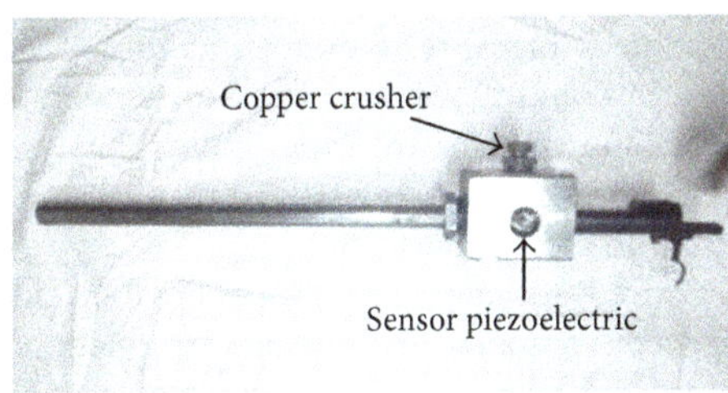

FIGURE 8: Barrel with 2 holes (for copper crusher and piezo pressure).

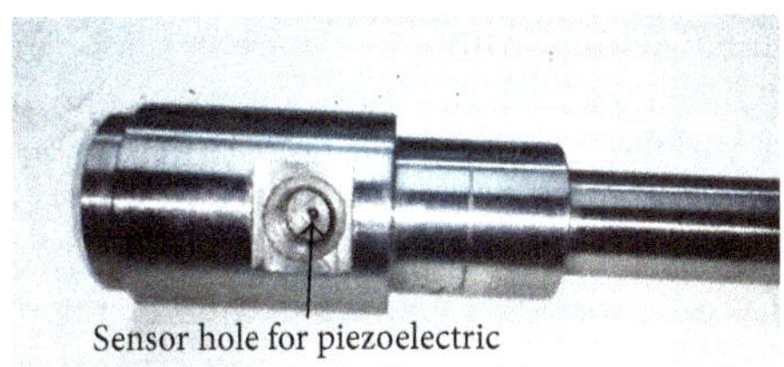

FIGURE 9: Barrel end with 1 hole for case mouth pressure.

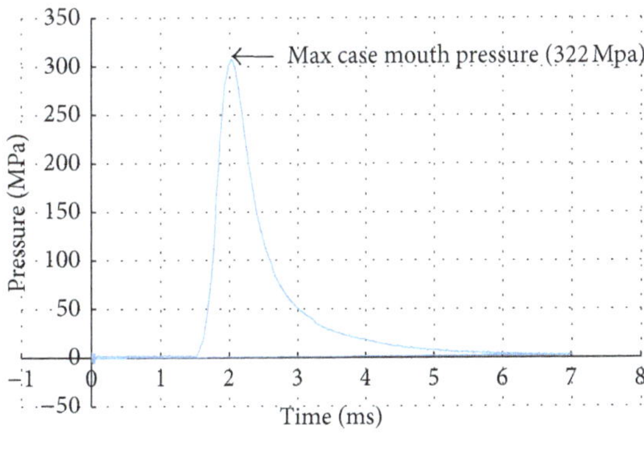

FIGURE 10: Pressure-time curve.

designed having provision for pressure measurement by copper crusher as well as piezo at the same distance on barrel. The salient features of the system are shown in Figure 7.

In the experimental set-up two (02) NATO proof barrels for the proofing of 7.62 × 51 mm ammo with 560 mm barrel length were manufactured: one barrel with two holes at 25 mm from chamber end (one hole for the copper crusher and the other for the piezoelectric transducer (Kistler 6203) installed at 90° to each other on proof barrel for pressure measurements as shown in Figure 8) and second barrel with single hole at the case mouth (54 mm from chamber end) as shown in Figure 9.

Figure 7 shows chamber pressure measurement mechanism where one piezoelectric transducer (Kistler 6203) was installed on proof barrel for pressure measurements and one charge amplifier (Kistler 5018A) was connected with piezoelectric transducers. The output from the sensor is fed into charge amplifier which is then passed to data acquisition system (TiePie Handy Scope HS-4, a four-channel input). Then two Light Gates (LS-06) are placed at specific distance from the muzzle end of proof barrel. When the bullet is fired two outputs signals (start and stop) are fed to the data acquisition system for velocity measurement which is used as reference in this experimental study. A data acquisition system (Handyscope HS-4) system simultaneously captures 3 signals (1 × pressure signal and 2 × velocity signals) and automatically saves raw data in specified folder as pressure-time curve as shown in Figure 10.

The customized Ballistic Data Measurement System (BDMS) software (developed in C# programing language

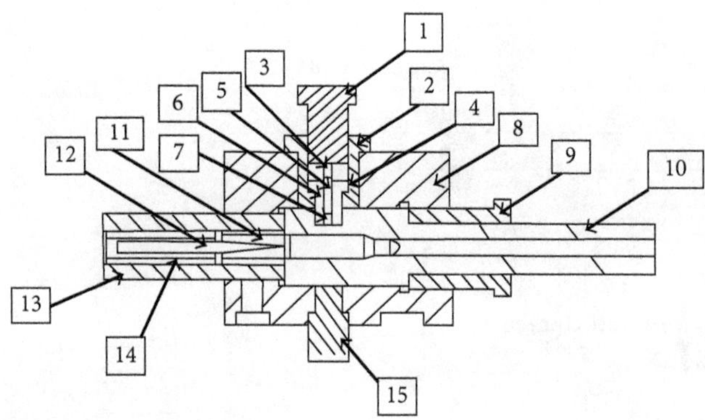

FIGURE 11: Mechanism for piezo and copper crusher at chamber (25 mm from chamber end) [15].

with MySQL databases) automatically detects saved raw data from a specific folder and calculates the following ballistics parameters:

(i) Case mouth pressure/chamber pressure (MPa)

(ii) Velocity (ft./s)

(iii) Energy (J)

(iv) The BDMS software automatically calculates corrected values of pressure based on the test sample and the standard rounds. The test report in PDF/MS Word and excel format is generated automatically and can be saved in a specified folder.

3.1. System Methodology. Once a round is fired for pressure measurement, the process starts with the burning of primer/ignition of propellant; hence burning of propellant generates gas pressure which is sensed by the piezoelectric transducer (Kistler 6203) as a charge signal. This signal is fed to charge amplifier (Kistler 5018A) to amplify the signal and for conversion into voltage. This voltage is fed to a smart and highly professional data accusation system and oscilloscope (HS-4 Handy scope) hence read by the computer in readable form. The setting of charge amplifier is 5000 bar/volts. So, if a signal of 0.644 v is captured, it gives pressure of 322 MPa (0.644 volts × 5000 bar/volts = 3220 bar = 322 MPa). Figure 11 shows an experimental set-up for piezoelectric and copper crusher.

4. Experimental Analysis

The results for various ammunition test fired for chamber pressure have been compared with two different experimental set-ups. The one involves a barrel having 2 holes, that is, one for copper crusher and another for piezoelectric sensor. The second manufactured barrel involves piezo only and can only measure ten rounds in each of two different categories of ammunition fired.

Experimentation was performed in order to determine material parameters and deformation behavior of copper crushers. Thereafter, extensive finite element analysis has been carried out using ANSYS after utilizing material property (elastic modulus) and density of crusher gauges. A set of tests (10 rounds in each test) were conducted by using the following parameters:

(i) Ammunitions of two different manufacturers:

(a) NATO specified standard ball ammunition labelled as NATO having propellant weight of 45.5 grams,

(b) nonstandard ammunition with cylindrical propellant labelled as NSTD also having propellant weight of 45.5 grams

(ii) Cylindrical copper crushers (4 × 6 mm)

(iii) Pressure measured against deformation of copper crusher.

Table 2 shows three different sets of experimental results as mean values for chamber pressure using 10 rounds in each test of standard NATO and non-NATO standard ammunition (ammo). The results are obtained through copper crusher using two different categories of ammunition and hence, marginal difference in values is observed.

As per NATO requirements, mean peak chamber pressure of any type of pressure shall not exceed 50,000 pounds per square inch (<345 MPa) (NATO 1997) and for case mouth mean peak pressure shall not exceed 380 MPa when ammunition is conditioned and fired at temperature 21°C (NATO 1997). On the basis of experimental analysis, deformations behavior was evaluated against the pressure given by manufacturer on a conversion table.

The pressure values noted by copper crushers during real time firing were compared with piezo transducer and it was analyzed that the pressure values observed through piezo transducer gave more precise values vis-à-vis specified (true) pressure. During the analysis, a number of firing results were obtained with the help of two different standard ammunitions fired ten rounds in each ammo test series.

The similar setting within the same barrel for piezo sensor located at the same distance from chamber end produced somewhat variable results as compared to copper crusher results. The difference in chamber pressure in megapascal (MPa) ranges from minimum 5% for NSTD ammo to maximum 9.58% for NSTD ammo. Similarly, the difference in chamber pressure for piezo and copper crusher for NATO ammo ranges from 3.64% to 7.15%.

The deformation achieved via copper crusher varies nonlinearly as per theory and the same becomes more and more evident for higher values of pressure. The results for current research however do not show a clear difference between deformations which is conclusive enough for nonlinear behavior of copper material. However, clear higher values for chamber pressure have been noted for piezo sensor as compared to copper crusher for both types of fired ammos. Mean values of fired test series are given in Table 3.

The data obtained for chamber pressure through copper crusher and piezo sensor can also be critically analyzed for relative spread in peak pressure values as described by Coghe et al. [3]. The statistical equivalence of measurement techniques can also be easily compared via calculating relative spread in peak pressures at gun chamber.

Table 4 depicts an interesting result regarding peak pressure at chamber using copper crusher. The values of standard deviation by using standard NATO ammunition show a clear difference of measured pressure compared to nonstandard local ammunition. The values can be taken as an indication of unreliability of copper crusher measurement technique which is clearly affected by type of ammunition as well as other multiple factors. The values of relative spread also indicate same trend of copper crusher methodology depending on type of ammunition. The consistent trend in stable values of chamber pressure is clearly visible for standard NATO ammunition.

Table 5 shows chamber pressure measured at 25 mm using piezo sensor. The values of standard deviation and relative spread both strongly stress previous observation of stable and repeatable values for both ammunitions using piezo sensor. The lower trend of standard deviation irrespective of ammunition type is a strong endorsement of our initial assessment regarding accuracy of piezo sensor compared to copper crusher method.

The spread and standard deviation values from aforementioned tables for both copper crusher and piezo sensor clearly indicate stable and marginally repeatable values for piezo sensor due to lower values of standard deviation (SD) and relative spread.

After having comparative analysis of pressure results through copper crusher and piezoelectric transducer at 25 mm (NATO specified for chamber pressure vide drawing 6 in NATO MANUAL MOPI as shown in Figure 7), having observed that piezo pressure is more accurate, the case mouth pressure (at 54 mm from chamber end vide drawing b-35 in NATO MANUAL MOPI as shown in Figure 3) as specified in NATO EPVAT system (MOPI NATO 7.62 mm ammunition) is even more accurate and closer to true pressure. A number of firing results were obtained with the help of two different standard ammunitions' firing of ten rounds in each ammo test series. Mean values of fired test series are shown in Table 6.

The results for case mouth pressure at 54 mm from chamber end indicate a variable difference in pressure as compared to chamber pressure which ranges from 3.1% to maximum of 15.3% higher values of case mouth pressure. The higher values of piezo pressure for case mouth indicate a clear pattern of higher values for NATO ammo. The corresponding values of case mouth pressure for second category of NSTD ammo indicates lower values of measured pressure.

5. Simulation Tools for Crusher Material Modeling

In order to obtain accurate finite element modeling results, it is essential to implement material models in conformance with the actual copper behavior. As the exact material characteristics of copper crushers were not known, an experimental study was performed in order to obtain necessary parameters for a suitable material model. As the copper crusher application does not involve working under extremely high pressures, the material characterization was limited to the determination of a strength model [3]. Extensive use of plastic stress models in simulation has been found under high deformation and elevated temperature conditions [17]. Two different strength models were considered for the copper material of the crushers.

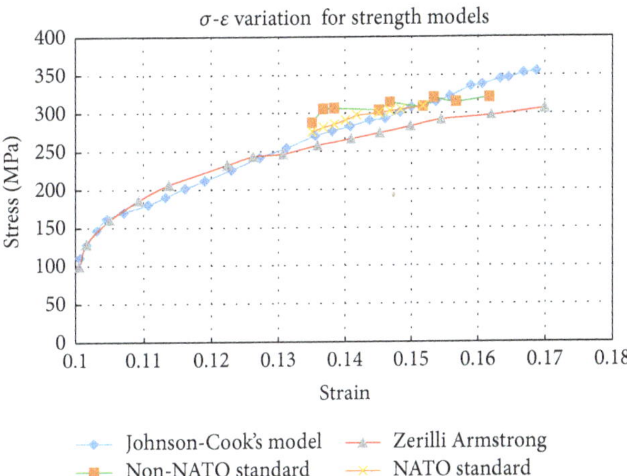

FIGURE 12: Curve fitting of experimental and model parameters.

The Zerilli-Armstrong model [5, 18] for face-centered cubic materials (FCC) like copper was given by

$$Y = A + C_0 \sqrt{\epsilon} e^{-C_1 + C_2 \ln \dot{\varepsilon}}. \tag{1}$$

The Johnson-Cook model [19] was developed to model the mechanical behavior of body-centered cubic materials (BCC), for example, steel, and declared as a very adaptable model which gives satisfactory results. Its mathematical expression is given by [5, 19] in

$$\sigma = \left(A + B\varepsilon_p^n\right)\left(1 + C \ln \frac{\dot{\varepsilon}}{\dot{\varepsilon}_0}\right)\left(1 - \frac{T - T_0}{T_m - T_0}^m\right), \tag{2}$$

where A, B, C, n, and m are the material parameters to be determined and $\dot{\varepsilon}$ is a reference strain rate (chosen as 1 per sec). T_m and T_0 are the melt temperature of the considered material and a reference temperature, respectively. Considering the melting temperature of pure copper, T_m was calculated as 1357 K, whereas, T_0 at room temperature was chosen as (294 K).

Figure 12 indicates the variation of σ-ε for copper which is nonlinear material. Analytical strength models Johnson-Cook and Zerilli-Armstrong have been compared with the results of real time copper crusher deformation. The results for both ammunitions (NATO standard and non-NATO standard) while using copper crusher show an agreement with Johnson-Cook model and hence, the same was used to get simulation results.

ANSYS workbench is used to carry out finite element analysis. The model is analyzed with real boundary conditions using Johnson-Cook model for copper material and a highly refined mesh. The pressure is applied mainly on the copper crusher to optimize the deformation with varying pressures. A fixture showing placement of copper crusher and piezoelectric transducer at chamber (25 mm from chamber end) is shown in Figures 13 and 11. Table 1 gives the components detail of Figure 11.

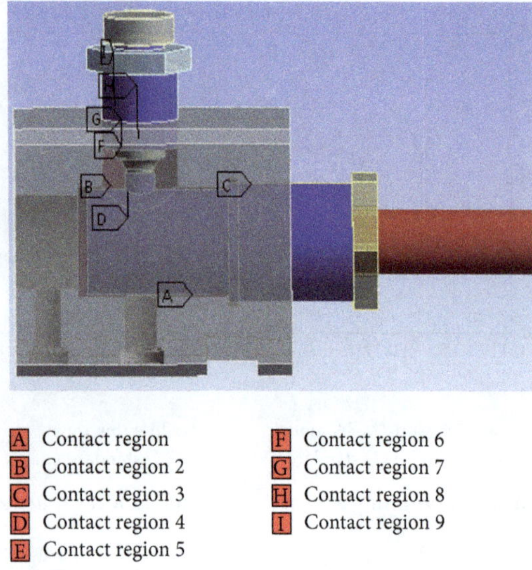

FIGURE 13: NATO proof barrel with copper crusher fixture.

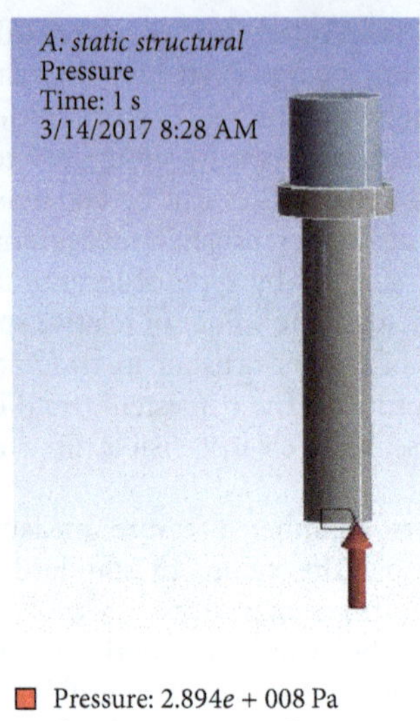

FIGURE 15: Applied pressure on copper crusher.

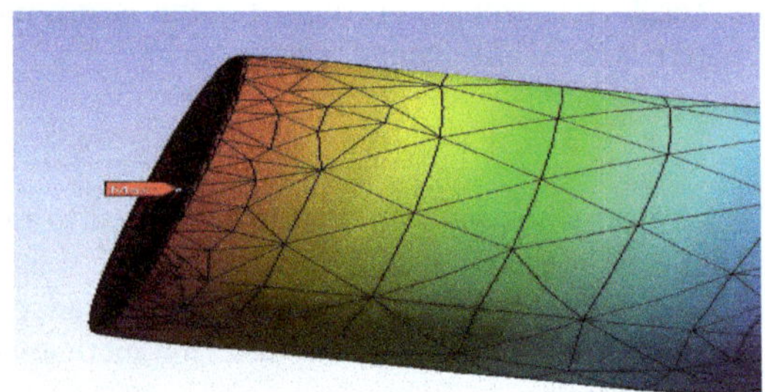

FIGURE 14: Refinement where max pressure occurs.

TABLE 1: Detail of labels used in Figure 11.

Label	Details
1	Thumb screw
2	Hollow nut
3	Copper crusher
4	Star washer
5	Piston
6	Piston guide
7	Brass sealing washer
8	Housing
9	Barrel locking sleeve
10	Barrel
11	Bolt head
12	Firing bolt
13	Bolt carrier
14	Firing pin
15	Barrel tightening nut

5.1. Meshing. To perform finite element analysis, our model is meshed such that the solution converges with more accuracy and real time results as shown in Figure 14. Refinement meshing is done on the regions of copper crusher chamber side to obtain and analyze the exact position from where the crusher deforms up to maximum limit.

5.2. Boundary Conditions for Copper Crusher. The material properties of copper are changed since copper crusher undergoes series of chemical processes from copper rod till miniature sized (4 mm × 6 mm) cylinders (copper rod, machining, blank, annealing, pickling, passivation, compression, and copper crushers finished product). Modulus of elasticity has been calculated through inverse numerical technique. The parameters used in the finite element analysis are as follows:

(i) Density = 8960 kg/m3 (constant value)

(ii) Modulus of elasticity = 1326 MPa (constant value)

(iii) Pressure = variable.

After setting up the constraints with certain mechanical material properties of pure copper crusher's density, ultimate tensile strength, yield strength, compressive strength, and modulus of elasticity. The model is simulated critically in 5 different scenarios and their corresponding deformation effects are shown in Figures 15–22.

5.2.1. Case 1

(i) Density = 8960 kg/m3

(ii) Modulus of elasticity = 1326 MPa

(iii) Pressure = 289.4 MPa

5.2.2. Case 2

(i) Density = 8960 kg/m3

(ii) Modulus of elasticity = 1326 MPa

(iii) Pressure = 296.9 MPa

TABLE 2: Experimental analysis of pressure at 25 mm copper crushers test.

Ammo test #	Mean of 10 fired rounds			
	Length of copper crusher		Deformation [mm]	Pressure [MPa]
	Before fire [mm]	After fire [mm]		
1-NATO	06	5.142	0.858	297.69
2-NATO		5.161	0.839	293.6
3-NATO		5.15	0.85	296.9
1-NSTD		5.148	0.852	296.24
2-NSTD		5.141	0.859	297.77
3-NSTD		5.126	0.874	300.93

TABLE 3: Comparative analysis of piezoelectric transducer and copper pressure at 25 mm.

Test ammo #	Mean of 10 fired rounds				
	Piezo pressure [MPa]	Copper crusher		Difference Piezo – copper	
		Deformation [mm]	Pressure [MPa]	[MPa]	%* A
1-NATO	317.28	0.858	297.69	19.59	6.17
2-NATO	316.2	0.839	293.6	22.6	7.15
3-NATO	308.125	0.85	296.9	11.22	3.64
1-NSTD	327.675	0.852	296.24	31.39	9.58
2-NSTD	319.305	0.859	297.77	21.53	6.74
3-NSTD	316.775	0.874	300.93	15.84	5.00

* A: percentage increased in piezo pressure with reference to copper crusher pressure.

TABLE 4: Standard deviation and relative spread using copper crusher method at 25 mm.

Test ammo #	Mean of 10 fired rounds				
	Chamber pressure [MPa]	Standard deviation [MPa]	Max value [MPa]	Min value [MPa]	Relative spread (%)
1-NATO	297.69	6.34	308.8	291.5	5.93
2-NATO	293.6	6.39	302.4	287.2	5.29
3-NATO	296.9	5.34	304.5	287.2	6.02
1-NSTD	296.24	9.80	321.2	287.2	11.83
2-NSTD	297.77	8.32	310.8	287.2	8.21
3-NSTD	300.93	11.81	321.2	287.2	11.83

TABLE 5: Standard deviation and relative spread using piezo pressure method at 25 mm.

Test ammo #	Mean of 10 fired rounds				
	Piezo pressure [MPa]	Standard deviation [MPa]	Max value [MPa]	Min value [MPa]	Relative spread (%)
1-NATO	317.28	4.73	326.8	310.95	5.09
2-NATO	316.2	3.75	322.7	309.55	4.24
3-NATO	308.125	4.54	317.8	304.3	4.43
1-NSTD	327.675	4.99	336.95	320.9	5.00
2-NSTD	319.305	5.50	326.25	307.5	6.09
3-NSTD	316.775	10.63	329.7	297.5	10.82

TABLE 6: Chamber pressure versus case mouth pressure.

Test ammo	Mean of 10 fired rounds (chamber versus case mouth)		
	Piezo pressure at		% increase in *CMP w.r.t. chamber [%]
	25 mm (chamber) [MPa]	54 mm (case mouth) [MPa]	
1-NATO	317.28	356.105	11.02
2-NATO	316.2	364.59	15.3
3-NATO	308.125	356.28	11.27
1-NSTD	327.675	337.83	3.1
2-NSTD	319.305	332.79	3.6
3-NSTD	316.775	335.53	4.71

*CMP: case mouth pressure.

TABLE 7: Cost benefit analysis (CBA).

(a)	Number of copper crushers used in proof firing during year 2012/13 (excluding production rejection)	15,400 crushers
(b)	Number of copper crushers used in proof firing during year 2013/14 (excluding production rejection)	13,582 crushers
(c)	Number of copper crushers used in proof firing during years 2012/13 & 2013/14 (excluding production rejection)	28,982 crushers
(d)	Cost of one radial copper crusher (4 × 6 mm)	2.96 USD
(e)	*Total cost of crushers used in two years*	*85,786.72 USD*
(f)	*Cost of one piezoelectric transducer*	*9,548 USD*
(g)	*Cost comparison between piezoelectric and copper crushers in two years*	*1 : 8.98424*
(h)	Saving (difference of cost in two years' proofing if piezo method is adopted)	76,241 USD
(i)	*Approx. average saving per year*	*≈38,120 USD*

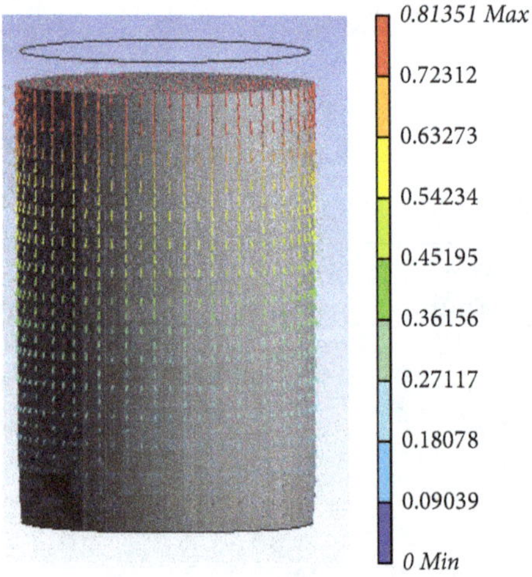

FIGURE 16: Linear deformation obtained.

5.2.3. Case 3

(i) Density = 8960 kg/m3

(ii) Modulus of elasticity = 1326 MPa

(iii) Pressure = 300.2 MPa

5.2.4. Case 4

(i) Density = 8960 kg/m3

(ii) Modulus of elasticity = 1326 MPa

(iii) Pressure = 304.5 MPa

5.2.5. Case 5

(i) Density = 8960 kg/m3

(ii) Modulus of elasticity = 1326 MPa

(iii) Pressure = 310.8 MPa

6. Cost Benefit Analysis

Copper crusher is used one time only as it is deformed during firing and has to be replaced for every subsequent fire. However, in case of piezoelectric transducer, a sensor once installed is used for a lifetime. In order to have a comparative analysis between copper crushers and piezoelectric transducer, cost of copper crushers used in two years during proof firing of 7.62 × 51 mm ammunition is compared with one piezoelectric transducer (considering warranty period of two years). The cost benefit analysis carried out for actual consumption of copper crushers during the last two years with piezoelectric transducer is shown in Table 7.

7. Results Validation through Simulations

In order to validate pressure values obtained from copper crushers through series of real time firing of ammunition, the pressure obtained through deformation of copper crusher was taken as reference and validated through ANSYS workbench. Comparative analysis on experimental and simulation results was carried out to validate experimental data by using a sample size of 10% test round results of a batch of 50 rounds. The same shows an encouraging agreement between values as the deformation achieved in simulation results is

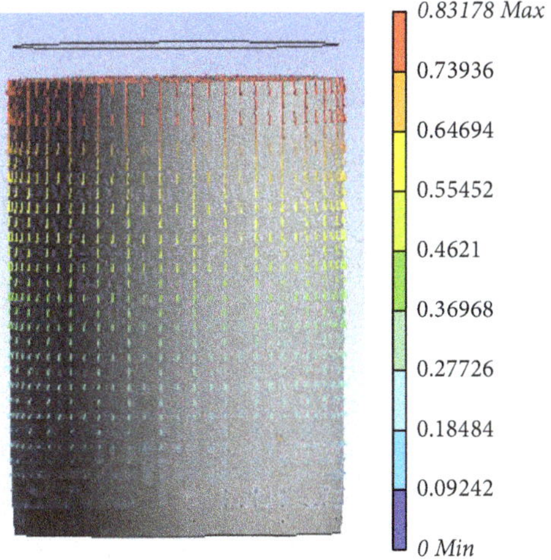

FIGURE 17: Linear deformation obtained.

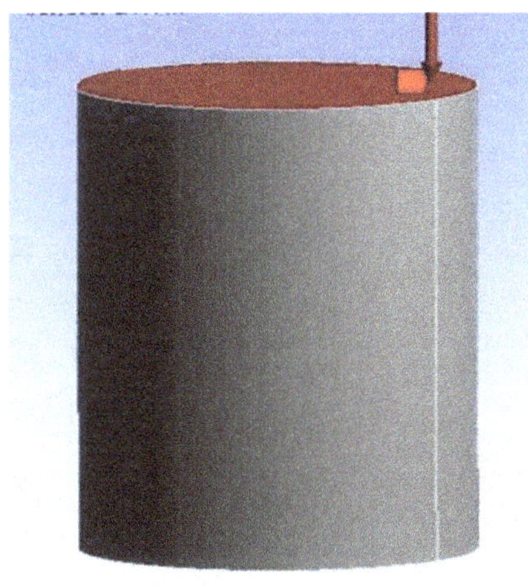

FIGURE 18: Applied pressure on crusher.

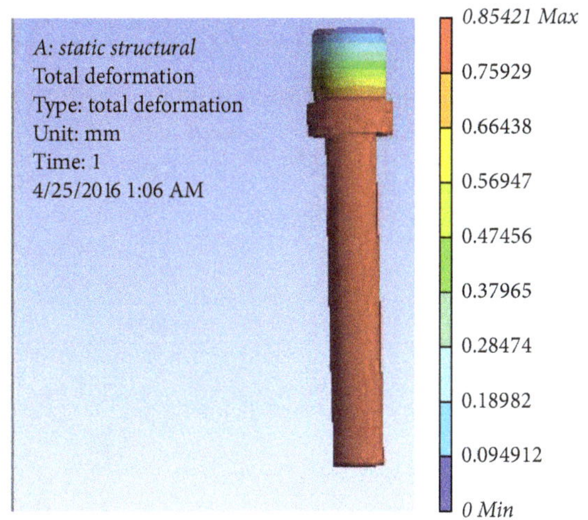

FIGURE 19: Linear deformation obtained.

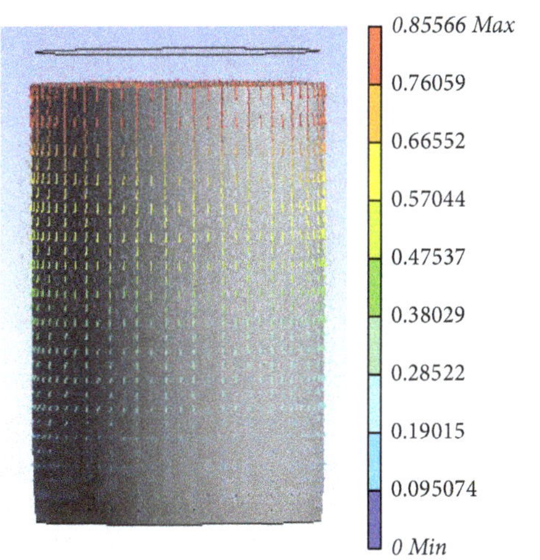

FIGURE 20: Linear deformation obtained.

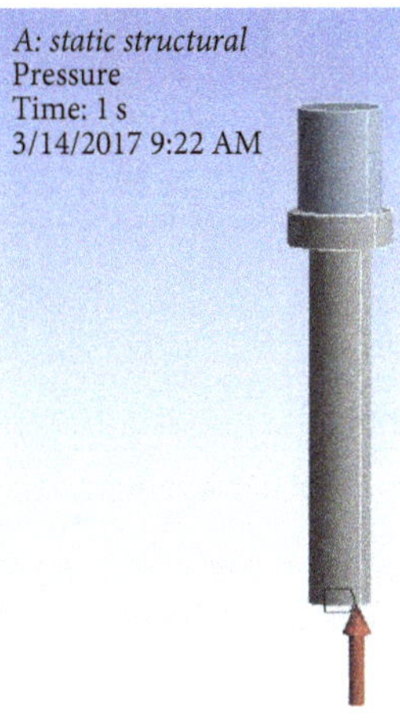

FIGURE 21: Applied pressure on crusher.

approximately the same as was actually achieved in proof firing via copper crusher. The values of peak pressure as converted from deformation of copper material using Terage tables were applied as load in simulation to check whether the same deformation is achieved through simulation. A close agreement of both deformation values is given in Figure 23.

Table 8 indicates the same comparative data in tabular form. The minor differences between two results are possibly due to the fact that few material properties were either assumed or unknown due to limited material property data from supplier.

8. Discussion

8.1. Discussion on Chamber Pressure Measurement. The comparison of results for non-NATO standard (NSTD) ammunition of local origin unveils clear flaws of copper crusher methodology in measuring chamber pressure. Figure 24

TABLE 8: Pressure results validation on ANSYS ® workbench.

Real time firing results obtained by copper crushers (original length = 6.0 mm)		Simulation results on copper crusher (6.0 mm) using ANSYS	
Deformation noted [mm]	Pressure obtained [MPa]	Applied pressure [MPa]	Deformation obtained [mm]
0.82	289.4	289.4	0.8135
0.85	296.9	296.9	0.8318
0.87	300.2	300.2	0.8542
0.89	304.5	304.5	0.8556
0.92	310.8	310.8	0.8734

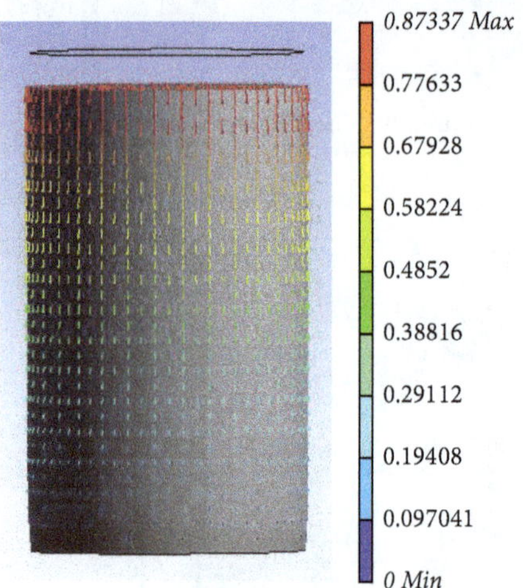

FIGURE 22: Linear deformation obtained.

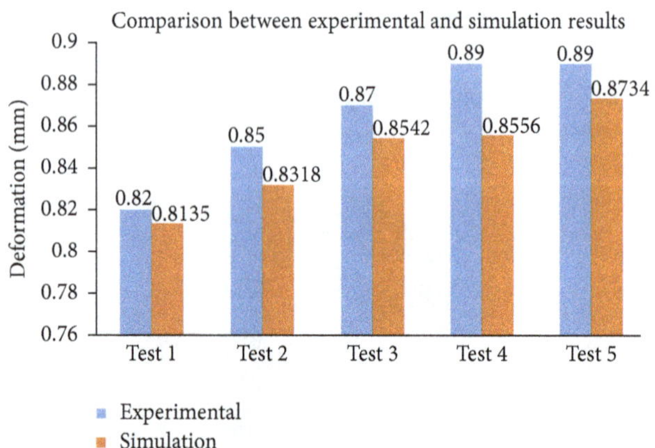

FIGURE 23: Comparison of experimental and simulation deformations.

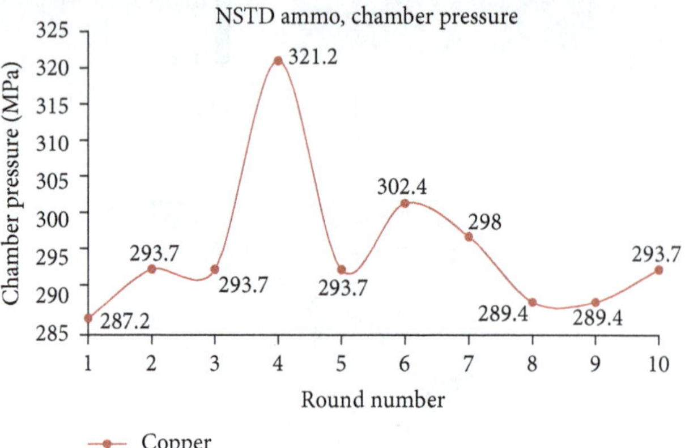

FIGURE 24: Chamber pressure measured with copper crusher using NSTD ammunition.

highlights a major difference in pressure between different NSTD rounds fired to measure chamber pressure via copper crusher and thus endorsing nonsuitability of method for sensitive applications. The chamber pressure results for various rounds vary from 287.2 MPa to highest 321.2 MPa and thus an erratic nature of more than 10% for NSTD ammunition is recorded. It could be further stated from test to test as there are numerous factors which come into play when using copper crusher as measurement technique.

The comparison between copper crusher and piezo sensor for chamber pressure values indicates another fact that copper crusher values for the same ammunition and simultaneous measurement with piezo sensor are not only erratic in nature but also on the lower side which validates our initial assumption of nonsuitability of copper crusher measurements. The comparison between copper crusher and piezo sensor for chamber pressure values indicates another fact that copper crusher values for the same ammunition and simultaneous measurement with piezo sensor are not only erratic in nature but also on the lower side which validates our initial assumption of nonsuitability of copper crusher measurements. The same has already been reported by Coghe et al. [3] where lower peak pressures have been observed for different caliber ammunition using copper crusher. The corresponding values from piezo sensor however are reported to be higher for the same ammunition. The same is on lower side as compared to true pressure due to presumable nonlinear deformation of copper material. Hence, the results obtained for nonstandard ammo clearly indicate a close agreement with existing observations that copper crusher technique does not present true barrel pressure. The results by piezo sensor on the contrary indicate a more smooth pattern of chamber pressure from round to round and erratic nature of results is nonevident as it was for copper crusher

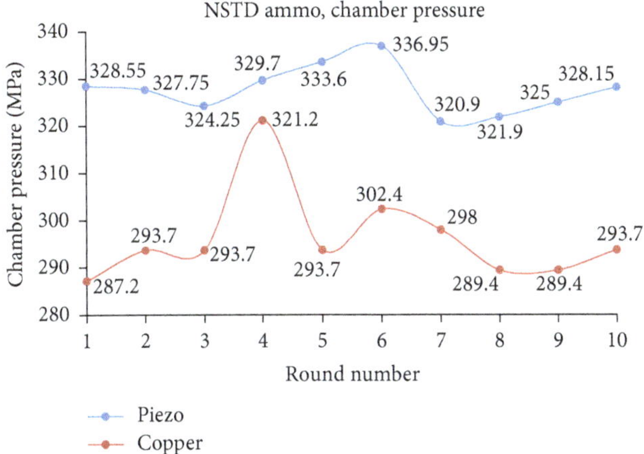

FIGURE 25: Comparison of chamber pressure measured with copper crusher and piezo sensor simultaneously using nonstandard ammunition.

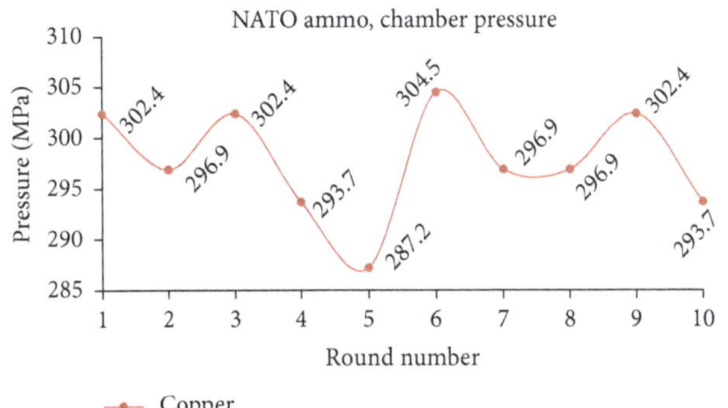

FIGURE 26: Chamber pressure measured with copper crusher using standard NATO ammunition.

results. Figure 25 presents a broad view of first series of proof fire using nonstandard ammunition to measure chamber pressure by both piezo sensor and copper crusher within the same barrel.

The results when extended to change of ammunition, from NSTD to NATO ammo, did not indicate any difference with regard to results as compared to nonstandard ammunition. The copper crusher method produced the same erratic nature of results for chamber pressure irrespective of ammunition. The copper crusher method produced the same erratic nature of results for chamber pressure irrespective of ammo type which is in agreement with existing research outlines by Kuokkala et al. [11]. However, for standard NATO ammo, comparatively higher values of chamber pressure have been recorded by copper crusher. The results are consistent from test series 1 to test series 3 for standard NATO ammo.

The values of chamber pressure by copper crusher using NATO ammo are presented in Figure 26. In order to judge the effect of ammunition type on chamber pressure measurement technique and the chamber pressure results for standard ammunition using both copper crusher and piezo sensor, an endorsement of previous finding is found; that is, piezo sensor method records higher pressure values as compared to copper

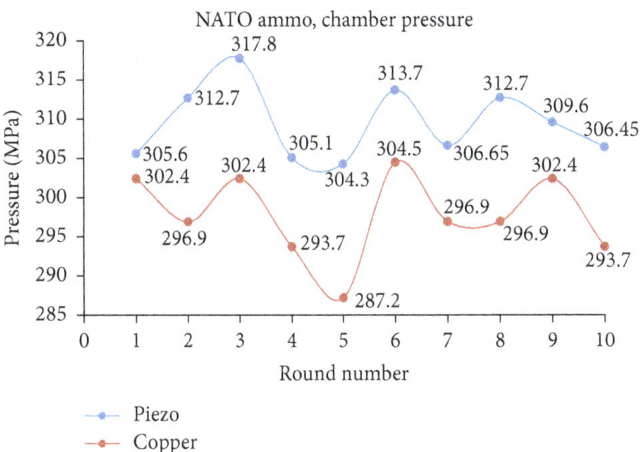

FIGURE 27: Chamber pressure measured with copper crusher and piezo sensor simultaneously using standard NATO ammunition.

crusher. The copper crusher on the other hand presents lower values of chamber pressure for the same experiment and the same standard ammo. It coincides with existing research output by Kuokkala et al. [11] who claim to have observed beyond 20% higher values in piezo sensor compared to copper crusher pressure values. The result remains consistent for three series of fires using NATO ammunition. The results obtained for standard ammunition once again stress the fact that copper crusher irrespective of ammunition type does not present true chamber pressure. Figure 27 presents a comparison of two measurement techniques installed simultaneously in a barrel using standard NATO rounds.

8.2. Discussion on Case Mouth Pressure Measurement. Case mouth pressure was measured using piezo sensor only as installation of copper crusher to measure the same is not possible. Furthermore, piezo sensor method for case mouth pressure presents even a simpler installation as compared to chamber pressure as no drilling in this case is required. The following results have been categorized as NSTD ammunition and NATO standard ammunition using piezo sensor.

The NSTD ammunition results for case mouth pressure as measured in three different test series indicate consistent pressure values from series 1 to series 3. Piezo sensor case mouth pressure values indicate smoother pattern even with NSTD ammunition. Figure 28 indicates different test series results for piezo sensor installed at case mouth position.

Case mouth pressure measurement using standard NATO ammunition with piezo sensor is also investigated for effect of ammunition type. The results present close agreement with previous discussions about piezo sensor method as reliable and consistent as compared to copper crusher method irrespective of ammo type. The result however indicates a close consistency in result for case mouth pressure and somewhat marginally deviates from round to round. Figure 29 presents three test series of fires for standard NATO ammunition.

8.3. Discussion on Simulation Results. With the help of reverse numerical technique, chamber pressure obtained

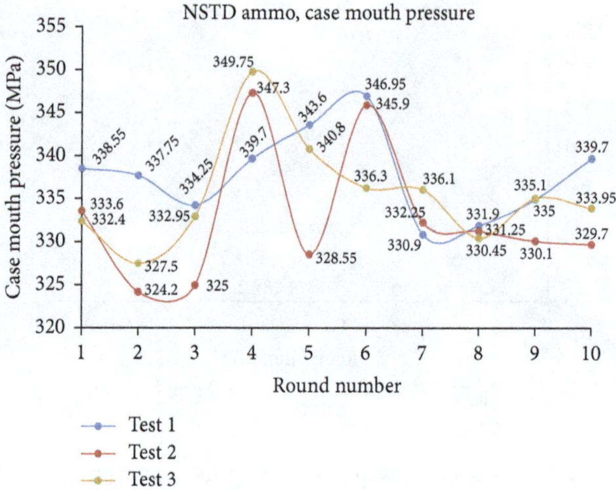

FIGURE 28: Case mouth pressure measured with piezo sensor using nonstandard ammunition.

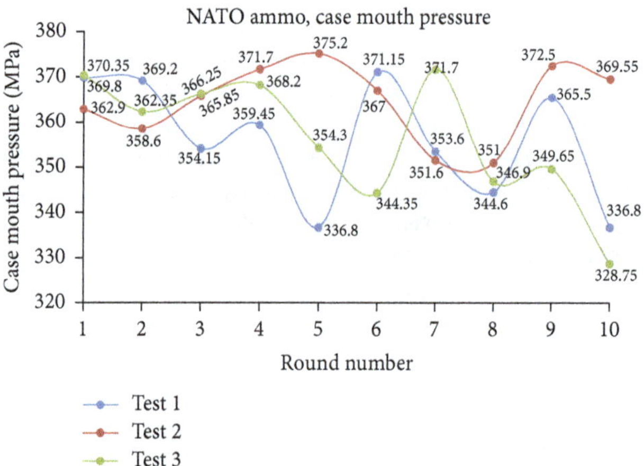

FIGURE 29: Case mouth pressure measured with piezo sensor using standard NATO ammunition.

through deformation of copper crusher was applied on copper crusher using ANSYS workbench to validate corresponding deformation, thus confirming results. On the basis of experimental and simulated results the following points are highlighted:

(i) The variation in pressure between piezoelectric sensor and copper crushers has been observed as 5–10%.

(ii) No significant difference in results of standard ammunition (NATO) and nonstandard (NSTD) ammunition is observed. Significant difference in CMP versus chamber pressure has been observed in case of NATO ball ammunition with 11–15% variation as compared to nonstandard ammunition.

(iii) It is a well-established fact that ANSYS is a very useful simulation tool for the validation of real time firing results (pressure values against deformation of copper crusher).

9. Conclusions

Based on experimental and numerical analysis, it is evident that the piezo pressure observed is higher (closer to true pressure) than copper crusher pressure at 25 mm distance from chamber end. Therefore, piezo pressure is more accurate and reliable to ensure effective ammunition proofing and due safety of weapon.

The results further reveal that the pressures recorded by piezo transducers are slightly higher compared to copper crusher pressure irrespective of the type of ammunition. The ammo of both categories revealed close agreement of persistent higher pressure values for piezo as compared to copper crusher. The same results have been validated by simulation and a close settlement between deformations is achieved through real time firing and simulation has been found, validating experimental results.

The results in current work indicate that copper crusher pressure and piezo pressure have close agreement at lower pressures and the same is not true for higher values of pressure due to nonlinear deformation of crusher material. Similarly, the pressure observed through EPVAT system at case mouth is even more accurate and closer to true pressure. Hence, the proofing of ammunition with NATO EPVAT system should be adopted instead of obsolete and unreliable pressure measuring by copper crusher method.

The confirmation of reliability and repeatability of results for piezo sensor method is also evident through standard deviation and relative spread values. Close agreement between values obtained for chamber pressure through piezo sensor with lower values of standard deviation and relative spread is conclusive enough for initial argument of piezo sensor being superior compared to copper crusher method.

The cost benefit analysis for a year of proofing clearly indicates a wide difference of financial impact if piezo sensor instead of conventional copper crusher method is adopted. The accuracy, repeatability, and reliability of ballistic results through EPVAT system supersede conventional and high variable method of ballistic measurement through copper crusher.

10. Recommendations

The analysis shows that, due to accuracy, cost, and quick results, piezoelectric transducers have an edge over copper crushers. This change in methodology will increase the acceptability of the product worldwide thus bringing it to the NATO standard. The ease of installation and automatic generation of ballistic parameters unlike copper crusher provides ammunition manufacturers and proofing personnel with an added advantage at a considerably lower cost. The accommodation of piezoelectric sensor method of ballistic measurement will enhance our capability to produce international quality standard ammunition and weapons. It is highly recommended that standard NATO ammunition should be manufactured and its testing be done by using EPVAT method of proof testing to ensure international quality, capturing customer satisfaction and thus reaching new markets.

Nomenclature

Abbreviations

A:	Yield strength, MPa
AVL:	Copper manufacturer, AVL Technology Inc.
B:	Hardening modulus, MPa
BDMS:	Ballistic Data Measurement System
C:	Strain rate sensitivity coefficient
CIP:	Permanent International Commission
CMP:	Case mouth pressure
C_0, C_2, C_3, C_4:	The material parameters
EPVAT:	Electronic Pressure, Velocity, and Action Time
M:	Thermal softening coefficient
MOPI:	Manual of Proof and Inspection
N:	Material constant
NATO:	North Atlantic Treaty Organization
NSTD:	Non-NATO standard
SAAMI:	Sporting Arms and Ammunition Manufacturers' Institute
T:	Temperature
US:	United States
USD:	Dollars (United States of America)
w.r.t.:	With reference to.

Units

ft./s:	Feet per second
g/cm3:	Gram per centimeter cube
J:	Joules
Mm:	Millimeters
MPa:	Megapascal.

Symbols

Σ: The stress, MPa
p: The plastic strain
Ė: The strain rate.

Conflicts of Interest

The authors declare that they have no conflicts of interest.

References

[1] A. Wilkinson, *Conventional Ammunition in surplus - A reference guide, Chapter 6 - Stockpile Management: Surveillance and Proof*, Small Arms Survey, Graduate Institute of International Studies, Geneva, Switzerland, 2008, ISBN 2-8288-0092-X.

[2] W. H. Ming, P. D. Xing, Z. Yu, and W. Yan, "Research of miniature internal electronic piezo gauge," *Electronic Design Engineering*, vol. 17, no. 12, pp. 29–31, 2009.

[3] F. Coghe, L. Elkarous, P. van de Maat, A. Khadimallah, and M. Pirlotl, "Comparison of piezoelectric and crusher measurements of Chamber pressure with finite element modeling results," in *International Symposium on Ballistics*, Freiburg, Germany, 2013.

[4] J. Z. Xin'e Li, T. Ma, and P. Xu, "Measuring chamber pressure of different caliber artilleries using a capacitive pressure sensor," *Sensors and Materials*, vol. 23, no. 5, pp. 277–292, 2011, MYU Tokyo.

[5] AEP-23 ED. 2, *Pressure Measurement by Crusher Gauges NATO Approved Tests For Crusher Gauges*, NATO Standardization Agreemens, 2005.

[6] NATO, "NATO piezo gauge replacement programme," 1997, AC/225(LG/3-SG/1)D/3(Rev.).

[7] NATO, "Combination electronic pressure, velocity and action time (EPVAT) test procedure," 2004, PFP (NAAG-LG/3-SG/1)D1 Chapter 12.

[8] International Permanent Commission (C.I.P), "Edition Synthétique des decisions C.I.P. en vigueur," in *Bureau Permanent C.I.P*, Brussels, Belgium, 2011.

[9] SAAMI/ANSI standards, *Sporting Arms and Ammunition Manufacturers' Institute (SAAMI) Technical Committee*, 2017, http://www.saami.org/index.cfm.

[10] D. Crescini, D. Marioli, E. Sardini, and A. Taroni, "Low-cost thick film sensor based on piezoelectric effect for ballistic application," *Sensors and Actuators*, vol. 87, pp. 131–138, 2001.

[11] V.-T. Kuokkala, J. Ramo, and T. Vuoristo, *Calibration of Crusher Pressure Gauges by High Strain Rate Testing*, Tampere University of Technology, Institute of Materials Science, Tampere, Finland, P.O.Box 589, FIN-33101.

[12] Ballistic Standardization of Gun Ammunition Part 8, "Crushers and Crusher gauges," 1991, Defense Standard 13–36 (PART 8), Issue 1, 8 February 1991.

[13] L. Elkarous, F. Coghe, M. Pirlot, and J. C. Golinval, "Experimental techniques for ballistic pressure measurements and recent development in means of calibration," *Journal of Physics: Conference Series*, vol. 459, Article ID 012048, 2013.

[14] NATO, "Manual of Proof and Inspection Procedures for NATO 7.62 mm Ammunition MOPI," 1997, AC/225(LG/-SG/1)D/9, p. 28.

[15] 1997, Proof and Inspection Procedures (MOPI) for 7.62mm Ammunition NATO Manual No. AC/225 (LG/3-SG/1) D/9.

[16] North Atlantic Treaty Organization (NATO), "Standardization agreement for small arms ammunition (7.62 mm)," in *STANAG No. 2310*, edition 3, Annex C, 1976.

[17] Kistler Group, "Operating instruction for high pressure quartz transducers 6203," in *No. B3.013e*, edition 10.85, pp. 1–33.

[18] F. J. Zerilli and R. W. Armstrong, "Description of tantalum deformation behavior by dislocation mechanics based constitutive relations," *Journal of Applied Physics*, vol. 68, pp. 1580–1591, 1990.

[19] G. R. Johnson and W. H. Cook, "A constitutive model and data for metals subjected to large strains, high strain rates and high temperatures," in *Proceedings of the 7th International Symposium on Ballistics*, 547 pages, The Hague, The Netherlands, 1983.

11

Shock Tube as an Impulsive Application Device

Soumya Ranjan Nanda, Sumit Agarwal, Vinayak Kulkarni, and Niranjan Sahoo

Department of Mechanical Engineering, Indian Institute of Technology Guwahati, Guwahati 781 039, India

Correspondence should be addressed to Niranjan Sahoo; shock@iitg.ernet.in

Academic Editor: Paul Williams

Current investigations solely focus on application of an impulse facility in diverse area of high-speed aerodynamics and structural mechanics. Shock tube, the fundamental impulse facility, is specially designed and calibrated for present objectives. Force measurement experiments are performed on a hemispherical test model integrated with the stress wave force balance. Similar test model is considered for heat transfer measurements using coaxial thermocouple. Force and heat transfer experiments demonstrated that the strain gauge and thermocouple have lag time of 11.5 and 9 microseconds, respectively. Response time of these sensors in measuring the peak load is also measured successfully using shock tube facility. As an outcome, these sensors are found to be suitable for impulse testing. Lastly, the response of aluminum plates subjected to impulsive loading is analyzed by measuring the in-plane strain produced during deformation. Thus, possibility of forming tests in shock is also confirmed.

1. Introduction

Hypersonic or hypervelocity flows bear complexities due to the thin shear layer, high temperature, and inviscid-viscous interaction. Missiles or launch vehicles encounter this flow regime and hence need special attention in their design phase. A paradigm shift can be noticed in their development over a short span of time due to the impetus of need for space exploration and faster global transportation. Moreover, major challenge ahead the research community lies in accounting impulsive mechanical and thermal loading for safer and cheaper space flights. Hence it is desirable to expose such configurations to the harsh environment as a part of design practice. Flight tests provide one such option, but these tests involve risk and are less cost effective. Therefore, realistic approach, like simulating flight conditions in the laboratories, should be considered for a practical design. Measurement of force for understanding the fuel requirement, heat transfer measurement for designing thermal protection system, and material testing for evaluating the structural strength are the prime motives behind the ground-based testing. In view of this, there is an immediate need to explore for force measurement, heat transfer measurement, and structural health monitoring of associated test configurations in the ground facilities. Various research groups have developed different techniques for the measurement of these design parameters.

Force measurement plays a vital role in design of the aerodynamic vehicles from perspective of its stability and estimation of load on propulsion system. But these experiments, when conducted in impulse facilities (like shock tubes, shock tunnels, or expansion tubes), demand for sophisticated instruments due to very small test flow duration. Typically, two force balance techniques, namely, inertia based [1] and stiffness based [2], have been reported in literature for force measurement. Out of these two techniques, stiffness based balance represents the system more accurately. It relies on measurement of force from strain signal measured either by semiconductor strain gauges or by piezofilm. Simmons [3] and Tuttle et al. [4] considered use of this force balance in free-piston shock tunnel. Another technique of strain measurement with help of polymer piezoelectric film is proposed by Matsumoto et al. [5] where specimens are subjected to uniaxial compressive loading. Calibration of the force balance has been mentioned in these strain based force measurement techniques, but it is also desirable to characterize the strain gauge used for its response time and lag time to justify its applicability in the short duration testing. However, these fundamental requirements are not attended in the open literature.

Prediction of the heat flux becomes very essential parameter to minimize the effect of aerodynamic heating of high-speed vehicles. There are many transient temperature measurement devices available, out of which thin-film and coaxial surface junction thermocouple (CSJT) are very potential candidates to measure the surface heat flux. Rose [6] worked on the development of heat transfer gauge for measuring extremely high heat flux under the quasi-transient conditions occurring in shock tubes. Mohammed et al. and Anderson et al. [7, 8] employed shock tube for carrying out the dynamic calibration and in turn performance assessment of the fast response coaxial thermal sensors. Other researchers also have noticed that coaxial thermocouples are more robust and durable as compared to thin-film sensors when it comes to the application of measuring the heat flux in the harsh environment [9, 10] Moreover, thermal sensors are routinely used in shock tunnels, expansion tubes, free-piston shock tunnel, and gun tunnels [11]. Calibration of the thermal sensor followed by measurement of temperature time history in ground test facilities is the standard procedure for estimation of local heat transfer rate. But this inference largely depends upon the response time and lag time of the sensor due to which special sensors like coaxial thermocouples or thin-film sensors are preferred for these experiments. Hence, it is highly essential to characterize the thermal sensor and test for its robustness before experimentation with the space vehicle model.

Properties like strength, damage tolerance, and crack propagation are the key issues that have to be addressed for the materials generally used for aerospace vehicles. In recent years, shock tube has been used to study the dynamic response of thin metallic plates subjected to varying levels of shock loading [12]. Shock tube, driven with explosives, has also been employed as a device to create a shock loading [13]. Diaphragmless shock tube was considered to study the response of the copper, brass, and aluminum plates [14]. Kumar et al. [15] implemented shock tube as the experimental set-up to study the effect of aluminum panel curvature on blast response where in-plane strain is measured on the back face of the panels. Gray III and Huang [16] investigated the response due to single and repeated shock loading using explosives on aluminum plates. These literature-based findings provide an initiative for application of shock tube in the field of impact testing of materials. Therefore, possibility of simultaneous experimentation for material testing and measurement for impact loading needs to be further explored.

From the open literature, it can be concluded that various researchers have implemented different impulse facilities such as shock tube and shock tunnel for measurement of force, heat transfer, or material testing. All these ground test facilities are evolved from shock tube fundamentals. Besides, shock tube testing is cheaper and effective when it is concerned with possibility of usage of a sensor, its robustness testing or characterization. Jagadeesh [17] had proved the usefulness of this impulse facility for industrial applications. Therefore, shock tube can be considered as the first experimental set-up to test sensors in the ambience involving highly transient compressible flow. In view of above facts, an attempt has been made to access the potential use of shock tube for determining fundamental constants of the sensors used in force and heat transfer measurements. These characteristic constants include "response time and lag time" of the sensors. Possibility of material testing against the impulse loading and associated measurement is also planned herein. With successful demonstration, such types of use of shock tube would become an integral part of sensor development. Hence, an initiative is taken to assess the usefulness of the shock tube in sensor and material testing. As a part of this drive, a shock tube is designed in-house, fabricated and calibrated successfully and, then, employed as an aerotest facility. In this phase, numerical simulations are also performed to get detailed insight of shock tube flow. Three major experiments are conducted in the shock tube, which include force measurement using strain gauge, heat transfer rate measurement using thermocouple, and deformation testing of aluminum sheets. The details of shock tube and associated experiments and the numerical simulations are explained in the subsequent sections.

2. Computational Investigation

2.1. Shock Tube Facility. The shock tube facility at "Indian Institute of Technology Guwahati (IITG), India", is made out of stainless steel material and has 2 m long driver section and 5 m long driven section. These sections are assembled using 1 m long tubes of inner diameter 55 mm and thickness 12 mm. For present studies, aluminum diaphragm of 1.2 mm thickness with V-grooves is considered to separate two sections of the shock tube. The driver section is equipped with high pressure digital pressure gauges, whereas vacuum pressure in the driven section is measured by Pirani gauge. Two pressure transducers (PCB make) have also been flush-mounted to the inner surface of the driven section for measuring the pressure jump in the presence of incident and reflected shock waves. Further, usage of these pressure signals is helpful in deducing their Mach numbers. The entire shock tube assembly with associated instrumentation is illustrated in Figure 1. In the most of the present studies, driver gas is chosen to be helium having temperature of 298 K and pressure of 19.85 bar, while air is maintained at 298 K and 0.18 bar as the driven gas.

2.2. Numerical Simulation. Before execution of the experiments in the actual test facility, numerical simulation of the same is necessary as it provides a clear visualization and better insight of the flow inside the domain. Also it offers essential information about magnitude of signals to be acquired and associated data acquisition settings. Hence, numerical simulation of complete shock tube is carried out by usage of the commercial CFD solver (ANSYS FLUENT 14.5), where governing equations of continuity, momentum, and energy are solved using finite volume approach. In order to incorporate the effect of different driver and driven gas species, transport equation is also solved herein. The computational domain along with its associated initial and boundary conditions is shown in Figure 2.

As the axisymmetric approach is sufficient to render an accurate description of the real flow configuration, so the shock tube is modeled as an axisymmetric body. Meshing of the flow domain is initially done by mapped facing method

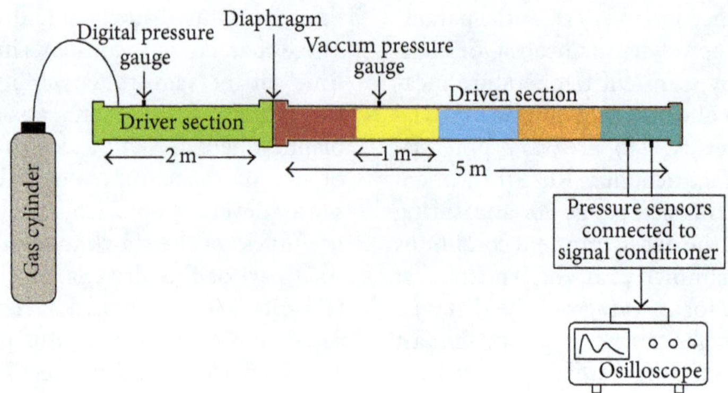

FIGURE 1: Shock tube facility present at IIT Guwahati.

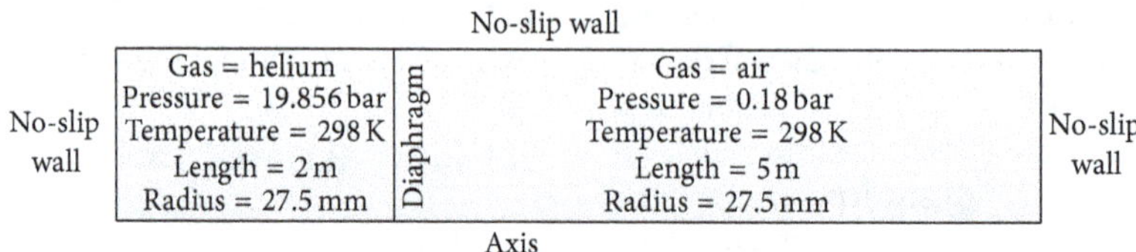

FIGURE 2: Computational domain and its boundary conditions.

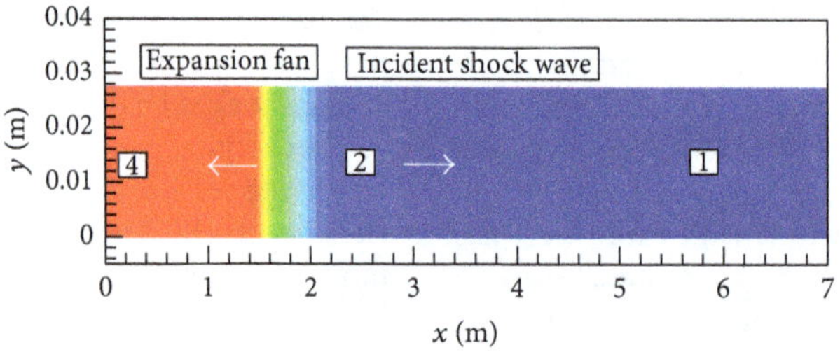

FIGURE 3: Pressure distribution after diagram rupture.

which is then modified by uniform quad method. Density based implicit solver with four-stage Runge-Kutta scheme has been chosen for transient analysis of the flow field. For the current numerical simulation, "advection upstream splitting method (AUSM)" flux vector splitting scheme has been adopted to compute the convective fluxes. User-defined function (UDF) has been incorporated in the present situation to initialize the computational domain into two different sections, namely, driver and driven sections. Bursting of the diaphragm is not the prime concern, so the diaphragm is assumed to be ruptured instantly at time $t = 0$. Initial conditions set in the computational domain are as given in Figure 2. This figure also describes the boundary conditions for the present problem. Simulations are performed for different time steps and mesh sizes so as to arrive at an optimum combination. Finally, the chosen mesh (320000 nodes) and time step have fetched the results, which are independent of the mesh size and time step with adequate resolution of the flow field. Pressure contours after 0.9 ms of the rupturing of the diaphragm are shown in Figure 3. It gives clear visualization of a right-running incident shock wave and a left-running expansion fan. The incident shock hits the end wall and reflects back creating high temperature and pressure zone behind it, which is evident in Figure 4.

Initial conditions of the shock tube are also used to obtain the x-t diagram using the tool of Wisconsin Shock Tube Laboratory [8]. Space-time details of different waves can be obtained using such x-t diagram. Essentially, it is an exact Riemann solver which employs a 2nd-order Muscl-Hancock method in finite volume formulation. Result of these simulations in the form of x-t diagram is shown in Figure 5. This figure clearly shows that the present operating conditions lead to tailored mode operation of the shock tube. Further, different wave speeds and the pressure change across the wave can also be evaluated using the slope of the x-t diagram.

It has been observed, from CFD simulations, that the primary shock wave requires 4.4 ms to reach the end wall. Similarly, x-t diagram suggests that the shock would take 3.96 ms to reach the driven section end after bursting of the diaphragm. Further, as both primary and reflected shocks are the concerned factors during experiments, so the parameter

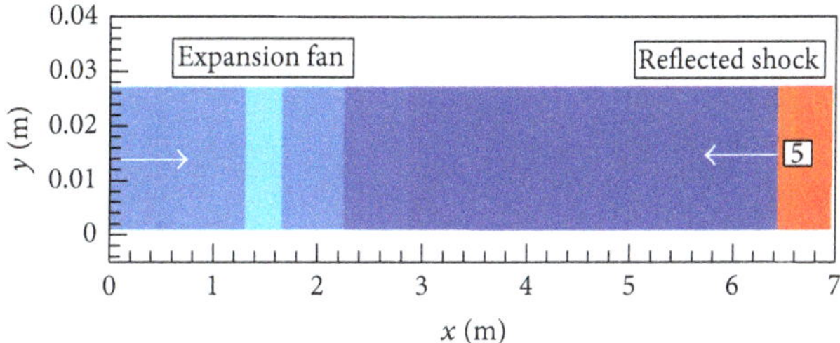

FIGURE 4: Pressure distribution after reflection from end wall.

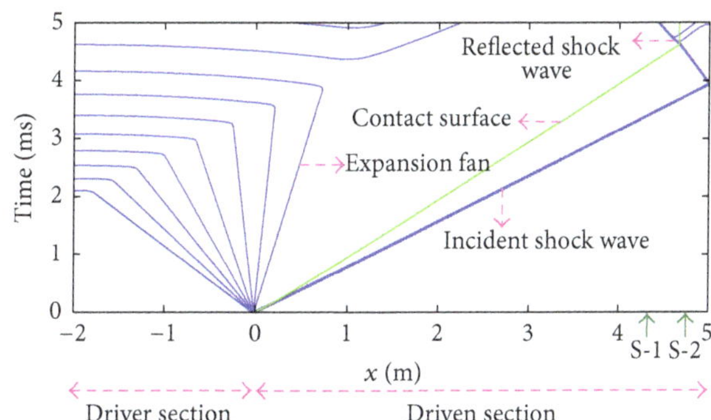

FIGURE 5: Representation of shock tube operating conditions in x-t diagram.

jump across these shocks is the critical information acquired as an outcome of computational efforts. Pressure and temperature expected behind the primary shock are 3.23 bar and 1029 K from the CFD simulation. Mach number of predicted by this simulation for primary shock is noted to be 3.6, while the induced mass motion behind this shock has Mach number of 1.49. Similarly, primary shock Mach number and pressure behind it, predicted from exact Riemann solver, are 3.65 and 2.75 bar, respectively. These results are found to be useful for designing and planning of the proposed experiments.

3. Calibration of Shock Tube

For the calibration experiments, driver and driven sections of the shock tube are separated by aluminum diaphragm. As considered for simulation, helium is chosen to be the driver gas and air is considered as the driven gas. In one of the experiments, initial pressure of the driven section is maintained at 0.18 bar, whereas pressure inside the driver section is increased till it reaches rupture pressure of the diaphragm. Before the actual experiments, driven section of the shock tube is evacuated to the known low pressure. Then driver gas is supplied in the driver section from the standard pressure cylinders till the diaphragm ruptures. The driver section pressure is measured to be 19.85 bar with an uncertainty of ±0.02%. Sudden rupture of the diaphragm resulted in a shock wave that propagates into the driven section which further induces a mass motion in the driven gas. This primary shock subsequently reflects from shock tube end wall. Typical pressure signal obtained during experiment from two pressure transducers is shown in Figure 6. Measured pressures can be used to evaluate the pressure ratios. Here, shock speed and onwards shock Mach number are calculated from the distance between both pressure transducers and time taken by the shock wave to travel the same distance. Calibration experiments of the shock tube showed reasonably good agreement with the theoretical and simulation based predictions. Experimentally measured pressure ratio across the primary shock is 16.73, while the same predicted from the shock tube theory is 15.59. Estimation of the pressure ratio across the primary shock from CFD simulations is 17.89, while the same ratio is predicted as 15.27 using exact Riemann solver. Similarly, experimental, theoretical, CFD, and exact Riemann solver based primary shock Mach numbers for the same shock tube driving conditions are 3.49, 3.68, 3.6, and 3.65, respectively. These results show that the experimental shock Mach number deviates 5.1% from the theoretical value, 2.1% from the CFD simulations, and 4.6% from estimation of exact Riemann solver. Along with the encouraging comparison, successful opening of metal diaphragm has been noticed as an outcome of these calibration experiments. Hence, these operating conditions are considered for experiments with force and heat transfer measurement sensors.

4. Applications of Shock Tube

4.1. Impulsive Force Measurement. Experimentation in impulse facilities needs sophisticated instrumentation since these facilities are characterized by small test duration (~ of 10 microseconds to 1 millisecond). Like other measurements, force measurement needs special attention due to difficulty in attaining steady state between the mounting system and test configuration. Therefore, shock tube testing is a viable

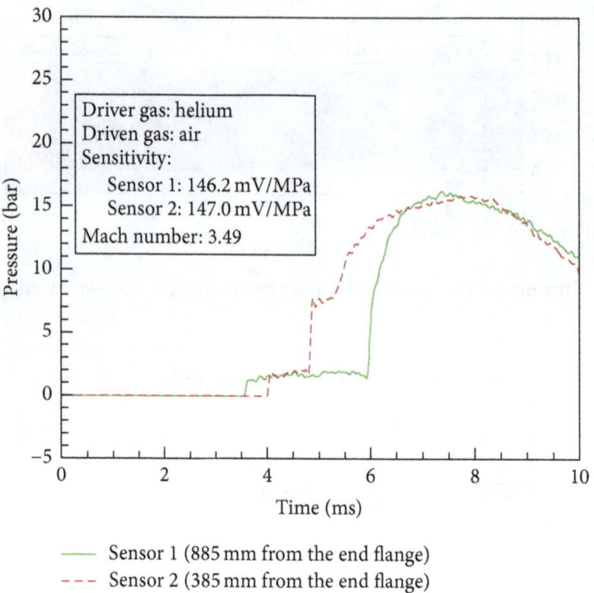

FIGURE 6: Pressure signal acquired in the shock tube using helium as the driver gas.

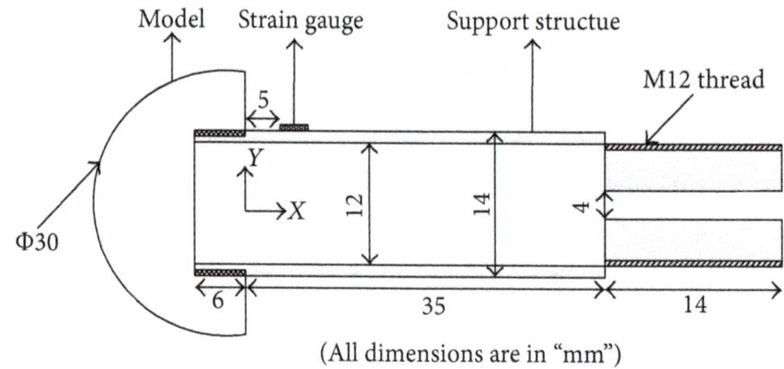

FIGURE 7: Schematic drawing of the hemispherical model.

alternative to assess the possibility of force measurement using a force balance since the test model experiences highly transient flow in this test facility.

The hemispherical test model (15 mm radius) made out of aluminum and attached with a brass stress bar (14 mm diameter and thickness 1 mm) is considered to be the test model for force measurement purpose (Figure 7). This model is equipped with stress wave force balance for the output response measurement in terms of strain. Here, strain is measured with the help of encapsulated semiconductor strain gauge having resistance 350 Ω and gauge factor 130. This strain gauge is mounted at a distance of 5 mm from the rear end of the hemispherical model. Strain gauge is further connected to Wheatstone bridge circuit for measuring the resistance change. The output of the Wheatstone bridge is of the order of few microvolts, while a voltage amplifier of gain factor 500 is used for amplifying the voltage response. Further, the output terminal of the amplifier is connected to the oscilloscope through BNC connecters for recording the signal during experiment.

The mentioned model in Figure 7 equipped with strain gauge is dynamically calibrated before testing in shock tube. During this dynamic calibration, model is fixed in a bench vice and impulse force is applied using the impulse hammer. Recorded strain signals and corresponding applied force time histories are shown in Figures 8 and 9, respectively. One such strain signal and corresponding force impulse signal are considered to obtain the system response function. Later, this response function and other strain signal from calibration experiment are employed to recover the applied force using MATLAB based FFT program. Encouraging recovery of force signals is evident in Figure 9.

Experiments are then carried out on the hemispherical test model with stress bar mounted with strain gauge in the shock tube. Mounting of the strain gauge is maintained at the same location as it was kept during the calibration experiments. For these experiments, model and stress bar assembly is fixed in the end flange located at the driven section end of the shock tube. Force measurement experiments are then conducted for the same initial conditions of shock tube calibration tests. Shock tube pressure signals are also recorded along with the strain signal in these experiments. Various sets of experiments have been carried out to verify the consistency and repeatability of measurement.

Strain time history along with the pressure signal of second pressure transducer is shown in Figure 10. Here, strain

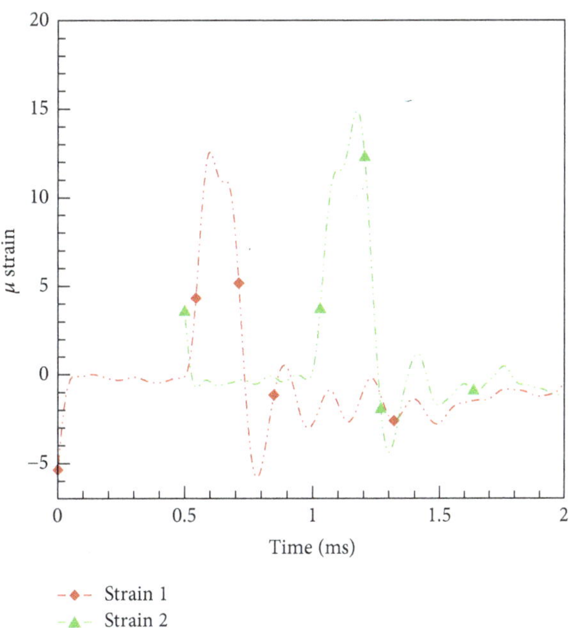

FIGURE 8: Strain signals recorded in two calibration tests.

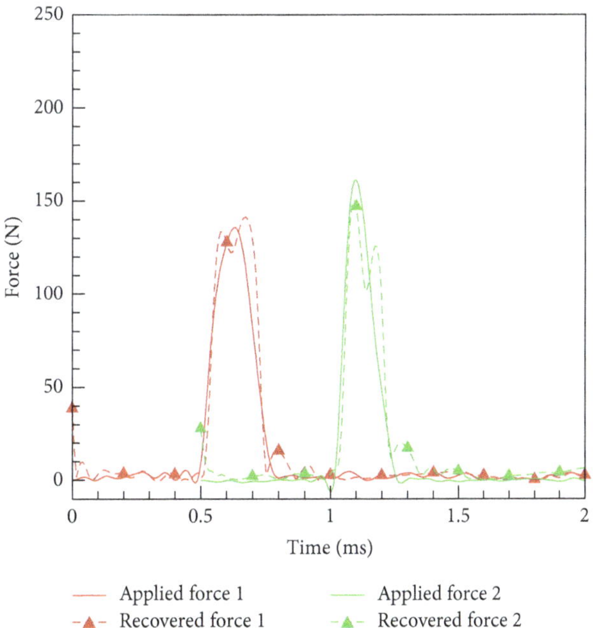

FIGURE 9: Applied and recovered force signals of two calibration tests.

signal has been filtered using "Butterworth low pass filter" having cut-off frequency of 12.5 kHz. Lag time of the strain gauge can be estimated by interpreting these signals. For this experimental condition, speed of the shock wave is 1108 m/s, so the theoretical time required for the shock wave to travel between second pressure transducer and the test model is 316.08 microseconds. The time difference between response of pressure transducer and strain gauge (Figure 10) is 327.59 microseconds. Thus, the lag time of the presently used strain gauge is calculated as, 11.51 microseconds. Such a small lag time of this sensor makes it perfectly applicable for force measurements in short duration impulse facilities. Further, this strain response is used to predict the drag force acting on the model by using deconvolution technique as illustrated in Figure 10. Response time of the sensor for force measurement is depicted in this figure (258 microseconds). It is evident that force measurement and strain gauge characterization is very much possible using shock tube. Thus, present investigations are found to be useful for implementation of shock tube in the area of force measurement on aerodynamic bodies.

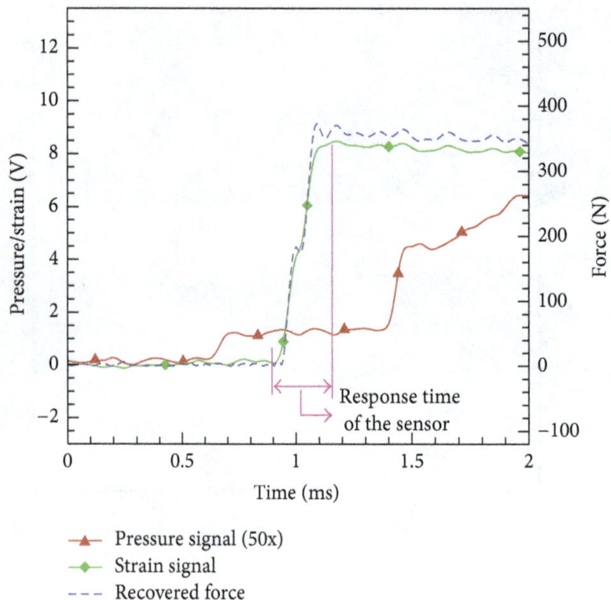

FIGURE 10: Experimentally obtained pressure and strain signal along with associated recovered force.

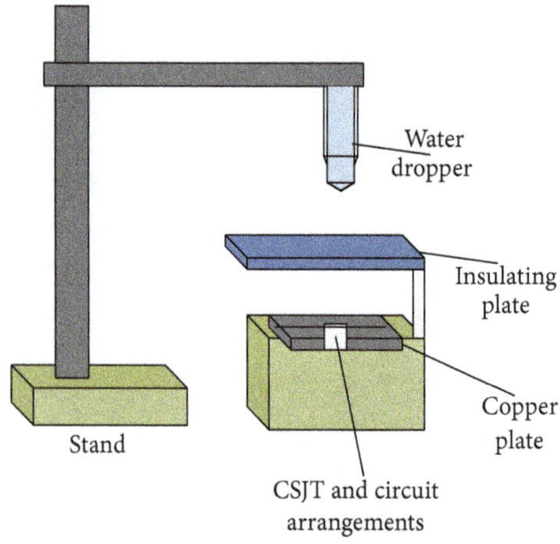

FIGURE 11: Schematic of the experimental set-up for estimation of thermal product.

4.2. Heat Flux Measurement. An attempt has been made to obtain specifications of thermal sensor and to measure the heat flux in the highly transient environment of the shock tube in order to review the application of shock tube as heat flux measurement device. Attainment of thermal equilibrium between the test object and the fluid flow is not possible due to small test duration. Therefore, the rate of temperature rise is invariantly measured for heat transfer rate prediction. On similar lines, heat transfer measurement experiments are carried out using a hemispherical aluminum model of radius 10 mm. An E-type (chromel-constantan) coaxial thermal sensor having diameter of 3.25 mm has been considered to measure the heat flux. This coaxial thermal sensor always gives the signal in the form of voltage as it works on the principle of Seebeck effect. Provision is made at the stagnation point of the model for mounting this thermocouple. The coaxial surface junction thermocouple (CSJT) has a surface junction where one element is swaged over the other element with a layer of electrical insulation in between them. The junction of the sensor is formed by slightly abrading one material over the other in turn creating plastic deformation of two materials. The sensor has a sensitivity of 58.96 $\mu V/^{\circ}C$, which is obtained from oil bath calibration of the sensor. The fabrication and calibration techniques are considered from [18].

An in-house experimental set-up is employed to measure the thermal product of the E-type coaxial surface junction thermocouple. This measurement procedure is generally termed as "water-droplet" technique for thermal product estimation. Schematic of the set-up is as shown in Figure 11. This arrangement is comprised of copper plate equipped with surface mounted E-type thermocouple. Water drop (at temperature of 26°C) from the dropper is allowed to fall on the thermocouple from a fixed location. Here, distance

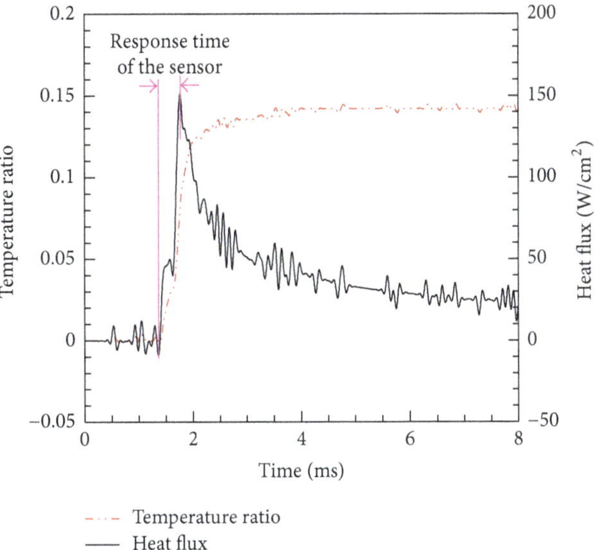

FIGURE 12: Temperature and corresponding heat flux signal for a water drop test.

between plate and the dropper is maintained using height adjustor. During the experiment, copper plate is heated using electric heater and its temperature is maintained at 50°C. An insulating plate is kept between the copper plate and the dropper to avoid the thermal influence of plate on the water drop.

It has been noticed that the water drop impact does not affect the surface junction of the thermocouple. Temperature time history, obtained during the experiment, is as shown in Figure 12, and (1) is used to compute the thermal product of the thermocouple:

$$\frac{T - T_s}{T_s - T_w} = \frac{\beta_w}{\beta_w + \beta}. \quad (1)$$

Here, required thermal product of water, β_w (=1643 J m^{-2} K^{-1} s$^{-0.5}$), is taken from the available literature [7, 8]. As an outcome of this experiment, presently, using E-type thermocouple is seen to have thermal product of 9493 J·m^{-2}·K^{-1}·s$^{-0.5}$. Generally, for such short duration time scale applications, heat transfer rates can be recovered from the transient temperature data, through the use of appropriate modeling of (2) with assumption of one-dimensional heat conduction [19, 20]:

$$q_L(t) = \frac{\beta}{\sqrt{\pi}}\left[\frac{T(t)}{\sqrt{t}} + \frac{1}{2}\int_0^t \frac{T(t) - T(\tau)}{(t-\tau)}\right]; \qquad \beta = \sqrt{\rho c k}. \quad (2)$$

Furthermore, as the thermal penetration depth during the experimental run times is small compared to the linear dimension of the thermal sensor, the system can be modeled by considering unsteady, linear conduction of heat in a one-dimensional semi-infinite solid [20]. In order to use (2), it is desirable to have a closed form solution of transient temperature data obtained from the captured experimental signals [10]. There are many discretization techniques available; but, for the present study, a polynomial based cubic-spline discretization technique is utilized to discretize the obtained temperature data (see (3) and (4)):

$$\{q_L(t)\}_{\text{spline}} = \left[2\sqrt{\frac{\rho_2 c_2 k_2}{\pi}}\sum_{i=1}^{M-1}\left\{V_i\left(P_i^{1/2} - R_i^{1/2}\right) - \frac{W_i}{3}\left(P_i^{3/2} - R_i^{3/2}\right) + \frac{a_{4,i}}{10}\left(P_i^{5/2} - R_i^{5/2}\right)\right\} + 2\sqrt{\frac{\rho_2 c_2 k_2}{\pi}}\left(V_M - P_M^{1/2} - \frac{W_M}{3}P_M^{3/2} + \frac{a_{4,i}}{10}P_M^{5/2}\right)\right] \cdot \sqrt{S_t}, \quad (3)$$

where

$$P_i = \tau_{M+1} - \tau_i;$$
$$R_i = \tau_{M+1} - \tau_{1+i};$$
$$F_i = a_{1,i} + a_{2,i}P_i + \frac{a_{3,i}}{2}P_i^2 + \frac{a_{4,i}}{6}P_i^3;$$
$$V_i = \frac{dF_i}{d\tau_{M+1}}; \quad (4)$$
$$W_i = \frac{d^2F_i}{d\tau_{M+1}^2};$$
$$S_t = \text{time scaling factor.}$$

The reduced heat flux signal processed with above technique from the temperature data is presented in Figure 12. This figure gives the information about response time of the sensor. It is evident here that the response time of the E-type CSJT in recovering peak heat flux is 375 microseconds.

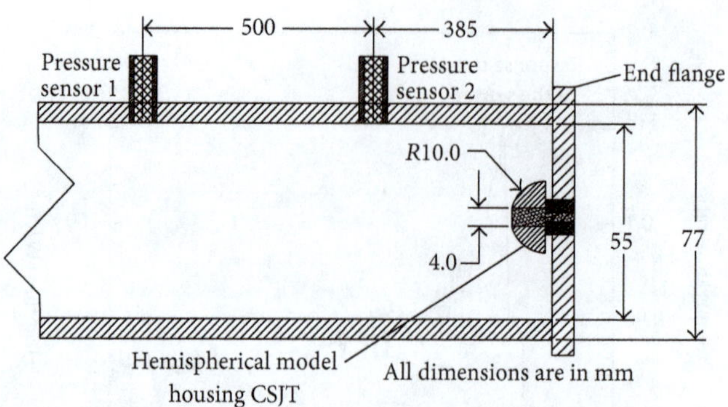

FIGURE 13: Schematic of the end-section of the shock tube showing the mounting of coaxial thermocouple on the hemispherical model at the end flange.

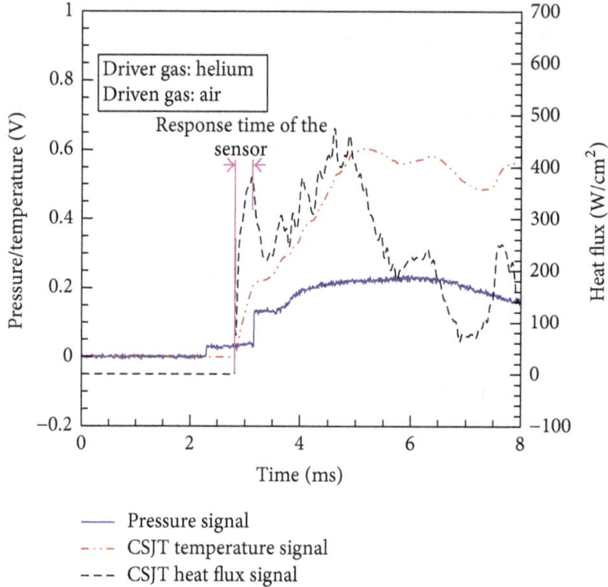

FIGURE 14: Temperature and heat flux variation for the spherical test model in shock tube test.

A test model integrated with thermocouple, flush-mounted at stagnation point, is then considered for measurement of stagnation heat flux in the shock tube. This spherical model of radius 10 mm is mounted on the end flange of the shock tube as shown in Figure 13. Same experimental settings are preferred as those used in force measurement experiments. The experimental temperature signal along with the pressure signal from second pressure transducer and heat flux histories (processed from temperature data by using (2)) is compared in Figure 14. Response time of the thermocouple can be estimated using the time varying heat transfer rate. These shock tube experiments depict that the response time of the E-type CSJT in recovering peak heat flux is 307 microseconds.

Separate experiment is conducted for lag time measurement of the in-house developed E-type CSJT. In this experiment, one coaxial thermocouple is flush-mounted in the end flange of the shock tube. Nitrogen is used here as driver gas to burst the same diaphragm of 1.2 mm thickness. Air pressure in the driven section is maintained as 0.18 bar. As an outcome of this experiment, temperature time history along with the pressure signal from second pressure transducer is shown in Figure 15. Lag time was calculated from these signals with known shock speed and it is found to be 9 microseconds for the in-house E-type CSJT. Fast reaction for the applied heat load makes this thermal sensor perfectly suitable for measurement in impulse facilities. Heat transfer rate is also estimated from the temperature time history as predicted in earlier cases (Figure 15). Response time for the peak heat flux is noted here as 352 microseconds. Thus, it is evident here that the shock tube can be used for measurement of lag time and response time of the thermocouple. Present efforts also portray that the E-type coaxial thermocouples have sufficiently low lag time and rise time in the range of 300–400 microseconds for predicting the peak heat flux.

4.3. Forming Response Study. Material strength and deformation characteristics under high pressure and temperature environment play a crucial role in designing aerodynamic

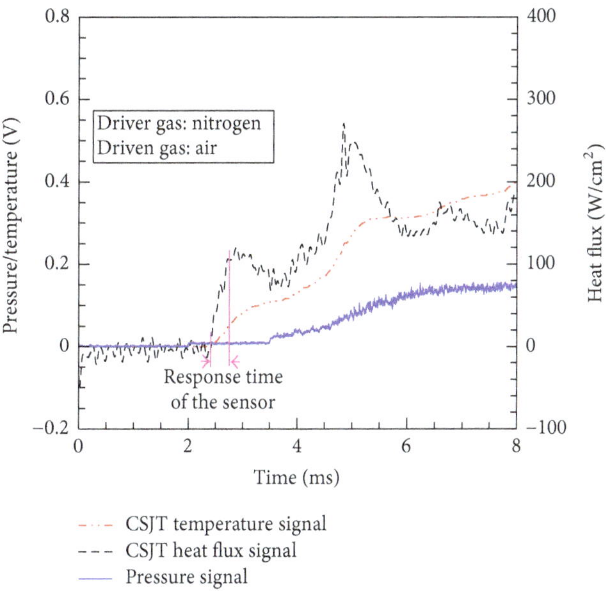

FIGURE 15: Temperature and heat flux variation for the end flange mounting of shock tube.

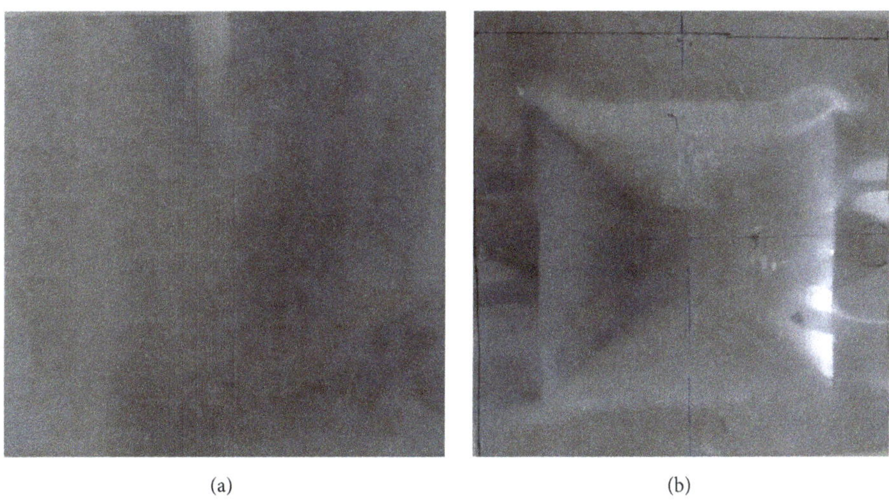

(a) (b)

FIGURE 16: (a) Initial test specimen. (b) Deformed shape due to shock loading.

vehicles. Generic aerospace applications make usage of aluminum and its alloys for fabrication purpose. Hence, to check the sustainability and stability of the materials subjected to impulse loading, preliminary forming study has been done. Shock tube is employed as a device for generation of high pressure impact loading. Same shock tube but with 2 m driver section and 2 m driven section is employed for the current investigation. Mylar sheets of 1 mm thickness have been chosen instead of metal diaphragm for the preliminary forming tests. The test specimen is considered as a flat plate of aluminum having 0.5 mm thickness. This plate is clamped at the end of the driven section during the experiment. The maximum in-plane strain in the specimen during experiment is measured with the help of semiconductor strain gauge having resistance of 7500 Ω and gauge factor 175. This strain gauge is glued at the centre of back surface of the sample along the horizontal axis. It is further connected to Wheatstone bridge circuit and voltage amplifier as a part of the data acquisition network. During the experiments, driven section is maintained at atmospheric pressure, whereas the driver section pressure is increased till rupture of the Mylar sheet. The bursting pressure is noticed to be 20 bar with an uncertainty of ±9%. From the pressure signal obtained through pressure transducers, it is evident that maximum incident pressure on to the plate is 2.73 bar. In fact, it is the pressure behind the reflected shock. Aluminum specimen is found to be deformed due to application of this impulse force. Initial and deformed plates are shown in Figure 16. The trend of deformation of the plate has reasonably good match with Kumar et al. [15]. Typical strain and pressure signal obtained has been shown in Figure 17. An instant rise in strain response can be seen from the figure as realised practically. Thus,

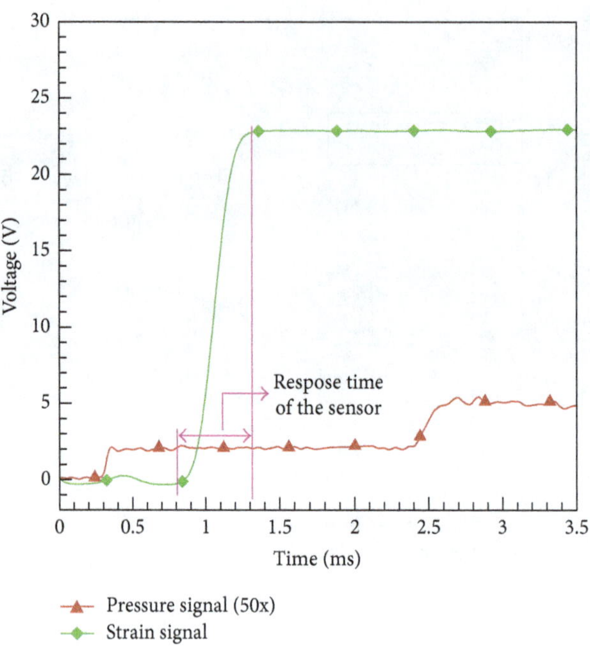

FIGURE 17: Obtained pressure and strain response during forming experiment.

present experiments demonstrate the use of shock tube for material response or forming studies as well.

5. Conclusion

A shock tube is successfully fabricated and calibrated and is effectively considered to demonstrate multiple applications. At the first phase, insight of the shock tube flow physics is earned through CFD simulations. Initially, this sophisticated experimental set-up is considered for force measurement experiments on a hemispherical model equipped with a high gauge factor strain gauge. From the output response, drag force is predicted using deconvolution technique through the system response function obtained during dynamic calibration. In this experiment, lag time and response time of the strain gauge are noticed to be 11.5 and 258 microseconds, respectively. Similarly, transient temperature change is recorded using a calibrated thermocouple for a hemispherical test configuration subjected to the shock tube flow conditions. Transient temperature time history has also been measured for the end flange mounting of the thermocouple in the shock tube. Further, using cubic-spline discretization method, stagnation point heat flux is calculated. These experiments portrayed that the E-type CSJT has lag time of 9 microseconds and rise time in the range of 300–400 microseconds. Since, in both the experiments, sensors could sustain the impulsive mechanical and heat load while recording the signal, these experiments in the shock tube showed that it is indeed possible to test the sensors for their characterization and robustness. Further experiments are also performed successfully on an aluminum plate subjected to impulse loading in shock tube for assessing its formability. During these experiments, in-plane strain is effectively measured using strain gauge. Thus, current study provides an ample scope of shock tube application in regard to force and heat flux measurement as well as structural behavior study for impulse loading.

Conflicts of Interest

The authors declare that there are no conflicts of interest regarding the publication of this paper.

Acknowledgments

The financial support received from "Aeronautical Research and Development Board (AR & DB)," and "Gas Turbine Materials and Processes (GTMAP)," Government of India (New Delhi), is highly acknowledged.

References

[1] N. Sahoo, D. R. Mahapatra, G. Jagadeesh, S. Gopalakrishnan, and K. P. J. Reddy, "Design and analysis of a flat accelerometer-based force balance system for shock tunnel testing," *Measurement*, vol. 40, no. 1, pp. 93–106, 2007.

[2] S. R. Sanderson and J. M. Simmons, "Drag balance for hypervelocity impulse facilities," *AIAA journal*, vol. 29, no. 12, pp. 2185–2191, 1991.

[3] J. M. Simmons, "Measurement techniques in high-enthalpy hypersonic facilities," *Experimental Thermal and Fluid Science*, vol. 10, no. 4, pp. 454–469, 1995.

[4] S. L. Tuttle, D. J. Mee, and J. M. Simmons, "Drag measurements at Mach 5 using a stress wave force balance," *Experiments in Fluids*, vol. 19, no. 5, pp. 336–341, 1995.

[5] E. Matsumoto, S. Biwa, K. Katsumi, Y. Omoto, K. Iguchi, and T. Shibata, "Surface strain sensing with polymer piezoelectric film," *NDT & E International*, vol. 37, no. 1, pp. 57–64, 2004.

[6] P. H. Rose, "Development of the calorimeter heat transfer gauge for use in shock tubes," *Review of Scientific Instruments*, vol. 29, no. 7, pp. 557–564, 1958.

[7] H. A. Mohammed, H. Salleh, and M. Z. Yusoff, "Dynamic calibration and performance of reliable and fast-response coaxial temperature probes in a shock tube facility," *Experimental Heat Transfer*, vol. 24, no. 2, pp. 109–132, 2011.

[8] M. H. Anderson, B. P. Puranik, J. G. Oakley, P. W. Brooks, and R. Bonazza, "Shock tube investigation of hydrodynamic issues related to inertial confinement fusion," *Shock Waves*, vol. 10, no. 5, pp. 377–387, 2000.

[9] R. Kumar, N. Sahoo, and V. Kulkarni, "Conduction based calibration of handmade platinum thin film heat transfer gauges for transient measurements," *International Journal of Heat and Mass Transfer*, vol. 55, no. 9-10, pp. 2707–2713, 2012.

[10] R. Kumar and N. Sahoo, "Dynamic calibration of a Coaxial thermocouples for short duration transient measurements," *Journal of Heat Transfer*, vol. 135, no. 12, Article ID 124502, 2013.

[11] K. Hariprakasham, M. D. Kumar, and T. Mukesh, "To develop a coaxial thermocouple sensor for temperature measurement in shock tube," *International Journal of Advanced Information Science and Technology*, vol. 27, pp. 2319–2682, 2014.

[12] N. Ray, G. Jagadeesh, and S. Suwas, "Response of shock wave deformation in AA5086 aluminum alloy," *Materials Science and Engineering A*, vol. 622, pp. 219–227, 2015.

[13] M. A. Louar, B. Belkassem, H. Ousji et al., "Explosive driven shock tube loading of aluminium plates: experimental study," *International Journal of Impact Engineering*, vol. 86, pp. 111–123, 2015.

[14] S. R. Nagaraja, J. K. Prasad, and G. Jagadeesh, "Theoretical-experimental study of shock wave-assisted metal forming process using a diaphragmless shock tube," *Proceedings of the Institution of Mechanical Engineers, Part G: Journal of Aerospace Engineering*, vol. 226, no. 12, pp. 1534–1543, 2012.

[15] P. Kumar, J. LeBlanc, D. S. Stargel, and A. Shukla, "Effect of plate curvature on blast response of aluminum panels," *International Journal of Impact Engineering*, vol. 46, pp. 74–85, 2012.

[16] G. T. Gray III and J. C. Huang, "Influence of repeated shock loading on the substructure evolution of 99.99 wt.% aluminum," *Materials Science and Engineering: A*, vol. 145, no. 1, pp. 21–35, 1991.

[17] G. Jagadeesh, "Industrial applications of shock waves," *Proceedings of the Institution of Mechanical Engineers, Part G: Journal of Aerospace Engineering*, vol. 222, no. 5, pp. 575–583, 2008.

[18] S. Agarwal, N. Sahoo, and R. K. Singh, "Experimental techniques for thermal product determination of coaxial surface junction thermocouples during short duration transient measurements," *International Journal of Heat and Mass Transfer*, vol. 103, pp. 327–335, 2016.

[19] D. L. Schultz and T. V. Jones, "Heat transfer measurements in short duration hypersonic facilities," Tech. Rep. AGARD-AG-165, 1973.

[20] J. Taler, "Theory of transient experimental techniques for surface heat transfer," *International Journal of Heat and Mass Transfer*, vol. 39, no. 17, pp. 3733–3748, 1996.

The Effect of Yaw Angle on a Compressible Rectangular Cavity Flow

Kuan-Huang Lee,[1] Kung-Ming Chung,[2] and Keh-Chin Chang[1]

[1] *Institute of Aeronautics and Astronautics, National Cheng Kung University, Tainan, Taiwan*
[2] *Aerospace Science and Technology Research Centre, National Cheng Kung University, Tainan, Taiwan*

Correspondence should be addressed to Kung-Ming Chung; kmchung@mail.ncku.edu.tw

Academic Editor: Saad A. Ahmed

Experiments are performed to determine the characteristics of a compressible flow over yawed rectangular cavities for Mach numbers of 0.64, 0.70, and 0.83. The cavity's length-to-depth ratio varies from 4.43 to 21.50 and the length-to-width ratio is unity. The yaw angle is 0°–45°. The upstream compression and downstream expansion near the front and rear corners of a cavity decrease when the value of the yaw angle increases. The amplitude of the fluctuating pressure is a maximum for an open cavity with a yaw angle of 15°. An increase in the yaw angle results in a reduction in the pressure fluctuations for both open and transitional cavities. In the span-wise direction, variations in the mean and fluctuating pressure are less significant than those in the chord-wise direction. The oscillating frequency of resonance varies slightly with the yaw angle, but the amplitudes for the power spectral density are significantly reduced when the yaw angle is larger than 30°. For lower Mach numbers, the lower mode plays an important role in self-sustained oscillations for an open cavity when there is an increase in the yaw angle.

1. Introduction

The presence of a cavity changes the mean and fluctuating pressure distributions inside and near a cavity [1, 2]. For compressible flow in a rectangular cavity (M = 0.2–0.95), the mean and fluctuation pressure distributions normal to the direction of the flow depend principally on the length-to-depth ratio, L/H [3–5]. When $L/H > 13$ (a closed cavity), the flow expands from the leading edge, attaches to the floor, and separates ahead of the rear face of the cavity. This results in a significant variation in the mean surface pressure in the stream-wise direction. The shear layer for an open cavity (L/H < 6–8) spans the cavity and impinges near the rear corner. Discrete acoustic tones are associated with a feedback loop between vortex shedding and acoustic disturbance, which is known as Rossiter resonance [6, 7].

For flow in a yawed rectangular cavity, the yaw angle, β, is defined as the angle between the freestream and the chord-wise direction of the cavity. An asymmetric flow pattern inside the cavity can be expected. Savory et al. (U_∞ = 7 m/s) [8] measured the drag of flow in a yawed cavity for L/H = 1.428–10.0 and β = 0°–90°, in which the maximum drag occurs for L/H = 2–2.5 and for β = 45–60 deg. They also noted that the drag for a square cavity (width-to-length, W/L = 1) is greater than that for a rectangular cavity with the same plan-form area. Czech et al. (U_∞ = 16 m/s) [9] demonstrated a critical value of β of 45°, based on the measurements of mean and fluctuating pressure in a wide cavity (W/L = 4.85), and showed that asymmetry is more apparent for a deep cavity (L/H = 1–3). The oil flow visualization and drag measurements by Gai et al. (U_∞ = 15 m/s) [10] showed that the most asymmetric flow pattern inside a cavity and the lowest drag occur at β = 45° for L/H = 6–16.

In terms of self-sustained oscillations, Bari and Chambers (U_∞ = 20–44 m/s) [11] showed that the yaw angle for a cavity need not significantly affect the resonant frequencies. However, there may be a switch in the dominant mode and the effective stream-wise length, $L/\cos(\beta)$, of a cavity at yaw is probably not a suitable characteristic length. A study by Lee et al. (L/H = 5.0, β = 0°–20°) [12] showed that the resonance switches from 2nd mode to 3rd mode at β = 15° for transonic and low supersonic flows (M = 0.84 and 1.10). The strength of

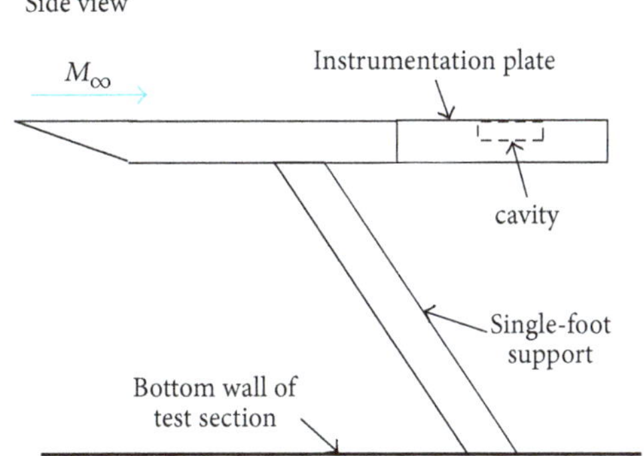

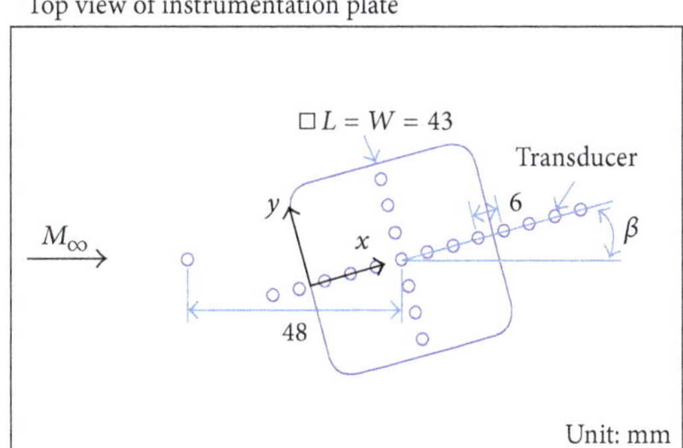

FIGURE 1: Test configuration.

the resonance is also significantly affected by the value of β. The overall fluctuation is increased at a critical yaw angle for a subsonic flow ($U_\infty = 25$ m/s) [13] and significantly reduced for a supersonic flow ($M = 2.0$) [14].

Since less work has been done for a compressible turbulent flow past rectangular cavities at yaw in the past, this experimental study aims to characterize the flow in detail. The chord-wise and span-wise distributions of the mean and fluctuating pressure are determined. The distribution of the power spectral density near the rear face is used to characterize self-sustained oscillations for both open and transitional cavity flows. The resonant frequencies are calculated and the corresponding Strouhal numbers were compared with those predicted using Rossiter's semiempirical formula, in which the empirical parameters are determined using a gradient-based searching method.

2. Experimental Techniques

2.1. The Transonic Wind Tunnel and Instrumentation. The transonic wind tunnel at the Aerospace Science and Technology Research Center in National Cheng Kung University is a blow-down type. The test section for this study had solid side walls and perforated top and bottom walls. It was 600 mm square and 1500 mm in length. Chung et al. [15] showed that perforated walls induce strong acoustic waves, for which the characteristic frequency is 4.2–4.8 kHz for $M = 0.64$–0.83. The stagnation pressure was controlled using a rotary perforated sleeve valve and, for subsonic flow, the test Mach number, M, was monitored using two choked flaps. The stagnation pressure and temperature were, respectively, 172 ± 1 kPa (25.0 ± 0.15 psia) and room temperature, for $M = 0.64$, 0.70, and 0.83 ± 0.01.

A National Instruments (NI-SCXI) system recorded the output signals from the dynamic pressure transducers (Kulite XCS-093-25A, B screen). The natural frequency of the transducers is 200 kHz, as quoted by the manufacturer. The transducers were powered by a DC power supply of 10.0 V (GW Instek PSS-3203) and Ectron amplifiers (753A), which had a roll-off frequency of approximately 140 kHz at a gain of 20, were used to improve the signal-to-noise ratio. The sample time was 5 μs and each sample record contained 131,072 data points. Each sample record was then divided into 32 subsets of 4096 data points for data analysis. The experimental results for the flat plate case show that the respective uncertainty in the values for the static pressure coefficient, C_p, and the surface fluctuating pressure coefficient, C_{σ_p}, is 2.4% and 0.4%. Each sample record was then divided into 31 segments with a 50% overlap and the corresponding frequency resolution was 24.4 Hz for each segment of 8192 data points. The power spectral density (PSD) was evaluated using a Hann window and a fast Fourier transform. Each spectrum was then generated by averaging 31 spectra for each test case. A factor of 8/3 for each spectrum was used to compensate for the loss that results from the Hann window [16].

2.2. Models and Test Conditions. The test model consisted of a flat plate (150 mm × 450 mm) that naturally develops a turbulent boundary layer and an instrumentation plate (150 mm square) with a yawed rectangular cavity, as shown in Figure 1. The pressure transducers were flush-mounted along the centerline of each cavity in the chord-wise ($y/L = 0$) and span-wise ($x/L = 0.5$) directions. The distance between the leading edges of the flat plate and the cavity's leading edge was approximately 480 mm. The origin of the Cartesian coordinates was set at the center of the leading edge of the cavity. The positive direction of the x-axis is in the chord-wise direction towards the trailing edge. The boundary layer thickness was approximately 7 mm, upstream of the cavity's leading edge [4]. The unit Reynolds numbers were 12.9–17.2 × 10^6 per meter for $M = 0.64$–0.83. The geometry of the cavities is summarized in Table 1, where 23 instrumentation plates were fabricated. For a fixed length ($L = 43$ mm) with different depths ($H = 2.0$–9.7 mm), the value for L/H ranges from 4.43 to 21.50 and $\beta = 5^\circ, 10^\circ, 15^\circ, 30^\circ$, and 45°. The data for a rectangular cavity that is normal to the flow direction ($\beta = 0^\circ$) that was gathered by Chung [17] is also included for comparison. Notably, the self-sustained oscillation corresponds to open and transitional-open cavities, for which the value of $L/H = 4.43$–8.60 ($H = 5.0$–9.7 mm) [4].

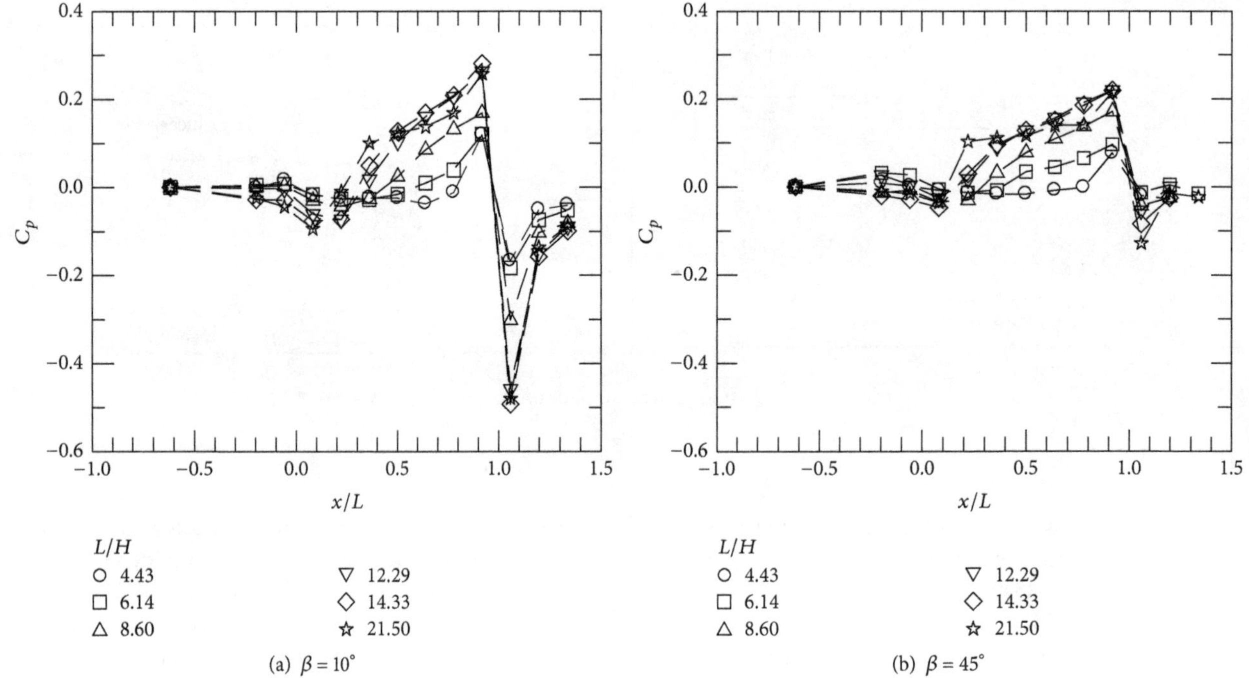

FIGURE 2: The chord-wise distributions of the mean surface pressure at $M = 0.83$: (a) $\beta = 10^\circ$ and (b) $\beta = 45^\circ$.

TABLE 1: The geometry of the yawed rectangular cavities.

β, deg.	H, mm						
	2.0	3.0	3.5	4.0	5.0	7.0	9.7
5	V	/	/	/	V	V	/
10	V	V	V	/	V	V	V
15	V	/	/	/	V	V	V
30	V	/	/	/	V	V	V
45	V	/	V	V	V	V	V

2.3. Empirical Constants in Rossiter's Formula. A semiempirical formula for a rectangular cavity flow was derived by Rossiter [6] as follows:

$$\mathrm{St}_n = \frac{f_n L}{U_\infty} = \frac{n-\alpha}{M + 1/k_c}. \quad (1)$$

The nth-mode Strouhal number, St_n, is calculated using the oscillation frequency, f_n, and the freestream velocity, U_∞. M is the freestream Mach number. The empirical parameter, α, corresponds to the lag time between the passage of a vortex and the emission of an acoustic pulse and k_c is the ratio of the convection velocity for the vortices to U_∞. Using a best fit to the measured data, Rossiter proposed values of α = 0.25 and $k_c = 0.57$ for rectangular cavities. However, Ünalmis et al. [18] showed that the empirical parameters depend on flow conditions and the value of L/H. For a cavity at yaw, the optimal values of the empirical parameters are evaluated by minimizing the difference between the experimental and the predicted Strouhal numbers. The steepest descent optimization algorithm is used [19] in this study and the details are given in [20].

3. Results and Discussion

3.1. Surface Mean Pressure Distributions. Examples of the C_p distributions for M = 0.83 are shown in Figure 2. For L/H = 12.29, 14.33, and 21.50 and $\beta = 10^\circ$ (Figure 2(a)), the flow expands near the leading edge and compresses towards the rear face. A reduction in the value of C_p is observed downstream of the rear face, following a recovery process. This corresponds to a transitional-closed cavity or a closed cavity [3, 18]. For L/H = 8.60, the expansion near the leading edge of the cavity abates. This is termed as transitional-open cavity. For open cavities (L/H = 4.43 and 6.14), a uniform C_p distribution inside the cavity is observed for values of x/L up to 0.5–0.7, following the formation of an adverse pressure gradient near the rear face. When the value of β increases (= 45°, Figure 2(b)), the C_p distribution exhibits a similar pattern for a given value of L/H. However, there is less significant

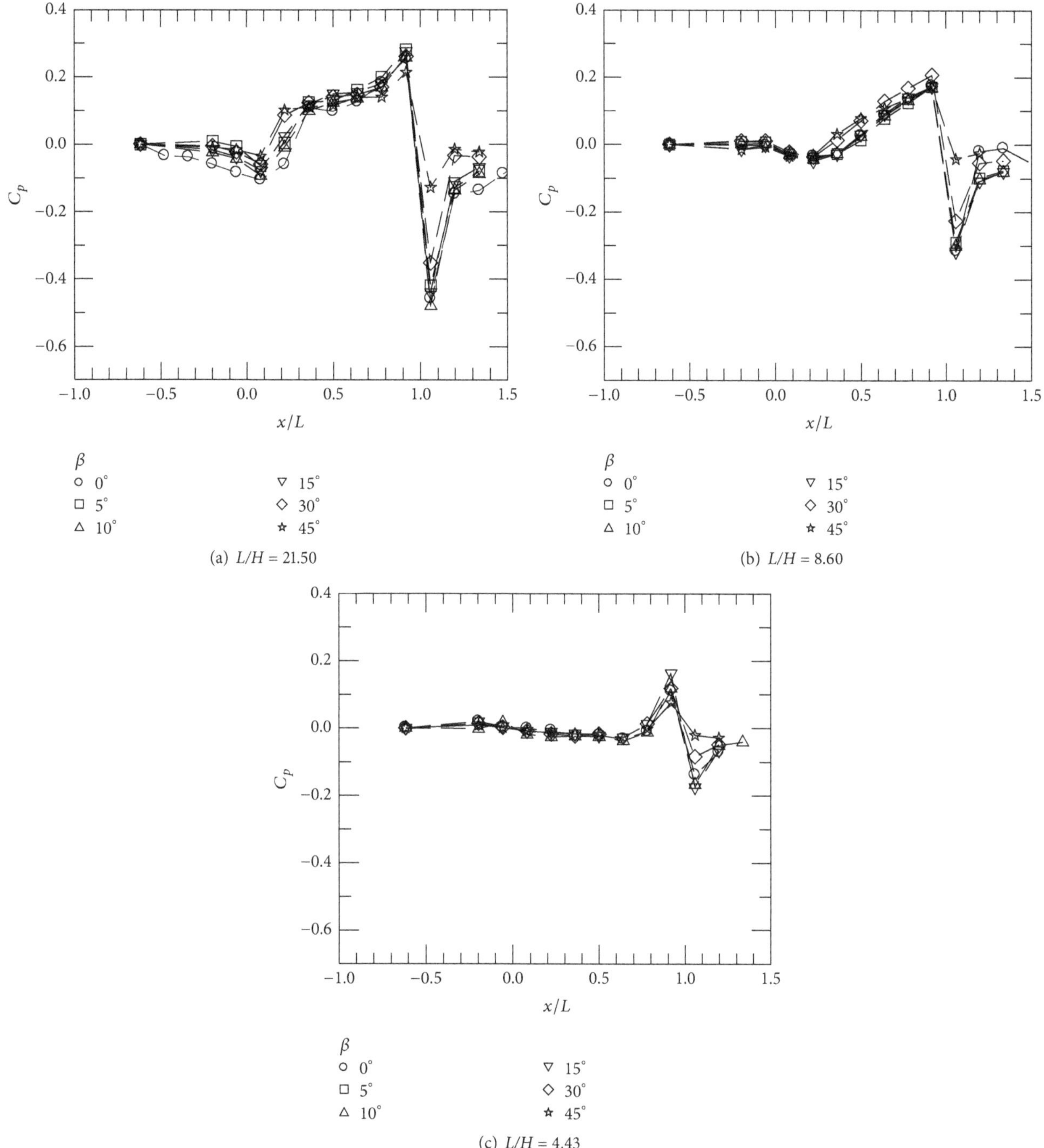

(a) $L/H = 21.50$

(b) $L/H = 8.60$

(c) $L/H = 4.43$

FIGURE 3: Chord-wise distributions of the surface mean pressure for $M = 0.83$: (a) $L/H = 21.50$, (b) $L/H = 8.60$, and (c) $L/H = 4.43$.

leading-edge expansion, compression on the cavity floor, and expansion near the rear face. The C_p distributions show that the boundaries for the flow type depend more on the value of L/H than on the value of β.

The effect of the value of β on a rectangular cavity flow is shown in Figure 3, where $M = 0.83$ and $L/H = 4.43$, 8.60, and 21.50. For a given value of L/H, the C_p distributions show that there is less expansion near the rear face when the value of β is increased. For $M = 0.64$ and 0.70, the C_p distributions show a similar feature. The mean surface pressure near the front and rear face of a cavity is also used to characterize the upstream and downstream influence. For $x/L = -0.058$, the value of C_p upstream of the cavities is shown in Figure 4. For a given value of M, the value of β has a minor effect on the amplitude of C_p. The Mach number effect is also not significant. Figure 5 shows the variation in C_p with the value

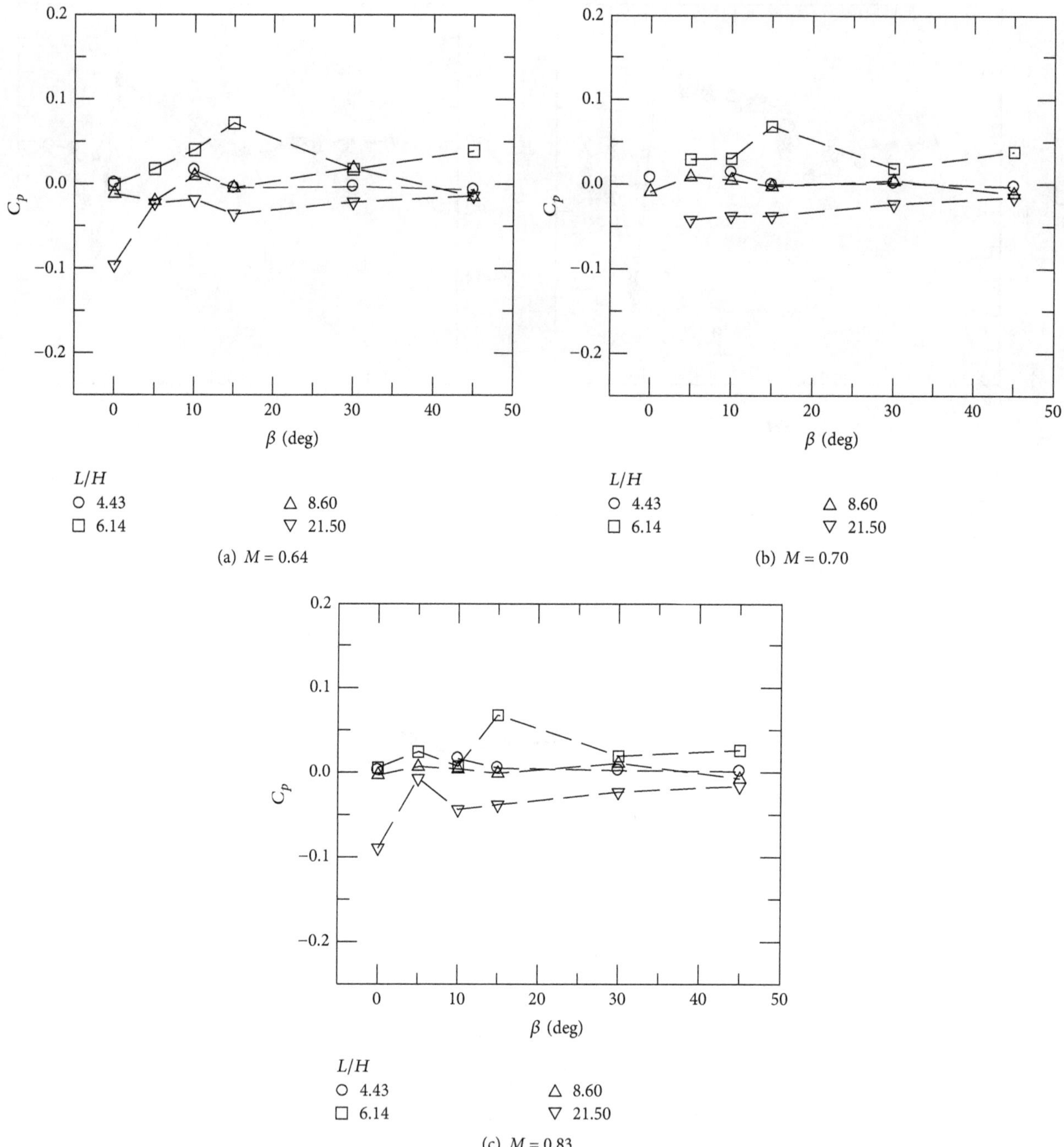

FIGURE 4: The pressure coefficient upstream of the cavity ($x/L = -0.058$): (a) $M = 0.64$, (b) $M = 0.70$, and (c) $M = 0.83$.

of β near the rear face ($x/L = 0.919$ and 1.058, which are, resp., labeled as hollow and solid symbols). The amplitude of C_p at $x/L = 0.919$ for a closed cavity ($L/H = 21.5$) is greater than that for a transitional-open ($L/H = 8.60$) or an open cavity ($L/H =$ 4.43 and 6.14). The value of β has a more significant effect on the amplitude of C_p for a value of $x/L = 1.058$ than for a value of $x/L = 0.919$. There is a minor variation in the amplitude of C_p for values of β up to 15°, following an increasing C_p as the value of β increases. For a closed cavity, there is also an increase in the pressure difference for $x/L = 0.919$ and 1.058 (greater expansion strength) near the rear face.

3.2. Surface Fluctuating Pressure Distributions. The C_{σ_p} distributions for $M = 0.83$ are shown in Figure 6. At a value of $\beta =$ 10°, the shear layer for an open cavity ($L/H = 4.43$ and 6.14) separates from the leading edge of a cavity and impinges near the rear face. The amplitude of C_{σ_p} increases gradually and reaches a peak value near the rear face. This corresponds to

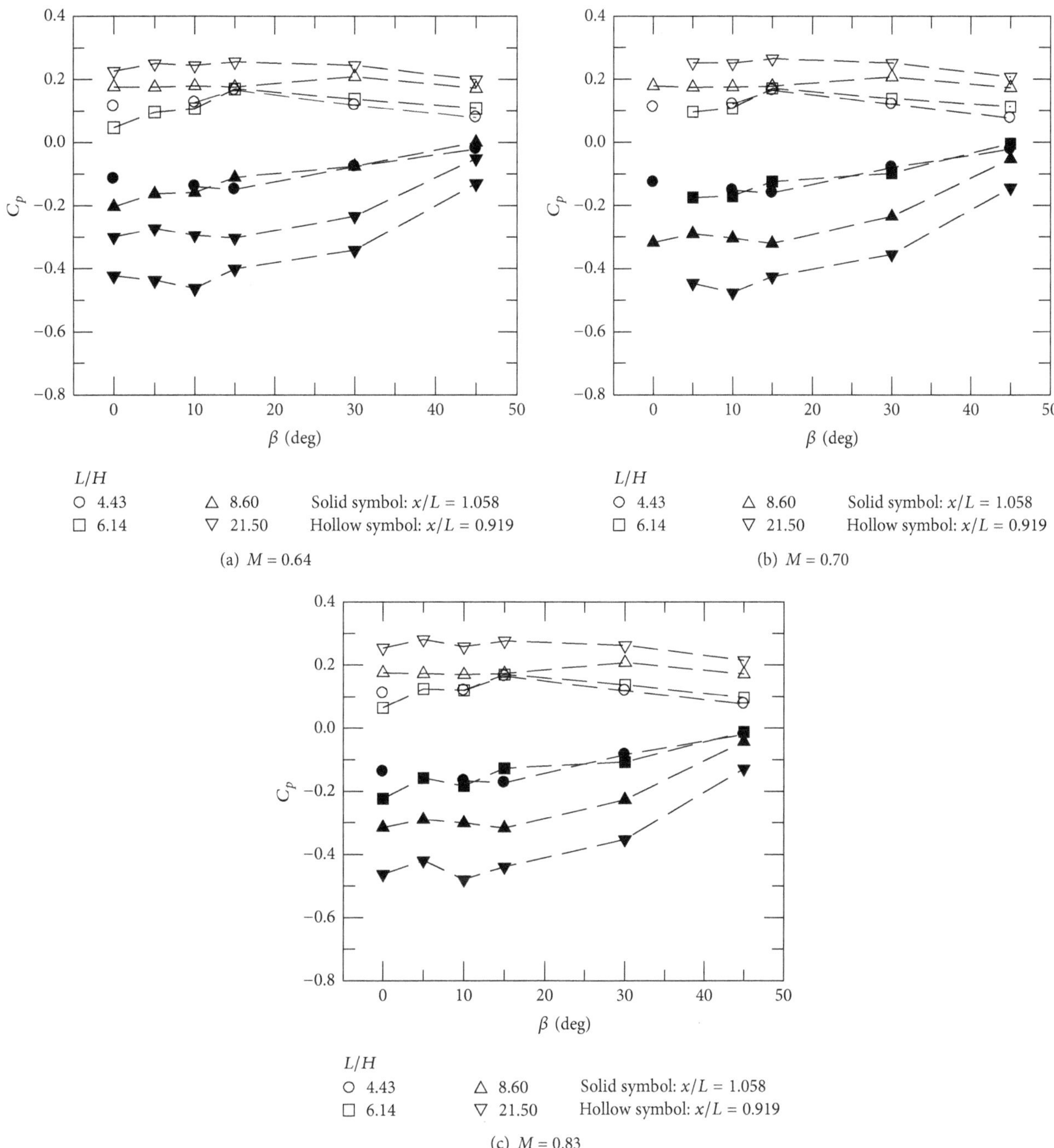

FIGURE 5: The pressure coefficients near the rear corner ($x/L = 0.919$ and 1.058): (a) $M = 0.64$, (b) $M = 0.70$, and (c) $M = 0.83$.

self-sustained oscillation [6]. For a transitional cavity ($L/H = 8.60$–14.33), there is an increase in the fluctuating pressure near the central region of the cavity ($x/L \approx 0.3$–0.7). For transitional-closed cavity and closed cavities, minor peak pressure fluctuations are observed for a value of $x/L \approx 0.36$, because of the deflection or the reattachment of shear layer. For an open cavity, the amplitude of C_{σ_p} increases ahead of the rear corner ($x/L = 0.919$) or when there is a decrease in the value of L/H. Downstream of the rear face ($x/L = 1.058$), Heller and Bliss [21] showed that the pressure fluctuations are associated with the balance between the energy that is supplied by the external flow and the energy that is dissipated by viscous losses and acoustic radiation. The peak pressure fluctuations, $C_{\sigma_{p,\max}}$, correspond to the unsteady process for the addition and removal of mass for open and transitional cavities ($L/H = 4.43$–14.33). For a value of $\beta = 45^\circ$, the C_{σ_p} distributions are similar to those for a value of $\beta = 10^\circ$. However, for a transitional cavity, the amplitude of C_{σ_p} at $x/L = 0.919$ is greater than that for an open cavity. Minor

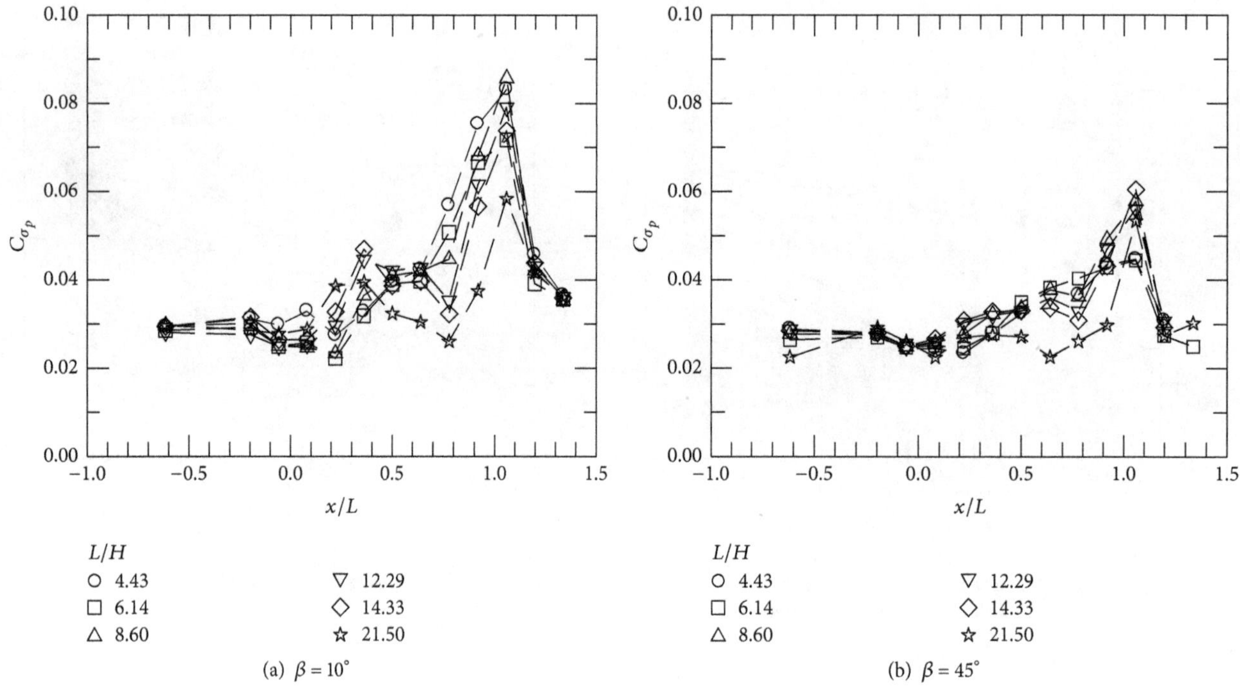

FIGURE 6: Chord-wise distributions of surface fluctuating pressure for $M = 0.83$: (a) $\beta = 10°$ and (b) $\beta = 45°$.

peak pressure fluctuations are observed at $x/L = 0.64$. The amplitude of $C_{\sigma_{p,\max}}$ at $x/L = 1.058$ also decreases significantly when there is an increase in the value of β.

The effect of the value of β on the C_{σ_p} distributions for $M = 0.83$ is shown in Figure 7. For a closed cavity (L/H = 21.50), the minor peak pressure fluctuations for $x/L \approx 0.36$ decrease as the value of β increases, as do the peak pressure fluctuations for $x/L = 1.058$. For a transitional cavity (L/H = 8.60), the effect of the yaw angle effect is minimal, except when there is a significant reduction in the amplitude of C_{σ_p} downstream of the rear face for $\beta = 45°$. For an open cavity (L/H = 4.43), the yaw angle has an evident effect on the amplitude of C_{σ_p} near the rear face. For $M = 0.64$ and 0.70, the distributions of C_{σ_p} are similar to those for $M = 0.83$. The effect of M and β on $C_{\sigma_{p,\max}}$ are shown in Figure 8. For a given value of M, the amplitude of $C_{\sigma_{p,\max}}$ at $\beta = 0°$ is the greatest for a transitional cavity and the least for a closed cavity. For a closed cavity (L/H = 21.50), there is a small variation in the amplitude of $C_{\sigma_{p,\max}}$ as the value of β varies and there is a reduction at $\beta = 45°$ for a transitional cavity ($L/H = 8.60$). For $L/H = 4.43$, an increase in the amplitude of $C_{\sigma_{p,\max}}$ is observed up to $\beta = 15°$, following a decrease as the value of β increases. However, the opposite trend is true for $L/H = 4.43$ when β = 0°–15°. It is also seen that, for an open cavity, the amplitude of $C_{\sigma_{p,\max}}$ at $\beta = 45°$ is less than that for a closed cavity. This demonstrates that the peak pressure fluctuations at $\beta = 45°$ mainly correspond to the unsteady process for the addition and removal of mass near the rear face and the self-sustained oscillation for an open cavity is attenuated.

3.3. Span-Wise Mean and Fluctuating Pressure Distributions. The mean and fluctuating pressure distributions in the span-wise direction are of interest. Figure 9 shows the C_p distributions for $M = 0.83$ at $\beta = 10°$ and 45°. For $x/L = 0.5$ and $\beta = 10°$, the C_p distributions for closed and transitional-closed cavities (L/H = 21.50–12.29) show small variations and transitional-open and open cavities (L/H = 8.60–4.43) show a slight increase from $y/L = -0.42$ to 0.42. For a value of $\beta = 45°$, the C_p distributions are asymmetric. This asymmetric feature is more significant for a transitional cavity (L/H = 8.60–14.33) and is less evident for closed and open cavities. The amplitude of C_p for open and transitional cavities increases when the value of β increases, which agrees with results that are shown in Figure 2(b) for $x/L = 0.5$. The span-wise fluctuating pressure distributions for $M = 0.83$ are shown in Figure 10, which shows a similar feature to that in Figure 6 for x/L = 0.5. The amplitude of C_{σ_p} is the least for a closed cavity (L/H = 21.50). There are small variations at $\beta = 10°$ and a gradual decrease from $y/L = -0.42$ to 0.42 at $\beta = 45°$. Notably, for a given value of L/H, the effect of the yaw angle on the amplitude of C_p and C_{σ_p} is more evident near the sidewalls of the cavity ($y/L = \pm 0.42$), particularly for the values of $\beta = 30°$ and 45°.

3.4. Power Spectra. Self-sustained oscillation in a cavity is a consequence of periodic vortex shedding and acoustic disturbance. The power spectral density (PSD) for $M = 0.83$ at $x/L = 0.919$ is shown in Figure 11. The plots are presented in terms of sound pressure level (SPL $= 20\log_{10}(p/p_s)$, $p_s = 2 \times 10^{-5}$ Pa) and are consecutively offset by 10 dB for clarity. The

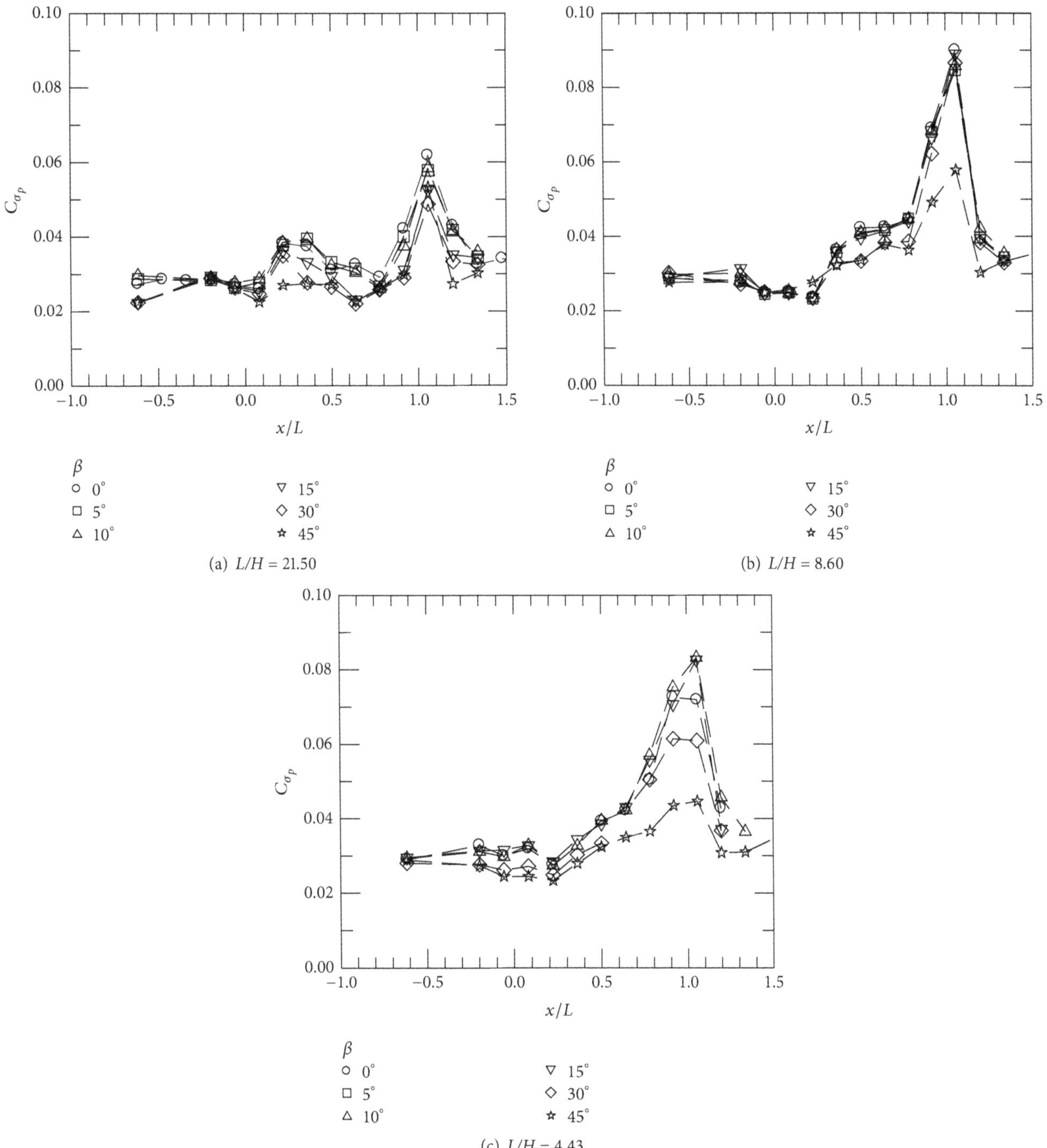

FIGURE 7: Chord-wise distributions of the surface fluctuating pressure for $M = 0.83$: (a) $L/H = 21.50$, (b) $L/H = 8.60$, and (c) $L/H = 4.43$.

uppermost plot has its original values. The PSD for a flat plate (FP) flow without the presence of a cavity is also shown, for reference. The peak frequency for this flow is approximately 4600 Hz and is induced by the perforated wall of the wind tunnel. For open cavities (L/H = 4.43 and 6.14) at $\beta = 0°$, discrete acoustic tones are observed at $f_1 \approx 1800$, $f_2 \approx 4100$, and $f_3 \approx 6300$ Hz. The values for the SPL for L/H = 4.43 are larger than those for L/H = 6.14. The 1st mode is not observed for a transitional-open cavity (L/H = 8.60) and the 2nd mode is less apparent. Taking the effect of the yaw angle into account, the frequency of the 1st mode for L/H = 4.43 and for $\beta = 10°$ and 15°, as shown in Figure 11(a), is slightly greater than that for $\beta = 0°$ and the values for the SPL decrease as the value of β increases. For the 2nd and 3rd modes, an increase in the value of β results in a reduction in their frequencies and their amplitudes. Only the 1st mode is evident for β = 30° and self-sustained oscillations are only just evident for β = 45°. Figure 11(b) shows that, for L/H = 6.14, there are

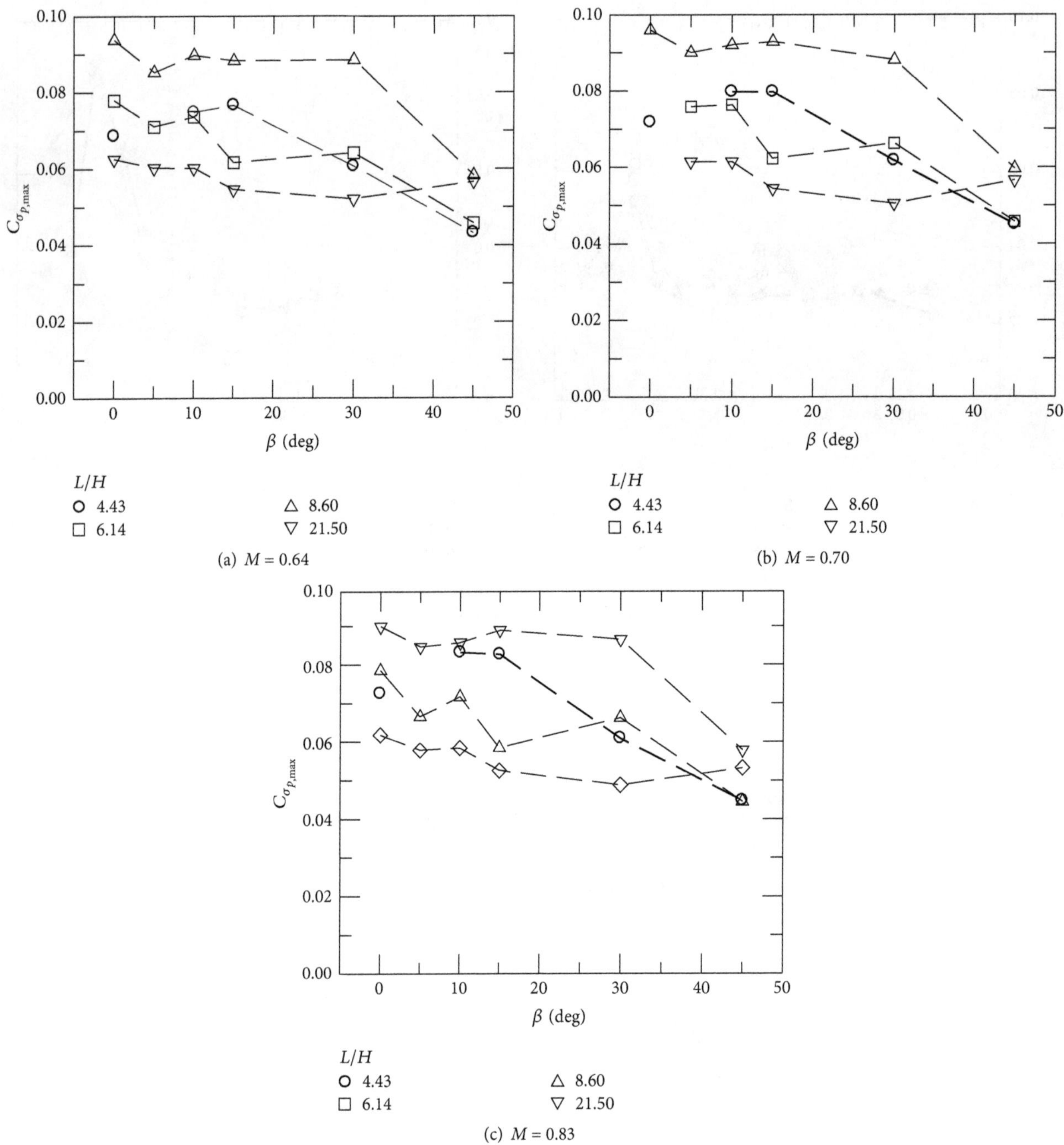

FIGURE 8: Peak pressure fluctuations: (a) $M = 0.64$, (b) $M = 0.70$, and (c) $M = 0.83$.

weaker oscillations than for $L/H = 4.43$. The 1st mode almost disappears at $\beta = 15^\circ$ and no modes are visible for $\beta = 30^\circ$ or 45°. For a transitional-open cavity ($L/H = 8.60$), Figure 11(c) shows that only the 2nd mode is evident for $\beta = 5^\circ$–15°. In summary, the 2nd mode dominates self-sustained oscillations for a cavity at yaw.

3.5. Self-Sustained Oscillations for a Cavity at Yaw. Previous studies have shown that, for rectangular cavity flow, self-sustained oscillations can be predicted using the semiempirical Rossiter's formula [22]. Figure 12 shows the amplitudes of the resonance for rectangular cavities at yaw, including open ($L/H = 4.43$ and 6.14) and transitional-open cavities ($L/H = 8.60$). In general, there is a decrease in the value of SPL as the value of β increases and the 2nd mode dominates. These values for the SPL are greater than those for the 1st and 3rd modes. However, for $M = 0.64$, the value for the SPL for the 1st mode is greater than that for the 2nd mode for $L/H = 4.43$ and 6.14, at $\beta = 15^\circ$ and 30°, and for $M = 0.70$ and 0.83 ($L/H = 4.43$) at $\beta = 30^\circ$. This shows that the dominant mode changes for different values of M and β. For $M = 0.83$ and $L/H = 8.60$, all three modes disappear at $\beta = 30^\circ$. It is also noted that an

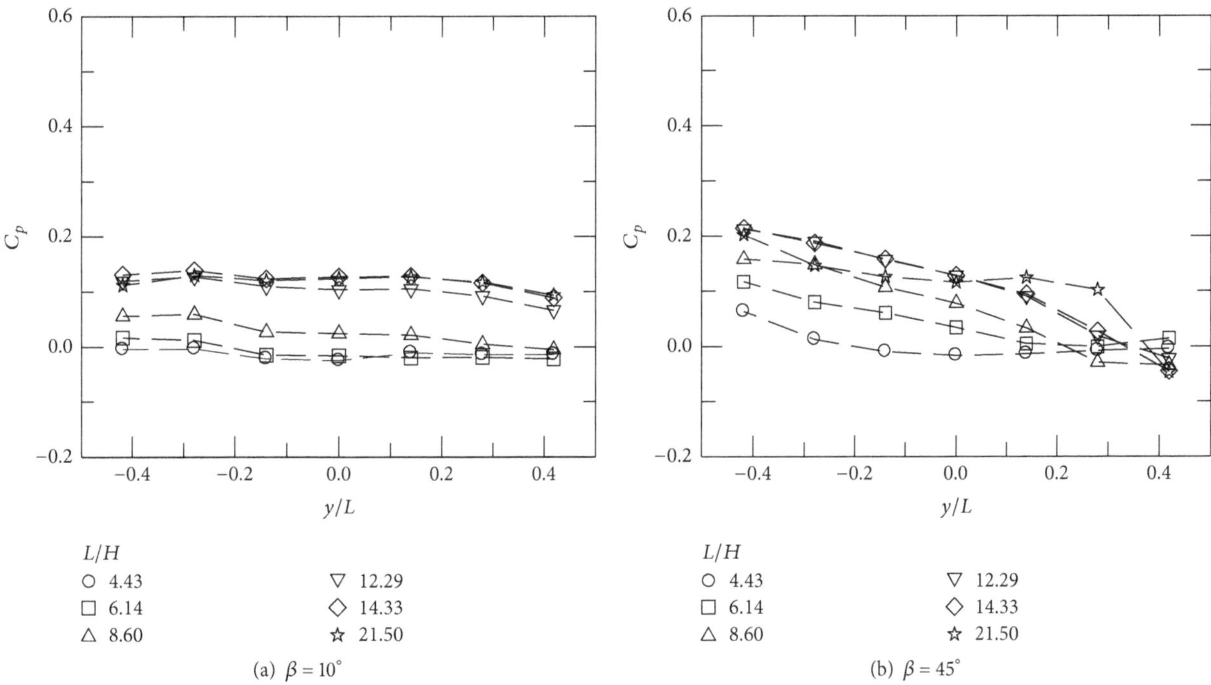

FIGURE 9: Span-wise mean surface pressure distributions for $M = 0.83$: (a) $\beta = 10°$ and (b) $\beta = 45°$.

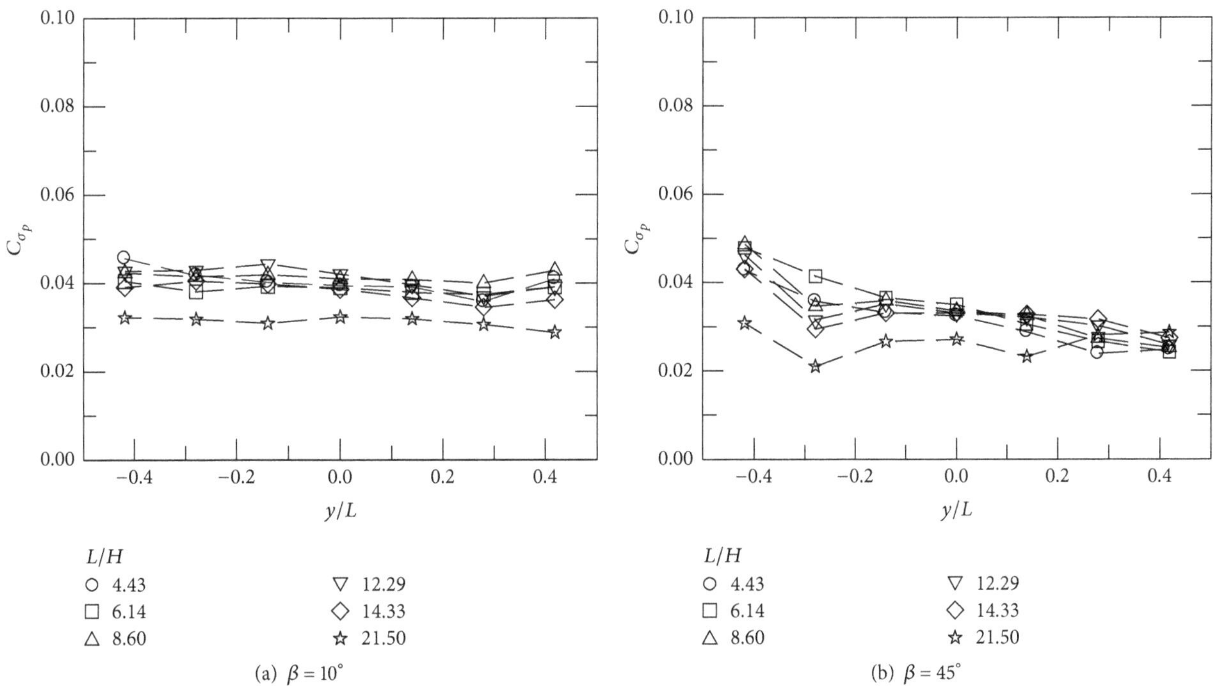

FIGURE 10: Span-wise fluctuating pressure distributions for $M = 0.83$: (a) $\beta = 10°$ and (b) $\beta = 45°$.

increase in the value of M results in an increase in the value of the SPL.

The variation in St_n with M for $L/H = 4.43$ and 6.14 is shown in Figure 13. There is also a prediction using the semiempirical Rossiter's formula ($\alpha = 0.25$ and $k_c = 0.57$ for a rectangular cavity). The uncertainty in St_n is estimated to be ±0.007, which is principally due to the resolution of the PSD. It is seen that $St_1 = 0.26$–0.35, $St_2 = 0.60$–0.68, and $St_3 = 0.97$–1.10. For a given value of M, there is a slight decrease in St_2 *and* St_3 as β increases but not in St_1. The empirical

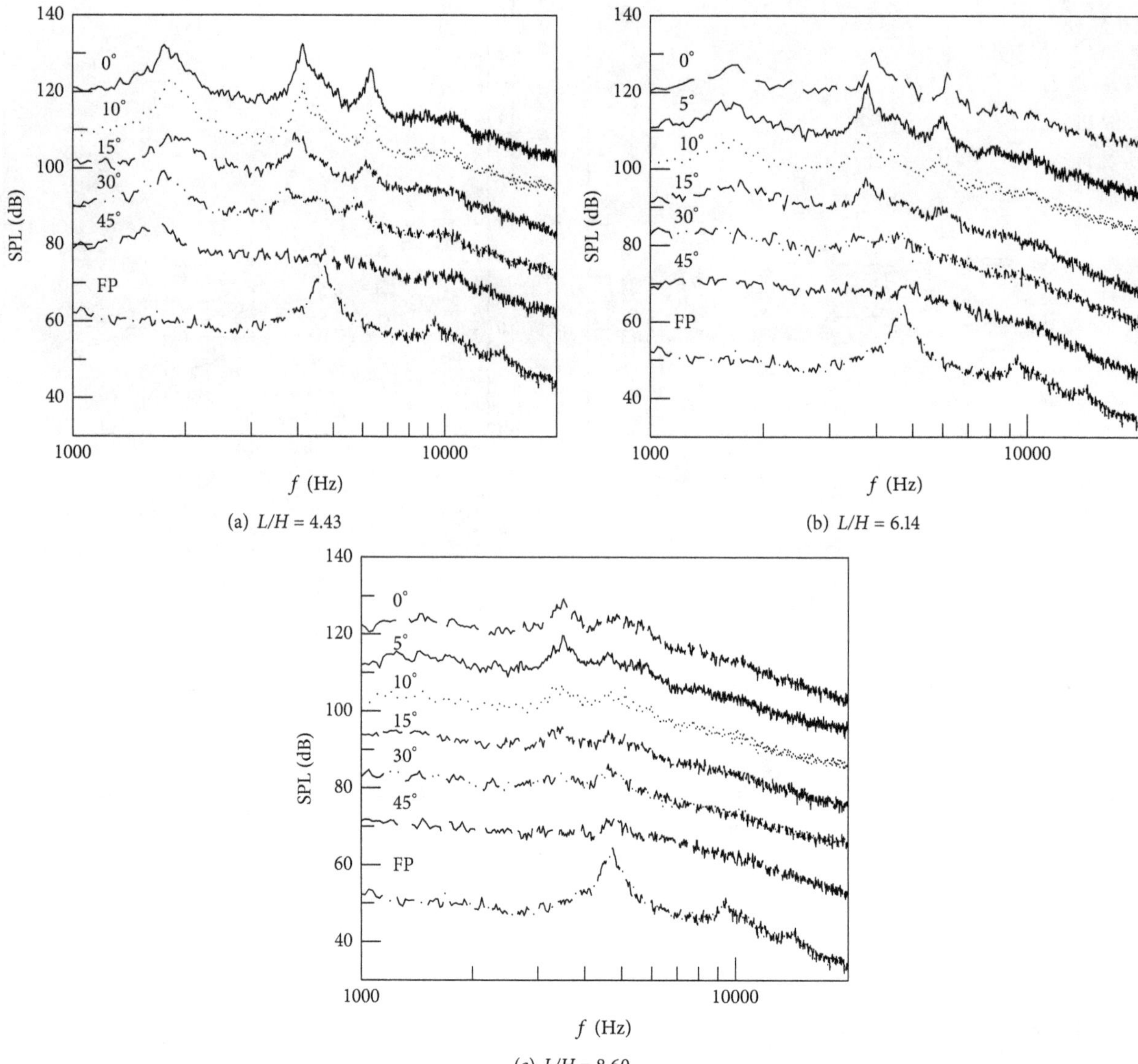

FIGURE 11: Distributions of the power spectral density for $M = 0.83$ and $x/L = 0.919$. (Each plot is offset by 10 dB along the amplitude axis to give greater clarity. The uppermost plot has its original values.)

constants for cavities at yaw are evaluated using the steepest descent method. The optimized values for α and k_c are 0.15 and 0.48, respectively. This demonstrates that a cavity at yaw has a smaller phase lag and a lesser convection velocity.

4. Conclusions

This experimental study determines the characteristics of a compressible, yawed rectangular cavity flow. The boundaries for the flow type correspond to L/H and the effect of the value of β is smaller. The mean surface pressure gradient in the chord-wise direction at the rear face decreases as the value of β increases. The peak amplitude of the fluctuating pressure is significantly less for a large value of β for open and transitional cavities. In the span-wise direction, there are asymmetric distributions for C_p and C_{σ_p}. These variations are relatively small, compared to those in the chord-wise direction. The resonant frequencies for an open cavity vary slightly with the value of β and there is a decrease in the amplitude of the PSD as β increases. The resonant modes disappear and the dominant mode changes for large values of β. Compared to the prediction using Rossiter's semiempirical formula, a cavity at yaw has less lag time and a smaller convection velocity.

Nomenclature

C_p: Static surface pressure coefficient, $(p_w - q_\infty)/q_\infty$

$C_{\sigma p}$: Fluctuating pressure coefficient, $(\sigma_p - \sigma_{p\infty})/q_\infty$

F: Objective function

f_n: Resonant frequency at mode n, Hz

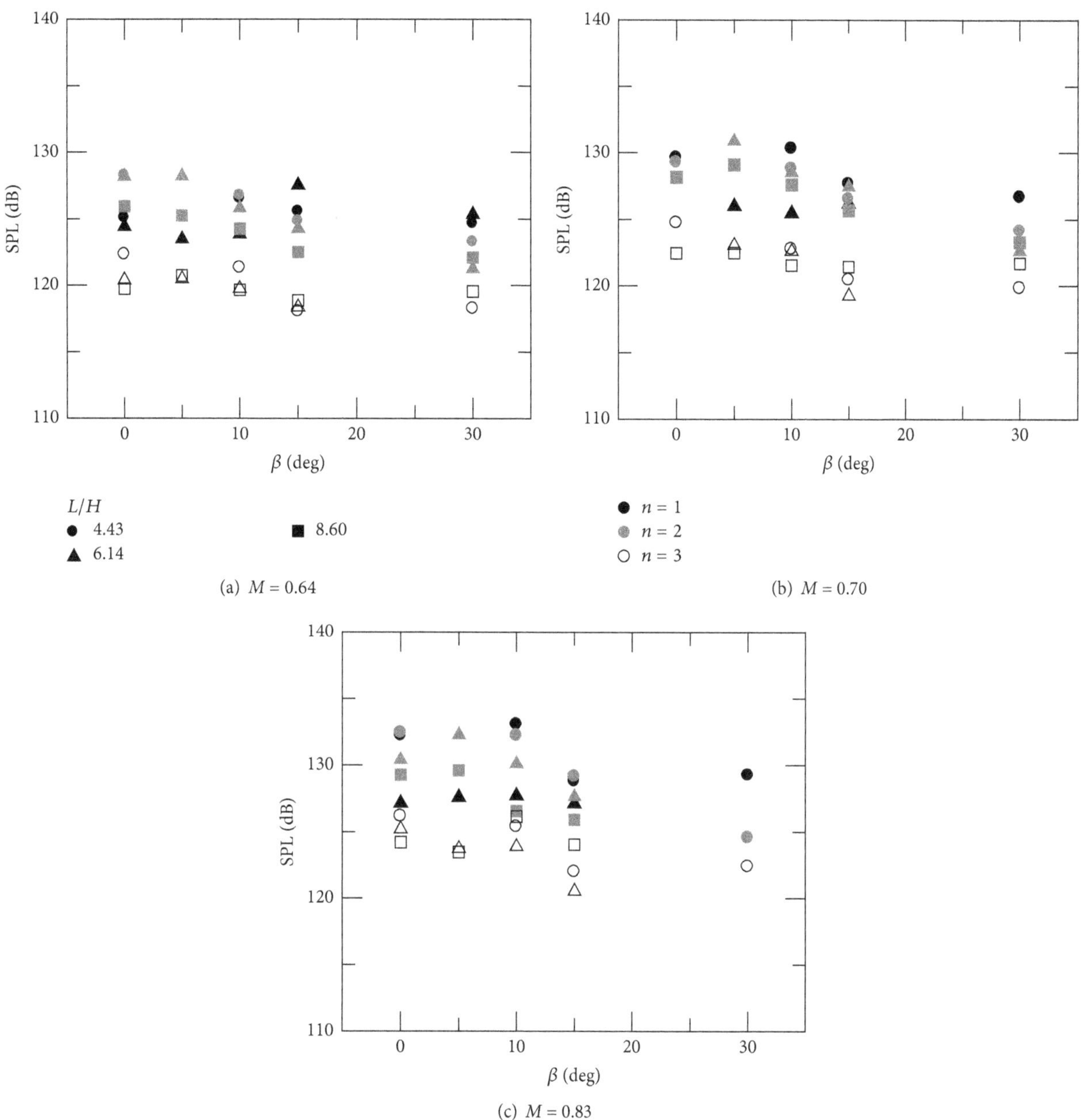

FIGURE 12: Amplitude of the resonant modes: (a) $M = 0.64$, (b) $M = 0.70$, and (c) $M = 0.83$.

k_c: Ratio of the convection velocity of vortices to freestream velocity
H: Cavity depth
L: Cavity length
M: Freestream Mach number
N: Number of test cases
n: Mode number
PSD: Power spectral density
p_w: Static surface pressure
q_∞: Freestream dynamic pressure
SPL: Sound pressure level, dB
St_n: Strouhal number at mode n, $f_n L/U_\infty$
U_∞: Freestream velocity, m/s
W: Cavity width
x: Chord-wise distance
y: Span-wise distance
α: Lag time
β: Yaw angle, degree
δ: Boundary thickness
σ_p: Standard deviation for the surface pressure signal.

Conflicts of Interest

The authors declare that they have no conflicts of interest.

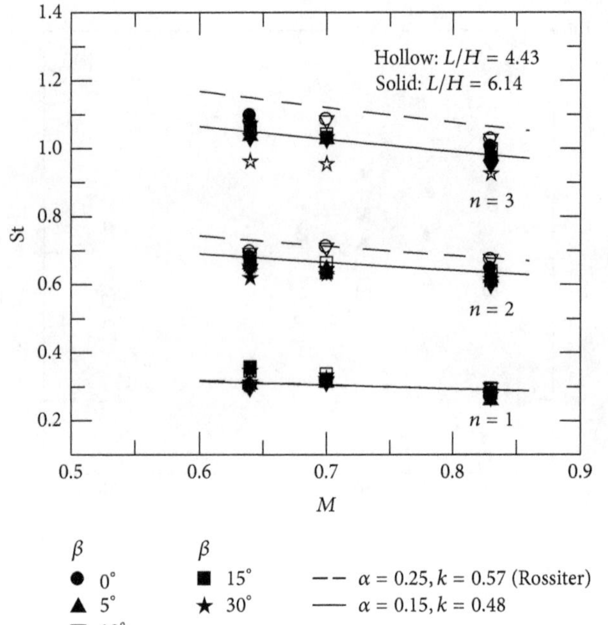

FIGURE 13: The effect of Mach number on the Strouhal number.

Acknowledgments

The authors acknowledge the support of Ministry of Science and Technology (MOST 104-2926-E-006-006-MY3), Taiwan.

References

[1] W. Li, "Suppression of supersonic cavity oscillations using pulsed upstream mass injection," *International Journal of Aerospace Engineering*, vol. 2016, Article ID 6702385, 6 pages, 2016.

[2] T. Handa, K. Tanigawa, Y. Kihara, H. Miyachi, and H. Kakuno, "Frequencies of transverse and longitudinal oscillations in supersonic cavity flows," *International Journal of Aerospace Engineering*, vol. 2015, Article ID 751029, 2015.

[3] M. B. Tracy and E. B. Plentovich, "Cavity unsteady-pressure measurements at subsonic and transonic speeds," NASA Technical Paper 3669, 1997.

[4] K.-M. Chung, "Characteristics of transonic rectangular cavity flows," *Journal of Aircraft*, vol. 37, no. 3, pp. 463–468, 2000.

[5] K.-M. Chung, "Three-dimensional effect on transonic rectangular cavity flows," *Experiments in Fluids*, vol. 30, no. 5, pp. 531–536, 2001.

[6] J. Rossiter, "Wind tunnel experiments on the flow over rectangular cavities at subsonic and transonic speeds," Royal Aircraft Establishment, Technical Report 64037, 1964.

[7] D. Rockwell and E. Naudascher, "Review—self-sustaining oscillations of flow past cavities," *Journal of Fluids Engineering*, vol. 100, no. 2, pp. 152–165, 1978.

[8] E. Savory, N. Toy, P. J. Disimile, and R. G. Dimicco, "The drag of three-dimensional rectangular cavities," *Applied Scientific Research*, vol. 50, no. 3-4, pp. 325–346, 1993.

[9] M. Czech, E. Savory, N. Toy, and T. Mavrides, "Flow regimes associated with yawed rectangular cavities," *Aeronautical Journal*, vol. 105, no. 1045, pp. 125–134, 2001.

[10] S. L. Gai, T. J. Soper, and J. F. Milthorpe, "Shallow rectangular cavities at low speeds including effects of yaw," *Journal of Aircraft*, vol. 45, no. 6, pp. 2145–2150, 2008.

[11] A. Bari and F. Chambers, "Shear layer resonance over open cavities at angles to the flow direction," in *15th Aeroacoustics Conference*, Long Beach, Calif, USA.

[12] B. H. K. Lee, D. M. Orchard, and F. C. Tang, "Flow past a yawed rectangular cavity in transonic and low supersonic flows," *Journal of Aircraft*, vol. 46, no. 5, pp. 1577–1583, 2009.

[13] P. J. Disimile, N. Toy, and E. Savory, "Pressure oscillations in a subsonic cavity at yaw," *AIAA Journal*, vol. 36, no. 7, pp. 1141–1148, 1998.

[14] P. J. Disimile and P. D. Orkwis, "Sound-pressure-level variations in a supersonic rectangular cavity at yaw," *Journal of Propulsion and Power*, vol. 14, no. 3, pp. 392–398, 1998.

[15] K.-M. Chung, P.-H. Chang, and K.-C. Chang, "Tunnel background noise on compressible convex-corner flows," *Journal of Aircraft*, vol. 50, no. 4, pp. 1011–1015, 2013.

[16] J. S. Bendat and A. G. Piersol, *Random Data: Analysis and Measurement Procedures*, John & Wileys, New York, NY, USA, 1986.

[17] K. Chung, "Characteristics of compressible rectangular cavity flows," *Journal of Aircraft*, vol. 40, no. 1, pp. 137–142, 2003.

[18] Ö. H. Ünalmis, N. T. Clemens, and D. S. Dolling, "Cavity oscillation mechanisms in high-speed flows," *AIAA Journal*, vol. 42, no. 10, pp. 2035–2041, 2004.

[19] J. A. Snyman, *Practical Mathematical Optimization: An Introduction to Basic Optimization Theory and Classical and New Gradient-Based Algorithms*, vol. 97 of *Applied Optimization*, Springer, New York, NY, USA, 2005.

[20] K. H. Lee, *Self-Sustained Oscillations of Compressible Cavity Flows [Ph.D. dissertation]*, National Cheng Kung University, Tainan, Taiwan, 2017.

[21] H. Heller and D. Bliss, "The physical mechanism of flow-induced pressure fluctuations in cavities and concepts for their suppression," in *Proceedings of the 2nd Aeroacoustics Conference*, Hampton, Va, USA.

[22] A. J. Bilanin and E. Covert, "Estimation of possible excitation frequencies for shallow rectangular cavities," *AIAA Journal*, vol. 11, no. 3, pp. 347–351, 1973.

13

Efficient Training Data Generation for Reduced-Order Modeling in a Transonic Flight Regime

Haojie Liu and Yonghui Zhao

State Key Laboratory of Mechanics and Control of Mechanical Structures, Nanjing University of Aeronautics and Astronautics, 210016 Nanjing, China

Correspondence should be addressed to Haojie Liu; liuhj@nuaa.edu.cn

Academic Editor: Zhiguang SONG

In this study, a time-dependent surrogate approach is presented to generate the training data for identifying the reduced-order model of an unsteady aerodynamic system with the variation of mean angle of attack and Mach number in a transonic flight regime. For such a purpose, a finite set of flight samples are selected to cover the flight range of concern at first. Subsequently, the unsteady aerodynamic outputs of the system under given inputs of filtered white Gaussian noise at these flight samples are simulated via CFD technique which solves Euler equations. The unsteady aerodynamic outputs, which are viewed as a time-dependent function of flight parameters, can be approximated via the Kriging technique at each time step. By this way, the training data for any combination of flight parameters in the range of concern can be obtained without performing any further CFD simulations. To illustrate the accuracy and validity of the training data generated via the proposed approach, the constructed data are used to identify the reduced-order aerodynamic models of a NACA 64A010 airfoil via a robust subspace identification algorithm. The unsteady aerodynamics and aeroelastic responses under various flight conditions in a transonic flight regime are computed. The results agree well with those obtained by using the training data of CFD technique.

1. Introduction

The techniques of computational fluid dynamics (CFD) have been widely used to simulate both linear and nonlinear flow fields for various flight vehicles. However, it is still time-consuming for any high-fidelity CFD techniques to simulate the unsteady aerodynamic loads due to the broad variation of parameters, such as different combinations of mean angle of attack and Mach number. For example, the linear doublet-lattice method has been frequently used for the aeroelastic analysis of aircraft although CFD techniques offer more accurate numerical simulations. In a transonic flight regime, however, the double-lattice method does not work properly because of the aerodynamic nonlinearities coming from shock waves and flow separations [1]. Thus, various CFD-based reduced-order models (ROMs) have been developed [2] so as to provide an effective way to simulate unsteady aerodynamic loads with a high level of accuracy.

In general, the unsteady aerodynamic ROMs can be classified into frequency-domain type and time-domain type. The frequency-domain ROMs mainly contain the proper orthogonal decomposition (POD) approach [3], balanced POD [4], and the harmonic balance approach [5]. On the other hand, the time-domain ROMs include the Eigensystem realization algorithm (ERA) [6], the Volterra theory [7], the auto regressive-moving-average (ARMA) model [8], surrogate models via artificial neural networks [9–11] and Kriging technique [12], and the Wiener-type cascade model [13].

Although a significant progress has been made for the unsteady aerodynamic ROMs, almost all aforementioned ROMs are only valid for a set of fixed flight parameters. It is still time-consuming to generate the unsteady aerodynamic ROMs for a range of flight parameters since the same procedure of computation has to be repeated under each flight condition in the range of concern. Recently, several reduced-order modeling approaches have been proposed to predict unsteady aerodynamic loads for a range of flight parameters. For example, a surrogate-based recurrence framework ROM was developed to model the unsteady aerodynamics on a rotating airfoil [14]. Yet, it is time-consuming to generate the training data for each combination of parameters in the parameter space. A ROM adaptation approach based on the interpolation in a tangent space to a Grassmann manifold was proposed to predict the aeroelastic characteristics for an F-16 configuration [15]. To obtain new basis vectors for a new flight condition via interpolation, the size of the basis vectors of different local ROMs should be the same so that the accuracy of the POD approach may be decreased. A ROM approach by combining linear convolution with a nonlinear correction factor was proposed to model the aerodynamic characteristics for multiple Mach regimes [16]. However, the approach gives a distinct phase shift at higher oscillation frequencies. A time-dependent surrogate model to fit the relationship between flight parameters and step functions was proposed to model the unsteady and nonlinear aerodynamic loads [17]. Only one kind of input, the pitching motion of the aircraft in their study, can be taken into account at each time. A Kriging surrogate model was proposed to model the unsteady aerodynamic forces with respect to a range of Mach numbers [18]. However, the computational cost for Kriging model increased significantly in order to take the Mach number into consideration. Recently, a reduced-order modeling approach based on recurrent local linear neuro-fuzzy models was developed to model the generalized aerodynamic forces over a range of Mach numbers [19]. To guarantee the accuracy of the local linear neuro-fuzzy model, the flight parameters for training data should be selected carefully in the flight range of concern. Very recently, a nonlinear interpolation method based on a set of local linear state-space models was proposed to model the generalized aerodynamic forces for an elastic wing with control surfaces [20]. Nevertheless, enough local linear state-space models were required to cover the flight range of concern and only Mach number variations were taken into consideration.

Almost all aforementioned unsteady aerodynamic ROMs, which take flight parameter variations into account, require enough training data to capture the dynamic characteristics of the unsteady aerodynamic systems. However, it is time-consuming to generate the training data via any direct CFD simulations when both Mach number and mean angle of attack are taken into consideration. The motivation of this study is to generate the training data efficiently for a range of flight parameters including mean angle of attack and Mach number in a transonic flight regime. For such a purpose, a time-dependent surrogate approach is proposed. Once the Kriging models are constructed via the proposed approach, the training data under an arbitrary flight condition in the range of concern can be obtained without performing any further CFD simulations. The remainder of the paper is organized as follows. In Section 2, the theoretical background of the training data generation approach and the robust subspace algorithm is presented. In Section 3, the unsteady aerodynamics and aeroelastic responses of a NACA 64A010 airfoil under different flight conditions in a transonic flight regime are investigated to validate the proposed approach. In Section 4, some conclusions are given.

2. Theoretical Background

In this section, the theoretical background of the training data generation approach and the robust subspace algorithm is presented. The flowchart of the reduced-order modeling approach is shown in Figure 1. The first step is to obtain the unsteady aerodynamic outputs of the system via direct CFD simulations at a finite set of flight samples selected to cover the flight range of concern. Then, a time-dependent surrogate method based on the Kriging technique is used to approximate the relationship between the unsteady aerodynamic outputs of the system and the flight parameters. Afterwards, with the training data obtained, the robust subspace algorithm is implemented to identify the discrete-time state-space model under an arbitrary flight condition.

2.1. Selection of Excitation Signal. As well known, the dynamic characteristics of an unsteady aerodynamic system of concern should be appropriately excited via the input signal in order to construct the training data for successfully identifying the corresponding time-domain ROMs. For this purpose, the filtered white Gaussian noise (FWGN) is used as the input signal to make CFD simulations. Under such a type of excitation, only the frequency response spectrum of concern can be excited and there is no need to repeat computations for each combination of frequency and amplitude.

Once the type of input signal is determined, it is necessary to select the sampling rate and the sampling length for the input signal. The sampling rate f_S determined by the time step Δt of the CFD solver is given as follows:

$$f_S = \frac{1}{\Delta t}. \tag{1}$$

According to the Nyquist sampling theorem, the Nyquist frequency f_N is $f_S/2$ which must exceed the maximal frequency of concern. Furthermore, the frequency resolution of the input signal with the sampling length N, given by f_N/N, should be adequate for the representation of the minimal frequency of concern. Hence, the time step Δt and the sampling length N for the input signal need to be adjusted carefully.

Although only one unsteady aerodynamic simulation under FWGN excitation is required under each flight condition, it is still time-consuming to generate the training data to capture the dynamic characteristics of an unsteady aerodynamic system in the mean angle of attack and Mach number space.

2.2. Time-Dependent Surrogate Approach for Training Data Generation. To generate the training data for a range of flight parameters efficiently, a time-dependent surrogate method based on the Kriging technique is proposed in this subsection. Similar to the previous study [17], the unsteady aerodynamic outputs of the system under the FWGN excitation and

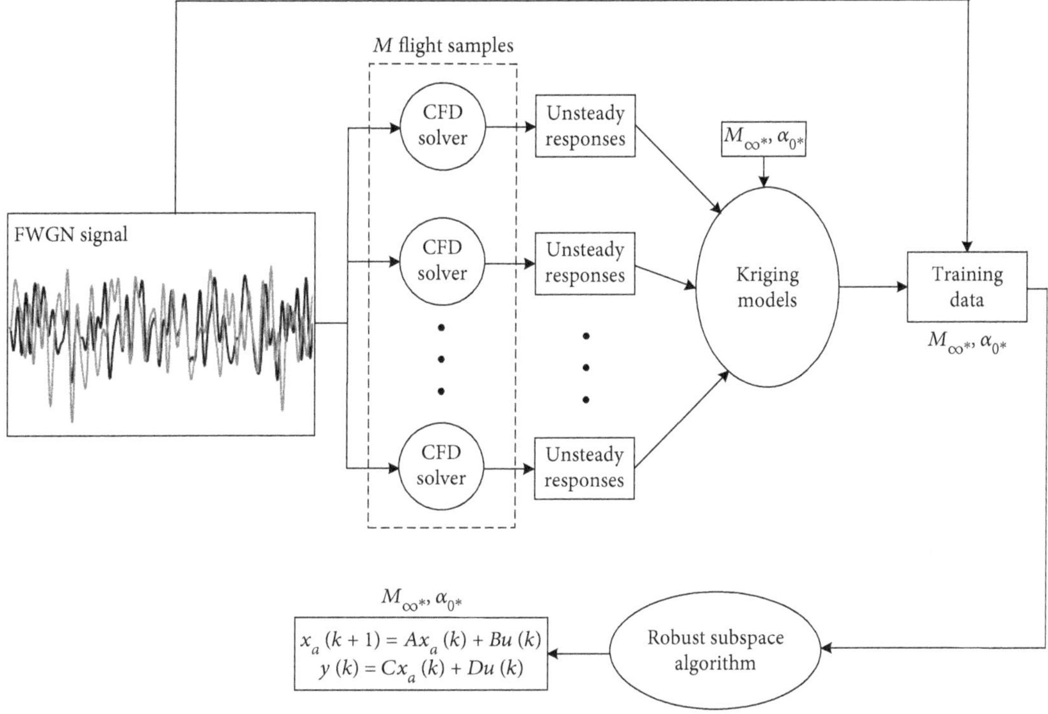

Figure 1: The flowchart of the reduced-order modeling approach.

a finite set of flight conditions are considered as a set of time-correlated spatial processes. Under such an assumption, the surrogate techniques can be used to model the time-dependent outputs of the aerodynamic system at each time step as a function of flight parameters. Here, the Kriging technique is implemented since it has good accuracy and robustness with small data sets [21].

To approximate the input-output relationship between the flight parameters and the unsteady aerodynamic outputs of the system, M flight samples need to be selected to cover the range of concern at first. The design of experimental studies [22] can be used to select these samples from the mean angle of attack α_0 and Mach number M_∞ space. It is worthy to note that the selection of the input parameters strongly depends on the flight regime of concern and the aircraft configuration. For example, more flight parameter samples need to be placed at the region where there exist strong aerodynamic nonlinearities in the flow field. Subsequently, the input vector $\mathbf{X}$ for the Kriging technique can be defined as

$$\mathbf{X} = \begin{bmatrix} M_{\infty,1} & \alpha_{0,1} \\ M_{\infty,2} & \alpha_{0,2} \\ \vdots & \vdots \\ M_{\infty,M} & \alpha_{0,M} \end{bmatrix}. \quad (2)$$

Under each flight condition defined by $M_{\infty,i}$ and $\alpha_{0,i}$, an unsteady aerodynamic computation needs to be performed via direct CFD simulation under the given FWGN excitation. The corresponding outputs of the aerodynamic system of the M flight samples can be written as

$$\mathbf{Y}_o = \begin{bmatrix} \mathbf{y}_{1,1} & \mathbf{y}_{2,1} & \cdots & \mathbf{y}_{N,1} \\ \mathbf{y}_{1,2} & \mathbf{y}_{2,2} & \cdots & \mathbf{y}_{N,2} \\ \vdots & \vdots & \ddots & \vdots \\ \mathbf{y}_{1,M} & \mathbf{y}_{2,M} & \cdots & \mathbf{y}_{N,M} \end{bmatrix}, \quad (3)$$

where $\mathbf{y}_{i,j}$ is the aerodynamic output vector and N is the number of time steps of the FWGN excitation. With the input vector $\mathbf{X}$ and the output vector $\mathbf{Y}_o$ known, the Kriging technique can be used to approximate the relationship between the flight parameter and the aerodynamic outputs of the system at each time step.

What follows is a brief description of the Kriging technique, where the Kriging model approximates the target function at an untried site $\mathbf{x}_*$ as

$$\Phi(\mathbf{x}_*) = \mathbf{f}(\mathbf{x}_*)\boldsymbol{\beta} + z(\mathbf{x}_*), \quad (4)$$

Here, $\mathbf{f}(\mathbf{x}_*)$ and $\boldsymbol{\beta}$ are the regression model and the regression coefficients, respectively. In (4), the first term is the mean value, which can be thought as a globally valid trend function. The second term $z(\mathbf{x}_*)$ given in (4) is a Gaussian distributed error term with zero mean and variance σ^2. To construct the Kriging model, the values of the regression coefficients $\boldsymbol{\beta}$ can be approximated by solving a least-square regression problem as follows:

$$\widehat{\boldsymbol{\beta}} = \left(\mathbf{F}^{\mathrm{T}}\left(\mathbf{R}^{-1}\right)\mathbf{F}\right)^{-1}\mathbf{F}^{\mathrm{T}}\mathbf{R}^{-1}\mathbf{h}, \tag{5}$$

where $\mathbf{R}$ is the spatial correlation matrix. $\mathbf{h}$ and $\mathbf{F}$ are the $M \times 1$ output vector and the $M \times M$ mapping matrix at the sample inputs, respectively. $\mathbf{h}$ and $\mathbf{F}$ are defined as

$$\mathbf{h} = \begin{bmatrix} y_1 \\ y_2 \\ \vdots \\ y_M \end{bmatrix}, \quad \mathbf{F} = \begin{bmatrix} f_1(\mathbf{x}_1) & f_2(\mathbf{x}_1) & \cdots & f_{r+1}(\mathbf{x}_1) \\ f_1(\mathbf{x}_2) & f_2(\mathbf{x}_2) & \cdots & f_{r+1}(\mathbf{x}_2) \\ \vdots & \vdots & \ddots & \vdots \\ f_1(\mathbf{x}_M) & f_2(\mathbf{x}_M) & \cdots & f_{r+1}(\mathbf{x}_M) \end{bmatrix}. \tag{6}$$

With the vector $\mathbf{r} = [R(\mathbf{x}_*, \mathbf{x}_1)R(\mathbf{x}_*, \mathbf{x}_2) \cdots R(\mathbf{x}_*, \mathbf{x}_M)]^{\mathrm{T}}$, the prediction at an unsampled location $\mathbf{x}_*$ can be obtained as

$$\widehat{\Phi}(\mathbf{x}_*) = \mathbf{f}(\mathbf{x}_*)\widehat{\boldsymbol{\beta}} + \mathbf{r}^{\mathrm{T}}\mathbf{R}^{-1}\left(\mathbf{h} - \mathbf{F}\widehat{\boldsymbol{\beta}}\right). \tag{7}$$

It has to be noted that one mapping function $\widehat{\Phi}$ needs to be constructed for each component of the output vector at each time step. Once the Kriging models are constructed, the unsteady aerodynamic outputs of the system under an arbitrary flight condition in the range of concern can be obtained without performing any further CFD simulations. For example, the corresponding aerodynamic outputs of the system under the same FWGN excitation for the mean angle of attack α_{0*} and Mach number $M_{\infty*}$ can be written as

$$\begin{aligned}&[y_{1*} \ \ y_{2*} \ \ \cdots \ \ y_{N*}] \\ &= \left[\widehat{\Phi}_1(M_{\infty*}, \alpha_{0*})\widehat{\Phi}_2(M_{\infty*}, \alpha_{0*})\cdots\widehat{\Phi}_N(M_{\infty*}, \alpha_{0*})\right].\end{aligned} \tag{8}$$

Hence, the training data under an arbitrary flight condition in the range of concern can be obtained efficiently via the proposed approach.

2.3. Robust Subspace Algorithm. With the training data obtained via the time-dependent surrogate approach, a time-domain ROM approach can be used to construct the unsteady aerodynamic ROM. In this study, the robust subspace algorithm [23] is implemented to identify the discrete-time state-space models. This algorithm is always numerically stable and convergent and has no identification problems, such as lack of convergence, slow convergence, or numerical instability.

A brief description of the algorithm is given as follows. The discrete-time state-space model assumes the following model structure:

$$\begin{aligned}\mathbf{x}_a(k+1) &= \mathbf{A}\mathbf{x}_a(k) + \mathbf{B}\mathbf{u}(k), \\ \mathbf{y}(k) &= \mathbf{C}\mathbf{x}_a(k) + \mathbf{D}\mathbf{u}(k),\end{aligned} \tag{9}$$

where $\mathbf{x}_a$, $\mathbf{u}$, and $\mathbf{y}$ are the vectors of state, input, and output variables, respectively. Given the training data, four constant matrices $\mathbf{A}$, $\mathbf{B}$, $\mathbf{C}$, and $\mathbf{D}$ are estimated such that the unsteady aerodynamic outputs of the system under the FWGN excitation can be reproduced.

At first, the input and output Hankel block matrices based on the training data are defined as

$$\begin{aligned}\begin{pmatrix}\mathbf{U}_p \\ \hline \mathbf{U}_f\end{pmatrix} &\overset{\text{def}}{=} \begin{pmatrix} \mathbf{u}_0 & \mathbf{u}_1 & \cdots & \mathbf{u}_{j-1} \\ \mathbf{u}_1 & \mathbf{u}_2 & \cdots & \mathbf{u}_j \\ \cdots & \cdots & \cdots & \cdots \\ \mathbf{u}_{i-1} & \mathbf{u}_i & \cdots & \mathbf{u}_{i+j-2} \\ \hline \mathbf{u}_i & \mathbf{u}_{i+1} & \cdots & \mathbf{u}_{i+j-1} \\ \mathbf{u}_{i+1} & \mathbf{u}_{i+2} & \cdots & \mathbf{u}_{i+j} \\ \cdots & \cdots & \cdots & \cdots \\ \mathbf{u}_{2i-1} & \mathbf{u}_{2i} & \cdots & \mathbf{u}_{2i+j-2} \end{pmatrix}, \\ \begin{pmatrix}\mathbf{Y}_p \\ \hline \mathbf{Y}_f\end{pmatrix} &\overset{\text{def}}{=} \begin{pmatrix} \mathbf{y}_0 & \mathbf{y}_1 & \cdots & \mathbf{y}_{j-1} \\ \mathbf{y}_1 & \mathbf{y}_2 & \cdots & \mathbf{y}_j \\ \cdots & \cdots & \cdots & \cdots \\ \mathbf{y}_{i-1} & \mathbf{y}_i & \cdots & \mathbf{y}_{i+j-2} \\ \hline \mathbf{y}_i & \mathbf{y}_{i+1} & \cdots & \mathbf{y}_{i+j-1} \\ \mathbf{y}_{i+1} & \mathbf{y}_{i+2} & \cdots & \mathbf{y}_{i+j} \\ \cdots & \cdots & \cdots & \cdots \\ \mathbf{y}_{2i-1} & \mathbf{y}_{2i} & \cdots & \mathbf{y}_{2i+j-2} \end{pmatrix},\end{aligned} \tag{10}$$

where the number j of columns is given as

$$j = N - 2i + 1. \tag{11}$$

In (11), N is the length of training data and i is the number of block rows in the Hankel block matrices. Then, based on the RQ decomposition of the Hankel block matrix, the weighted oblique projection $o_i\Pi_{\mathbf{U}_f^\perp}$ can be computed. The singular value decomposition of the weighted oblique projection is given as follows:

$$o_i\Pi_{\mathbf{U}_f^\perp} = \mathbf{U}\mathbf{S}\mathbf{V}^T. \tag{12}$$

The order of state-space model can be determined by inspecting the singular values in $\mathbf{S}$. Afterwards, $\mathbf{U}_1$ and

$\mathbf{S}_1$ can be obtained according to the order and the matrices. The extended observability matrix Γ_i can be computed as

$$\Gamma_i = \mathbf{U}_1 \mathbf{S}_1^{1/2}. \quad (13)$$

Here, Γ_{i-1} denotes the matrix Γ_i without the last p rows. To this end, the system matrices **A**, **B**, **C**, and **D** can be obtained with the matrices Γ_i, Γ_{i-1}, and the decomposition matrix of the Hankel block matrix. For further details, one can refer [23].

In summary, the efficiency of the algorithm enables one to identify the unsteady aerodynamic ROM quickly under an arbitrary flight condition.

3. Case Studies

This section presents the study on the unsteady aerodynamic problem and aeroelastic problem of a NACA 64A010 airfoil, respectively, for the mean angle of attack α_0 ranging from 0 to 5 deg and the Mach number M_∞ ranging from 0.75 to 0.85 so as to validate the proposed approach.

The airfoil of concern has two degrees of freedom, as shown in Figure 2, where $\bar{x}_\alpha$ is the dimensionless distance between the center of gravity and the stiffness center, $\bar{r}_\alpha$ is the gyration radius of the airfoil around the stiffness center, ω_α and ω_h are the uncoupled natural frequencies of the pitch and plunge, and b, μ, V_∞, and V_∞^* are the half-chord length, mass ratio, free-stream velocity, and dimensionless velocity, respectively. With the dimensionless time $\tau = \omega_\alpha \cdot t$, the dynamic equation of the airfoil can be established in the following state-space form.

$$\dot{\mathbf{x}} = \mathbf{A}_s \cdot \mathbf{x} + \mathbf{B}_s \cdot \mathbf{f}_a(\mathbf{x}, t). \quad (14)$$

The vectors and matrices in (14) can be written as

$$\begin{aligned}
\mathbf{A}_s &= \begin{bmatrix} \mathbf{0}_{2\times2} & \mathbf{I}_{2\times2} \\ -\mathbf{M}^{-1}\mathbf{K} & \mathbf{0}_{2\times2} \end{bmatrix}, \\
\mathbf{B}_s &= \begin{bmatrix} \mathbf{0}_{2\times2} \\ \mathbf{M}^{-1} \end{bmatrix}, \\
\mathbf{f}_a &= \frac{(V_\infty^*)^2}{\pi} \begin{Bmatrix} -C_l \\ 2C_m \end{Bmatrix}, \\
\boldsymbol{\xi} &= \begin{Bmatrix} \dfrac{h}{b} \\ \alpha \end{Bmatrix}, \\
\mathbf{x} &= \begin{Bmatrix} \boldsymbol{\xi} \\ \dot{\boldsymbol{\xi}} \end{Bmatrix},
\end{aligned} \quad (15)$$

where the mass and stiffness matrices are defined as

$$\begin{aligned}
\mathbf{M} &= \begin{bmatrix} 1 & \bar{x}_\alpha \\ \bar{x}_\alpha & \bar{r}_\alpha^2 \end{bmatrix}, \\
\mathbf{K} &= \begin{bmatrix} \left(\dfrac{\omega_h}{\omega_\alpha}\right)^2 & 0 \\ 0 & \bar{r}_\alpha^2 \end{bmatrix}.
\end{aligned} \quad (16)$$

The structural parameters of the airfoil are chosen to be $\bar{x}_\alpha = 1.8$, $\bar{r}_\alpha^2 = 3.48$, $\bar{a} = -2.0$, $\omega_h = 100$ rad/sec, $\omega_\alpha = 100$ rad/sec, and $\mu = 60$, respectively.

3.1. Training Data for the Unsteady Aerodynamic System. The CFD technique which solves the Euler equations [13, 18] is used to compute the unsteady aerodynamics over the airfoil in transonic regime. The spatial discretization of the CFD solver is based on the cell-centered finite-volume approach. The widely used central scheme with artificial dissipation is employed for convective flux calculation. The dual time-stepping approach is employed for unsteady simulations. After performing mesh and time step sensitivity analyses, the unstructured mesh with 4098 cells and dimensionless fluid time step 0.2 are used in the present study. Figure 3 presents the CFD grid consisting of 4098 triangle elements and 2145 nodes. The high-fidelity CFD solver needs to be validated at first. In order to validate the Euler solver used in the present study, the test case CT6 of [24] is studied. For such case, the free stream Mach number is 0.796 and the 64A010 airfoil pitches around its quarter-chord point and the pitching motion are governed as

$$\alpha(t) = \alpha_0 + \alpha_A \sin(\omega t), \quad (17)$$

where $\alpha_0 = 0$ deg and $\alpha_A = 1.01$ deg are the mean angle of attack and the amplitude of pitching oscillation, respectively. The relationship between the angular frequency ω and the reduced frequency $k_r = 0.202$ is described as

$$k_r = \frac{\omega c}{2V_\infty}, \quad (18)$$

where c and V_∞ are the chord length and the free-stream velocity, respectively. As shown in Figure 4, the CFD lift coefficients agree well with the experimental data. However, there are apparent differences for the moment coefficients between the numerical and experimental results. Such discrepancy may be due to the high sensitivity of the moment coefficients to the location of shock wave.

In the numerical simulations, the dimensionless bandwidth of the FWGN excitation signal is chosen to be 0.01~0.4. Figure 5 presents the time histories of the two FWGN excitation signals selected to drive the pitch and plunge motions of the airfoil, respectively. The length of the input signal in CFD simulations is taken as 2000. Subsequently, 20 flight samples are selected for the mean angle of attack α_0 in [0, 5] deg and for the Mach number M_∞ in [0.75, 0.85], respectively, as shown in Figure 6. For each flight

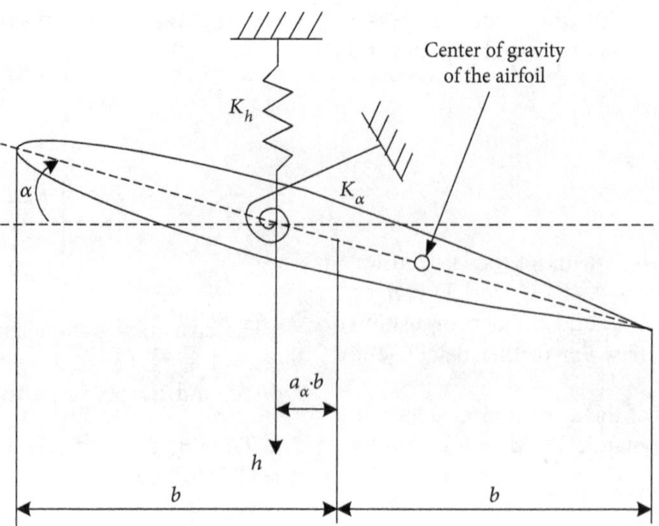

FIGURE 2: An airfoil of two degrees of freedom.

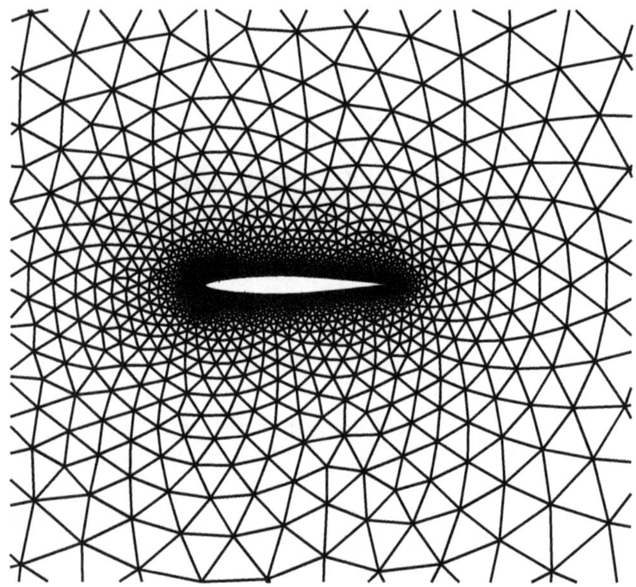

FIGURE 3: The CFD grid for the NACA64A010 airfoil.

parameter sample, an unsteady aerodynamic simulation is performed via the CFD solver.

According to the time-dependent surrogate approach, 4000 Kriging models are required since the length of the FWGN excitation is 2000 for two unsteady aerodynamic outputs of the system, that is, the lift coefficient C_l and the moment coefficient C_m. To demonstrate the accuracy of the training data, C_l and C_m are predicted via the Kriging models under two flight conditions in a transonic flight regime, one with $M_\infty = 0.76$ and $\alpha_0 = 1$ deg and the other with $M_\infty = 0.82$ and $\alpha_0 = 4$ deg. As shown in Figures 7 and 8, the results predicted via the Kriging models agree well with those obtained via the direct CFD technique.

The CPU time to construct the Kriging models is shown in Table 1, where all the computations are performed by using a laptop with two CPUs of 2.4 GHz. Even though the total CPU time cost associated with the construction of the Kriging models is a little bit expensive, the training data under an arbitrary flight condition in the range of concern can be obtained efficiently without performing any further CFD simulation. For example, the computational time of the direct CFD approach which solves Euler equations is about 0.284 h in order to obtain the training data at $M_\infty = 0.76$ and $\alpha_0 = 3$ deg, while the computational time of the Kriging models for the same case is 0.006 h only.

3.2. Unsteady Aerodynamic Prediction. To demonstrate the application of the training data in unsteady aerodynamic prediction, the constructed data are used to identify the state-space models via the robust subspace algorithm. The

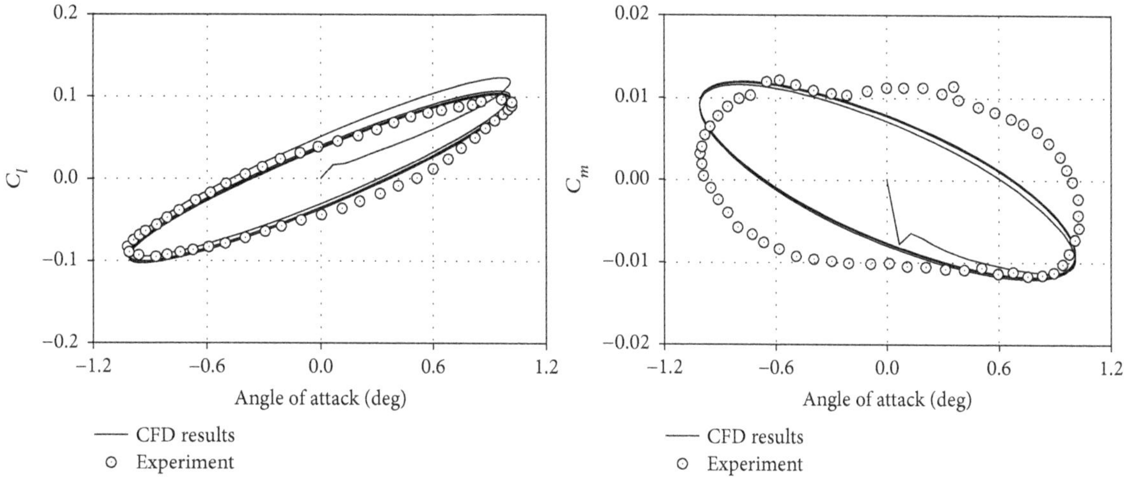

Figure 4: Lift and moment coefficient loops for a NACA 64A010 airfoil at Mach number 0.796.

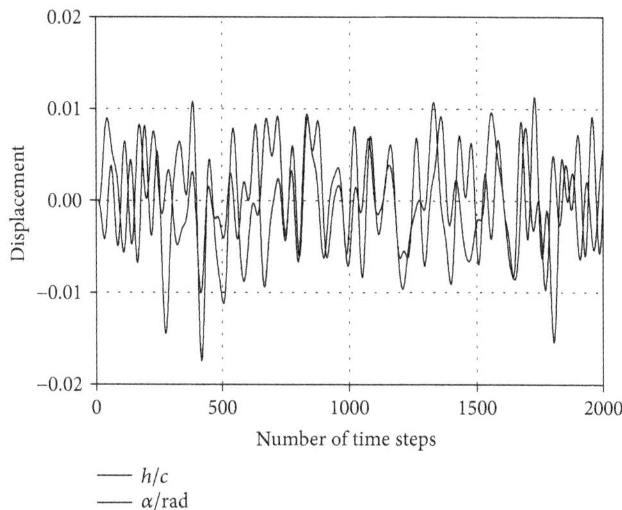

Figure 5: The FWGN excitation signals to drive pitch and plunge motions.

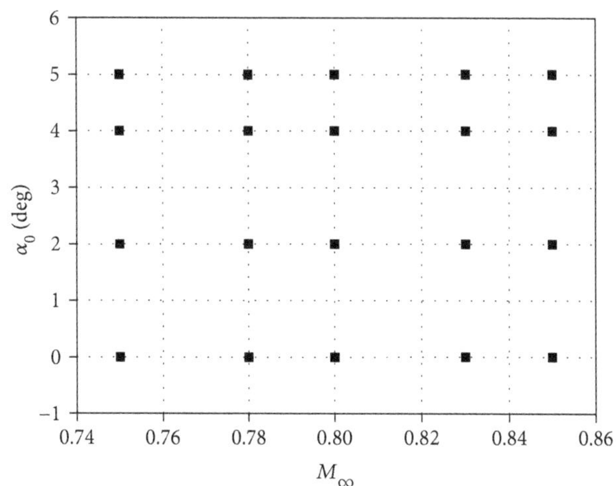

Figure 6: The flight parameter samples to cover the range of concern.

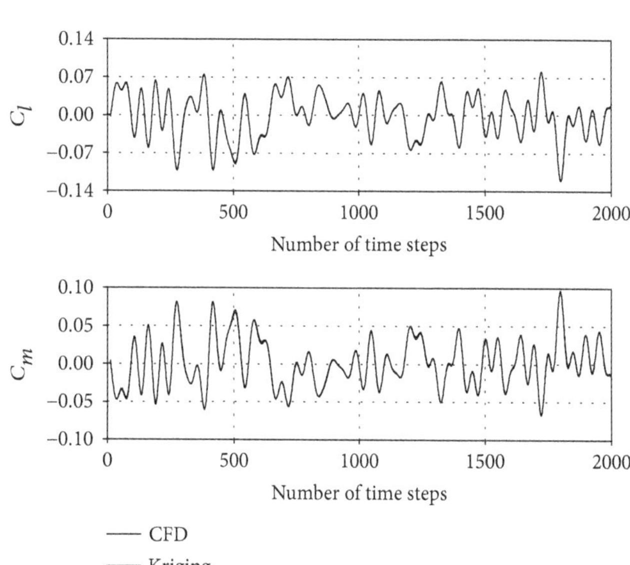

Figure 7: The training data validation of a NACA 64A010 airfoil at $M_\infty = 0.76$ and $\alpha_0 = 1$ deg.

unsteady aerodynamic outputs of the system are computed under different excitation signals, and the results are compared with those obtained by using the CFD training data.

At first, the FWGN excitation signals, not the same as the training signal during the ROM construction, are used to drive the airfoil at $M_\infty = 0.76$ and $\alpha_0 = 3$ deg. The pitch and plunge motions of the airfoil are shown in Figure 9. The corresponding unsteady aerodynamic coefficients are shown in Figure 10. For further comparison, the flight condition with $M_\infty = 0.82$ and $\alpha_0 = 5$ deg is also selected. The excitation signals and the corresponding unsteady aerodynamic results are presented in Figures 11 and 12, respectively. In addition, a harmonic oscillation is chosen as the input signal at $M_\infty = 0.84$ and $\alpha_0 = 0$ deg. The unsteady aerodynamic results are presented in Figure 12.

As can be seen from Figures 10, 12, and 13, the time histories of C_l and C_m obtained via the ROM2s identified

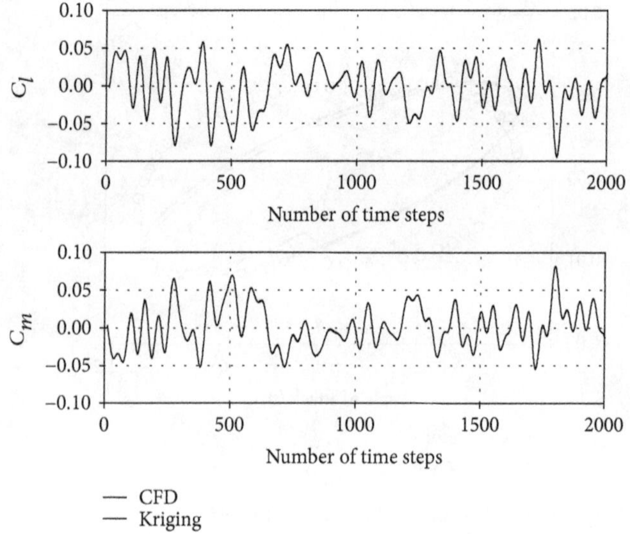

FIGURE 8: The training data validation of a NACA 64A010 airfoil at $M_\infty = 0.82$ and $\alpha_0 = 4$ deg.

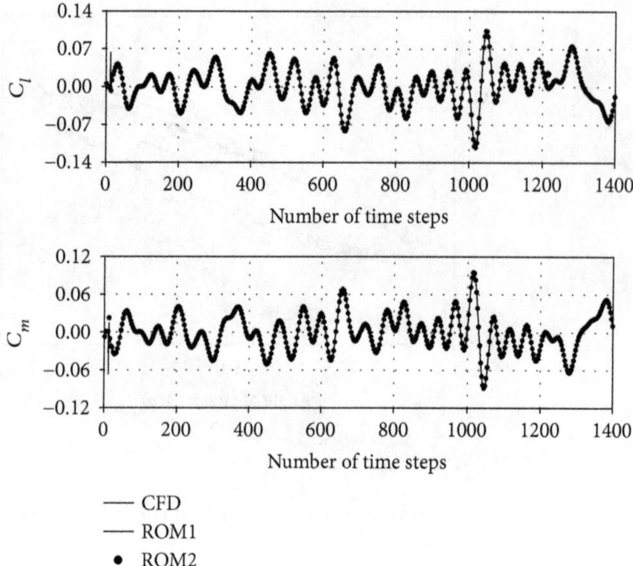

FIGURE 10: The ROM validation under the FWGN excitation at $M_\infty = 0.76$ and $\alpha_0 = 3$ deg (ROM1-training data via CFD; ROM2-training data via the proposed method).

TABLE 1: The computational cost associated with the time-dependent surrogate approach.

	CPU time
Computation of the 20 training cases	4.98 h
Generation of the Kriging models	0.01 h
Total CPU time	4.99 h

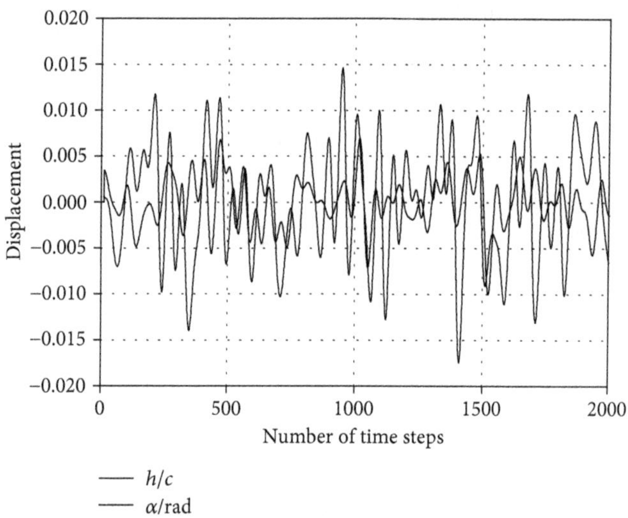

FIGURE 11: The FWGN excitation signals for ROM validation at $M_\infty = 0.82$ and $\alpha_0 = 5$ deg.

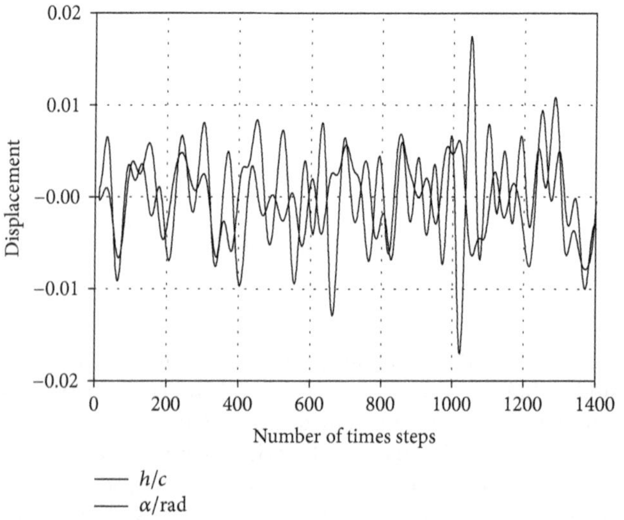

FIGURE 9: The pitch and plunge motions under the FWGN excitation for ROM validation at $M_\infty = 0.76$ and $\alpha_0 = 3$ deg.

by using the time-dependent surrogate approach agree well with those of ROM1s identified by using the direct CFD simulation data. That is, the proposed approach is able to generate good training data for ROMs in a transonic flight regime.

3.3. Aeroelastic Time-Marching Simulation. To further demonstrate the performance of the proposed approach in the flutter prediction in a transonic flight regime, the ROM2s are coupled with the structural model to make time-marching simulations. At first, the airfoil is forced to make three complete cycles of harmonic motion in pitch at the frequency $\omega_\alpha/2$. Then, the aeroelastic system is allowed to evolve by its own self-induced loads.

In the present case, the aeroelastic responses of the airfoil predicted via the ROM2s and ROM1s are compared under two flight conditions. As shown in Figure 14, the aeroelastic responses of the airfoil with $M_\infty = 0.81$ and $\alpha_0 = 0$ deg are computed for $V^*_\infty = 0.73$. Both ROM2 and ROM1 can predict a neutrally stable oscillation. Afterwards, the aeroelastic responses of the airfoil with $M_\infty =$

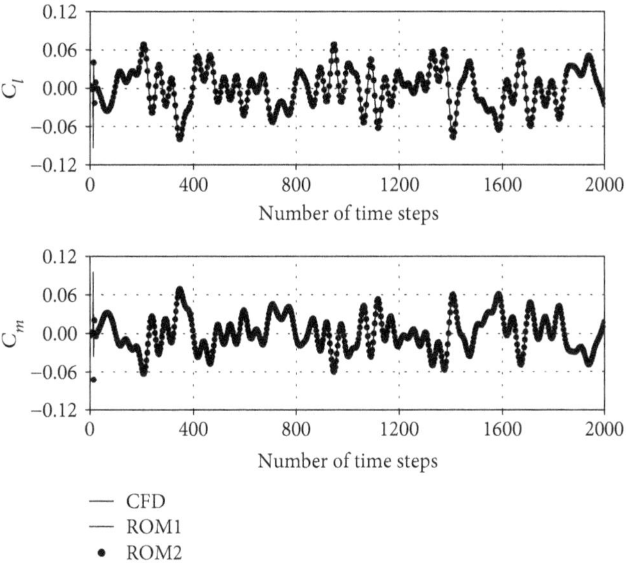

Figure 12: The ROM validation under the FWGN excitation at $M_\infty = 0.82$ and $\alpha_0 = 5$ deg (ROM1-training data via CFD; ROM2-training data via the proposed method).

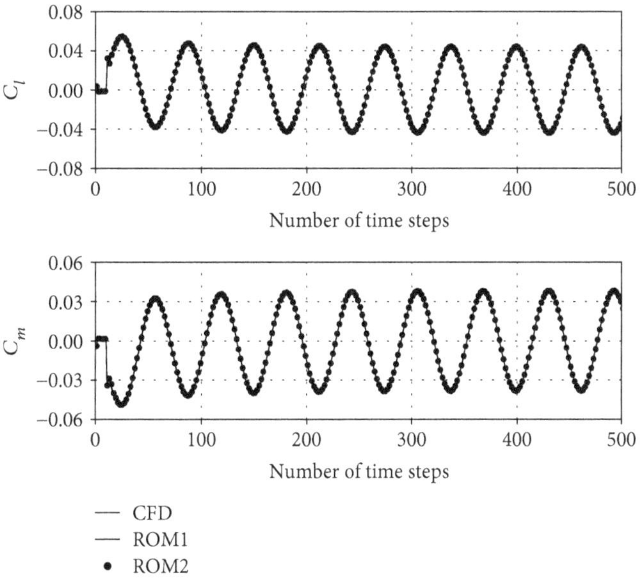

Figure 13: The ROM validation under the harmonic excitation at $M_\infty = 0.84$ and $\alpha_0 = 0$ deg (ROM1-training data via CFD; ROM2-training data via the proposed method).

0.85 and $\alpha_0 = 1$ deg are evaluated. As shown in Figure 15, the results of ROM2 and ROM1 present a neutrally stable oscillation at $V^*_\infty = 0.57$.

Finally, Figures 16 and 17 give the flutter boundaries of the aeroelastic model with the mean angle of attack $\alpha_0 = 1$ deg and $\alpha_0 = 3$ deg, respectively. As shown in these two figures, the computational results based on ROM2s and ROM1s get a good agreement in the flutter boundaries. Nevertheless, the computation of ROM2 is much more efficient than that of ROM1 since no further CFD simulations are required once the Kriging models are constructed.

As can be seen from the flutter boundaries of the airfoil at different mean angle of attack, the aeroelastic characteristics of the model are very sensitive to the mean angle of attack in a transonic flight regime. Hence, the effects of the variations of the mean angle of attack have to be taken into account for aeroelastic analysis.

4. Conclusions

In this study, an efficient approach is proposed to generate the training data for constructing the reduced-order model of an

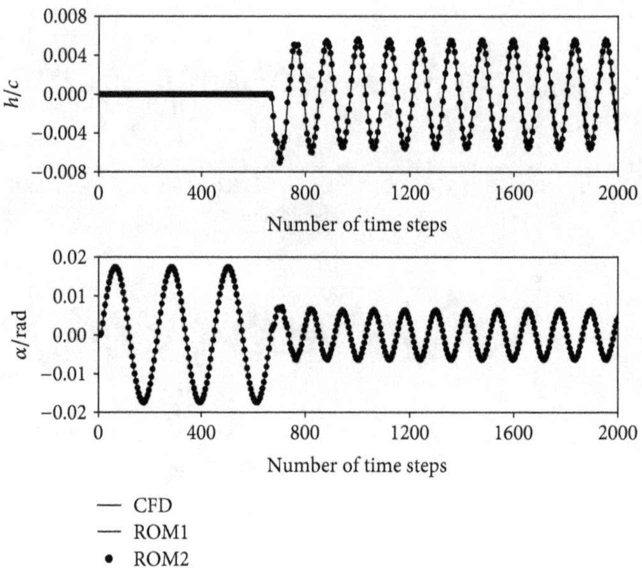

FIGURE 14: The aeroelastic responses of the airfoil for $V_{\infty}^{*} = 0.73$ at $M_{\infty} = 0.81$ and $\alpha_0 = 0$ deg (ROM1-training data via CFD; ROM2-training data via the proposed method).

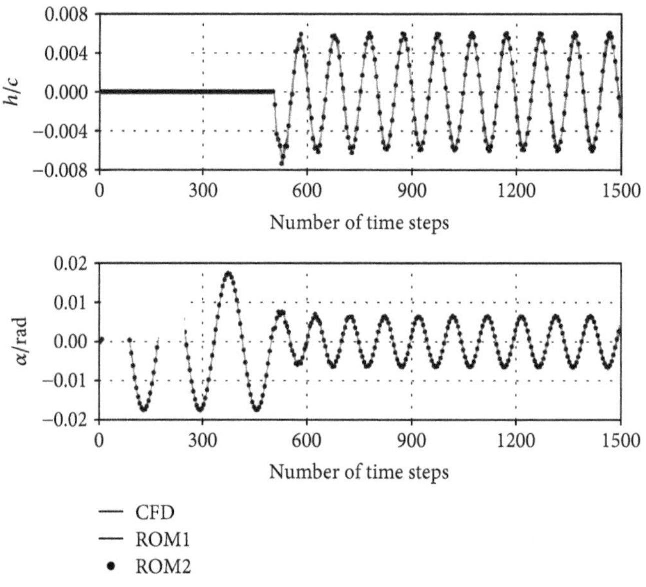

FIGURE 15: The aeroelastic responses of the airfoil for $V_{\infty}^{*} = 0.57$ at $M_{\infty} = 0.85$ and $\alpha_0 = 1$ deg (ROM1-training data via CFD; ROM2-training data via the proposed method).

unsteady aerodynamic system with the variation of flight parameters in a transonic flight regime. The Kriging technique is used in the approach to approximate the relationship between the unsteady aerodynamic outputs of the system and the flight parameters. No further CFD simulations are required to generate the training data once the Kriging models are constructed. The training data obtained via the proposed approach are used to identify the reduced-order aerodynamic models of a NACA 64A010 airfoil via a robust subspace identification algorithm. The unsteady aerodynamics and aeroelastic responses of the airfoil under various flight conditions in a transonic flight regime are computed. The results agree well with those obtained by using the direct simulations of computational fluid dynamics.

When the time-dependent surrogate approach is applied to a three-dimensional aeroelastic model in a transonic flight regime, it is expected to save more computation time compared with two-dimensional cases. Furthermore, the unsteady aerodynamics generated via the approach can be used as the validation data to determine the validation region of an unsteady aerodynamic ROM. Based on the time-dependent surrogate approach, a method by assembling the

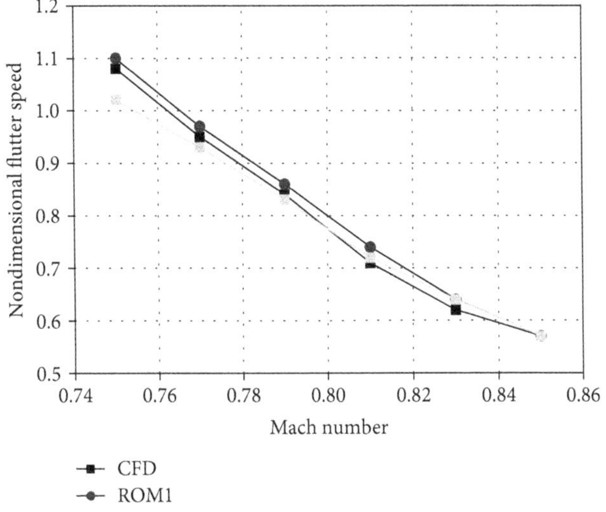

FIGURE 16: The flutter boundary of the airfoil with the mean angle of attack $\alpha_0 = 1$ deg (ROM1-training data via CFD; ROM2-training data via the proposed method).

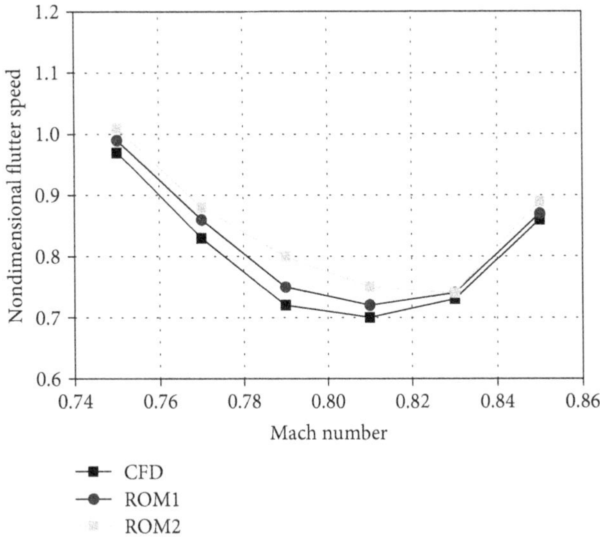

FIGURE 17: The flutter boundary of the airfoil with the mean angle of attack $\alpha_0 = 3$ deg (ROM1-training data via CFD; ROM2-training data via the proposed method).

aerodynamic outputs of a set of precomputed local linear models will be investigated to construct the aerodynamic ROM for a range of flight parameters.

Conflicts of Interest

The authors declare that they have no competing interests.

Acknowledgments

This work was supported by the National Natural Science Foundation of China under grant 11472128 and grant 11502106, National Science Foundation of Jiangsu Province under Grant BK20150736, and Aeronautical Science Foundation of China under Grant 2015ZA52004.

References

[1] O. O. Bendiksen, "Review of unsteady transonic aerodynamics: theory and applications," *Progress in Aerospace Sciences*, vol. 47, no. 2, pp. 135–167, 2011.

[2] D. J. Lucia, P. S. Beran, and W. A. Silva, "Reduced-order modeling: new approaches for computational physics," *Progress in Aerospace Sciences*, vol. 40, no. 1-2, pp. 51–117, 2004.

[3] K. C. Hall, J. P. Thomas, and E. H. Dowell, "Proper orthogonal decomposition technique for transonic unsteady aerodynamic flows," *AIAA Journal*, vol. 38, no. 10, pp. 1853–1862, 2000.

[4] C. W. Rowley, "Model reduction for fluids, using balanced proper orthogonal decomposition," *International Journal of Bifurcation and Chaos*, vol. 15, no. 03, pp. 997–1013, 2005.

[5] J. P. Thomas, E. H. Dowell, and K. C. Hall, "Modeling viscous transonic limit cycle oscillation behavior using a harmonic balance approach," *Journal of Aircraft*, vol. 41, no. 6, pp. 1266–1274, 2004.

[6] J. N. Juang and R. S. Pappa, "An eigensystem realization algorithm for modal parameter identification and model reduction," *Journal of Guidance, Control, and Dynamics*, vol. 8, no. 5, pp. 620–627, 1985.

[7] D. E. Raveh, "Reduced-order models for nonlinear unsteady aerodynamics," *AIAA Journal*, vol. 39, no. 8, pp. 1417–1429, 2001.

[8] T. J. Cowan, A. S. Arena, and K. K. Gupta, "Accelerating computational fluid dynamics based aeroelastic predictions using system identification," *Journal of Aircraft*, vol. 38, no. 1, pp. 81–87, 2001.

[9] A. Mannarino and P. Mantegazza, "Nonlinear aerodynamic reduced order modeling by discrete time recurrent neural networks," *Aerospace Science and Technology*, vol. 47, pp. 406–419, 2015.

[10] F. D. Marques and J. Anderson, "Identification and prediction of unsteady transonic aerodynamic loads by multi-layer functionals," *Journal of Fluids and Structures*, vol. 15, no. 1, pp. 83–106, 2001.

[11] W. Zhang, B. Wang, Z. Ye, and J. Quan, "Efficient method for limit cycle flutter analysis based on nonlinear aerodynamic reduced-order models," *AIAA Journal*, vol. 50, no. 5, pp. 1019–1028, 2012.

[12] T. J. Mackman, C. B. Allen, M. Ghoreyshi, and K. J. Badcock, "Comparison of adaptive sampling methods for generation of surrogate aerodynamic models," *AIAA Journal*, vol. 51, no. 4, pp. 797–808, 2013.

[13] R. Huang, H. Hu, and Y. Zhao, "Nonlinear reduced-order modeling for multiple-input/multiple-output aerodynamic systems," *AIAA Journal*, vol. 52, no. 6, pp. 1219–1231, 2014.

[14] B. Glaz, L. Liu, and P. P. Friedmann, "Reduced-order nonlinear unsteady aerodynamic modeling using a surrogate-based recurrence framework," *AIAA Journal*, vol. 48, no. 10, pp. 2418–2429, 2010.

[15] D. Amsallem, J. Cortial, and C. Farhat, "Towards real-time computational-fluid-dynamics-based aeroelastic computations using a database of reduced-order information," *AIAA Journal*, vol. 48, no. 9, pp. 2029–2037, 2010.

[16] T. Skujins and C. Cesnik, "Toward an unsteady aerodynamic ROM for multiple Mach regimes," in *53rd AIAA/ASME/ASCE/AHS/ASC Structures, Structural Dynamics and Materials Conference*, Honolulu, Hawaii, April 2012.

[17] M. Ghoreyshi, R. M. Cummings, A. D. Ronch, and K. J. Badcock, "Transonic aerodynamic load modeling of X-31 aircraft pitching motions," *AIAA Journal*, vol. 51, no. 10, pp. 2447–2464, 2013.

[18] H. Liu, H. Hu, Y. Zhao, and R. Huang, "Efficient reduced-order modeling of unsteady aerodynamics robust to flight parameter variations," *Journal of Fluids and Structures*, vol. 49, pp. 728–741, 2014.

[19] M. Winter and C. Breitsamter, "Neurofuzzy-model-based unsteady aerodynamic computations across varying freestream conditions," *AIAA Journal*, vol. 54, no. 9, pp. 2705–2720, 2016.

[20] H. Liu, R. Huang, Y. Zhao, and H. Hu, "Reduced-order modeling of unsteady aerodynamics for an elastic wing with control surfaces," *Journal of Aerospace Engineering*, vol. 30, no. 3, pp. 1–19, 2016.

[21] T. W. Simpson, J. D. Poplinski, P. N. Koch, and J. K. Allen, "Metamodels for computer-based engineering design: survey and recommendations," *Engineering with Computers*, vol. 17, no. 2, pp. 129–150, 2001.

[22] N. V. Queipo, R. T. Haftka, W. Shyy, T. Goel, R. Vaidyanathan, and P. Kevin Tucker, "Surrogate-based analysis and optimization," *Progress in Aerospace Sciences*, vol. 41, no. 1, pp. 1–28, 2005.

[23] P. Van Overschee and B. De Moor, *Subspace Identification for Linear Systems: Theory, Implementation, Applications*, Kluwer Academic, Dordrecht, the Netherlands, 1996.

[24] S. S. Davis, "NACA64A010 oscillatory pitching. Compendium of unsteady aerodynamics measurements," *AGARD Report-702*, 1982.

14

Characterization of Flow Interactions in a One-Stage Shrouded Axial Turbine

Adel Ghenaiet

Laboratory of Energetic Mechanics and Conversion Systems, Faculty of Mechanical Engineering, University of Sciences and Technology Houari Boumediene, BP32 El-Alia, Bab-Ezzouar, 16111 Algiers, Algeria

Correspondence should be addressed to Adel Ghenaiet; ag1964@yahoo.com

Academic Editor: Giacomo V. Iungo

The aim of this paper is to characterize the steady and unsteady flow interactions through a one-stage high-pressure (hp) shrouded axial turbine with a tip cavity. The vane and blade passages were reduced based on the scaling technique, and the domains of compromise were identified and used in the flow computations. The flow structures are mainly in the form of vanes' wakes and vortices inducing circumferential distortions and interacting with the rotor blades. Fast Fourier transform (FFT) of the static pressure fluctuations recorded at the selected points and lines through the turbine stage revealed high unsteadiness characterized by a space-time periodic behavior, and described by the double Fourier decomposition. The vane-rotor interactions (VRI) appeared in the form of a potential flow field about the blades extending both upstream and downstream and correlated with the rotational speed. The other sources of unsteadiness are induced in the rotor blades by the vanes' wakes and referred to as the wake interaction, in addition to the secondary flows and vortices in endwall regions.

1. Introduction

High-pressure axial turbines are designed at high loading factors, leading to inherently complex flows which are in essence unsteady. The phenomena of vane/rotor interaction (VRI) arise from the displacement of the rotating blades against the stationary vanes, thereby complicating the mechanism of losses generation. One most important source of unsteadiness is the potential effect in which the pressure field associated with the leading edge of a blade sweeps past the trailing edge of a vane [1]. The other main contributor to unsteadiness is the vanes wakes swept into the blade row due to periodic chopping of wakes [2], added to the secondary flows and vortices convected from an upstream row [3]. Moreover, large variations in the size and strength of the secondary flows and vortices are observed as the rotor blades passages sweep through the flow distortions generated by upstream vanes [4]. Early experimental studies of wake/rotor interactions in axial turbines were performed by Hodson [2] and Binder et al. [5]. According to Sharma et al. [4, 6] and Arndt [7] the phenomena of VRI are presumably one of the difficult investigations to carry out both experimentally and numerically. Miller et al. [8] investigated the effect of the neighboring blade row on the pressure field around an hp rotor blade and showed, in addition to upstream and downstream interactions, a new interaction between downstream vane and upstream vane potential fields. There are only a few publications reporting the vortex-blade interaction mechanism for the shrouded axial turbines compared to unshrouded low aspect ratio axial turbines, such as the ones published by Chaluvadi et al. [9] and Schlienger et al. [10], which indicate that the unsteady secondary flows are primarily dominated by the rotor hub passage vortex and the secondary flows from the upstream vane. For the consideration of unsteady interactions between the leakage flows and the adjacent vanes/blade row, publications are rare; the only one noticed is that due to Qi and Zhou [11] who claimed that an upstream wake may reduce the strength of the tip leakage vortices.

Despite the possession of CFD codes providing acceptable flow details, the computation of unsteady flows in entire blade rows is still prohibitive. Therefore, there is a need for reducing the computational domain to save the

Figure 1: Hp shrouded axial turbine stage.

computing resources and time without affecting too much the flow physics. Periodic parts of a blade row may serve for such simplifications, but this is not usually possible owing to different vane/blade counts. Therefore, several developed techniques may account for the nonintegral blade counts without modeling the entire blade rows, such as the phase-lagged approach modified by Giles [12] to eliminate the assumption of temporal periodicity. To overcome the problem of nonintegral blade counts while maintaining a constant solidity, Rai and Madavan [13] developed the scaling technique. Arnone and Pacciani [14] based on the scaling method performed 2D unsteady flow analysis in the first stage of a transonic turbine consisting of 22 vanes and 38 blades and showed that the configuration 1:2 (1 vane and 2 blades) led to a premature choking of rotor blades passages and a lowering of exit vane velocity with an alteration of pressure distribution, which were nonexistent in the configuration 4:7. Also, Clark et al. [15] completed 3-D unsteady simulations in a 1.5-stage hp axial turbine of vane/blade counts 36, 56, and 36, through a quarter of an entire domain and a simplified domain 1:18 consdiering the scaling technique. As a result, the time-averaged pressure distributions were not affected, contrary to the pressure fluctuations. Furthermore, Yao et al. [16] simulated 3-D unsteady flows in a 1.5-stage axial turbine of blade counts 36, 41, and 36 and arrived to the same conclusions.

The aim of this paper is to investigate the aerodynamic performance and the VRI phenomena through an hp shrouded axial turbine with a tip cavity. The scaling technique allowed creating several reduced multichannel domains. Following the analyses of expansion properties and flow structures as well as the computing cost, the optimal computational domain was identified and adopted. After studying the steady aerodynamic performance and flow structures, the unsteady flow simulations served to characterize the fluctuations of static pressure, both in time and space, at different monitoring points and lines through the components. By means of FFT analysis, the different frequencies and spatial modes of components interactions and their origins were revealed.

2. CFD Model

The single-stage hp axial turbine (Figure 1) is a component from a mixed turbofan consiting of an NGV of 46 vanes and a rotor with 64 shrouded blades with a tip cavity. The main geometry data are summarized in Table 1. The reproduced CAD models of NGV and rotor are shown by Figure 2. At the takeoff operating conditions, the engine parameters correspond to a rotational speed of 12,250 rpm, total air mass flow of 146 kg/s, bypass ratio of 1.03, pressure ratio of 15.5, and turbine inlet temperature of 1335.15 K [17].

2.1. Solver and Setting. ANSYS CFX-solver [18] is used to solve URANS equations (as below) in the finite volume environment. The computational domain has an upstream duct connecting the combustion chamber to NGV and a shrouded rotor blade with a tip cavity (Figure 2):

$$\frac{\partial \rho}{\partial t} + \frac{\partial}{\partial x_j}(\rho U_j) = 0, \tag{1a}$$

$$\frac{\partial}{\partial t}(\rho U_i) + \frac{\partial}{\partial x_j}(\rho U_j U_i) = -\frac{\partial P}{\partial x_i} + \frac{\partial}{\partial x_j} \cdot \left[\tau_{ij} - \rho \overline{u_i u_j}\right] + S_i, \tag{1b}$$

$$\frac{\partial}{\partial t}(\rho h_t) - \frac{\partial P}{\partial t} + \frac{\partial}{\partial x_j}(\rho U_j h_t) = \frac{\partial}{\partial x_j}\left(\lambda \frac{\partial T}{\partial x_j} - \rho \overline{u_j h}\right) + \frac{\partial}{\partial x_j}\left[U_i(\tau_{ij} - \rho \overline{u_i u_j})\right] + S_E, \tag{1c}$$

where the stress tensor $\tau = \mu(\nabla U + (\nabla U)^t - (2/3)\delta \nabla U)$, $\vec{S} = -\rho(2\vec{\Omega} \times \vec{U} + \vec{\Omega} \times (\vec{\Omega} \times \vec{r}))$: Coriolis and centrifugal force terms. $h_t = h + U_i U_i/2 + k$ total enthalpy and $k = (1/2)\overline{u_i^2}$, k turbulence kinetic energy, S_E energy source term. These equations are integrated over each control volume defined by joining the edge centres and the element centres surrounding each node. The volume integral is evaluated by considering the flow properties as constant and equal to the central value (mesh node), whereas the surface integral is evaluated at the integration points located at the centres of surface segments.

The boundary conditions, summarized in Table 2, include the total pressure and temperature (takeoff operating conditions [17]) imposed at the inlet of NGV, while downstream of the rotor blade, a static pressure was varied to sweep the totality of the mass flow rate for different rotational speeds. Moreover, a free stream turbulence intensity $\mathrm{Tu} = (2k/3)^{1/2}/V$ was imposed at the inlet, where V is a representative velocity and k denotes the turbulence kinetic energy of incoming flow which the intensity depends on the inlet vane length scale. The turbulent viscosity is evaluated from Wilcox's model [19].

The steady flow solution (RANS equations) used the high resolution scheme. To prevent pressure field oscillations due to a nonstaggered collocated grid, a coupled solver solves the flow equations as a single system using the SIMPLE algorithm in conjunction with the interpolation technique for the momentum. The solution method is fully implicit, and the

Table 1: Geometry parameters.

Parameters		NGV	Rotor
Vane/blade count		46	64
Inlet minimal diameter	(mm)	490.5	487.5
Inlet maximal diameter	(mm)	644	647.5
Outlet minimal diameter	(mm)	487.5	476
Outlet maximal diameter	(mm)	647.5	656.3/665.5 (with cavity)
Tip clearance	(mm)	—	1.45
Axial chord (min, max)	(mm)	48.5–50.6	32.4/35.7 (with cavity)
Twist angle at hub at shroud	(deg)	40	24
	(deg)	37.5	40
Camber angle	(deg)	67	115

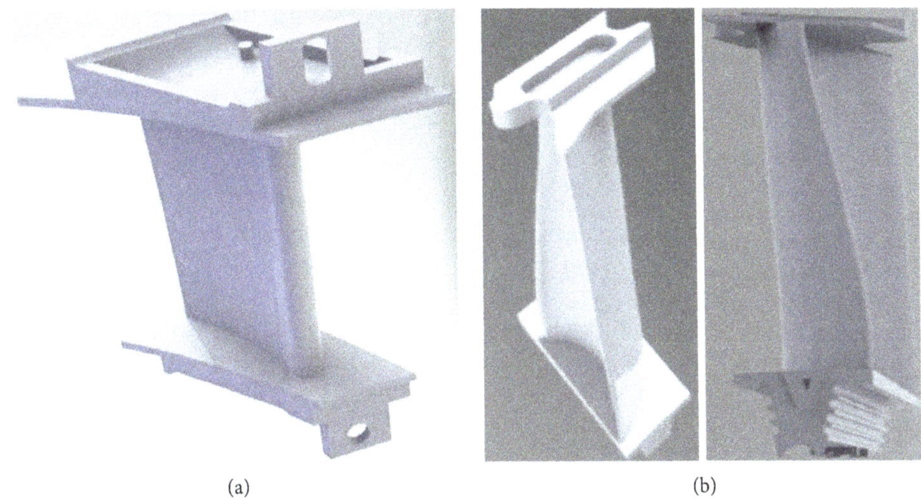

(a) (b)

Figure 2: CAD view of (a) NGV and (b) rotor blade with tip cavity.

Table 2: Boundary conditions.

Fluid		Combustion gas mixture		
Inlet conditions		Total pressure 1.539 MPa & total temperature = 1335.15 K Turbulence intensity~5% & length scale = $0.1 \times h$		
Outlet conditions		Static pressure: Pa		
Wall conditions		NGV		Rotor
	Vane	No slip stationary wall	Blade	No slip rotating wall
	Hub	No slip stationary wall	Hub	No slip stationary wall
	Shroud	No slip stationary wall	Shroud	Counterrotating wall
Lateral surfaces		Periodicity		
Interface in steady simulations		Frozen and stage interface		
Interface in unsteady simulations		Transient rotor/stator interface		

time step acts as a relaxation parameter for a convergence residual inferior to 10^{-6}.

Multiple reference frames differ in the way the interface between moving and nonmoving cell zones is treated. In the frozen rotor model, the coupling between the cell zones is done by just switching the absolute velocity between relative and absolute frames, and thus, a snapshot of the flow field is obtained with the consideration of wakes between different cell zones. Presently, the frozen rotor interface was used to characterize the flow details of the turbine stage. In the mixing plane model (stage interface), the steady state solution is calculated for each fluid zone, and the two

Table 3: Fraction of laminar flow of different components.

	C (mm)	V or W (m/s)	ρ (kg/m^3)	μ (Pa·s) $\times 10^{-5}$	K (J/kg)	Tu (%)	$\mathrm{Re}_c \times 10^6$	Fraction of laminar flow (%)
NGV	49.55	525	3.128	1.83	350	2.91	4.446	2.936
Rotor	34.05	395	2.613	—	750	5.66	1.920	3.495

adjacent frames are coupled by an exchange of the flow field variables at the interface. The flow field data are averaged circumferentially for both frames at the interface and passed to the adjacent zone as boundary conditions. This spatially averaging technique at the interface removes any unsteadiness that would arise due to variations in the zone-to-zone flow field. Herein, the mixing plane model is used to determine the steady aerothermodynamic performance. The transient rotor/stator interface model accounts for the frame of reference transfer and pitch changes, and assumes that the flow field is unsteady, thus can model the interactions between vanes and blades. The relative motion is simulated through the implemented general grid interface (GGI) connection permitting the nonmatching of node locations, element type, and surface overlap among others.

Unsteady flow solution (URANS equations) used the second order backward Euler as a transient scheme with a high-resolution advection scheme. The transient simulations consist in physically advancing the flow in a real time, and because it was not possible to use the final time step's flow field alone to assess the convergence in terms of RMS/MAX residuals and imbalances, some sort of average was required over an appropriate timescale. The number of iterations per time step corresponed to 10 iterations for one inner loop to achieve the accuracy of the residual (RMS $\leq 10^{-6}$) using the selected time step which is small enough to get the necessary time resolution of unsteady VRI depending on the speed of rotation. The different characteristic timescales are estimated as follows:

(i) Necessary time to accomplish one rotor round is $\Delta t_{\mathrm{round}} = 60/n$

(ii) Apparent blade passing period is $\Delta t_i = \Delta t_{\mathrm{round}}/N_{3-i}$, $i = 1, 2$, where N_{3-i} is the blade number of row $(3 - i)$

(iii) Necessary time to cover geometrical coincidence is given by $\Delta t_{\min} = \Delta t_{\mathrm{round}} \mathrm{GCD}(N_i, N_{3-i})/N_i N_{3-i}$, where $\mathrm{GCD}(N_i, N_{3-i})$ is the greatest common divisor

For this hp single-stage axial turbine consisting of 46 vanes and 64 blades and operating at a nominal speed of rotation of 12,242 rpm, $\Delta t_{\mathrm{round}} = 4.901$ ms and thus $\Delta t_{\min} = 3.3295 \mu s$ equivalent to 0.2445 degrees.

2.2. Turbulence Model. The conventional turbulence models offer a good balance between grid resolution requirement and flow field complexity; on the other hand, the large eddy simulation is still prohibitive and probably too detailed to be used for engineering design purposes [20]. Zhou et al. [21] used several turbulence models such as Spalart-Allmaras, standard k-ε, kω-SST, and transition kω-SST in studying the effect of the trailing edge of ultrahigh lift lp turbine blades and showed that the two last models provided a good prediction of aerodynamic performance compared with experiments. Moreover, the kω-SST model has proven certain reliability in studying the flow interactions with a good revelation of pressure oscillations [22]. The transition modeling represents a critical matter for industrial applications; however, according to the literature, this requires the solution of two extra equations with an additional CPU time about 18%, as well as fine grids with a maximum value $y^+ \leq 1$, to resolve the laminar boundary layers and sufficient grid points in the streamwise direction to resolve the transitional region. Herein, the dimensionless parameter $y^+ = (\Delta y/v)U_t$ considers the friction law of a flat plate $U_t = V_\infty \sqrt{C_f/2}$ and $C_f = 0.037\,\mathrm{Re}_c^{-1/5}$ with $\mathrm{Re}_c = \rho V_\infty c/\mu$, where c is the chord. In general, the fraction (percentage) of laminar flow can be estimated from $\mathrm{Re}_{xt}/\mathrm{Re}_c = 380{,}000\,(100\mathrm{Tu})^{-5/4}/\mathrm{Re}_c$, based on the empirical correlation for the transition onset (Mayle [23]), where Re_{xt} is the transition Reynolds number. In an hp axial turbine, the fraction of laminar flow is small, and since the aerodynamic losses are mostly attributed to the turbulent flow after transition, the effect of transition on losses is not significant [23]. The estimations of the fractions of laminar flow are based on the mid vane/blade chord and flow properties averaged at the exit. Table 3 shows small fractions of laminar flow, because of high flow velocities, and as consequence the adopted *kω*-SST model was selected, which allows switching between the standard wall function (for $20 < y^+ < 100$) and the low Reynolds number model (for $y^+ < 2$) to resolve the details of boundary layers.

2.3. Mesh Generation. Unstructured blocks of H-grids and O-grids (Figure 3) were used, with refinements around the vane and blade, whereas the tip cavity (Figure 4) was meshed separately. The parameter y^+ was used to check the distance of the first node from the vane or blade surfaces. To cover regions of high Reynolds number, the exit flow velocities from the vane or rotor blade correspond to the takeoff operating conditions.

The aerodynamic performances were assessed for seven grids of different sizes. The expansion performances show stabilization (Figure 5) above the sixth grid size, and hence, the overall grid size per one channel of the vane/blade is 667,088 : 261,018 nodes for one vane passage, and 250,486 nodes for one blade passage added the tip cavity of 155,584 nodes.

The consideration of the entire channels is very expansive and limited by the resolution of the grids, and this is why the computational domain was reduced by the scaling method. As shown by Table 4, four configurations of computational

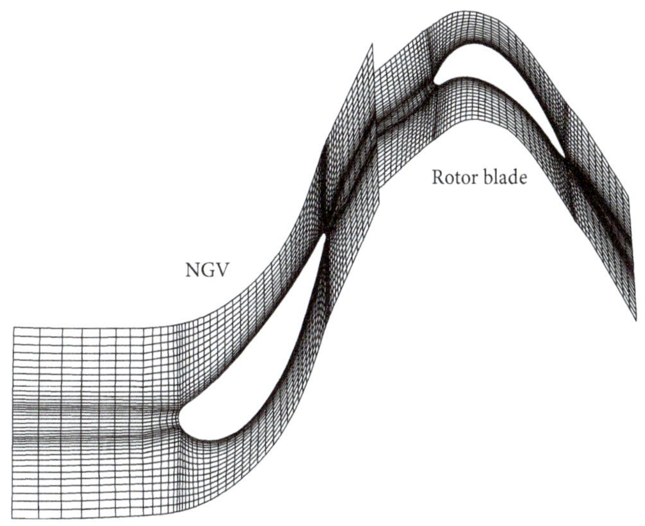

FIGURE 3: Midspan grids per one vane and one blade passages.

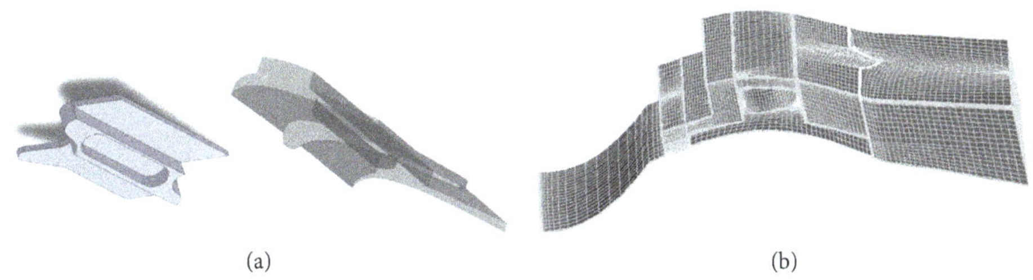

(a) (b)

FIGURE 4: Tip clearance cavity: (a) geometry and (b) grid.

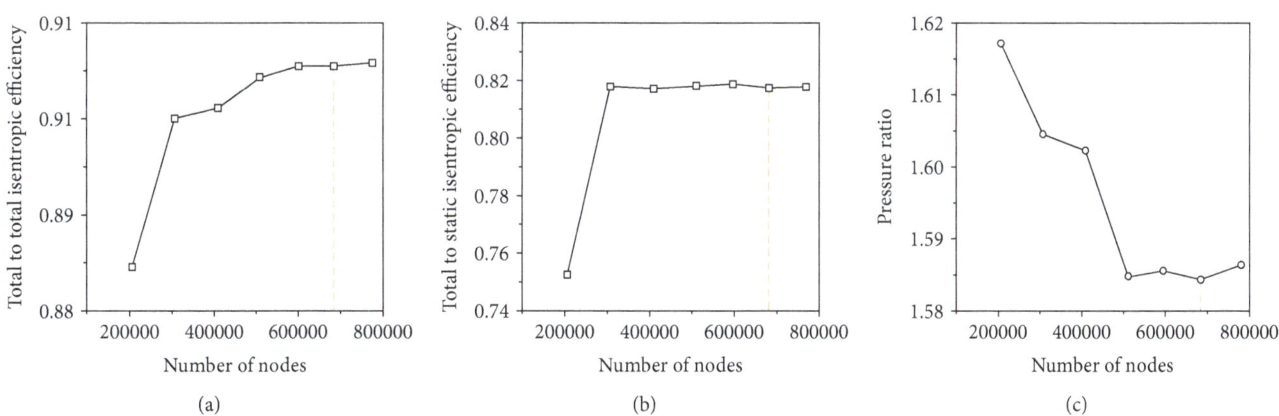

FIGURE 5: Grid dependency analysis: (a) total-to-total isentropic efficiency, (b) total-to-static isentropic efficiency, and (c) pressure ratio.

domains (Figure 6) were evaluated including: the baseline configuration consisting of a half of blade rows of 23 vanes and 32 blades, the configuration "NGV 11 : 16" consisting of 11 vanes and 16 blades (vane chord reduced by 4.5%), the configuration "NGV 3 : 4" consisting of 3 vanes and 4 blades (vane chord reduced by 4.2%), and the last configuration "rotor 2 : 3" consisting of 2 vanes and 3 blades (blade chord reduced by 7.25%). The different grid sizes are listed in Table 4 depicting a scaling factor near unity which in turn affects slightly the blade pressure distribution and avoids an excessive grid size which would require longer simulation time.

2.4. Comparison between Reduced Domains. The three scale-reduced domains are compared with the baseline in terms of entropy (obtained from the frozen rotor simulations) as depicted in Figures 7 and 8, and expansion properties and performance (based on the stage interface simulations) as depicted in Figures 9 and 10. The plots of static entropy in blade-to-blade from hub to tip at spans 10%, 50%, and 90%

Table 4: Scaled computational domains.

	Total vane number	Total blade number	Domain vane number	Domain blade number	Scaling ratio	Chord variation	Grid size × 10^6
Baseline	46	64	23	32	—	—	19.5
NGV 3:4	48	64	3	4	0.958	−4.2%	2.4
NGV 11:16	44	64	11	16	1.045	4.5%	9.3
Rotor 2:3	46	69	2	3	0.9275	−7.25%	1.7

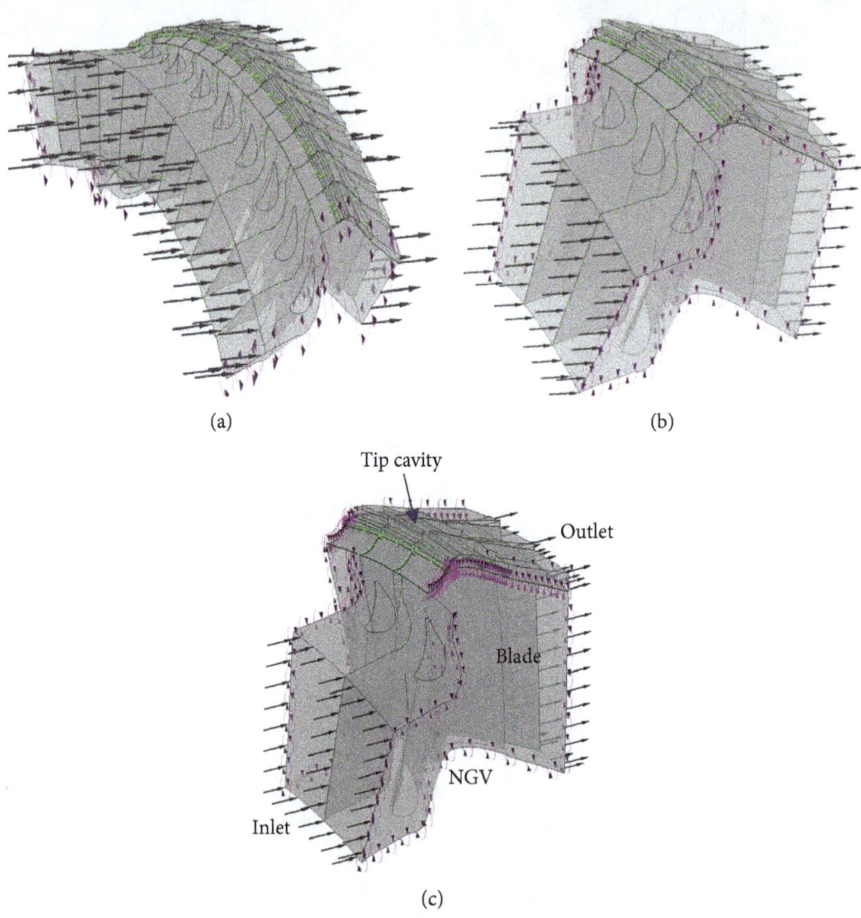

Figure 6: Scaled computational domains: (a) NGV 11:16, (b) NGV 3:4, and (c) rotor 2:3.

and in the tip cavity at spans 96% and 98.5% are shown by Figure 7. It is evident that the vanes wakes convected inside the rotor blades differ by comparing between the baseline and the other reduced domains. The baseline (Figure 7(a)) produced better resolution of flow details whereas the configuration rotor 2:3 produced Mach number over the suction side bit higher, especially at 50% and 90% of the span, explaining the difference in aerodynamic losses and loading. As the flow structure in the tip clearance and tip cavity of the configuration rotor 2:3 differs drastically, this in turn translates into a higher entropy generation at spans 96% and 98.5%.

Figure 9 depicts streamwise variation of the averaged expansion properties. The configurations adopting scaling factors close to unity i.e. 0.958 for NGV 3:4 and 1.045 for NGV 11:16 seem not affecting too much the expansion properties. At the nominal point, there is almost the same trend compared with the baseline, confirming the good choice of the current scaling factors. However, with the configuration rotor 2:3, the expansion properties differ beyond 70% of the blade chord, showing that the scaling of the rotor blade induces more losses since there are more blades and tip cavities in addition to the modification of blade throat and its critical condition.

Furthermore, to show the circumferential flow perturbations downstream of vanes and blades, the contours of entropy are plotted in Figure 8. At the exit of vanes, the radial equilibrium of flow translates into a positive

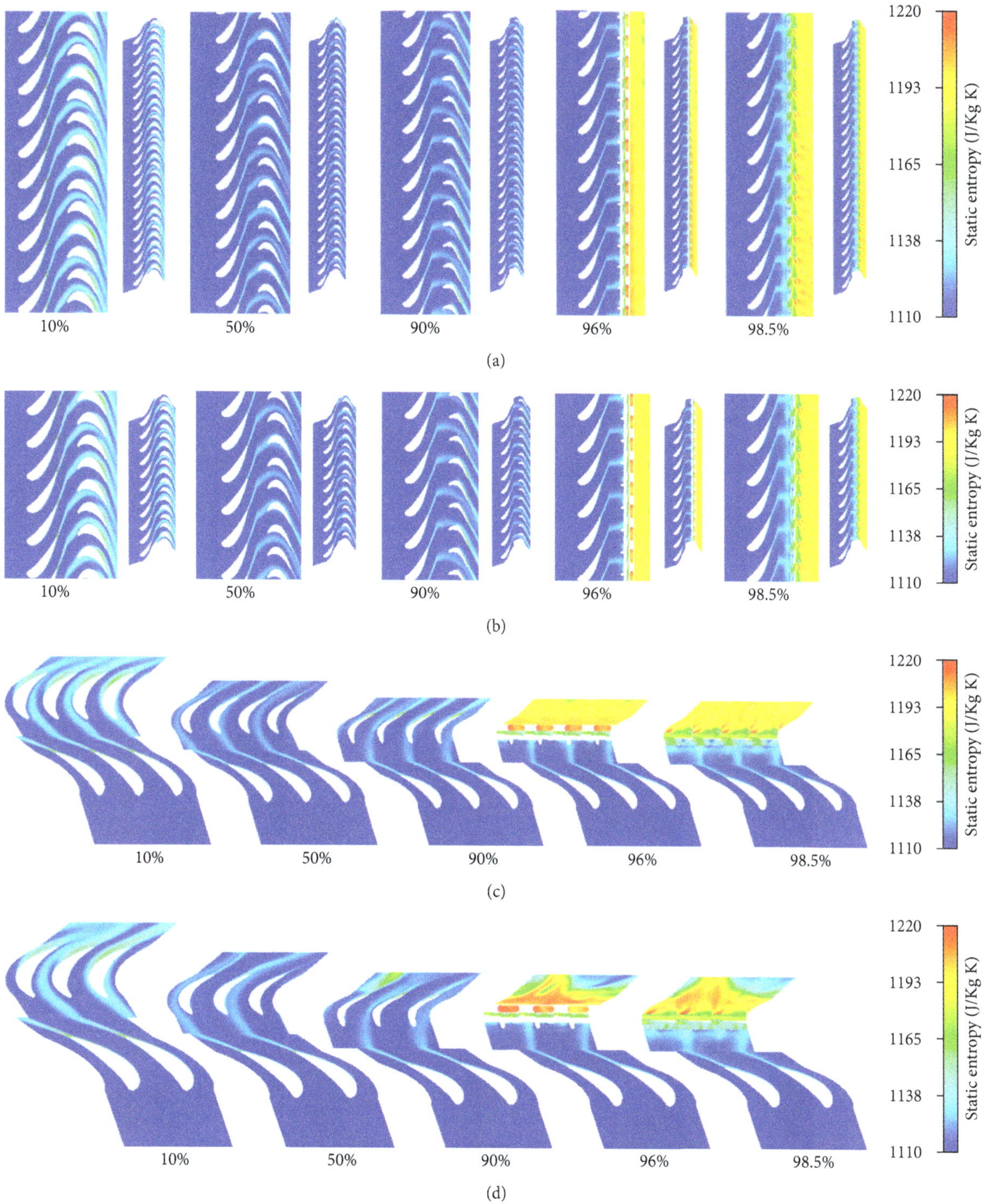

FIGURE 7: Entropy contours (at nominal point) at spans 10%, 50%, 90%, 96%, and 98% (from left to right), successively: (a) baseline, (b) NGV 11 : 16, (c) NGV 3 : 4, and (d) rotor 2 : 3.

gradient of pressure from hub to shroud, and a radial distribution of entropy. Downstream of the rotor blades and due to different aerodynamic loadings, the radial equilibrium is affected differently and hence the flow structures. Moreover, the incoming vanes wakes and vortices are swept differently through the blade row, and this is why the structure of secondary flows and vortices near the hub and shroud corners differ completely.

The turbine stage performance computed for 100% rotational speed shows that the expansion ratio (Figure 10(a)) for all the configurations are aligned with the baseline, but at the high mass flow rate the choke is reached earlier for rotor

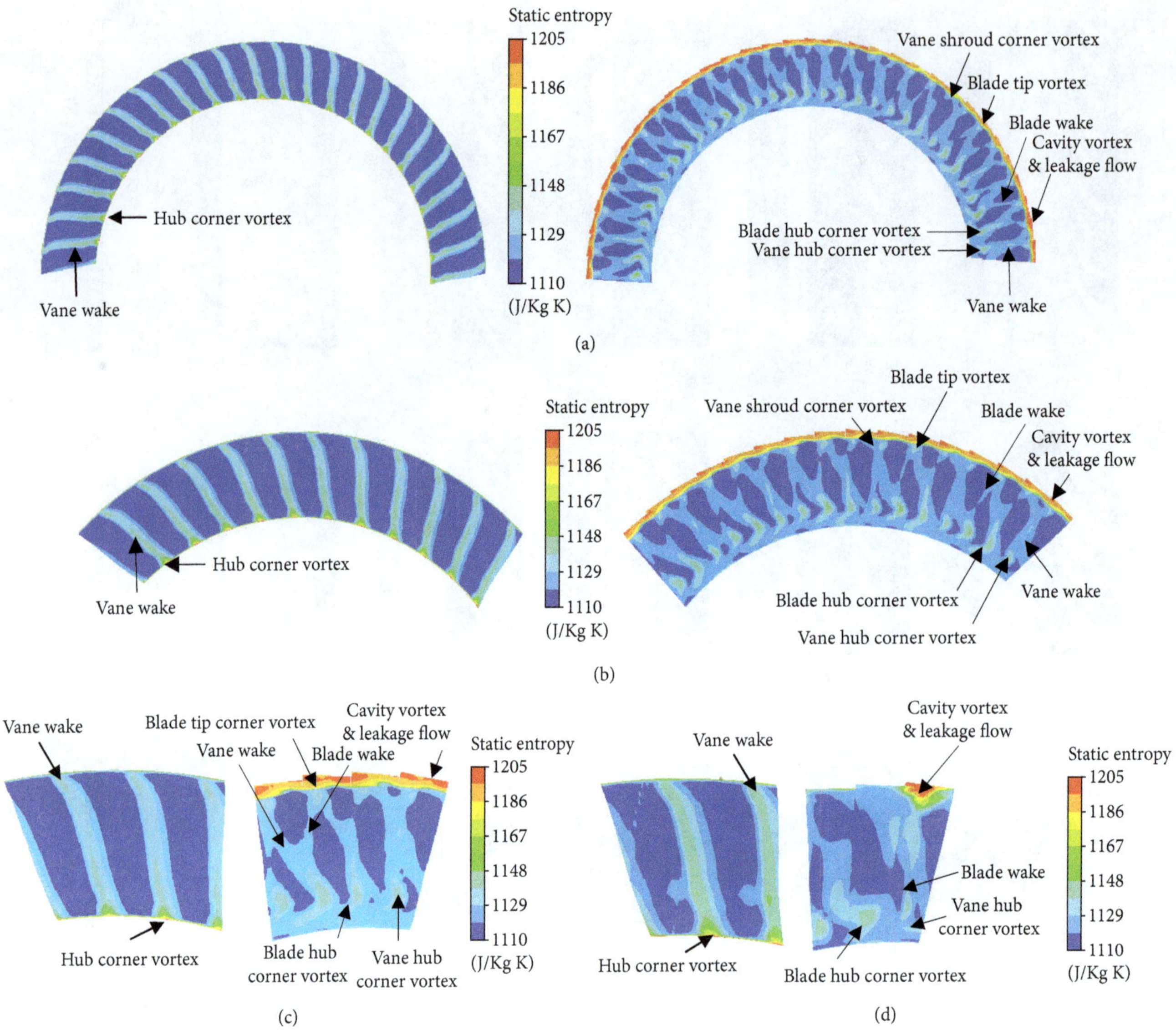

Figure 8: Entropy contours at exit planes of NGV and rotor blade (at nominal point) for (a) baseline, (b) NGV 11 : 16, (c) NGV 3 : 4, and (d) rotor 2 : 3.

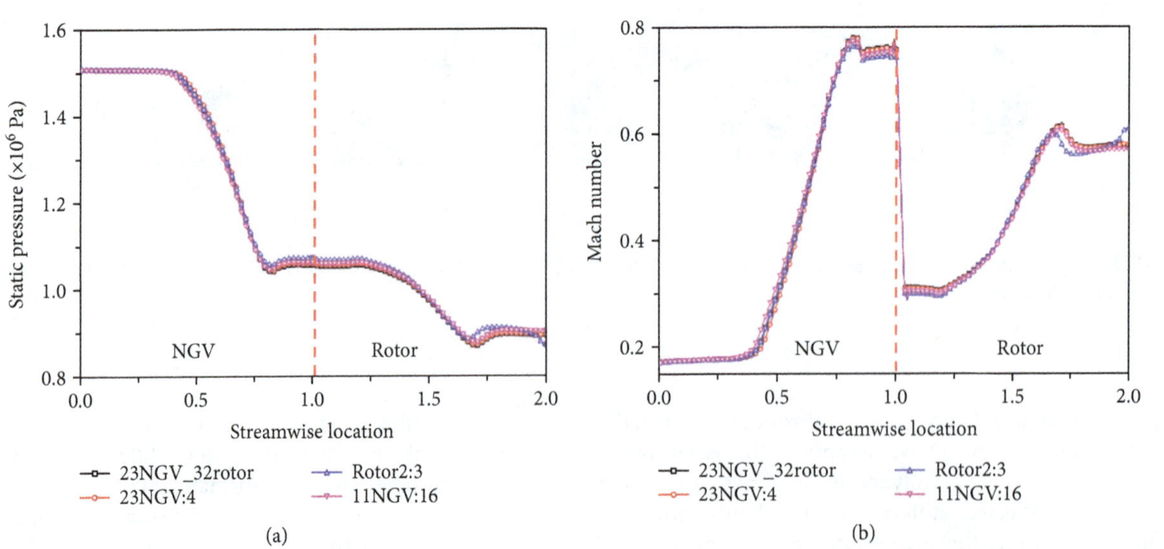

Figure 9: Streamwise variation of (a) static pressure and (b) relative Mach number.

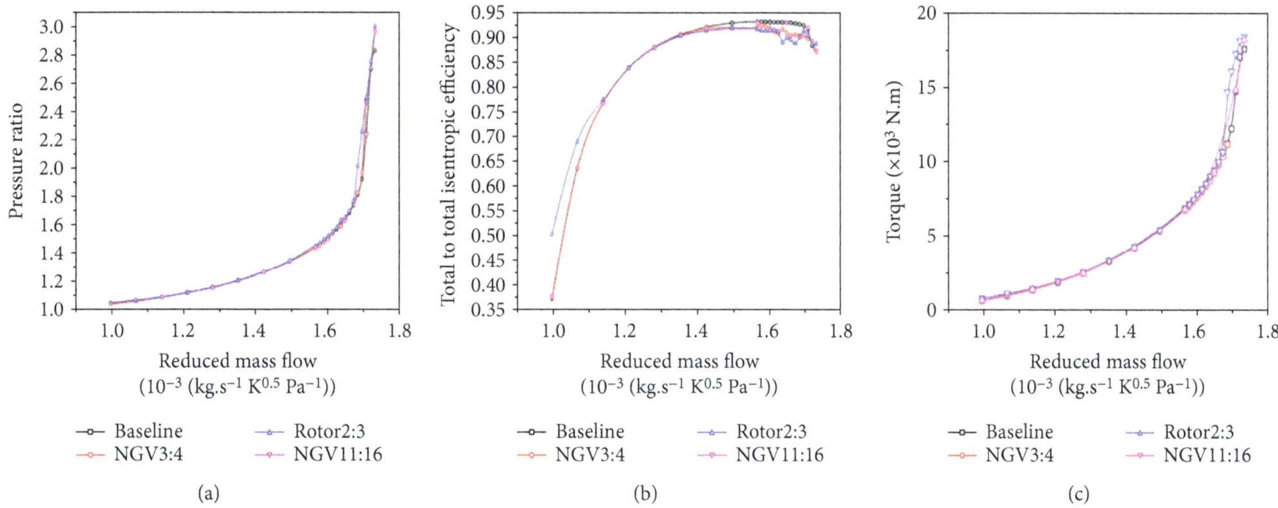

FIGURE 10: Turbine stage performance: (a) total-to-total pressure ratio, (b) total-to-total isentropic efficiency, and (c) torque.

2:3 due to reduction in the throat area of the blade. The total-to-total isentropic efficiency curves (Figure 10(b)) follow practically the same trend at low mass flow rates but deviate slightly at high mass flow rates except for the configuration rotor 2:3 which the efficiency undulates at near choking, since the scaling affected the losses. The torque of the configuration rotor 2:3 differs (Figure 10(c)) at high mass flow rates since there is no kinematic similarity maintaining the same speed of sound in the scaled configurations, and hence, the pressure distributions and aerodynamic loading are affected. Globally, it seems that there is a good agreement between NGV 11:16, NGV 3:4, and the baseline.

2.5. Computational Gain. The approximated number of nodes of the baseline configuration (23 vanes and 32 blades) is 19.5×10^6, for the upscaled (4.5%) NGV 11:16 configuration is 9.3×10^6, for the downscaled (4.2%) NGV 3:4 configuration is 2.4×10^6, and for the downscaled (7.25%) configuration rotor 2:3 is 1.7×10^6. The preliminary unsteady flow simulations at the nominal operating point permitted evaluating the gain in the computing power. For the configurations of baseline and upscaled NGV 11:16, the average computing time was 14 min and 9 min, respectively, per time step so that the total time for each simulation could reach 2184 days and 1404 days, respectively. Because of the cluster capacity did not allow writing down the data files, it was decided to limit the unsteady simulations to the configuration NGV 3:4 of a total nodes' number of 2,407,334 which constituted a good compromise. In this instance, the computations used a workstation of 7 processors and 16 GB of RAM, and each time step took about 3 min so that the total simulation time was 20 days.

3. Steady Flow Results

The steady performance maps are obtained based on the computational domain NGV 3:4 and by adopting the stage interface as the mixing plane. As depicted in Figure 11 for each rotational speed, the total-to-total expansion ratio (π_{tt}) initially varied moderately, increases rapidly to reach the choke limit which shifts with the speed of rotation. The limit of the expansion ratio reaches a maximum of 2.97 at the rotational speed 100% and a reduced mass flow rate of $1.73 \times 10^{-3}[\text{kg} \cdot \text{s}^{-1}\text{K}^{0.5}\text{Pa}^{-1}]$ equivalent to a mass flow rate of 73 kg/s. The curves of efficiency (η_{ttis}) reveal a maximum of efficiency that increases slightly with the rotational speed. The nominal point corresponds to a maximum of 93.05% of total-to-total isentropic efficiency and 85.63% of total-to-static isentropic efficiency at the nominal speed and reduced mass flow rate of $1.638 \times 10^{-3}[\text{kg} \cdot \text{s}^{-1}\text{K}^{0.5}\text{Pa}^{-1}]$ which corresponds to a mass flow of 69.5 kg/s. For this high aspect ratio hp turbine, the operating range is broad even at high rotational speeds.

3.1. Flow Structures. To characterize the flow structures at the nominal operating point, the configuration NGV 3:4 and the frozen rotor interface were used for the steady flow computations. The expansion through NGV and rotor occurs with a significant drop in static pressure and the flow accelerates beyond the throat in the form of expansion waves. After the throat and near the vane hub, an eddy is formed owing to interaction between the boundary layer and the expansion waves. At the rotor leading edge, the flow incidence increases from hub to shroud due to high peripheral speed and blade twist; hence, the stagnation point moves towards the blade pressure side near the hub and towards the suction side near the shroud to cause flow acceleration around the leading edge. Over the blade suction side from the hub and under the tip cavity, flow deviations are seen which, lead to the formation of secondary flows interacting with the main flow. As also noticed, the hub passage vortex mixes with the core flow up to 10% span. Inside the tip cavity, the flow is very slow with the formation of a trapped vortex, in addition to another region of decelerated flow due to mixing between leakage flow and free step vortex behind the tip cavity. At the 90%

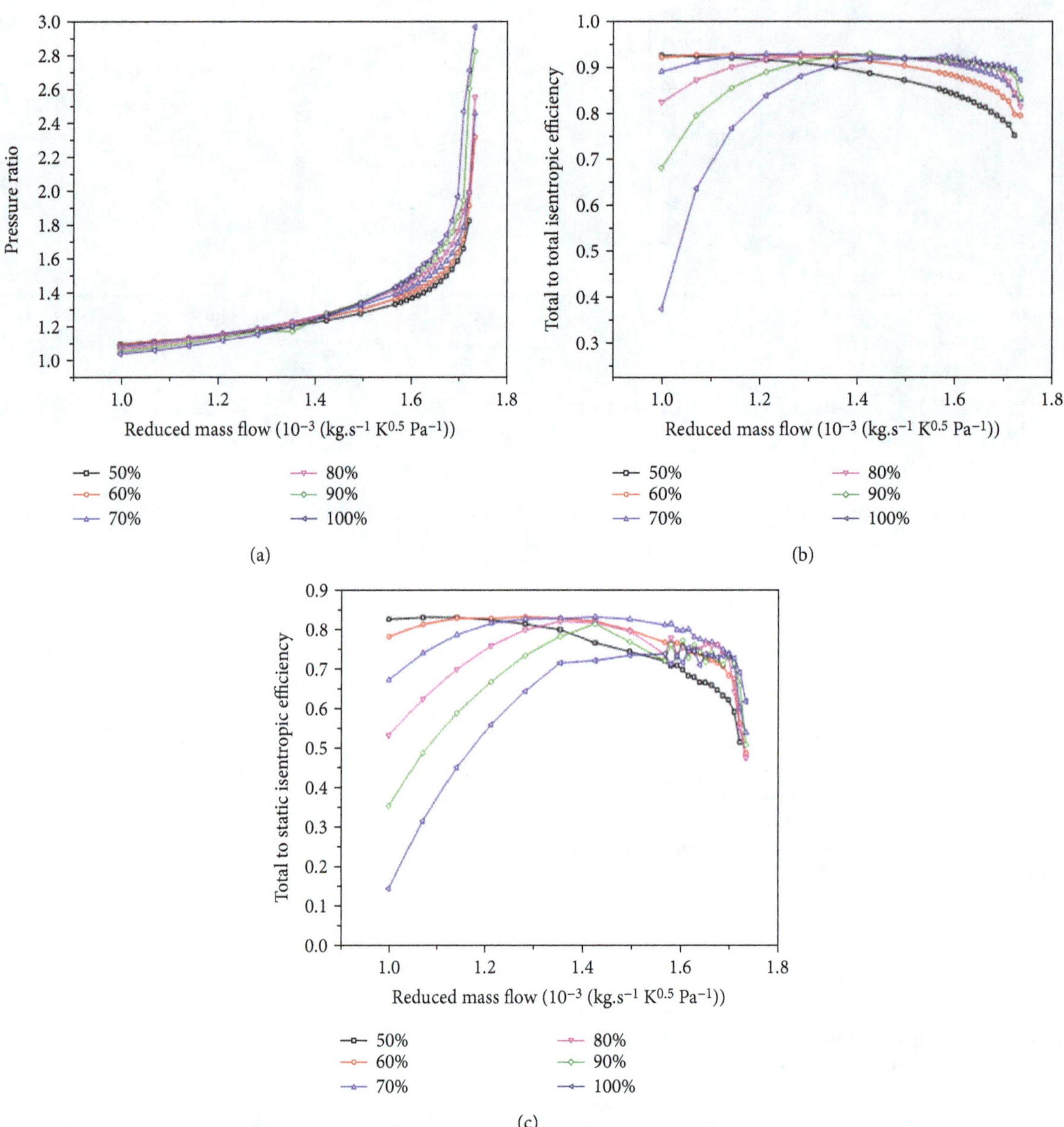

Figure 11: Turbine stage performance maps: (a) total-to-total pressure ratio, (b) total-to-total isentropic efficiency, and (c) total-to-static isentropic efficiency.

span, there is a jet-pointing band located on the suction side transmitted by the under cavity vortex but vanishes due to mixing with the main flow. Underneath the tip cavity, the negative incidence creates a slight flow separation on the pressure side just after the leading edge, and its interaction with the under cavity boundary layer leads to a vortex creation. The effect of tip cavity is clearly illustrated above the 96% span where the uniformly low static pressure is related to the leakage flow reduction. The highest value of static entropy is marked in the tip cavity as characterized by flow recirculation and mixing with leakage flow, shroud passage vortex, and free step cavity vortex. The under cavity vortex, passage vortex, and leakage flow meet the blade wake, hence intensifying the losses. It seems that by incorporating a tip cavity, the leakage flow is reduced drastically about 25%, as seen from Figure 12(a) compared to the case of a flat tip. Moreover, the flow recirculation in the tip cavity seems to reenergize the flow by a ventilation near the shroud, and subsequently the total-to-total and total-to-static isentropic efficiencies are increased by 5.4% and 13.26% as seen from Figure 12(b) and (Figure 12(c)), respectively.

3.2. Impacts of the Rotor Blade Positions. The effects of varying the pitchwise positions of the rotor blade on the expansion are assessed in terms of the total-to-total and total-to-static isentropic efficiencies and the pressure coefficient. The trigger positions (4°, 2°, 0°, −2°, and −4°) are indicated

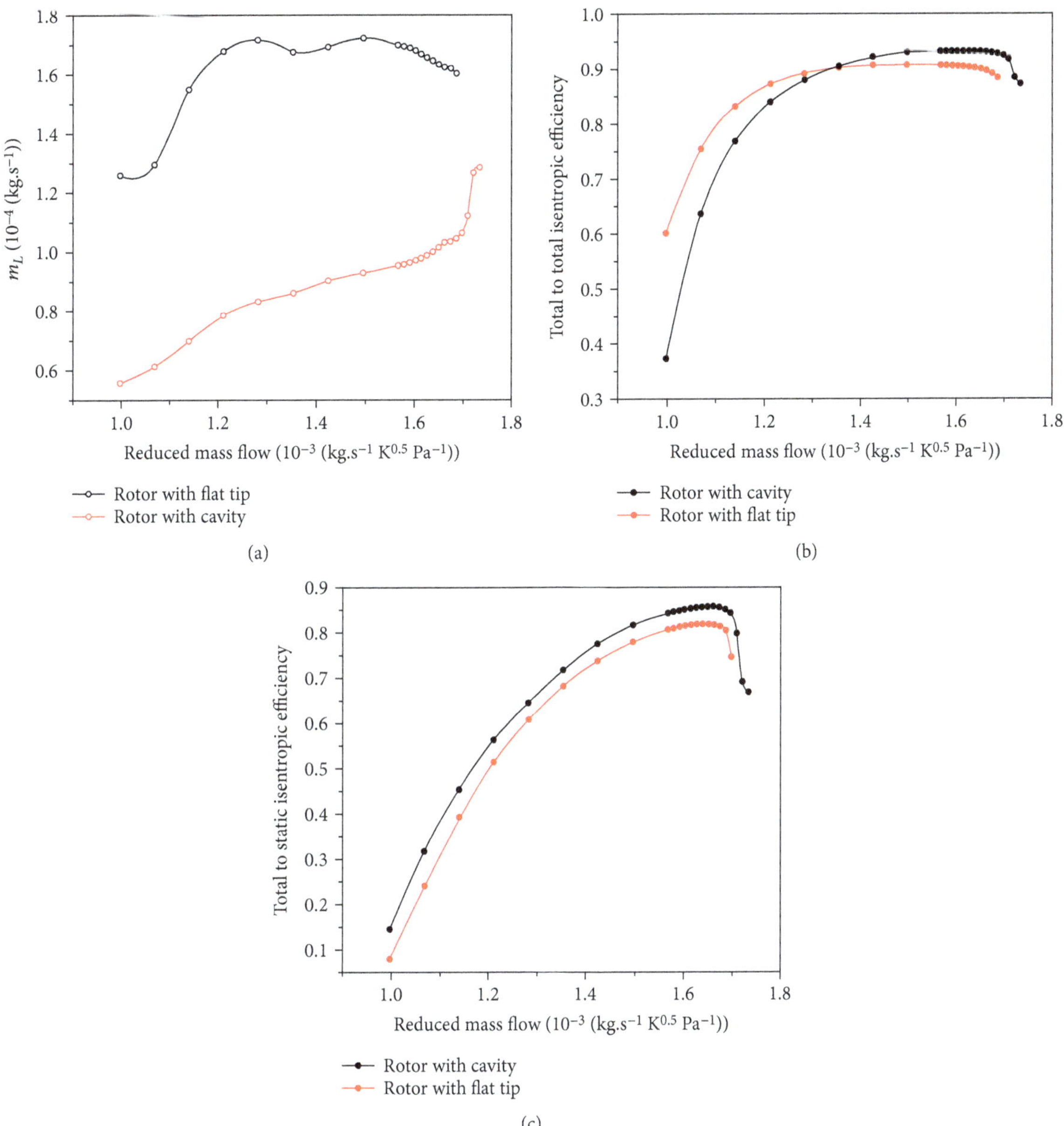

FIGURE 12: Comparison between flat tip and tip cavity: (a) leakage flow rate, (b) total-to-total isentropic efficiency, and (c) total-to-static isentropic efficiency.

by Figure 13. The relative variations of the total-to-total efficiency and total-to-static efficiency are defined as $100 \times (\eta/\eta_0 - 1)$, where η_0 is the efficiency at the trigger position R0, and presented in Figure 14 for different trigger positions and operating points. The peak values are reached for the positions of R1 and R4. The shift of efficiency curves is explained by the vane wake interacting with the blade leading edge. As observed, the minimum of isentropic efficiency occurs at the nominal operating point. When the wake meets the blade leading edge, and this represents the position of minimum losses.

The pressure coefficient $C_p = (P - P_{\text{exit}})/(1/2)\check{\rho}\check{W}^2_{\text{exit}}$ over the vanes is practically invariable with the pitchwise positions, except near the trailing edge owing to the back pressure generated from the rotor blades. The plot of C_p along the second blade exhibits (Figure 15) a predominant variation of pressure loading at the leading edge caused by incoming vane wake. Also, the sudden change at 50% of the blade chord seen from the suction side is referred to as a local lambda shock wave, which is more intense at the maximum mass flow as revealed by a drop in C_p shifted towards the trailing edge from 75% of the chord. At the minimum mass flow rate, the noticeable variation in C_p occurs with an opposite pressure gradient over the fore of the blade up to 20% of the chord. This can be explained by the displacement in stagnation point from the blade leading edge over the suction side, causing a negative incidence and creating a bubble recirculation over

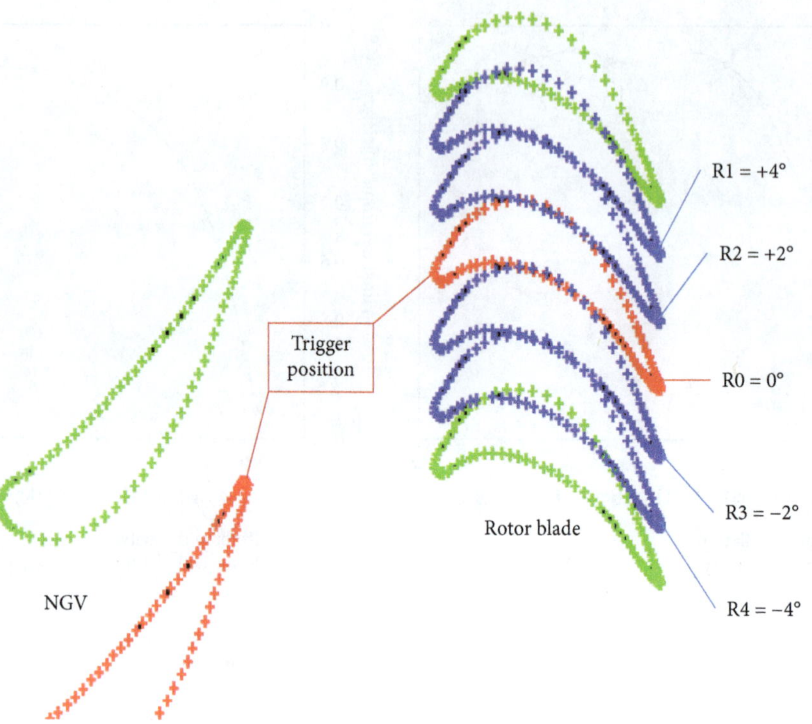

Figure 13: 2nd blade trigger position.

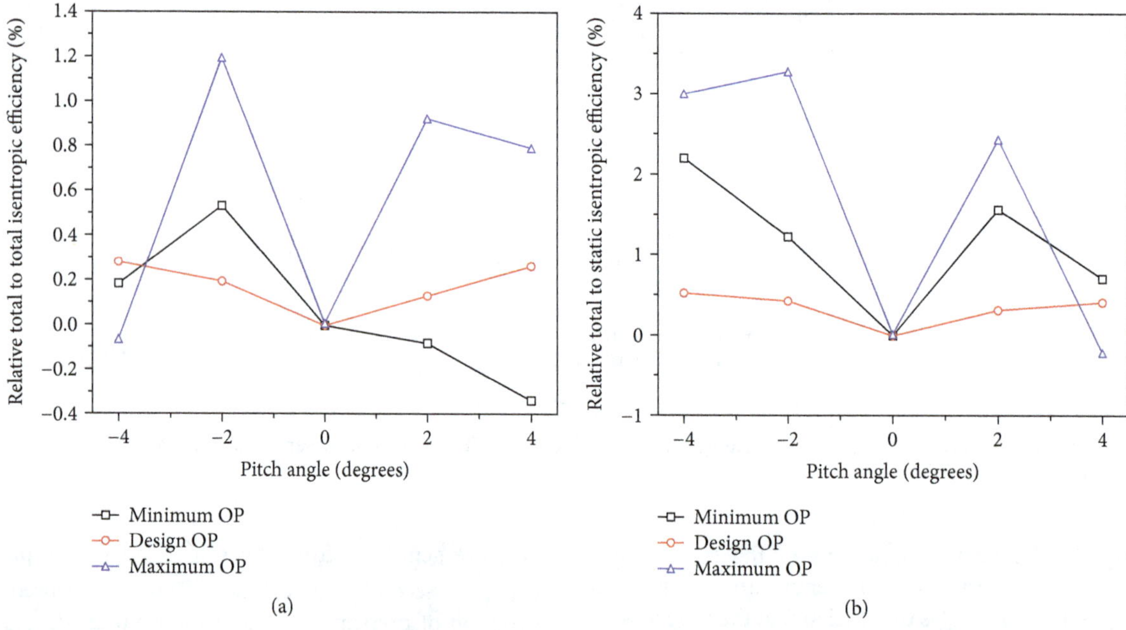

Figure 14: Effect of the pitchwise positions of rotor blade in terms of (a) total-to-total isentropic efficiency and (b) total-to-static isentropic efficiency.

the pressure side where an adverse pressure gradient occurs. As far as the pressure loading distribution is concerned, over the totality of rotor blade, the position leading to the maximum of turbine efficiency presents a wider area of static pressure distributions compared to that leading to the minimum of turbine efficiency, added to high C_p at the leading edge related to the wakes segments impinging locally.

4. Unsteady Flow Results

The unsteady flow results are presented and discussed for the nominal operating point (12,242 rpm, $m = 69.5$ kg/s). The time step was $\Delta t_{\min} = 3.329\mu s$, and the total simulation time equal to 9.802 ms corresponding to two rounds of the rotor. However, by respecting the Shannon criteria [24], the period $\Delta t_{\min}$ was divided by 3 (equal to $1.019\mu s$) allowing detecting

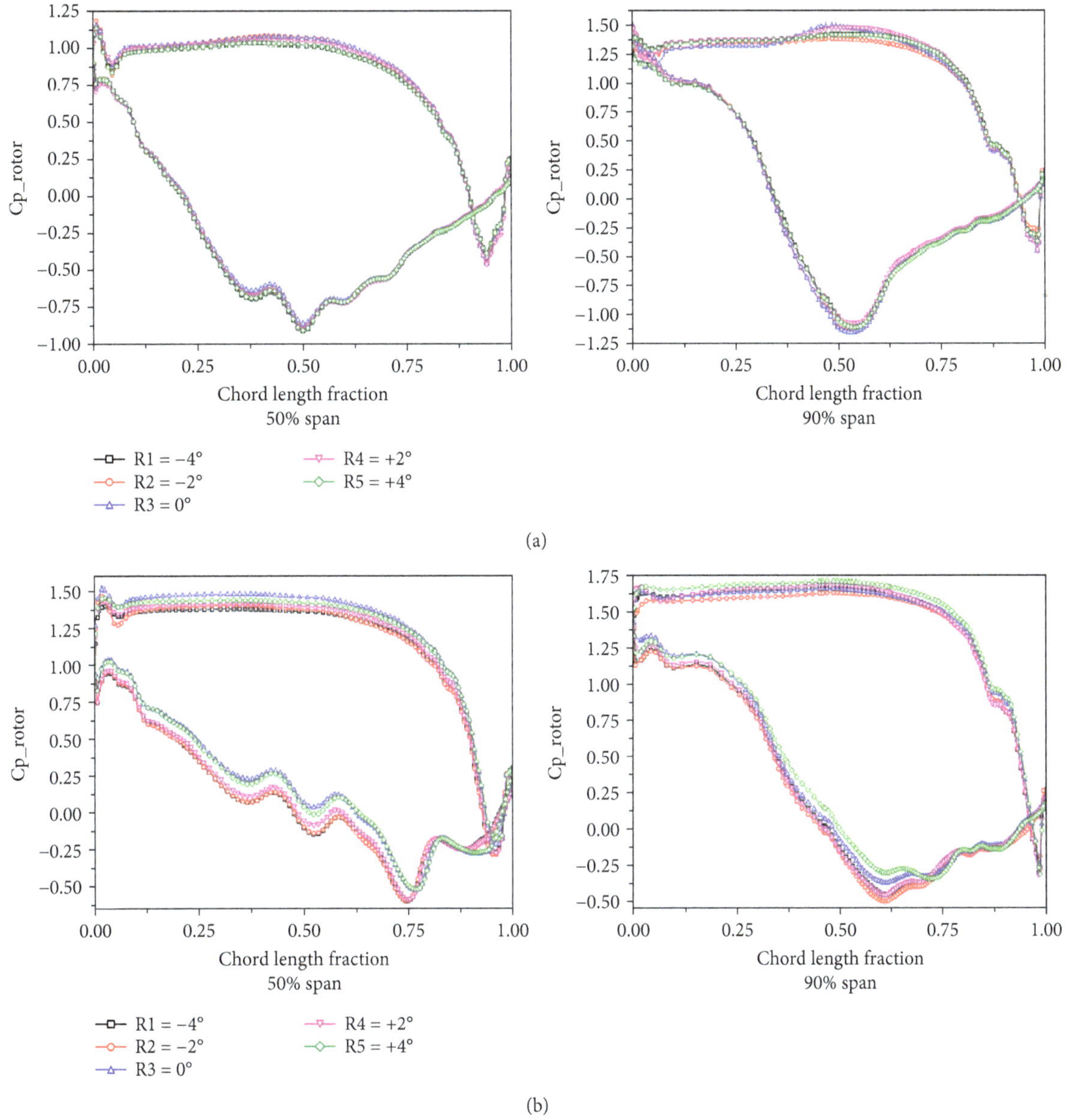

(a)

(b)

FIGURE 15: Pressure coefficient over the 2nd blade at the different pitch positions and the (a) nominal operating point and (b) maximum mass flow rate operating point.

the small fluctuations of pressure. The recording points of temporal static pressure were at midpitch of first, second, and third vanes from 10%, 50%, and 96% spans of the exit plane, whereas those upstream and downstream of the rotor blade, were set at the midpitch of first, second, third, and fourth blades. These positions were selected to better highlight the effect of flow structures: interactions of the vanes wakes and rotor blades and the vane shroud secondary flows with the blade tip leakage vortex, in addition to the vane hub secondary flow with the rotor hub secondary flow.

It was necessary to eliminate the transitory period from the temporal pressure signals as revealed from Figure 16. The recorded pressure signal is characterized by a first frequency of 100 Hz corresponding to the time of simulation which the magnitude is related by the fluctuations curve pace until its stabilization. Also, there are several peaks of

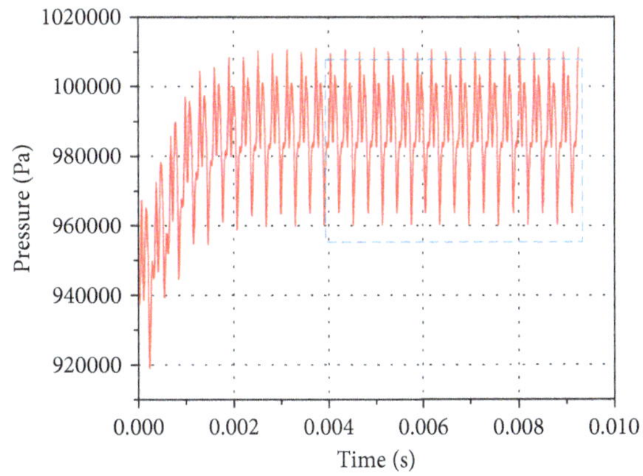

FIGURE 16: Sample of a temporal pressure signal. Stabilization: [4 ms, 9.8 ms].

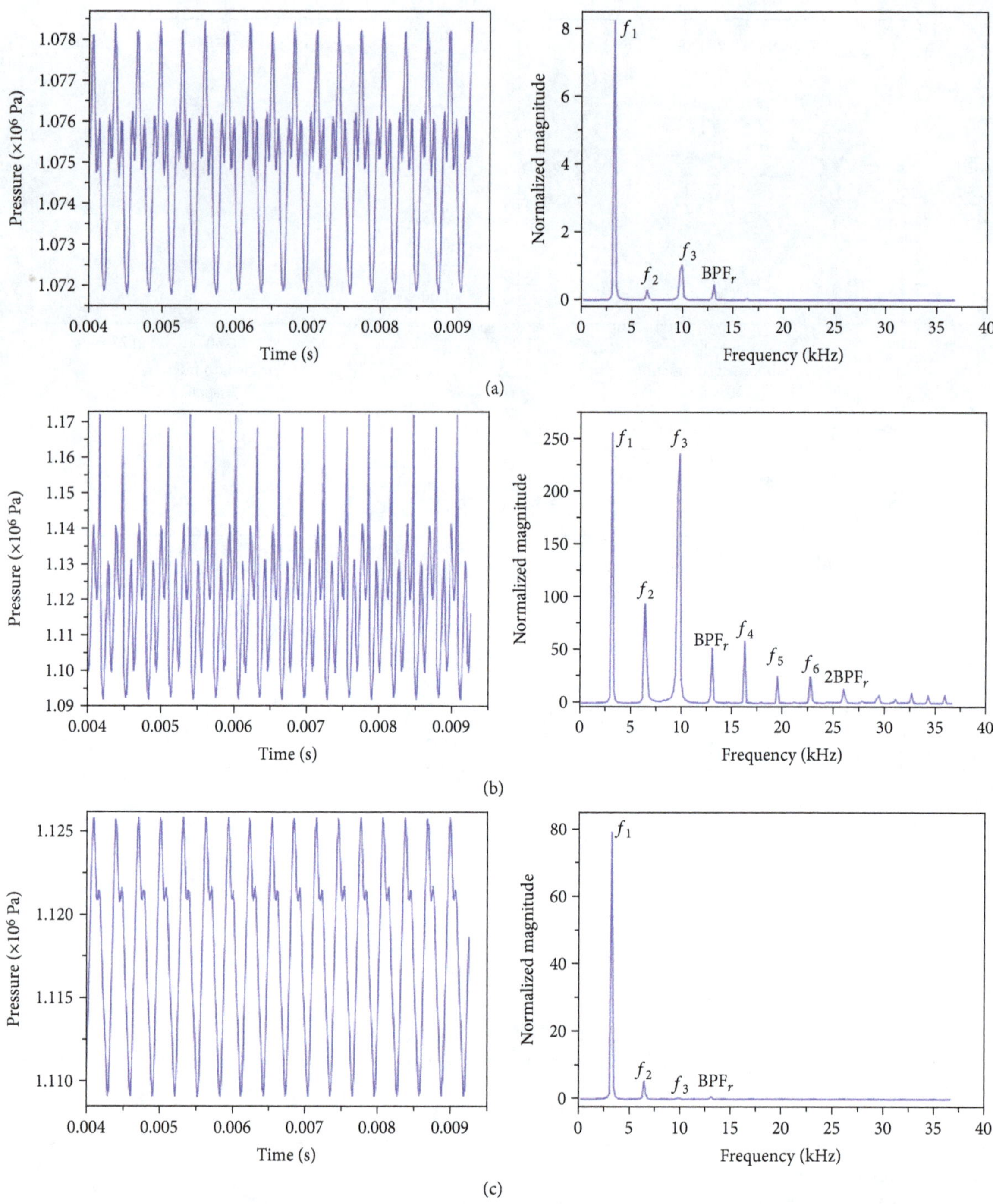

FIGURE 17: Temporal pressure fluctuations and FFT, recorded at the exit of NGV at midspan and midpitch of (a) first, (b) second, and (c) third vanes.

frequencies due to the transitory fluctuations of irregular variations with time (no fluctuation periodicity). The treated region in the time interval 4 ms–9.8 ms, seems to have a regular periodicity in time and characterized by frequencies describing physical phenomena.

FFT analyses applied to the spatial and the temporal static pressures recorded at different positions and lines produced the spectrums in Figures 17–20, depict a fundamental blade passing frequency $\text{BPF} = (\Omega/60)N_r = 13{,}066.68$ Hz related to the potential effect of the rotor blades. The pressure waves characterized by this harmonic and its multiples propagate both upstream and downstream of blades and cause flow perturbations through the blade passage, but are attenuated throughout. The magnitudes of the principal peaks vary depending on the recording points as noticed for the same spanwise location. The wave amplitude seems to increase starting from the suction side until a maximum in the middle of interblade and then reduces until the pressure side. For this

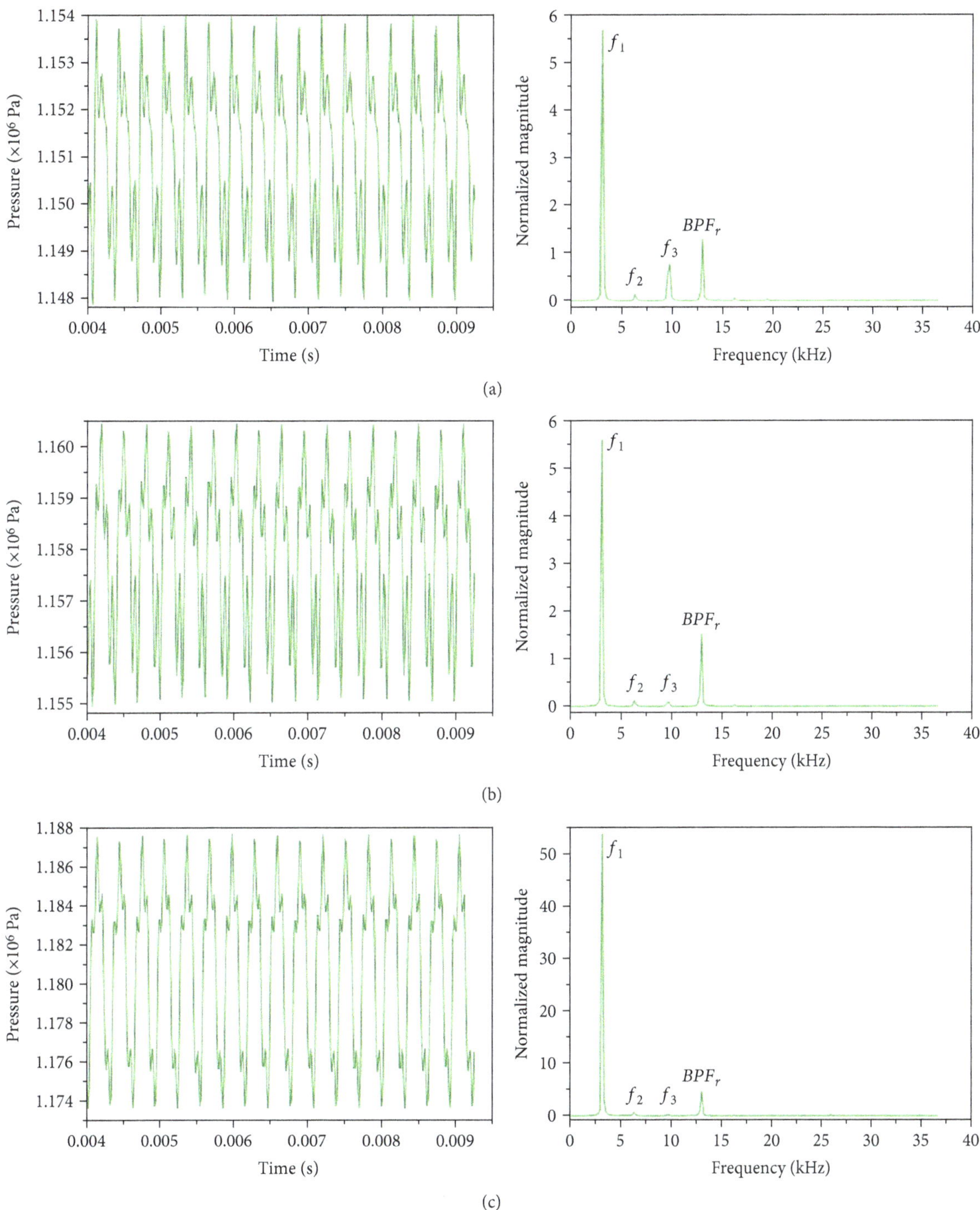

FIGURE 18: Temporal pressure fluctuations and FFT, recorded at the exit of NGV at the 96% span and midpitch of (a) first, (b) second, and (c) third vanes.

reason, the temporal pressure signals were taken at the midpitch where they have the highest amplitudes. At the exit of vanes, the pressure fluctuations recorded at mid-span and near shroud, specifically at the midpitch of first, second, and third vanes, and their associated spectrums are shown by Figures 17 and 18. Also appear other frequencies f_1, f_2, f_3, f_4, f_5, and f_6 of higher amplitudes attributed to the embedded rotor flow disturbances, predominately affected by upstream flow distortions caused by the vane wakes and the downstream potential disturbances related to the rotor blade, shroud ring and tip cavity. In principle, when two primary harmonics say, f_1 and f_2, coexist, the flow responses may contain subharmonics with frequencies like $nf_1 \mp mf_2$, where n and m are integers. From one recording point to another, there is emergence of different frequency peaks of varying magnitudes which

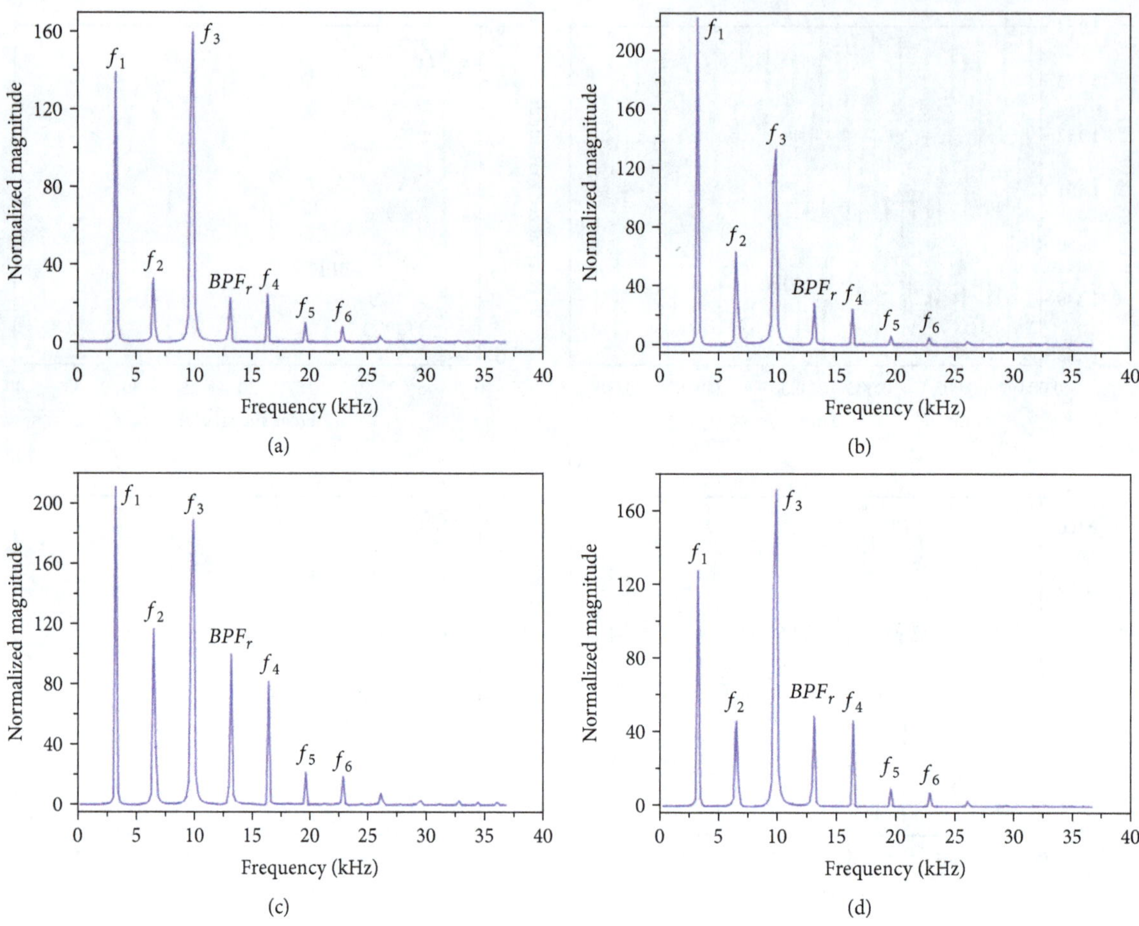

FIGURE 19: FFT of pressure fluctuations recorded upstream of the blade at (50%) span and midpitch of (a) first, (b) second, (c) third, and (d) fourth blades.

depends on the relative blade position, since the blades do not coincide with the vanes' wakes at the same instant. At the recording point just behind the trailing edge of the vane, the magnitude of frequency f_1 reaches the maximum but lesser when the recording point is not aligned with the trailing edge. The magnitude of the frequency f_2 is lower than f_1 due to backward potential effect which has a magnitude weaker than that implied by the vanes' wakes. The frequencies f_2 and f_3 are more pronounced for the second vane recording point, added to other important frequencies f_4, f_5, and f_6. Figure 18 shows that when the static pressure was recorded near the shroud, the first frequency is still important whereas the others are less compared with BPF.

Figures 19 and 20 show the spectrums of pressure fluctuations recorded just upstream the blade leading edge, which indicate that the main frequencies of the temporal evolutions are stronger than downstream of vanes. The strength of the potential effects is such that BPF is felt upstream the blade leading edge; however, its magnitude is lesser than the first frequencies f_1 and f_2. The pressure fluctuations recoded upstream the rotor blade is characterized by more peaks of frequencies as described by the chorochronic theory and revealed by FFT analysis. The magnitudes of the fundamental frequencies depend on the recoding locations which are more significant at the midspan than hub and shroud. Moreover, the frequencies such as $f_1, f_2, f_3, f_4, f_5, f_6$, and BPF have the highest amplitudes for the recording point of the third blade channel. Downstream of rotor blade, the temporal evolutions of static pressure are dominated by the composition of the frequencies f_1 and f_2. The vanes wakes are chopped by the rotor blades and convected downstream until the rotor exit and afterwards decay. The frequency f_1 has a smaller magnitude near the hub as might be explained by the formation of hub vortex convected towards the rotor exit. Moreover, high amplitudes frequency peaks are seen near the shroud for the blades 3 and 4, and hence, the pressure fluctuations induced by the tip cavity vortices are of higher amplitudes than the convected wakes or the hub passage vortices.

The distributions of static pressure and entropy at the instant of 5.17 ms are shown by Figures 21 and 22 for different normalized axial distances. These parameters are independent of the frame of reference, and their azimuthal evolutions are continuous across the interfaces. The static pressure contours (Figure 21) highlight the potential effect of the rotating blades as confirmed by the shape of the pressure wave, which increases close to the blade leading edge.

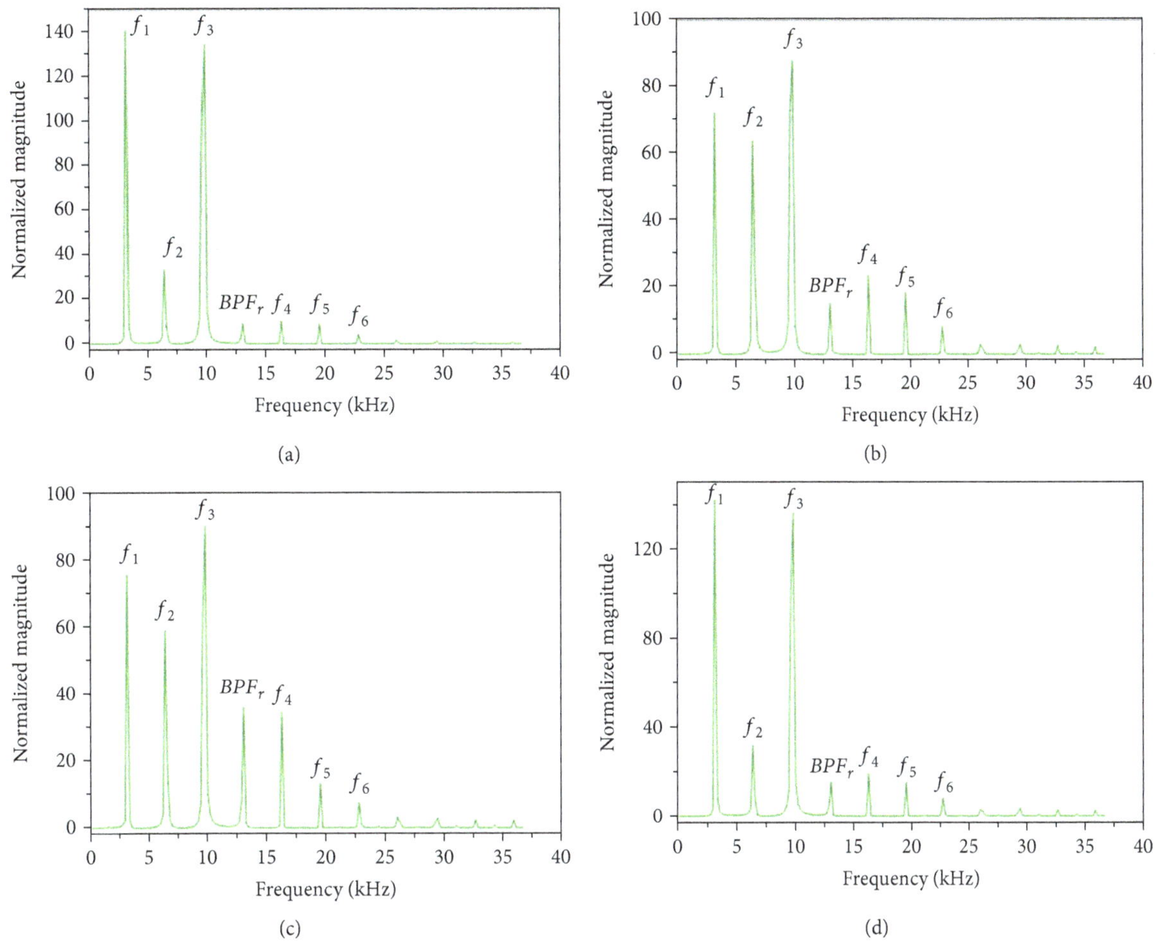

(a) (b) (c) (d)

Figure 20: FFT of pressure fluctuations recorded upstream of the blade at (96%) span and midpitch of (a) first, (b) second, (c) third, and (d) fourth blades.

Figure 21: Static pressure contours at different normalized streamwise cross-sections, at the instant of 5.17 ms.

Downstream of rotor blade, there are 64 lobed structures with different forms of pressure waves, related to the vane wakes chopped by the rotor blades, of a decaying amplitude owing to mixing. These results are also confirmed by the contours of entropy (Figure 22) highlighting regions of vanes wakes in addition to secondary flows (hub and shroud passage vortices ...) convected downstream. The entropy distributions near the blade leading edge show evidence of losses related to VRI which produce a circumferential distortions with a clear deficit in flow velocity. Downstream of rotor blade, the dominant losses are those related to the leakage flow and the convected wakes and secondary flows.

FFT analyses of spatial pressure fluctuations at the instant of 5.17 ms obtained from the recorded lines passing near hub, midspan, and shroud (96%) are depicted in

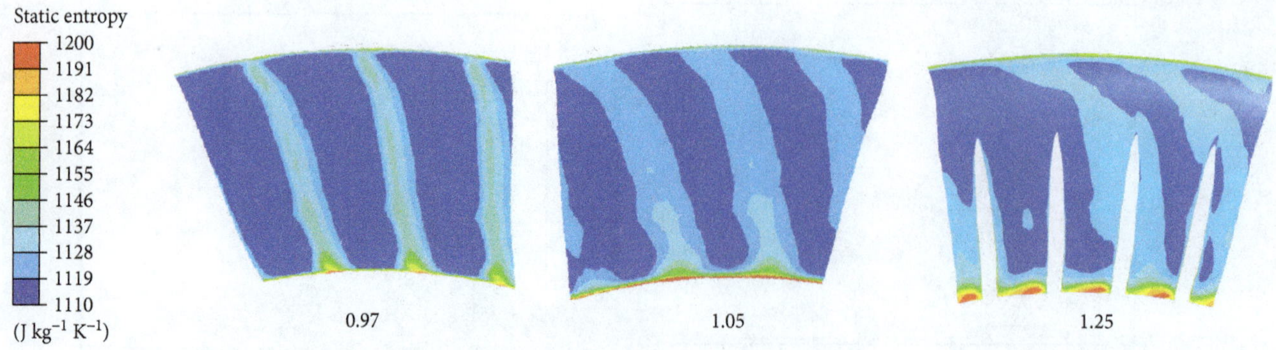

FIGURE 22: Entropy contours at different normalized streamwise cross-sections, at the instant of 5.17 ms.

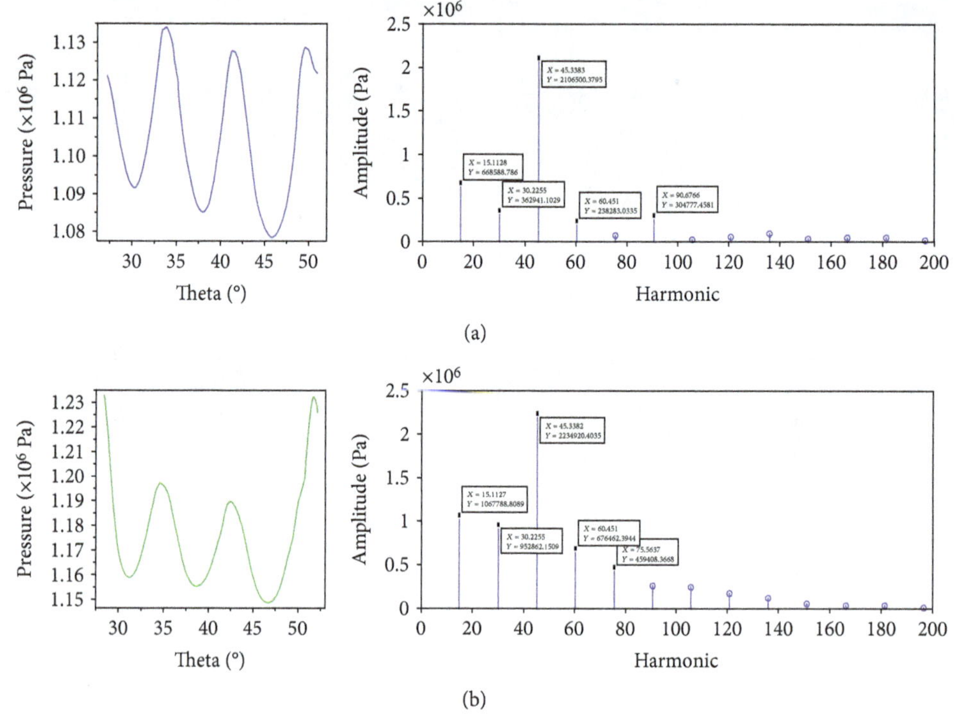

FIGURE 23: Spatial pressure distributions and FFT, recorded at the vane exit: (a) 50% span and (b) 96% span.

Figures 23–25. Indeed, the harmonic zero was eliminated since it represents the meridional distribution (average field) and qualifies the decay in the static pressure amplitude from inlet to outlet. The flow interactions are composed by an infinite number of rotating lobed structures described by Tyler and Sofrin [25], and given by $k = mN_r + nN_s$ and $n = \cdots, -1, 0, 1, \ldots$ of rotating speeds $\Omega_k = (mN_r/k)\Omega_r$. The interaction produced by the run of rotor blades in front of NGV produces the space modes resulting from the combination of vane count and blade count. As noticed from the respective FFT spectrums, the harmonics of $k = 15.031, 15.112, 15.765, \ldots$ are approximated by the fundamental harmonic 16 and its multiples related to the stroboscopic effect induced by the passing of the blade row in the front of NGV. In this picture, the dominant unsteady fluctuation is due to the rotor trace itself. A physical illustration of this mechanism can be made, for example, the wakes coming from the vanes are characterized circumferentially by the spatial periodicity of $k = N_n$, and the potential effect of the rotating row is represented by the spatial periodicity of $k = N_r$. Thus, the wake/potential interaction is being translated into nonlinear terms of $k = mN_r \mp nN_n$.

The spatial static pressure recorded along the lines passing by the midspan rear the vane trailing edge and their subsequent spectrums are depicted in Figure 23 showing clearly the dominant harmonic 48 corresponding to NGV wakes. The vane and blade counts present a common divider, explaining the appearance of the harmonic 16 related to the first order combination between the vane and blade counts. The presence of the spatial harmonic 64 is the revelation of the potential effect which is of smaller amplitude compared with the fundamental harmonics downstream of vanes. On the contrary, close to the rotor blade (Figure 24), the amplitude of the harmonic 64 increases as the potential effect

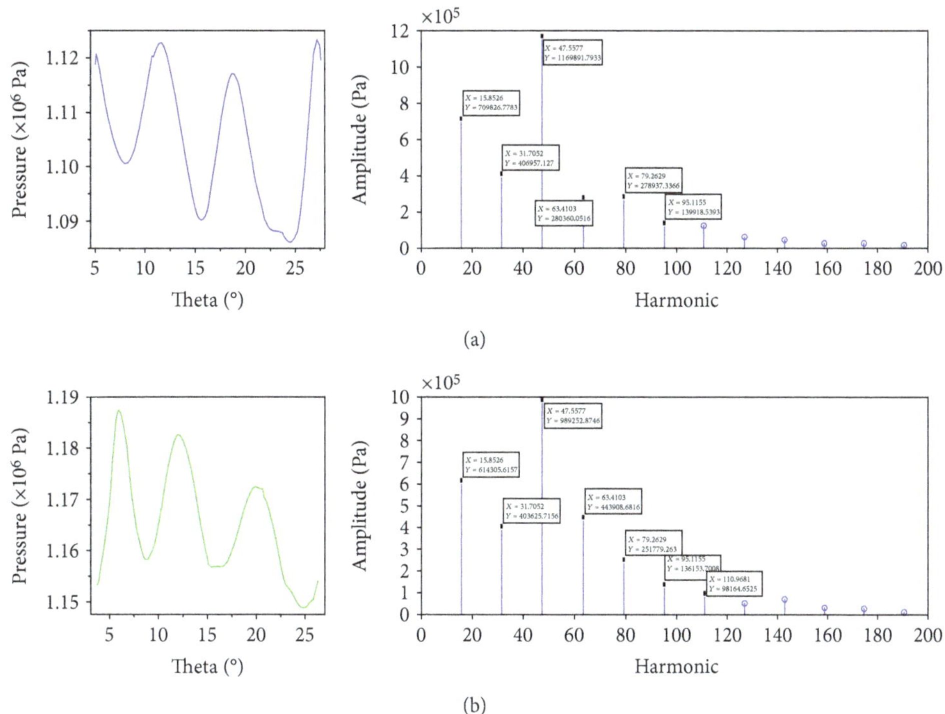

FIGURE 24: Spatial pressure and FFT, recorded upstream rotor at lines: (a) 50% span and (b) 96% span.

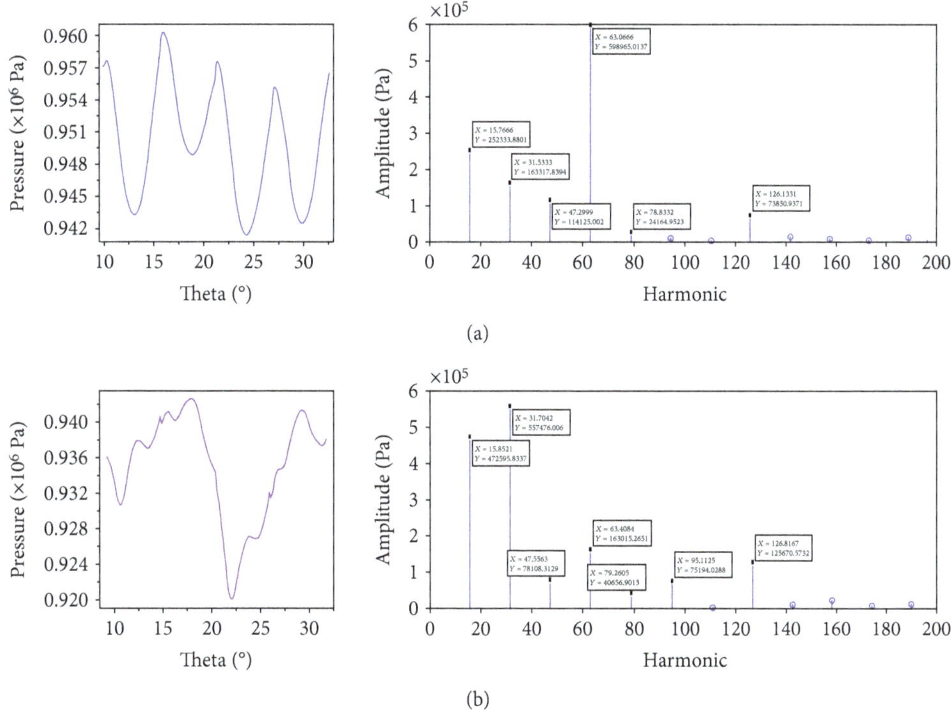

FIGURE 25: Spatial pressure and FFT, recorded downstream rotor blades at lines: (a) 50% span and (b) 96% span.

becomes more significant. Through the blade row, the flow undergoes sudden azimuthal and radial pressure gradients and sustains the effects of vanes' wake segments interacting with the potential effect. The accumulation of these various mechanisms contributes to a radial mixing and an exchange of energy. Downstream of rotor blade, the spectrums of the recorded pressure fluctuations (Figure 25) exhibit the fundamental harmonic 64 and its multiples which are the spatial modulation of the rotor blades and represent in the first order the rotor blade wakes but later they decay due to mixing. A second spatial harmonic is superimposed which is a combination of the first order between the vane and blade counts. The incoming wakes and vortices are shown to persist downstream of the rotor but of attenuated amplitudes. Moreover, the fluctuations induced by the tip cavity vortices are more significant than the wakes or the hub passage vortices.

The simulations based on the reduced domain NGV 3 : 4 allowed detecting the modes with harmonics less than the vane and blade counts. The harmonic 48 represents the effect of the vanes' wakes chopped by rotor blades, whereas the harmonic 64 represents the potential effect of rotating blades, while the harmonic 16 represents the VRI. Downstream of NGV, the effects of wakes are dominant, but the potential effect increases close to the rotor blades. Their effects are pronounced along the spacing between vanes and blades but attenuate downstream. From other unsteady simulations which considered a single channel by a component, in addition to overestimating the pressure fluctuation the frequencies of interactions related to the effects of wake segments chopped by rotor blades and other vortex structures were not detected.

5. Conclusion

The reduced computational domain NGV 3 : 4 seems a compromised choice to perform both the steady and unsteady flow computations. The flow structures in the vanes and the shrouded rotor with a tip cavity were characterized in details. As shown the wakes and secondary flows from the vanes persist downstream the rotor. Unsteady flow simulations based on the transient rotor/stator interface permitted investigating the unsteady VRI which are the valuable data in analyzing the forced vibrations. The amplitudes of the temporal pressure fluctuations due to rotor blade passing events and their harmonics vary significantly with the probing locations. The circumferential distribution of static pressure in each component reveals the existence of a unique harmonic and its multiples, related to the vane/blade count. The main interaction contibution is that due to the potential effect which propagates in the form of pressure wave upstream and downstream of the blade row. The other forms of VRI seem attributed to the embedded rotor flow disturbances predominately affected by the wake distortions impinging on the rotor blade suction surface, added to the shroud and tip cavity disturbances and the vortices formation of lesser importance.

Nomenclature

C: Blade chord (m)
C_p: Pressure coefficient
f: Frequency (Hz)
h: Enthalpy (J/kg)
k: Turbulence kinetic energy (m^2/s^2)
$\dot{m}$: Mass flow rate (kg/s)
n: Rotating speed (rpm)
N: Vane/blade count
P: Static pressure (Pa)
Re: Reynolds number
Δt: Time of simulation (ms)
$\Delta t_{\min}$: Time step (μs)
T: Temperature (K)
Tu: Turbulence intensity (%)
V_∞: Free stream velocity (m/s)
W: Relative flow velocity (m/s)
y^+: Dimensionless wall distance
U_i: Average flow velocity component (m/s)
u_i: Turbulent flow velocity component (m/s)
τ: Stress tensor (N/m^2).

Greek

ε: Turbulence rate of dissipation (m^2/s^3)
η: Efficiency
μ: Dynamic viscosity (kg/m·s)
ρ: Density (kg/m^3)
ω: Specific turbulence dissipation rate.

Subscripts

is: Isentropic
n: Nominal
t: Total
tt: Total-to-total
red: Reduced
xt: Transition.

Acronyms

BPF: Blade passing frequency
hp: High pressure
NGV: Nozzle guide vane
VRI: Vane-rotor interactions
SST: Sheer stress transport.

Conflicts of Interest

The author declared no potential conflicts of interest with respect to the research, authorship, and/or publication of this article.

References

[1] M. B. Giles, "Calculation of unsteady wake/rotor interaction," in *25th AIAA Aerospace Sciences Meeting*, Reno, Nevada, USA, March 1987.

[2] H. P. Hodson, "Measurements of wake-generated unsteadiness in the rotor passages of axial flow turbines," *Journal of Engineering for Gas Turbines and Power*, vol. 107, no. 2, pp. 467–475, 1985.

[3] R. E. Walraevens, H. E. Gallus, A. R. Jung, J. F. Mayer, and H. Stetter, "Experimental and computational study of the unsteady flow in a 1.5 stage axial turbine with emphasis on

the secondary flow in the second stator," in *ASME 1998 International Gas Turbine and Aeroengine Congress and Exhibition: Volume 1: Turbomachinery*, Stockholm, Sweden, June 1998.

[4] O. P. Sharma, E. Renaud, T. L. Buttler, K. Milsaps Jr., R. P. Dring, and H. D. Joslyn, "Rotor-stator interaction in multistage axial-flow turbines," in *24th Joint Propulsion Conference*, Boston, MA, USA, July 1988.

[5] A. Binder, W. Förster, H. Kruse, and H. Rogge, "An experimental investigation into the effect of wakes on the unsteady turbine rotor flow," *Journal of Engineering for Gas Turbines and Power*, vol. 107, no. 2, pp. 458–465, 1985.

[6] O. P. Sharma, T. L. Butler, H. D. Joslyn, and R. P. Dring, "Three-dimensional unsteady flow in an axial flow turbine," *Journal of Propulsion and Power*, vol. 1, no. 1, pp. 29–38, 1985.

[7] N. Arndt, "Blade row interaction in a multistage low-pressure turbine," *Journal of Turbomachinery*, vol. 115, no. 1, pp. 137–146, 1993.

[8] R. J. Miller, R. W. Moss, R. W. Ainsworth, and N. W. Harvey, "The development of turbine exit flow in a swan-necked interstage diffuser," in *ASME Turbo Expo 2003, collocated with the 2003 International Joint Power Generation Conference: Volume 6: Turbo Expo 2003, Parts A and B*, Atlanta, Georgia, USA, June 2003.

[9] V. S. P. Chaluvadi, A. I. Kalfas, H. P. Hodson, H. Ohyama, and E. Watanabe, "Blade row interaction in a high-pressure steam turbine," *Journal of Turbomachinery*, vol. 125, no. 1, pp. 14–24, 2003.

[10] J. Schlienger, A. I. Kalfas, and R. S. Abhari, "Vortex-wake-blade interaction in a shrouded axial turbine," *Journal of Turbomachinery*, vol. 127, no. 4, pp. 699–707, 2005.

[11] L. Qi and Y. Zhou, "Turbine blade tip leakage flow control by unsteady periodic wakes of upstream blade row," *Procedia Engineering*, vol. 80, pp. 202–215, 2014.

[12] M. B. Giles, "Calculation of unsteady wake/rotor interaction," *Journal of Propulsion and Power*, vol. 4, no. 4, pp. 356–362, 1988.

[13] M. M. Rai and N. K. Madavan, "Multi-airfoil Navier-Stokes simulations of turbine rotor-stator interaction," *Journal of Turbomachinery*, vol. 112, no. 3, pp. 377–384, 1990.

[14] A. Arnone and R. Pacciani, "Rotor-stator interaction analysis using the Navier-Stokes equations and a multi-grid method," *Journal of Turbomachinery*, vol. 118, no. 4, pp. 679–689, 1996.

[15] J. P. Clark, G. M. Stetson, S. S. Magge, R. H. Ni, C. W. Haldeman Jr., and M. G. Dunn, "The effect of airfoil scaling on the predicted unsteady loading on the blade of a 1 and 1/2 stage transonic turbine and a comparison with experimental results," in *ASME Turbo Expo 2000: Power for Land, Sea, and Air: Volume 1: Aircraft Engine; Marine; Turbomachinery; Microturbines and Small Turbomachinery*, Munich, Germany, May 2000.

[16] J. Yao, R. L. Davis, J. J. Alonso, and A. Jameson, "Massively parallel simulation of the unsteady flow in an axial turbine stage," *Journal of Propulsion and Power*, vol. 18, no. 2, pp. 465–471, 2002.

[17] JT8D-17A, *Turbofan Engine, SPEC. NO. 6281, Date: 6-27-80*, Pratt & Whitney Aircraft group, 1980.

[18] Ansys-CFX, *Theory Guide. Release 16.0*, ANSYS, Inc., Canonsburg, USA, 2015.

[19] D. A. Wilcox, "Simulation of transition with a two-equation turbulence model," *AIAA Journal*, vol. 32, no. 2, pp. 247–255, 1994.

[20] P. Adami and F. Martelli, "Three-dimensional unsteady investigation of HP turbine stages," *Proceedings of the Institution of Mechanical Engineers, Part A: Journal of Power and Energy*, vol. 220, no. 2, pp. 155–167, 2006.

[21] C. Zhou, H. Hodson, and C. Himmel, "The effects of trailing edge thickness on the losses of ultrahigh lift low pressure turbine blades," *Journal of Turbomachinery*, vol. 136, no. 8, article 081011, 2014.

[22] Z. Li, Z. Wang, X. Wei, and D. Qin, "Flow similarity in the rotor-stator interaction affected region in prototype and model Francis pump-turbines in generating mode," *Journal of Fluids Engineering*, vol. 138, no. 6, article 061201, 2016.

[23] R. E. Mayle, "The 1991 IGTI scholar lecture: the role of laminar-turbulent transition in gas turbine engines," *Journal of Turbomachinery*, vol. 113, no. 4, pp. 509–536, 1991.

[24] C. E. Shannon, "Communication in the presence of noise," *Proceedings of the IEEE*, vol. 86, no. 2, pp. 447–457, 1998.

[25] J. M. Tyler and T. G. Sofrin, "Axial flow compressor noise studies," *SAE Transactions*, vol. 70, pp. 309–332, 1962.

Modified Regression Rate Formula of PMMA Combustion by a Single Plane Impinging Jet

Tsuneyoshi Matsuoka,[1] Kyohei Kamei,[1] Yuji Nakamura,[1] and Harunori Nagata[2]

[1] *Department of Mechanical Engineering, Toyohashi University of Technology, Toyohashi, Japan*
[2] *Division of Mechanical and Space Engineering, Hokkaido University, Sapporo, Japan*

Correspondence should be addressed to Tsuneyoshi Matsuoka; matsuoka@me.tut.ac.jp

Academic Editor: Corin Segal

A modified regression rate formula for the uppermost stage of CAMUI-type hybrid rocket motor is proposed in this study. Assuming a quasi-steady, one-dimensional, an energy balance against a control volume near the fuel surface is considered. Accordingly, the regression rate formula which can calculate the local regression rate by the quenching distance between the flame and the regression surface is derived. An experimental setup which simulates the combustion phenomenon involved in the uppermost stage of a CAMUI-type hybrid rocket motor was constructed and the burning tests with various flow velocities and impinging distances were performed. A PMMA slab of 20 mm height, 60 mm width, and 20 mm thickness was chosen as a sample specimen and pure oxygen and O_2/N_2 mixture (50/50 vol.%) were employed as the oxidizers. The time-averaged regression rate along the fuel surface was measured by a laser displacement sensor. The quenching distance during the combustion event was also identified from the observation. The comparison between the purely experimental and calculated values showed good agreement, although a large systematic error was expected due to the difficulty in accurately identifying the quenching distance.

1. Introduction

1.1. Design of Fuel Grain for CAMUI-Type Hybrid Rocket Motor. A conventional hybrid rocket, which typically uses a solid as fuel with a liquid or gas as oxidizer, has several disadvantages and the critical issue among those is the low regression rate [1, 2]. In the hybrid rocket, the combustion of oxidizer gas and pyrolyzed vapor from the fuel grain takes place in the boundary layer and the diffusion flame is established. The mixing and combustion process results in slightly lower overall combustion efficiency and eventually causes a lower regression rate. In fact, the regression rate of HTPB, which is a conventional fuel, is typically an order of magnitude lower than solid propellants. To overcome this disadvantage and enhance the fuel regression rate, several ideas have been proposed: ideas based on the grain shape and/or the injector [3–6] and employing novel fuels [7–10].

The Cascaded Multistage Impinging-jet type hybrid rocket (so-called "CAMUI" hybrid rocket) developed in Hokkaido University is also one of new types of hybrid rockets [11]. The concept is based on a unique configuration of the fuel grain which allows the regression rate to be enhanced without changing fuel materials or injectors. The fuel grain consists of multiple combustible blocks having two ports and they are shifted by 90 degrees each in an axial arrangement. The oxidizer and/or burnt gas flow through the ports and repeatedly impinge to the next fuel block. The regression with and without combustion take place at the front-end and back-end surfaces of the fuel block by the impinging jet as well as the inner port surfaces. The impinging jet and the multiple burning surfaces accelerate the regression; thus they can improve thrust density at least three times larger than a conventional hybrid rocket [11]. Recently, the group has developed 15 kN thrust level CAMUI-type hybrid rocket and successfully performed a static firing test in 2014 [12].

A key parameter for the design of a hybrid rocket motor is the regression rate. Marxman and Gilbert developed a well-known theory which provides the basic understanding of the regression behavior of hybrid rocket combustion [13, 14]. Their model is based on combustion over a flat fuel surface in a turbulent boundary layer and the regression rate is governed by convective heat transfer from the flame to the fuel surface. Considering the convective heat transfer as well, the following regression rate formula for each burning surface of CAMUI-type motor, that is, the upstream end face, downstream end face, and port inner wall, has been developed [15–17]:

$$\dot{r} = c\left(G_p\right)^m\left(\frac{H}{D}\right)^n, \quad (1)$$

where G_p [kg/m^2·s] is the mass flux of propellant, H [mm] is separation distance between the fuel blocks, and D [mm] is the port diameter. c, m, and n are empirical constants which vary depending on the local O/F ratio. Since the propellant mass flux is a function of not only the oxygen mass introduced but also the evaporated fuel mass, the propellant mass flux varies along the flow. Thus, the regression behavior of the uppermost stage is critically important. In other words, the regression rate of the uppermost stage may change the overall performance of CAMUI hybrid rocket motor.

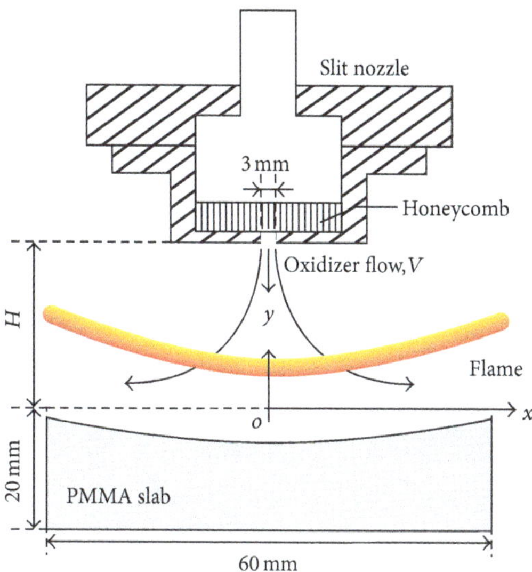

FIGURE 1: Schematic image of experimental apparatus.

1.2. Regression Rate Formula for Uppermost Stage. It is noticed that the above regression rate formula (i.e., (1)) predicts well except for the front-end surface of the uppermost fuel [16, 17]. According to Nagata et al. [17], it is because the pure oxygen gas impinges to the fuel surface in the uppermost stage, while the high temperature burnt gas is considered as working fluid in the following stages. The gasified oxygen reacts with the gaseous fuel and the flame establishes very close to the fuel surface there; the conduction is a rather dominant heat transfer. Therefore, the regression rate formula based on the convective heat transfer from the burnt gas may not be applied in the existence of the flame.

In addition to its applicability, several ambiguities remain in the conventional regression rate formula. The first one is that the formula requires empirical constants. In order to determine the values, preliminary tests must be carried out before starting the development of new motors which, in turn, creates high cost. In engineering point of view, to minimize the fuel residue is another important issue because it potentially blocks the nozzle and results in serious accidents such as explosion. However, it is impossible to predict the "local" regression rate by the conventional regression rate formula which provides only the spatial averaging regression rate over the whole combustion surface. To overcome these issues, the regression rate formula which can predict the local regression rate without any empirical constants and is applicable to the uppermost stage is needed for optimal design of CAMUI hybrid rocket motor.

The aim of this study is to develop a modified regression rate formula and validate it experimentally. First, burning tests of PMMA with a planar impinging oxidizer jet are performed and the local regression rate along the fuel surface is investigated for various impinging flow velocities and impinging distance. Although the experimental system is different from the actual CAMUI-type hybrid rocket motor, it is expected to reproduce the fundamental physics of the regression behavior of the uppermost stage. The modified regression rate formula which allows the calculation of the local regression rate with only the quenching distance is then derived considering a one-dimensional energy balance for a control volume near the fuel surface. The regression rate calculated is compared with the measurement data for validation.

2. Experimental Methods

A schematic image of the experimental model is shown in Figure 1. A PMMA was used as a fuel and its height, width, and depth were 20 mm, 60 mm, and 20 mm, respectively. The sample specimen of fuel was held by a stainless holder and installed below the nozzle. An oxidizer gas was ejected through a slit nozzle (width 3 mm and depth 30 mm) and a planer jet impinged to the sample. The location of sample specimen and the impinging distance, H [mm], between the sample and the nozzle exit were adjusted by a traverser. The oxidizer gas was supplied through the nozzle. The flow velocity and the oxygen concentration were adjusted by a gas flow control system. In this study, the pure oxygen or O_2/N_2 mixture gas containing 50 vol.% oxygen was used as an oxidizer. The oxygen concentration of the mixture gas was preliminary confirmed by the gas chromatography. The flow velocity at the nozzle exit, V [m/s], was set in the range of 1.9 to 11 m/s. A honeycomb that is 10 mm thick, having 100 cells per 100 cm^2, is built in the nozzle to rectify the oxidizer flow, although measurement of the velocity profile of the ejecting jet is not attempted in this study. Instead, the mean flow velocity was calculated from calibrated volume flow rates by means of the continuity equation. According to the test condition of oxygen concentration and flow velocity, several volume flow meters (variable-area flow meters) were used. For the pure oxygen test, we used three different flow meters properly (all of them were MODEL RK1200 SERIES available from KOFLOC). In addition to them, the MODEL RK1700 SERIES was used for 50 vol.% oxygen test. The MODEL RK1200 SERIES has the precision accuracy of less than 2%

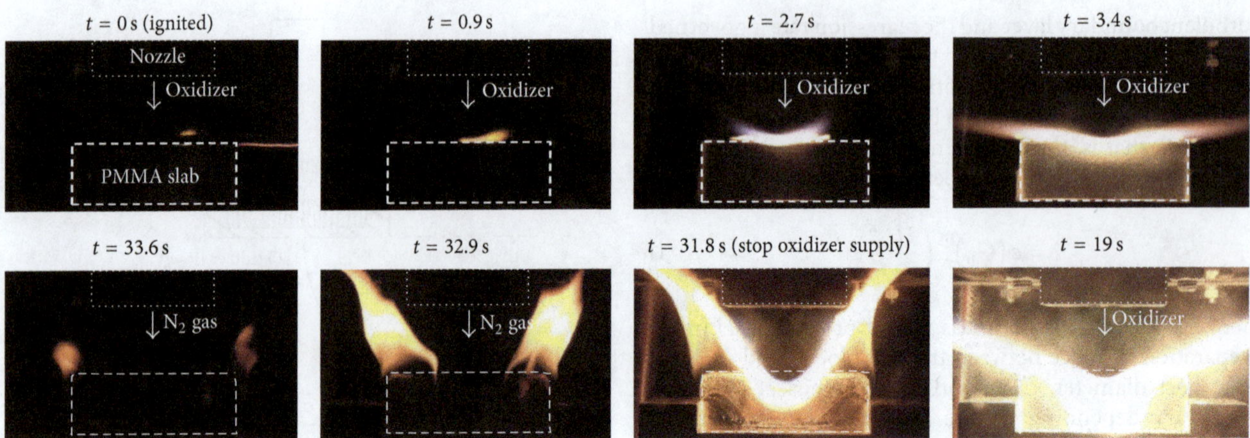

FIGURE 2: Time sequential images of PMMA combustion for $V = 5.6$ m/s and $H = 24$ mm.

FS, while that of RK1700 is less than 7% of FS. The volume flow rate was preliminarily calibrated. The calibration tests were performed as follows. The volume flow rate was set to a certain value and the actual flow rate was measured using a dry gas meter. Then, the set value was varied in a range of each flow meter. The test at the certain flow rate was repeated 12 times. The averaged data was plotted against the setting value and then fitted to a line by using the square least method. The calibration tests were done for all the flow meters used. The corresponding Reynolds number (Re) based on the slit width was between about 340 and 2100. It is noted that the experiments were conducted for much lower Re than the actual CAMUI-type hybrid rocket motor to validate the modified regression rate under the existence of the flame avoiding extinguishment.

Ignition to the sample specimen was introduced by an electric wire after confirming the obtained constant flow rate. Followed by successful ignition, the igniter was quickly removed. The combustion was forcibly stopped by introducing nitrogen gas. The entire combustion event was recorded by digital video camera (iVIS HF G20, CANON) or CCD camera (XC-EU50, SONY). The elapsed time until the flame experiences extinction, t [s], was obtained from the recorded movie.

After the combustion test, the regression depth along the horizontal axis, $r(x)$ [mm], defined as a distance from unburnt (initial) surface to burnt surface of the sample, was measured by a laser displacement sensor. The laser displacement was calibrated to output ±5 V when detecting ±20 mm. The burnt sample was fixed on a linear slider and the regression data was acquired. The sample edge has a large influence on the combustion phenomenon and the phenomenon cannot be assumed as two-dimensional. In order to eliminate the edge effect, all the measurement points were located at the center of depth direction in the range of $-20 < x < 20$ mm. As discussed later, assuming the regression phenomenon is steady, the time-averaged local regression rate, $\dot{r}(x)$, was determined as the regression amount per unit of the elapsed time, $\dot{r}(x) = r(x)/t$.

3. Results and Discussion

3.1. Measurement of Regression Rate. Figure 2 shows time sequential images of PMMA burning for $V = 5.6$ m/s. A reference time, $t = 0$, is set to the time at which the ignition is successfully confirmed. The flame spreads in the horizontal direction for the first few seconds, and then the flame covered the entire region above the fuel. The flame was extinguished at a certain time by introducing nitrogen gas through the nozzle. The elapsed time until the flame experiences extinction, t [s], was derived from the obtained movie. Figure 3(a) shows the regression amount from the initial unburnt surface plotted against the elapsed time at the representative locations. The regression amount linearly increases, suggesting that steady combustion is successfully achieved except for the initial 10 seconds after the ignition. Hence, the time-averaged regression rate, defined as the regression amount divided by the elapsed time, was evaluated as shown in Figure 3(b). The regression rate increases as the test proceeds during the initial duration. Later, a nearly constant regression rate was obtained within the entire region of the fuel. Accordingly, the elapsed time was set to about 30 s to eliminate the ignition disturbance on the regression rate in the following experiments. Kaneko et al. measured the regression rate of polyethylene of which thermal diffusivity is larger than that of PMMA by ultrasonic pulse-echo technique and obtained the steady regression rate within 100 s before the extinction treatment [15]. Thus, the time-averaged regression rate used in this study is considered as an adequate parameter for the evaluation. Since the experiment was performed only once, it is impossible to evaluate the uncertainty for these experiments. However, it may be deduced to the same extent as those obtained for 50 vol.% oxygen test (see Figure 7).

Figure 4 shows the local regression rate along the horizontal axis for the impinging distance equal to 24 mm. Note that the local regression rate seems symmetric to the center axis ($x = 0$), and the figure shows averaged regression rates on both $|x| > 0$ sides. The maximum regression rate is obtained in the vicinity of the center and it gradually decreases outwardly. For the solid combustion

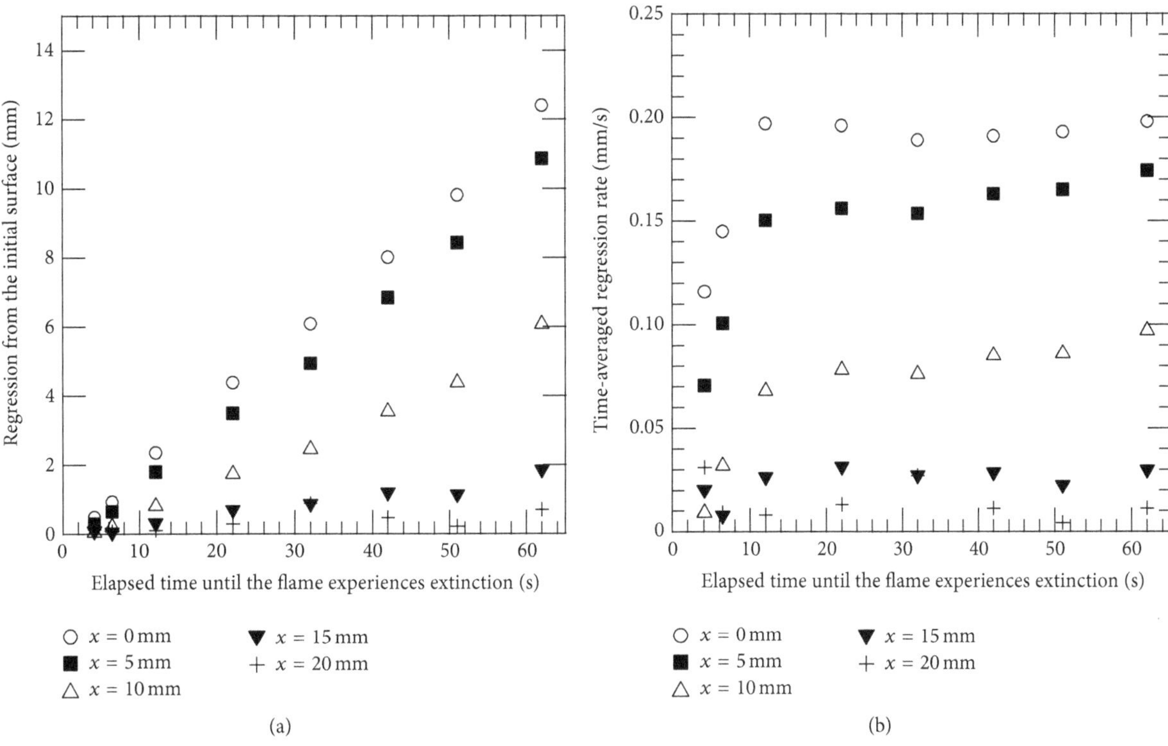

FIGURE 3: (a) The regression from the initial surface and (b) the time-averaged regression rate plotted against the elapsed time until the flame experiences extinction for $V = 5.6$ m/s and $H = 24$ mm.

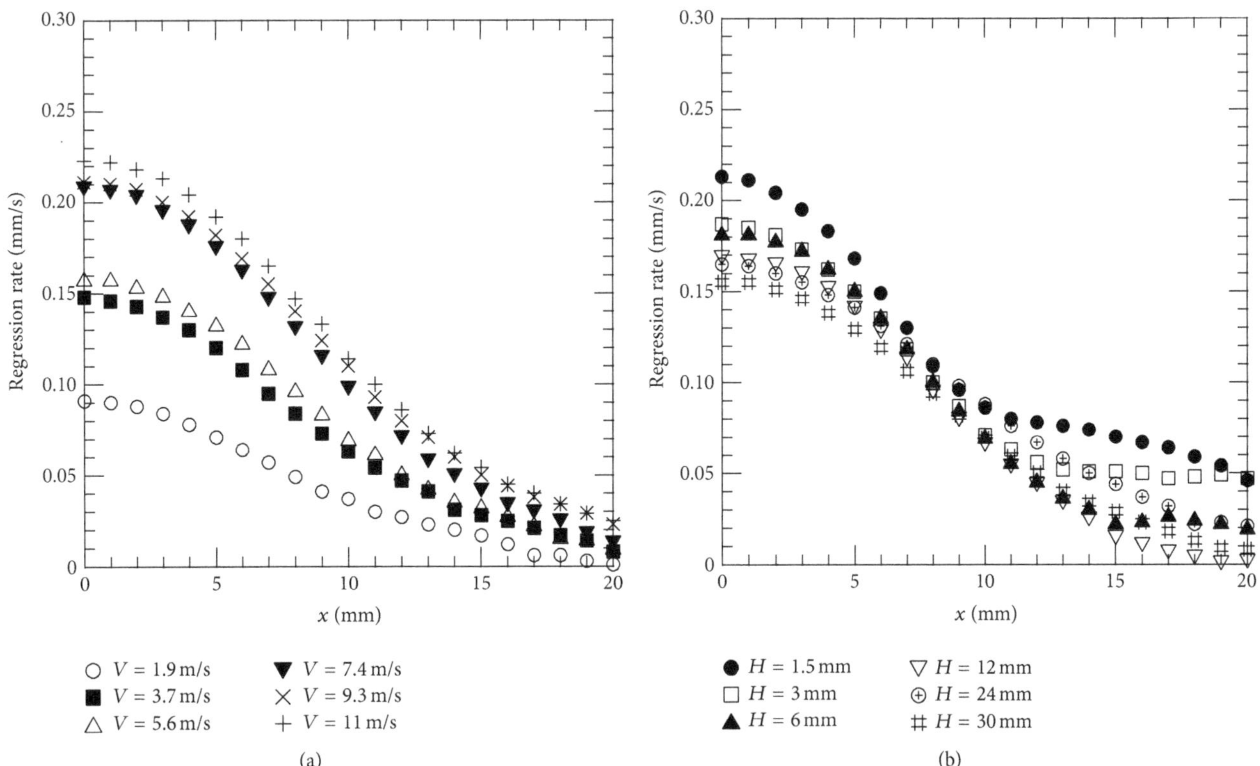

FIGURE 4: Local regression rate along the center line ($x = 0$). (a) The experiments were performed at a constant impinging distance of 24 mm and (b) at a constant impinging jet velocity of 5.6 m/s.

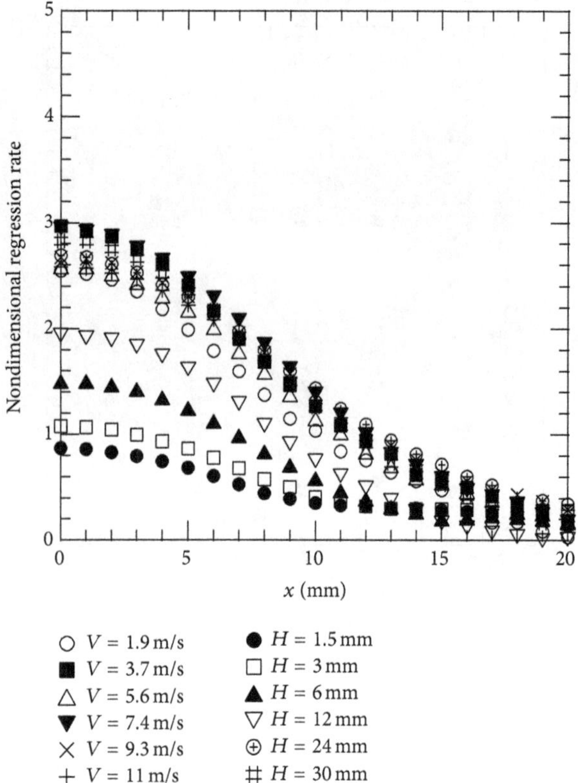

FIGURE 5: Nondimensional regression rate.

in stagnation flow, Matsui et al. [18, 19] showed that the nondimensional regression rate, defined as the regression rate divided by the characteristic velocity, is the suitable parameter to correlate the kinematic effect with the hydrodynamic effects and obtained well-summarized correlation between the Damköhler number and the nondimensional regression rate. Such nondimensional treatment is expected to work as well. Here, the characteristic velocity is expressed as $(a\nu)^{1/2}$ for the two-dimensional jet, where $a\,[\mathrm{s}^{-1}]$ is the so-called stagnation velocity gradient, and $\nu\,[\mathrm{m^2/s}]$ is the kinematic viscosity of the gas at the surface. The residence time is inversely proportional to the velocity gradient and the amount of momentum transfer of the evaporated fuel is evaluated by the kinematic viscosity. Hence, the velocity $(a\nu)^{1/2}$ physically characterizes the velocity of the fuel evaporated at the surface. As expected, the nondimensional regression rate, $\dot{r}/\sqrt{a\nu}$, for the various velocities and the impinging distances collapses into a single line except for $H < 12$ mm (Figure 5). This indicates that the kinetic of the gas-phase reaction needs to be taken into account due to the relatively short residence time.

The obtained regression rate shown in Figure 4 shows a similar trend to the heat transfer characteristic of a nonreactive impinging slot jet [20]. When no combustion takes place, it is known that the Nusselt number is proportional to $\mathrm{Re}^{0.5}$ in the stagnation region, while $\mathrm{Re}^{0.8}$ in the wall-jet region [21]. However, since the convective heat transfer coefficient and hence the Nusselt number depend on the flow properties such as velocity, viscosity, and other flow and temperature dependent properties, remodeling the regression behavior based on the conductive heat transfer from the flame to the solid may clarify the phenomenon and give a better regression rate formula.

3.2. Modified Regression Rate Formula for Uppermost Stage. The observation reveals that the diffusion flame covers the entire fuel surface for the condition far from the extinction limit. Ignoring the flame curvature, it is considered that the positional relation between the flame and regression surface is one-dimensional. Then, assuming the one-dimensional, quasi-steady regression motion, the heat balance equation of the control volume near the fuel surface is expressed as follows:

$$\dot{q}_{\mathrm{in}} = \dot{q}_1 + \dot{q}_2 + \dot{r}\rho_s L, \tag{2}$$

where $\dot{q}_{\mathrm{in}}\,[\mathrm{W/m^2}]$ is the heat flux transferred from the flame to the fuel, $\dot{q}_1\,[\mathrm{W/m^2}]$ is the conductive heat flux through the solid, $\dot{q}_2\,[\mathrm{W/m^2}]$ is the radiative heat loss from the fuel, $\rho_s\,[\mathrm{kg/m^3}]$ is the density of the fuel, and L [J/kg] is the latent heat of gasification. Here, the temperature profile in the condensed phase is given by the following equation [23]:

$$T(y) = T_\infty + (T_s - T_\infty)\exp\left(\frac{\dot{r}}{\alpha}y\right). \tag{3}$$

$\alpha\,[\mathrm{m^2/s}]$ is the thermal diffusivity and T_s [K] and T_∞ [K] are surface and bulk temperature of the fuel, respectively. Let δ [mm] be the quenching distance between the flame and the surface, and the regression rate is expressed by the following equation:

$$\dot{r}(x) = \frac{(\lambda_g/\delta(x))(T_f - T_s) - \varepsilon_s\sigma(T_s^4 - T_\infty^4)}{\rho_s[L + c_s(T_s - T_\infty)]}, \tag{4}$$

where c_s [J/kg·K] is the specific heat of solid, λ_g [W/m·K] is the thermal conductivity of gas phase, T_f [K] is the flame temperature, ε_s is the emissivity, and $\sigma\,[\mathrm{W/m^2{\cdot}K^4}]$ is the Stefan-Boltzmann constant.

Equation (4) shows that it is necessary to know the quenching distance for calculating the regression rate. In this study, the quenching distance was measured from the image during the combustion although it is also predictable as discussed later. Figure 6 shows front views of PMMA captured 20 seconds after ignition for $V = 1.9$ and 5.6 m/s and $H = 24$ mm with CCD camera. In order to make observation and specification of the flame easily, the glass plates of 20 mm height and 59 mm width were installed at the front and rear faces. The upper ends of both the glass plate and fuel specimen were flush against each other. In this experiment, the oxygen concentration was reduced to 50 vol.% to avoid thermal damage of the glass. When PMMA is exposed to the flame, surface degradation takes place due to linear pyrolysis. Figure 6 visualizes two distinctive regions: the black region corresponding to the condensed phase and the gray region where small bubbles are observed. Although the bubble layer adhered to the front and back walls inhibit clear observation from the side, it is considered that the luminous zone, which

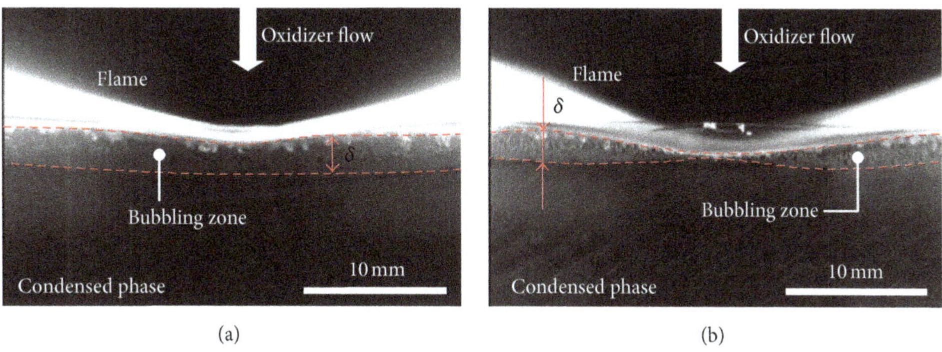

FIGURE 6: Photographs of PMMA burning for (a) $V = 1.9$ m/s and (b) 5.6 m/s. The oxygen concentration was 50 vol.%.

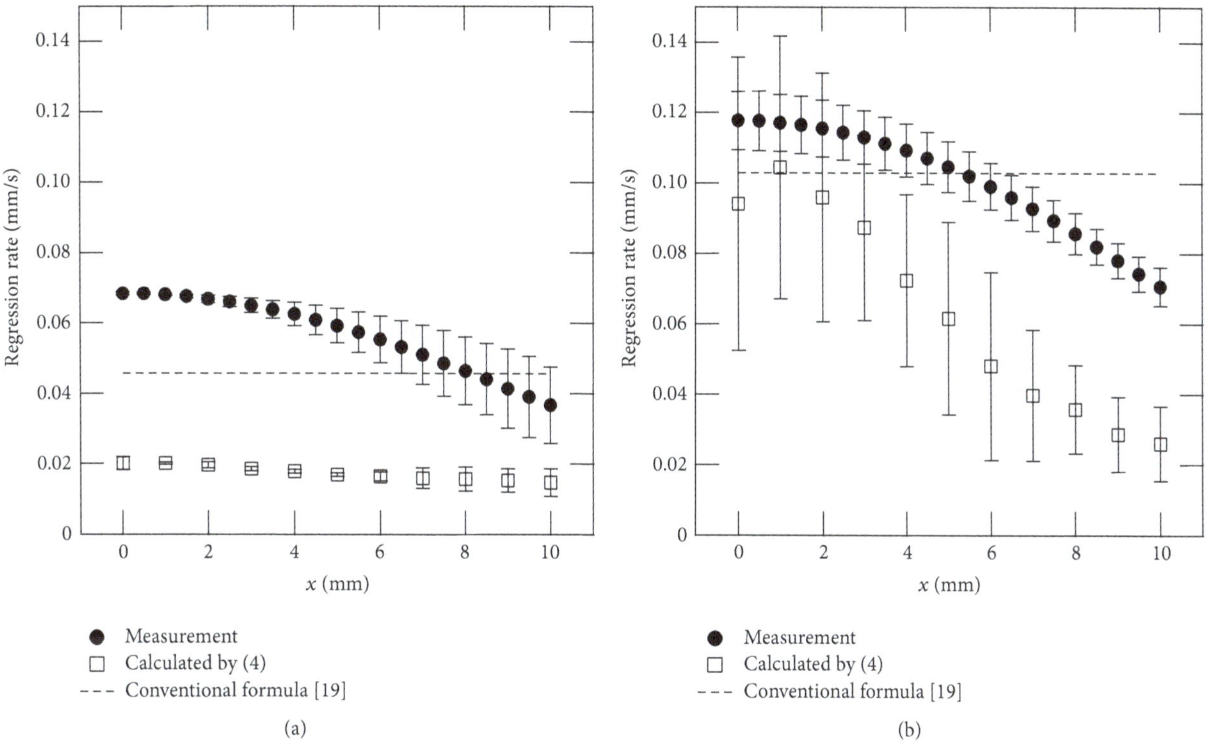

FIGURE 7: Comparison of local regression rates for (a) $V = 1.9$ m/s and (b) 5.6 m/s.

appears to overlap on the boiling layer, is corresponding to the area where the luminous flame is established. The quenching distance seemed slightly changed along a wide direction for $V = 1.9$ m/s. Meanwhile, the flame formed closely to the surface for $V = 5.6$ m/s and the quenching distance became a minimum in the vicinity of the center. Note that the visible distance between the flame and the bubbling layer (the distance between the solid and dashed line in the figure) gives the quenching distance at the central position in depth direction. Hence, in the following section, the regression rate evaluated by (4) with the measured quenching distance is that along the center line.

The regression rates were calculated by (4) and compared with the experimentally measured regression rate by the laser displacement sensor, as shown in Figure 7. Table 1 shows physical and chemical constants used in the calculation. In order to evaluate the uncertainty, the experiments were repeated for three ($V = 1.9$ m/s) or four times ($V = 5.6$ m/s), respectively. Assuming the obtained regression rate follows the normal distribution, the uncertainty of the regression rate within 95% confidence interval was then evaluated as 1.96 times the standard deviation of the data. The regression rate calculated by the conventional regression rate formula [17], $\dot{r} = c\,\mathrm{Re}^m (H/B)^n$, is also shown in the figure. The values of c, m, and n are shown in Table 2. Although the empirical constants obtained for the CAMUI-250 motor are applied in spite of the difference of flow fields, the conventional formula reasonably agrees with the measured regression rate. However, it cannot provide the *local* regression rate in principle. On the other hand, the modified regression rate formula gives the *local* regression rate. Although the difference between the calculated and measured regression rate is relatively large,

TABLE 1: Physical and chemical constants used for (4).

T_f	[K]	2850	Evaluated for 50 vol.% of O_2 concentration [22]
T_s	[K]	763	[23] p. 311
T_∞	[K]	300	
σ	[W/m^2·K^4]	5.67×10^{-8}	
ε_s	[—]	0.9	
λ_g	[W/m·K]	0.1077	For 1800 K of air [24]
L_v	[J/kg]	159×10^6	[25]
ρ_s	[kg/m^3]	1190	[26]
c_s	[J/kg·K]	2942	Evaluated for T_s = 763 K [26]

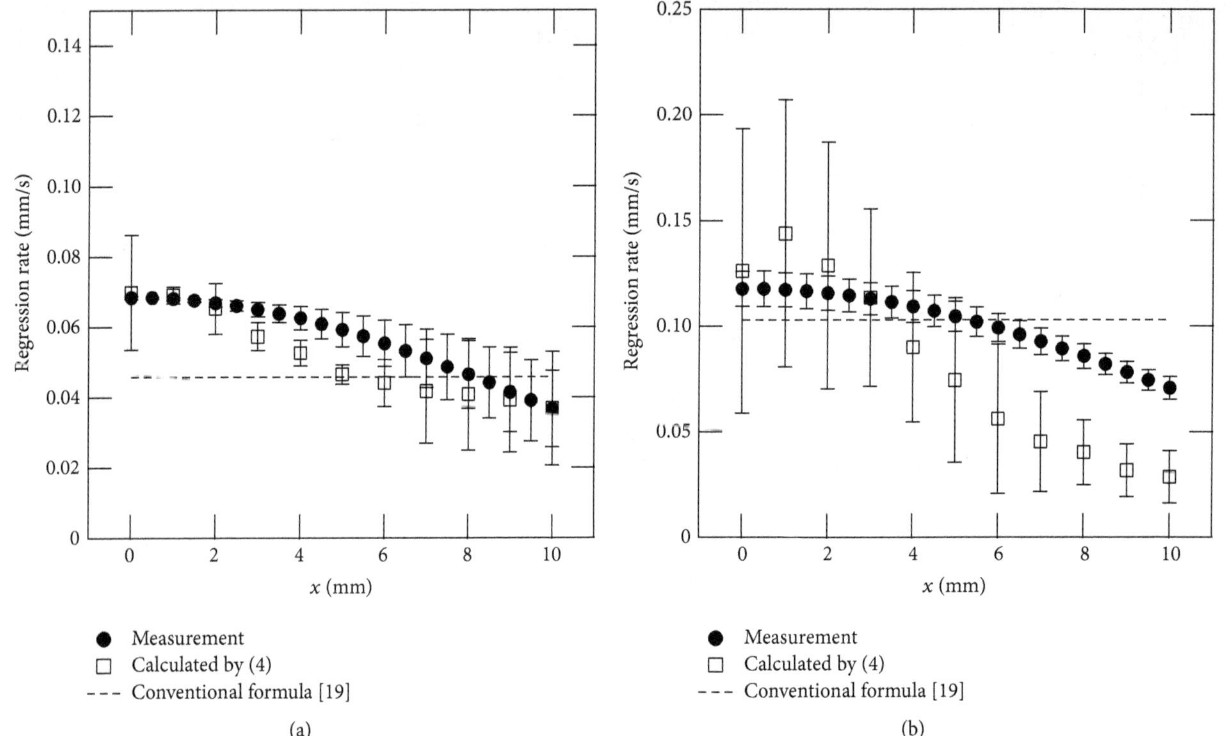

FIGURE 8: Comparison of local regression rates for (a) V = 1.9 m/s, offset value of δ = 1.7 mm, and (b) 5.6 m/s, offset value of δ = 0.15 mm.

TABLE 2: Empirical constants [17].

c	0.028
m	0.75
n	−0.09

especially for V = 1.9 m/s, the qualitative agreement of the trend indicates that the proposed heat balance equation reasonably predicts the regression rate once the quenching distance is identified. A possible reason to cause a large systematic error is the difficulty of accurate specification of the quenching distance. In addition, manual detection of the luminous flame edge and the interface between the bubbling zone and the condensed phase may also cause a further error. Considering the resolution of the image is about 0.05 mm/pixel, several tens pixels of offset greatly changes the regression rate. As shown in Figure 8, if the quenching distance is 1.7 mm (about 34 pixels) offset above the surface for V = 1.9 m/s, the modified regression rate formula is in good agreement with the measured regression rate, while the 0.15 mm offset (about 3 pixels) seems suitable for V = 5.6 m/s.

The present model is based on the energy balance between the flame and the fuel surface. Thus, the condition where the equation is applicable must be far from the extinction limit. The basic concept is applicable to the actual motor design. However, the actual CAMUI hybrid rocket motor is usually operated in much higher Reynolds number; for example, for 2500 N thrust class CAMUI motor, the Reynolds number ranges from about 33000 to 75000 [17]. In addition, the obtained regression rate formula requires one-dimensional arrangement of the regression surface and the flame. When the velocity gradient is large, the flame becomes stretched and thus the regression rate formula may not be applicable.

It is expected that the obtained modified regression rate formula has a potential to predict the regression rate more precisely once the precise quenching distance is given. This is a big advantage because the flame location from the surface is predictable from the initial and ambient parameters. It is well-known that for the droplet combustion the flame location is theoretically predicted with the Spalding mass transfer number, which represents the ratio of the driving force for vaporization of the fuel to the resistance to vaporization [27]. In turn, since the order of the regression rate is about 0.1 mm/s for general cases performed in this study, the characteristic time for changes in the gas-phase is considered shorter than the regression behavior. Thus, the combustion phenomenon is in quasi-steady state. Therefore, once the velocity field is determined independently, the flame location is expected to be obtained by solving the governing equations in a similar manner to the droplet combustion. Further modification to predict the local regression rate without performing any experiments is expected as a future work.

4. Conclusions

The regression behavior of PMMA combustion with the impinging oxidizer jet was experimentally investigated and the modified regression rate formula based on a quasi-steady, one-dimensional energy balance equation was proposed. An experimental setup which simulated the uppermost stage of CAMUI hybrid rocket was developed and burning tests with 100 vol.% and 50 vol.% oxygen were conducted. The quasi-steady combustion of PMMA was successfully confirmed immediately after the ignition process. Results show that the regression rate for various flow velocities and impinging distance was well-summarized by using the characteristic velocity of evaporated fuel at the surface. For a certain condition, the obtained regression rate formula was validated by comparing the calculated regression rate with the measured regression rate, although they did not match quantitatively. It is expected that the accuracy of the calculated regression rate can be improved by precise measurement or theoretical prediction of the quenching distance. The latter way suggests a big advantage over the conventional regression formula because the present model can predict the regression rate without performing any experiments once the quenching distance is given from the condition parameters. Though this study is initiated to predict the regression rate for optimal design of the CAMUI-type hybrid rocket motor, it is believed that the basic concept is applicable to general solid combustion with impinging jet.

Nomenclature

a: Velocity gradient, s^{-1}
B: Width of slit nozzle, mm
c: Empirical constant for (1)
c_s: Specific heat of solid, J/kg·K
D: Port diameter, mm
G_p: Mass flux of propellant, kg/m^2·s
H: Separation distance or impinging distance, mm
L: Latent heat of gasification, J/kg
m: Empirical constant for (1)
n: Empirical constant for (1)
Nu: Nusselt number
$\dot{q}_{\text{in}}$: Total heat flux transferred to fuel, W/m^2
$\dot{q}_1$: Conductive heat flux through the solid phase, W/m^2
$\dot{q}_2$: Radiative heat loss from the fuel, W/m^2
r: Regression amount from initial grain surface, mm
$\dot{r}$: Regression rate, mm/s
Re: Reynolds number
T_f: Flame temperature, K
T_s: Surface temperature of fuel, K
T_∞: Ambient temperature, K
V: Mean flow velocity at the nozzle exit, m/s
x: Horizontal axis, mm
y: Vertical axis, mm
α: Thermal diffusivity, m^2/s
δ: Quenching distance, mm
ε_s : Emissivity of fuel
λ_g: Thermal conductivity of gas phase, W/m·K
ν: Kinematic viscosity, mm^2/s
ρ_s: Density of the fuel, kg/m^3
σ: Stefan-Boltzmann constant: W/m^2·K^4.

Competing Interests

The authors declare that they have no competing interests.

Acknowledgments

This work was supported by JSPS KAKENHI Grant no. JP24860002. It was also supported by The Naito Science & Engineering Foundation and Tatematsu-Foundation.

References

[1] M. J. Chiaverini, "Review of solid-fuel regression rate behavior in classical and nonclassical hybrid rocket motors," in *Fundamentals of Hybrid Rocket Combustion and Propulsion*, M. Chiaverini, K. K. Kuo, and M. J. Chiaverini, Eds., chapter 2, pp. 37–125, American Institute of Aeronautics and Astronautics, Reston, Va, USA, 2007.

[2] D. Pastrone, "Approaches to low fuel regression rate in hybrid rocket engines," *International Journal of Aerospace Engineering*, vol. 2012, Article ID 649753, 12 pages, 2012.

[3] S. Yuasa, K. Yamamoto, H. Hachiya, K. Kitagawa, and Y. Oowada, "Development of a small sounding hybrid rocket with a swirling-oxidizer-type engine," in *Proceedings of the 37th Joint Propulsion Conference and Exhibit*, AIAA 2001-3537, AIAA, Salt Lake City, Utah, USA, July 2001.

[4] W. H. Knuth, D. J. Gramer, M. J. Chiaverini, and J. Arthur Saue, "Development and testing of a vortex-driven, high-regression rate hybrid rocket engine," in *Proceedings of the 34th AIAA/ASME/SAE/ASEE Joint Propulsion Conference and Exhibit*, AIAA 1998-3507, Cleveland, Ohio, USA, July 1998.

[5] W. H. Knuth, M. J. Chiaverini, J. A. Sauer, and D. J. Gramer, "Solid-fuel regression rate behavior of vortex hybrid rocket engines," *Journal of Propulsion and Power*, vol. 18, no. 3, pp. 600–609, 2002.

[6] J. R. Caravella Jr., S. D. Heister, and E. J. Wernimont, "Characterization of fuel regression in a radial flow hybrid rocket," *Journal of Propulsion and Power*, vol. 14, no. 1, pp. 51–56, 1998.

[7] P. G. Carrick and C. W. Larson, "Lab scale test and evaluation of cryogenic solid hybrid rocket fuels," in *Proceedings of the 31st Joint Propulsion Conference and Exhibit*, AIAA 1995-2948, San Diego, Calif, USA, July 1995.

[8] M. E. DeRose, K. L. Pfeil, P. G. Carrick, and C. W. Larson, "Tube burner studies of cryogenic solid combustion," in *Proceedings of the 33rd Joint Propulsion Conference and Exhibit*, AIAA 1997-3076, AIAA, Seattle, Wash, USA, July 1997.

[9] G. Risha, E. Boyer, R. Wehrman, and K. Kuo, "Performance comparison of HTPB-based solid fuels containing nano-sized energetic powder in a cylindrical hybrid rocket motor," in *Proceedings of the 38th AIAA/ASME/SAE/ASEE Joint Propulsion Conference and Exhibit*, AIAA 2002-3576, Indianapolis, Ind, USA, July 2002.

[10] M. Karabeyoglu, B. Cantwell, and D. Altman, "Development and testing of paraffin-based hybrid rocket fuels," in *Proceedings of the 37th Joint Propulsion Conference and Exhibit*, AIAA 2001-4503, Salt Lake City, Utah, USA, July 2001.

[11] H. Nagata, M. Ito, T. Maeda et al., "Development of CAMUI hybrid rocket to create a market for small rocket experiments," *Acta Astronautica*, vol. 59, no. 1–5, pp. 253–258, 2006.

[12] L. Uematsu Electric, *15kN Thrust Class CAMUI-Type Hybrid Rocket Motor Static Firing Test*, 2014, https://www.youtube.com/watch?v=jMSsNrPiLA0.

[13] G. Marxman and M. Gilbert, "Turbulent boundary layer combustion in the hybrid rocket," *Proceedings of the Combustion Institute*, vol. 9, no. 1, pp. 371–383, 1963.

[14] G. A. Marxman, "Combustion in the turbulent boundary layer on a vaporizing surface," *Proceedings of the Combustion Institute*, vol. 10, no. 1, pp. 1337–1349, 1965.

[15] Y. Kaneko, M. Itoh, A. Kakikura et al., "Fuel regression rate behavior of CAMUI hybrid rocket," *Transactions of the Japan Society for Aeronautical and Space Sciences, Space Technology Japan*, vol. 7, no. ists26, pp. 77–80, 2009.

[16] H. Nagata, S. Hagiwara, Y. Kaneko, M. Wakita, and T. Totani, "Development of regression rate formulas for CAMUI type," in *Proceedings of the 46th AIAA/ASME/SAE/ASEE Joint Propulsion Conference and Exhibit*, AIAA 2010-7117, Nashville, Tenn, USA, July 2010.

[17] H. Nagata, S. Hagiwara, N. Wakita, and T. Totani, "Optimal fuel grain design method for CAMUI type hybrid rocket," in *Proceedings of the 47th AIAA/ASME/SAE/ASEE Joint Propulsion Conference and Exhibit*, AIAA 2011-6105, San Diego, Calif, USA, July 2011.

[18] K. Matsui, A. Kôyama, and K. Uehara, "Fluid-mechanical effects on the combustion rate of solid carbon," *Combustion and Flame*, vol. 25, pp. 57–66, 1975.

[19] H. Tsuji and K. Matsui, "An aerothermochemical analysis of combustion of carbon in the stagnation flow," *Combustion and Flame*, vol. 26, pp. 283–297, 1976.

[20] M. Zukowski, "Heat transfer performance of a confined single slot jet of air impinging on a flat surface," *International Journal of Heat and Mass Transfer*, vol. 57, no. 2, pp. 484–490, 2013.

[21] M. Kumada, T. Nakatogawa, and K. Hirata, "Heat and mass transfer by Impining Jet," *Journal of Japan Society of Mechanical Engineering*, vol. 76, no. 655, pp. 822–830, 1973 (Japanese).

[22] S. Bhattacharjee, "The Expert System for Thermodynamics," http://www.thermofluids.net.

[23] N. A. Khalturinskii and A. A. Berlin, "High-temperature pyrolysis of thermoplastic polymers," in *Degradation and Stabilization of Polymers: A Series of Comprehensive Reviews*, H. H. G. Jellinek, Ed., vol. 1, chapter 6, pp. 289–336, Elsevier Science Publishers B.V., Amsterdam, The Netherlands, 1983.

[24] K. Kadoya, N. Matsunaga, and A. Nagashima, "Viscosity and thermal conductivity of dry air in the gaseous phase," *Journal of Physical and Chemical Reference Data*, vol. 14, no. 4, pp. 947–970, 1985.

[25] N. Hashimoto, H. Nagata, T. Totani, and I. Kudo, "Determining factor for the blowoff limit of a flame spreading in an opposed turbulent flow, in a narrow solid-fuel duct," *Combustion and Flame*, vol. 147, no. 3, pp. 222–232, 2006.

[26] Y. Nakamura and T. Kashiwagi, "Effects of sample orientation on nonpiloted ignition of thin poly(methyl methacrylate) sheet by a laser: 1. Theoretical prediction," *Combustion and Flame*, vol. 141, no. 1-2, pp. 149–169, 2005.

[27] J. C. Yang, "Heterogeneous combustion," in *Environmental Implications of Combustion Processes*, I. K. Puri, Ed., chapter 4, pp. 97–107, CRC Press, Boca Raton, Fla, USA, 1993.

16

Data Transfer Schemes in Rotorcraft Fluid-Structure Interaction Predictions

Young H. You, Deokhwan Na, and Sung N. Jung

Department of Aerospace Information Engineering, Konkuk University, Seoul 143-701, Republic of Korea

Correspondence should be addressed to Sung N. Jung; snjung@konkuk.ac.kr

Academic Editor: Vaios Lappas

For a CFD (computation fluid dynamics)/CSD (computational structural dynamics) coupling, appropriate data exchange strategy is required for the successful operation of the coupling computation, due to fundamental differences between CFD and CSD analyses. This study aims at evaluating various data transfer schemes of a loose CFD/CSD coupling algorithm to validate the higher harmonic control aeroacoustic rotor test (HART) data in descending flight. Three different data transfer methods in relation to the time domain airloads are considered. The first (method 1) uses random data selection matched with the timewise resolution of the CSD analysis whereas the last (method 2) adopts a harmonic filter to the original signals in CFD and CSD analyses. The second (method 3) is a mixture of the two methods. All methods lead to convergent solutions after a few cycles of coupling iterations are marched. The final converged solutions for each of the data transfer methods are correlated with the measured HART data. It is found that both method 1 and method 2 exhibit nearly identical results on airloads and blade motions leading to excellent correlations with the measured data while the agreement is less satisfactory with method 3. The reason of the discrepancy is identified and discussed illustrating CFD-/CSD-coupled aeromechanics predictions.

1. Introduction

Recently, with the advancement of modern software skills and innovative computer hardware technologies, mathematically challenging problems such as rotorcraft blade vortex interaction-(BVI-) induced noise and vibration have been tackled in a level that no one anticipated previously. One of the major breakthroughs in this regard is reached with the introduction of CFD method for loads and vibration prediction of a rotor. The classic CSD approach is combined with a CFD code in a form of loose coupling such that the blade elastic motions computed using CSD method and the aerodynamic forces and moments obtained using CFD method are exchanged at specified instant of time. This CFD/CSD coupling algorithm has been originally devised by Tung et al. [1] for simple lift coupling, developed separately by Beaumier [2] for lift and pitching moment coupling and later finalized by Potsdam et al. [3] using the delta airloads technique leading to close agreement with the measured data of UH-60A rotor. The CFD/CSD approach enables one to benefit the broad spectrum of CSD analysis power with the aid of first-principle-based CFD analysis capability, with the cost of computational efficiency. Even though the prediction capability has been improved drastically with CFD/CSD approach, there is no standard protocol or straightforward methodology for transferring data between CFD and CSD codes where significant gaps exist in solution methods and grid resolutions.

For rotor prediction capability, acquiring highly reliable test data plays crucial roles for cross-validating analytical results and implementing mathematical models to enhance physical understanding of an engineering phenomenon. Among other cases, HART data [4] offer unique opportunity to exploit and challenge high-frequency BVI-intensive signals in any advanced rotorcraft analysis system. The HART experiment was performed in the open-jet anechoic chamber of the German-Dutch wind tunnel (DNW) in 1994. The goal was to measure rotor loads, blade motions, acoustic signature, and flow fields in various flight conditions, with and without higher harmonic control (HHC) pitch control inputs. The

Table 1: Comparison of instrumentations for airloads and structural loads.

	HART	HART II
Number of pressure transducers	124	51
Number of measured air stations	3 ($r/R = 0.75, 0.87, 0.97$)	1 ($r/R = 0.87$)
Number of strain gages	34 (13 flap, 12 lag, and 9 torsion)	6 (3 flap, 2 lag, and 1 torsion)

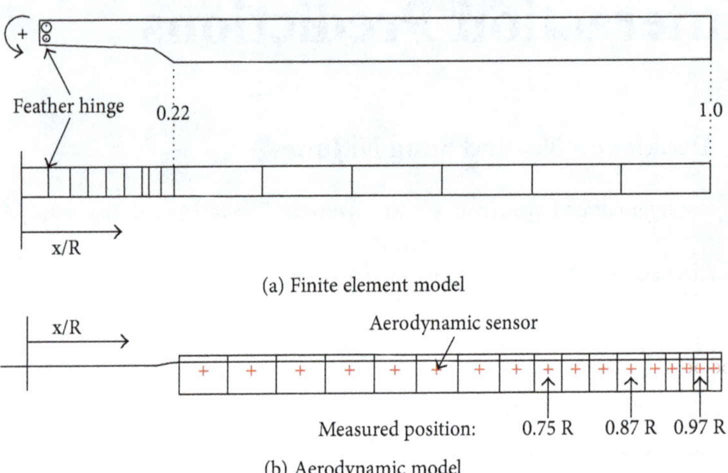

Figure 1: HART blade models used for CSD analysis.

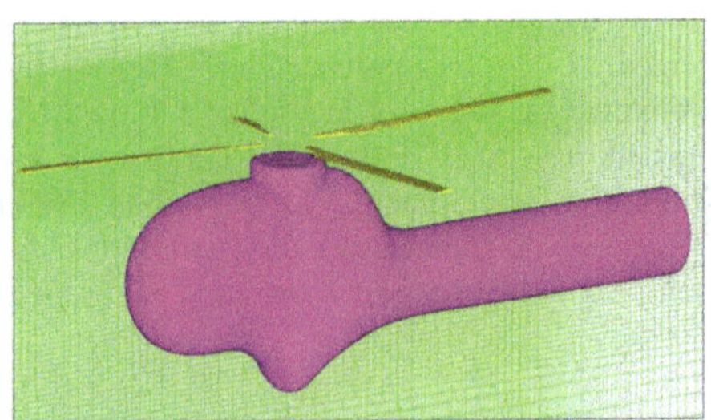

Figure 2: Computational grids used for CFD analysis.

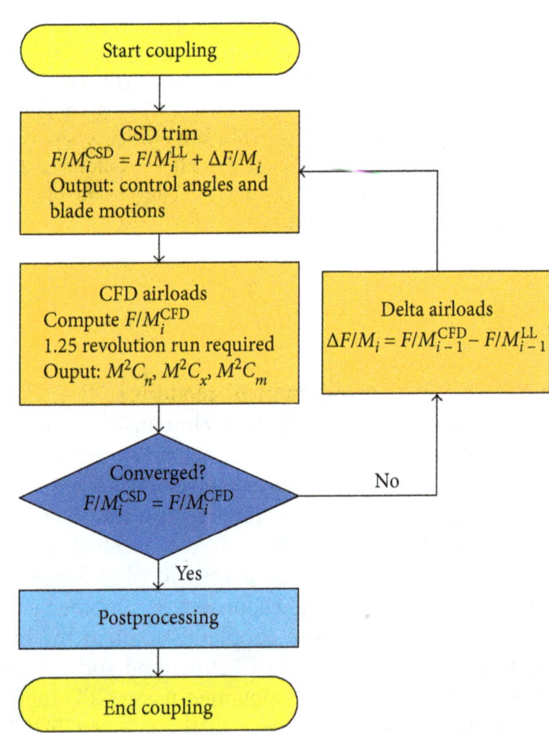

Figure 3: Flow diagrams for a loose CFD/CSD coupling approach.

international collaborative research results in a remarkable success [5], particularly obtaining high-precision test data set, advancing the smart rotor concept, gaining detailed knowledge of BVI phenomenon and its reduction mechanism due to the HHC technique, and leading to a follow-on test program HART II in 2001 [6]. A number of validation activities of HART rotor have been reported in the literature [7–11]. However, the published records are significantly less in volume as opposed to the postdecessor program HART II where the measured data have been extensively validated by researchers worldwide [12, 13]. Furthermore, very limited work has been carried out via CFD/CSD coupling except the work of Lim et al. [9] where a significant improvement on airloads prediction is reached over the conventional CSD approaches. The effect of fuselage is however neglected in the analysis of Lim et al. [9] which may lead to a phase shift problem on airloads along with the underprediction of BVI peak oscillations as observed by Jung et al. [14]. In addition, the update of the blade structural properties obtained by the recent measurement activity [15, 16] is lacking in most of the published works.

The present study aims at conducting a refined computation using CFD/CSD coupling for HART rotor in descent flight. A three-dimensional (3D), compressible RANS (Reynolds averaged Navier-Stokes) solver is employed as a CFD code. Considering the missing gap in the literature associated with the data exchange algorithm between CFD and CSD codes, this study is focused to investigate several data transfer

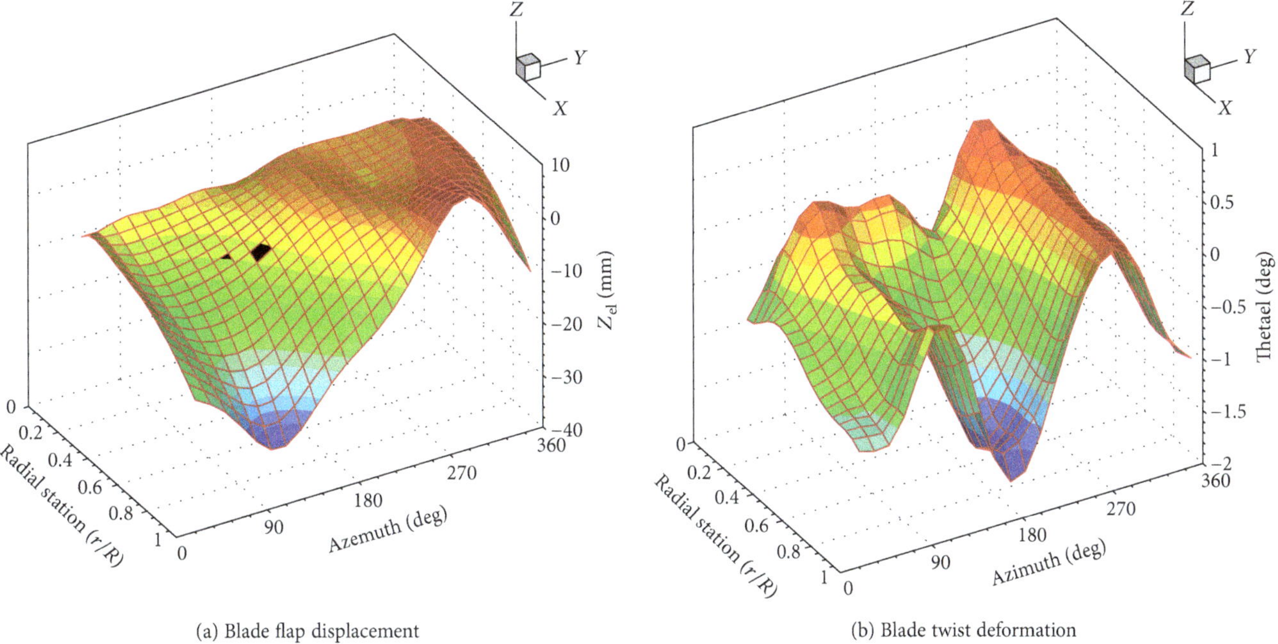

(a) Blade flap displacement

(b) Blade twist deformation

FIGURE 4: Blade elastic motions interpolated from CSD results.

TABLE 2: Summary of data synthesis methods.

	Method 1	Method 2	Method 3
CFD airloads (timewise)	Every 5 deg ($\Delta\psi = 5$ deg)	Every 5 deg ($\Delta\psi = 5$ deg)	Low-pass filtered (up to 10/rev)
CSD airloads (timewise)	Every 5 deg ($\Delta\psi = 5$ deg)	Low-pass filtered (up to 10/rev)	Low-pass filtered (up to 10/rev)
Spanwise airloads (CFD & CSD)	Cubic spline	Cubic spline	Cubic spline
Blade motions (spanwise)	Polynomial with nth order	Polynomial with nth order	Polynomial with nth order
Blade motions (timewise)	Fourier series with mth components	Polynomial with mth components	Polynomial with mth components

schemes in a loose coupling approach. Three different methods classified associated with the time domain airloads are examined. The first uses specified timewise data selection matched with the CSD analysis whereas the last adopts a harmonic filter to the raw data. The second one is a mixture of both methods. The accuracy of the final converged solutions for each of the data transfer schemes is evaluated by comparison with the measured HART data which include section airloads, blade elastic motions, and structural moments of the rotor. In addition, the newly measured blade properties are used for more realistic analysis.

2. HART Experiment

The HART rotor was tested at the DNW by an international joint team in 1994 [4]. A four-bladed, 40% Mach-scaled hingeless BO-105 model with 2 m radius and 0.121 m chord length is used for the test. The blade is constructed of E-glass spar and skin and has a rectangular planform with a modified NACA 23012 airfoil with trailing-edge tab. The test condition considered in the present study is a descending forward flight with an advance ratio $\mu = 0.15$, shaft tilt angle $\alpha_s = 4.5$ deg aft (after the wall correction), and thrust level $C_T = 0.0044$. The rotor is trimmed to match the target values

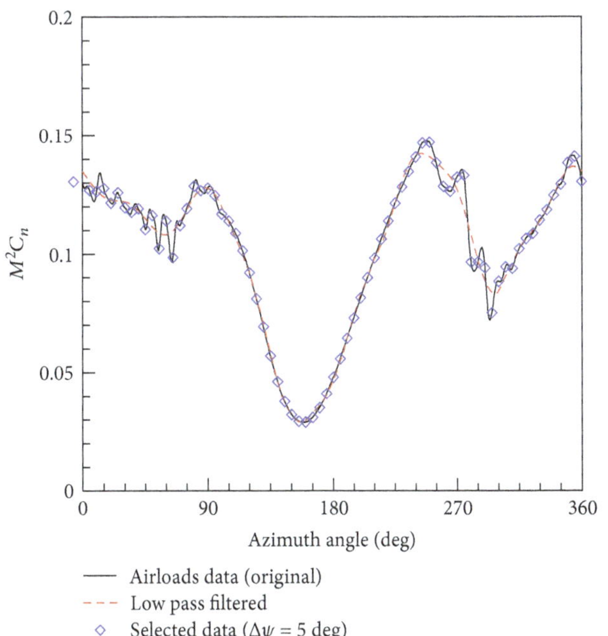

FIGURE 5: Comparison of data selection schemes on airloads signals.

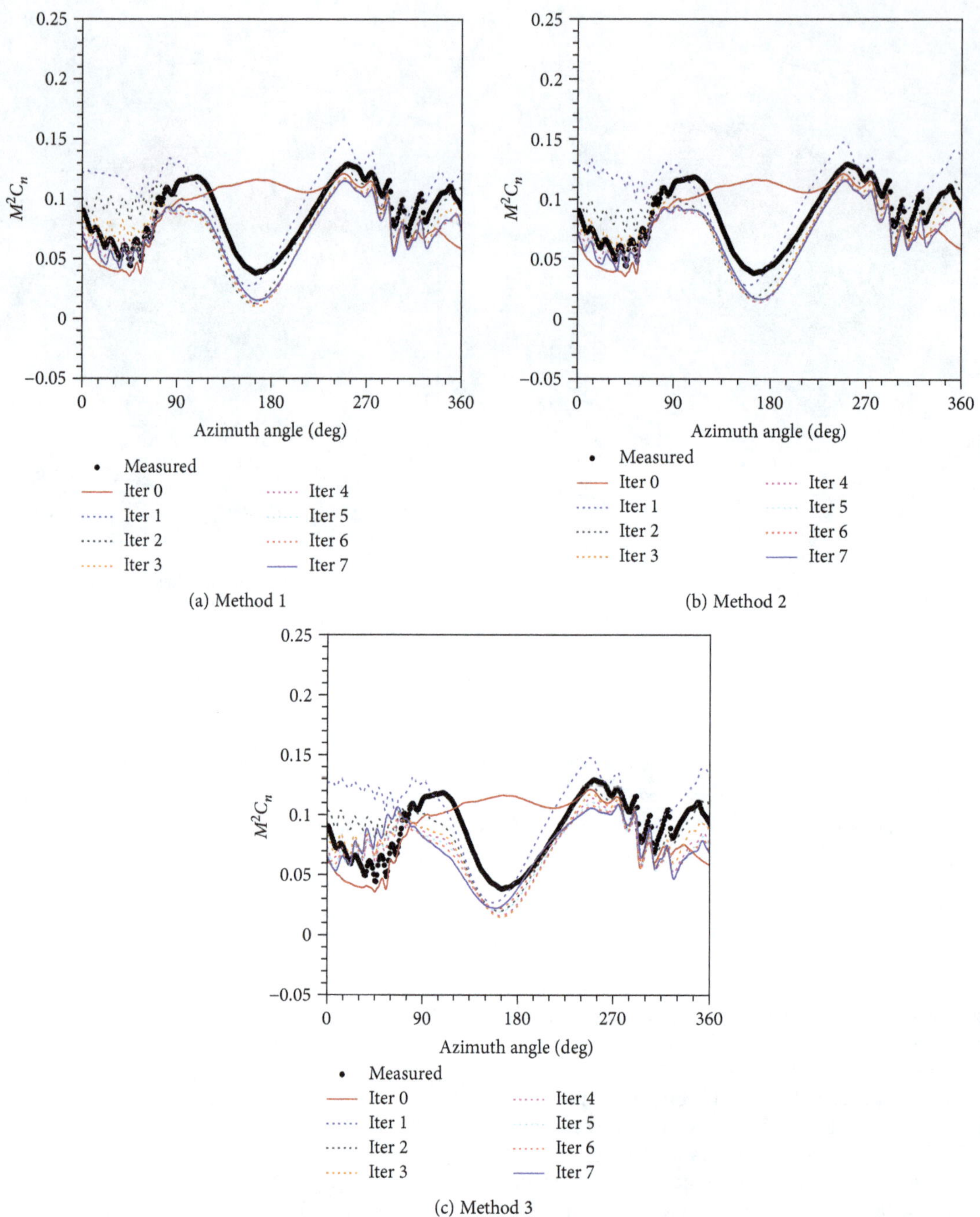

(a) Method 1

(b) Method 2

(c) Method 3

FIGURE 6: Comparison of convergence on section airloads M^2C_n at $r/R = 0.87$ with coupling iterations.

for the thrust, hub roll, and pitching moment as 3100 N, 11.2 N-m, and −20 N-m, respectively. The sign conventions used are positive when the advancing side goes up and a pitch-up is induced for the moments. The pressure measurements are made at three radial stations ($r/R = 0.75$, 0.87, and 0.97) of the reference blade (H1Y) for the complete pressure distribution along the blade chord. A total of 32 strain gauges are attached on the blade surface, distributed between $r/R = 0.14$ and 0.83, to measure the structural loads as well as the elastic deformation of the blade. An alternative optical technique was used to measure the blade tip motions. It is noted that HART rotor allows wider spectrum for airloads and structural load data than the follow-on program HART II (see Table 1). The blade structural properties are measured recently using the original set of blades tested in the wind tunnel [15, 16], and these are implemented in the present analysis.

3. Analysis Methodologies

A loose CFD/CSD coupling is employed to validate the measured HART data. To this end, a CSD analysis code CAMRAD II [17] is combined with a RANS flow solver KFLOW

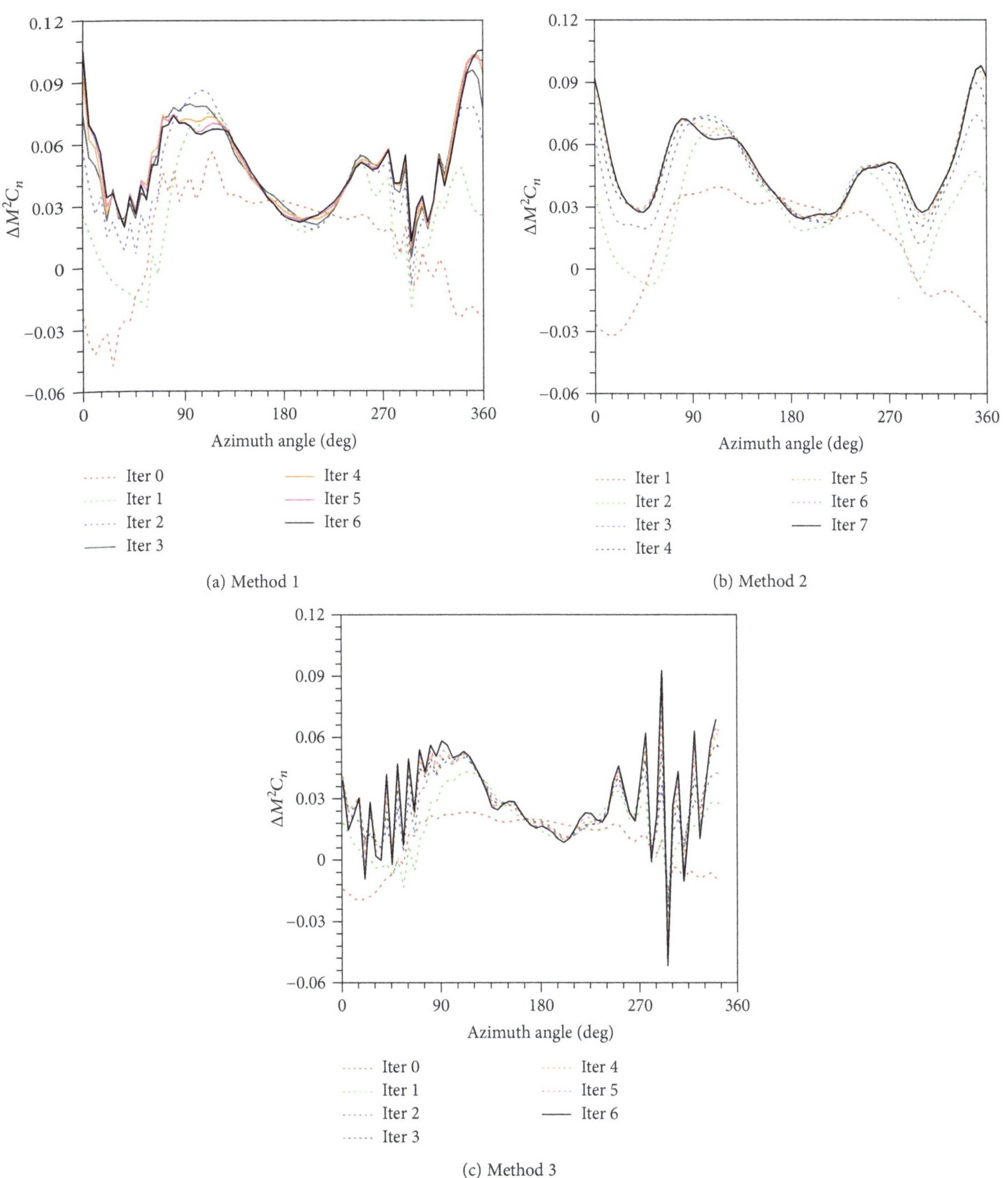

FIGURE 7: Comparison of convergence on delta airloads ΔM^2C_n at $r/R = 0.87$ with coupling iterations.

[18]. The essential features and modeling details adopted in the present approach are summarized in this section.

3.1. CSD Approach. CAMRAD II is a comprehensive aeromechanical analysis tool that is characterized by multibody dynamics, nonlinear finite elements, and various level of rotorcraft aerodynamics [17]. For the structural analysis, the blade motion is composed of the rigid body motion and the elastic deformation. The rigid body motion describes the motion of one end of a beam element, and the elastic motion is measured relative to the rigid motion. The beam elements are represented by 6 degrees of freedom (DOF) for the rigid motion and 9 DOF for the elastic motion (3 axial, 2 flap, 2 lag, and 2 torsion) that results in a 15 DOF for each beam finite element. The aerodynamic model is based on the ONERA-EDLIN unsteady airfoil theory combined with C81 airfoil table look-up. For the vortex wake model, the free wake geometry is used to compute the nonuniform-induced inflow around the rotor disk. The formation of the tip vortices is modeled using a free rolled-up wake model. The rolled-up wake model is based on the feature that a tip vortex forms at the blade tip. Both single and dual peak models are available considering the distribution of bound circulation peaks over the blade length.

In the present analysis, the blade structure is modeled using 15 beam finite elements with the discretization of finer elements in the blade inboard portion to counter a large variation in the structural properties of the blade, as can be seen in Figure 1(a). The airfoil blade region is divided into 17 nonuniform aerodynamic panels with finer lengths toward the blade tip, as shown in Figure 1(b). Specifically, the center of the aerodynamic panels is aligned to coincide with the measured airloads stations to minimize the discretization error.

3.2. CFD Approach. A 3D compressible flow solver KFLOW [18] is used for the CFD analysis. The KFLOW is a structured, parallelized multiblock, RANS solver that can compute time-accurate moving body problems. A second-order accurate, dual-time stepping scheme combined with a diagonalized alternating-directional implicit method is used to compute the unsteady flow fields around a rotor. The inviscid fluxes are calculated using the fifth-order weighted essentially nonoscillatory (WENO) scheme, while the central differencing technique is applied to the viscous fluxes. The k-ω Wilkox-Durbin (WD+) scheme is adopted for the turbulence model. The characteristic boundary conditions using the Riemann invariant are applied to the far-field boundary, whereas a no-slip condition is used at the solid wall surface.

A moving overlapped Chimera grid system with the near body and the Cartesian off-body grid are employed. Either C-mesh topology grids or O-mesh-based grids are formed, respectively, for the blade and the fuselage. The blade grids extend 1.5 times of a chord length (c) in the normal direction, measured from the blade surface. The cell spacing for the first grid point from the wall boundary used is $1.0 \times 10^{-5}c$. The off-body grids consist of an inner region that extends $4c$ upward, $3c$ below from the blade, and $1.5c$ away from the blade tip. The far-field boundary is stretched up to $5R$ (blade radius), centered at the rotor hub. The Cartesian off-body grids have a uniform spacing of $0.1c$. Figure 2 shows the overall computational grid system used for the HART rotor. The CFD grids result in about 38 million (M) cells in total, 6.4 M for the blade grid, 29.1 M for the off-body grid, and another 2.5 M for the fuselage grid.

3.3. CFD/CSD Coupling Approach. A loose coupling between CAMRAD II and KFLOW codes is adopted for the analysis. The basic idea of the coupling is to benefit the strength of the other code which requires an exchange of information between CSD-computed blade motions and CFD-computed airloads, per revolution base. Figure 3 illustrates a loose coupling strategy. The coupling iteration begins with CSD analysis using a low-fidelity aerodynamic model. The resulting blade motions along with trim control angles are transferred to the CFD code to compute refined aerodynamic forces and moments (F, M). The difference in airloads between the two codes (i.e., delta airloads) is calculated and superposed to the CSD airloads for the updated blade motions and trim controls for the next iteration stage. This process continues until the airloads and trim control angles show a convergence.

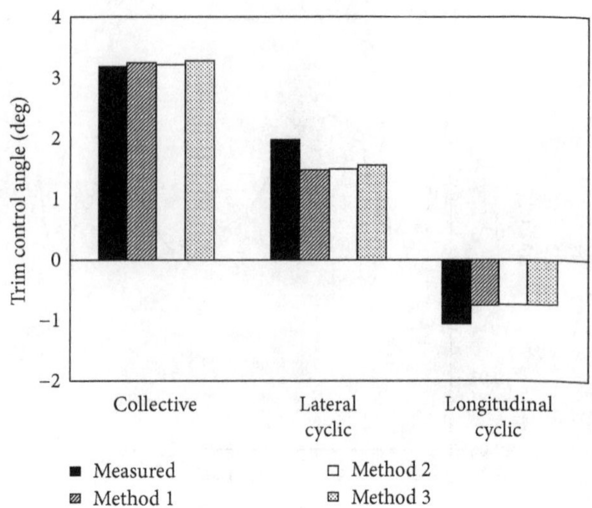

FIGURE 8: Comparison of trim control angles.

The CFD/CSD coupling requires adequate data regression schemes for blade motions and rotor airloads results. The blade motions are interpolated to allow sufficiently smooth curves in both radial and azimuthal directions, as described in van der Wall [19] and Sa et al. [20]. This method is straightforward and based on the best fit of the measured deformations of HART II rotor [12]. In this study, the radial interpolation is represented using polynomials with the seventh order while the time domain is interpolated using a Fourier series up to the eleventh components. The curve-fitted deformation on flap and lag motions is depicted in a 3D format in Figure 4. The CSD-synthesized data are smooth enough to be taken at any desired stations in a CFD code.

The airloads in spanwise direction are mostly monotonic, and the classical interpolation schemes such as the cubic spline are suitable to fit the data. The timewise airloads signals are arbitrary in general, and no systematic method or standard protocol is set to adopt universally. We employ three different data transfer schemes in association with the time domain airloads: method 1 uses random data selection matched at every azimuth angle of 5 deg for both CFD and CSD analyses; method 2 adopts a low pass filter containing up to 10/rev (per revolution) for both CFD and CSD data; and method 3 is a mixture of the two methods. The details of the three data synthesis methods for CFD/CSD coupling are summarized in Table 2. Figure 5 illustrates how the data are synthesized for the respective methods considered. As can be seen, the original CFD airloads data represented by the continuous line have an azimuth resolution of 0.2 deg ($\Delta\psi = 0.2$ deg) and appear to contain high-frequency signals (due to BVI events). The dashed line shows a low-pass-filtered (up to 10/rev) signal indicating a significant loss of information, particularly for the high-frequency signal. The random data selection made at every 5 deg azimuth angle are represented using hollow diamonds which indicates also some loss of information particularly over highly oscillating data zones. The influence of the candidate data synthesis methods on CFD-/CSD-coupled aeromechanics solutions is examined in the following section.

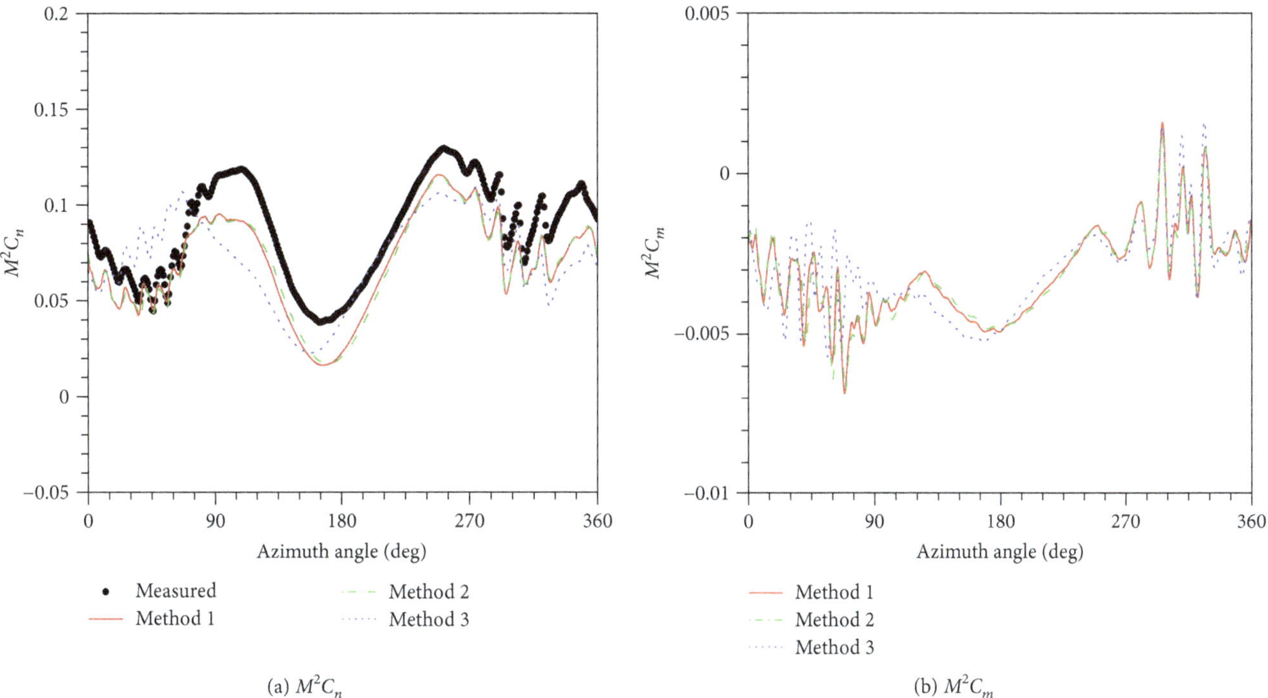

(a) M^2C_n

(b) M^2C_m

FIGURE 9: Comparison of section airloads at 87% radial station.

4. Results and Discussion

The three data transfer schemes are investigated first to evaluate the validity and overall performance of the techniques with respect to the coupling. To this end, the measured data of HART rotor are used as the referendum to evaluate the accuracy of the coupled predictions according to the data transfer methods. Next, the present CFD-/CSD-coupled results using the desired transfer method are compared with the measured data and also with CSD-alone predictions. Only the baseline (BL; Dpt. 140) condition of HART rotor is considered. It is noted that a pitch-bearing stiffness of 1706 N-m/rad is used to match the first torsion frequency and to represent the control system characteristic of HART rotor [21].

4.1. Parametric Investigation of Data Transfer Methods. Following the procedures given in Figure 3, the CFD-/CSD-coupled trim iterations are marched using the methods proposed in this study. The trim is handled by CSD code to match the trim targets specified as 3100 N, 11.2 N-m, and −20 N-m, respectively, for thrust, roll, and pitching moment. Figures 6 and 7 show the convergence behavior of CFD airloads M^2C_n (Mach number-scaled section normal forces) and the corresponding delta airloads with coupling iterations, respectively, for the three data transfer methods. It is clearly indicated that all methods lead to a convergence after 6 to 7 coupling iterations are stepped. The obvious differences according to the data transfer methods are represented in delta airloads results (Figure 7). In methods 1 and 3, some of the BVI peaks are preserved and passed over to the next coupling iterations while no such peaks are inherited to the next cycle in method 2 because of the low-pass-filtering operation performed at each step. It should be remarked that, in method 3, the BVI oscillatory peaks in the first and fourth quadrants of the rotor are gaining strengths to become very strong spikes as the coupling iterations are stepped further, in comparison with the other methods. Figure 8 shows the predicted control trim angles according to the three methods, as compared with the measured data. All methods indicate good correlation for the collective pitch settings with a significant underprediction in the cyclic pitch angles by up to 28.5%. However, there appear no remarkable deviations on predicted trim control angles between the data transfer methods.

Figure 9 shows the comparison of converged CFD airloads results on section normal forces and pitching moments, respectively, at 87% radial station of the rotor. Since no measured pitching moments are available for HART rotor, only the predicted results are compared with each other. As can be seen in both plots, methods 1 and 2 indicate nearly identical results leading to good agreements with the measured data. The peak-to-peak intensity of the airloads signal along with the down-up pattern near the front edge of the rotor disk is captured correctly. In addition, both BVI events apparent in the first and fourth quadrants of the rotor disk are predicted nicely in terms of magnitudes and phases of the signals. However, method 3 indicates significant deviations for section normal forces and section pitching moments over most of the azimuthal domain, as compared with the other methods, resulting in a considerable offset with the measured data. For the section normal forces, the predicted mean in the first quadrant increases significantly and the down-up pulse leads the measured data by about 15 deg. The BVI peaks in the pitching moment signal increase somewhat also. The higher BVI peak magnitudes of method 3 are

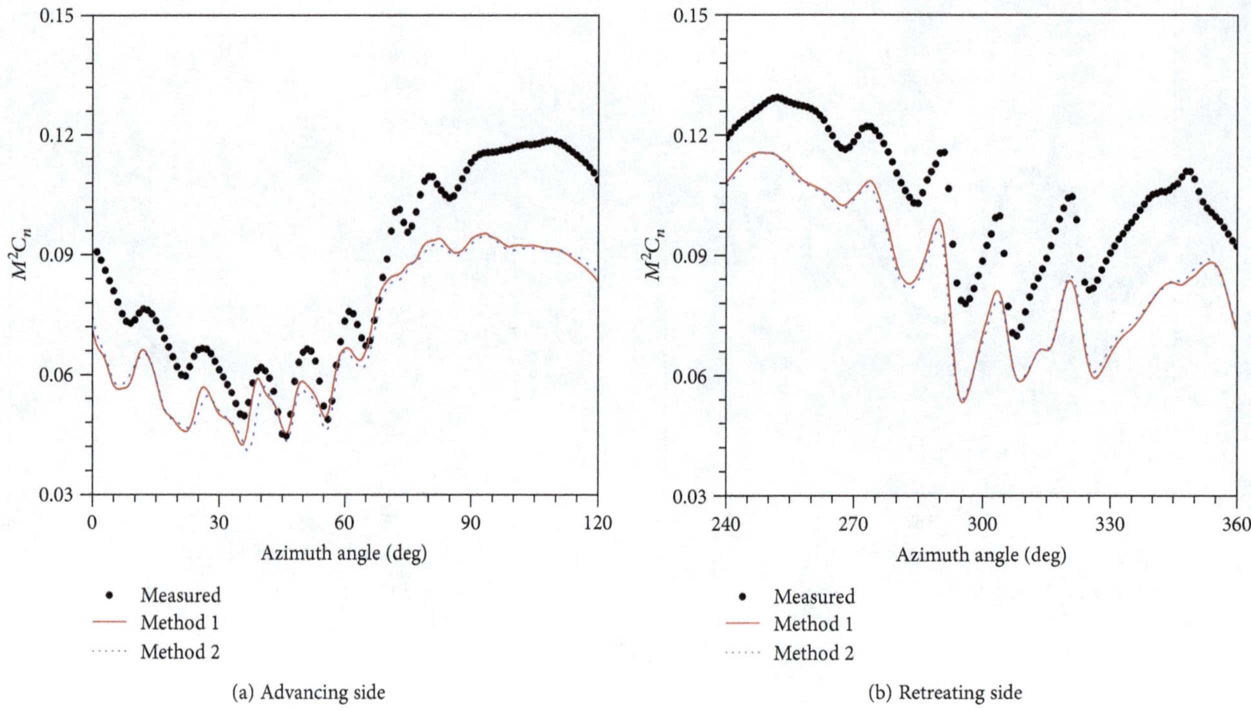

FIGURE 10: Comparison of enlarged view of section airloads M^2C_n at 87% radial station.

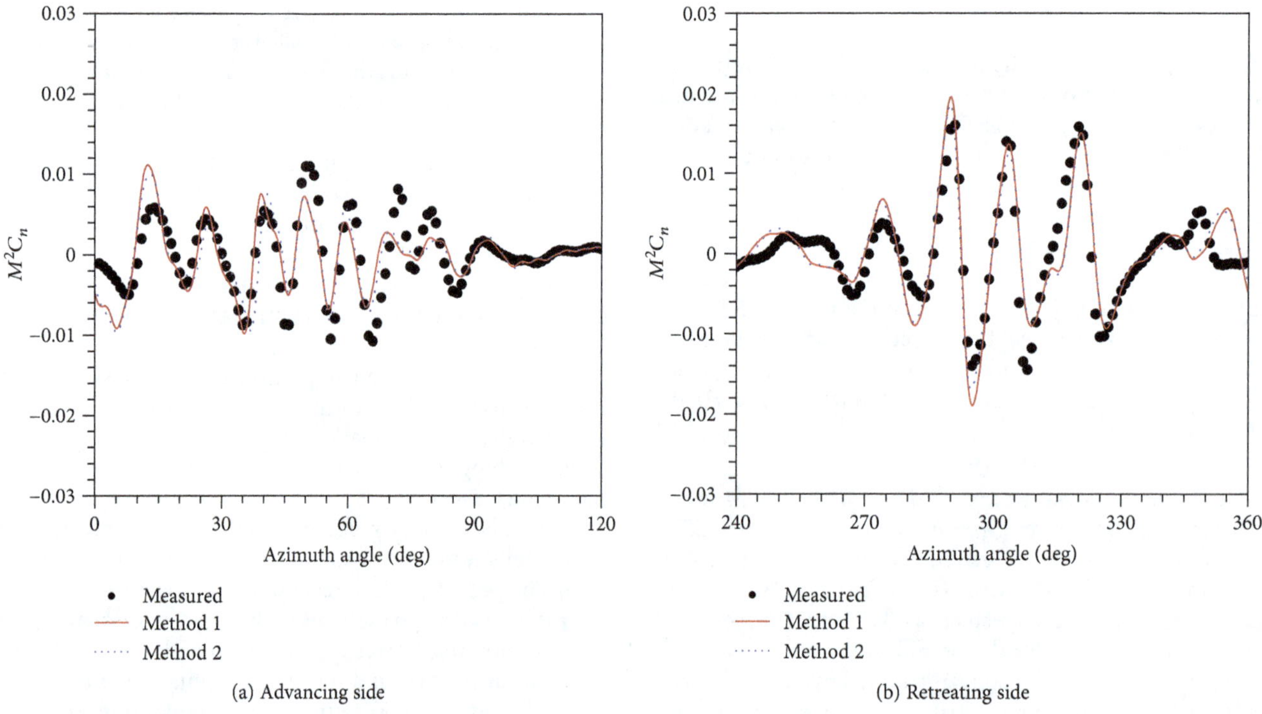

FIGURE 11: Comparison of higher harmonic contents (11/rev and higher) of section airloads M^2C_n at 87% radial station.

due to an increase in peaks accumulated during the coupling iterations before the convergence is reached, as observed in Figure 7. This can also be explained considering the way the coupling computations are managed. It is recalled that method 2 is a mixture method since CFD airloads are interpolated in the same way adopted in method 1 (i.e., random selection at every azimuth angle of 5 deg), and CSD airloads are low-pass filtered as method 2. A natural consequence of the mixed method is that the differences between CFD and CSD airloads (i.e., delta airloads) may be gaining in strengths, particularly for the case when highly oscillatory signals (larger than 10/rev) are present. More specifically, some of the spikes and kinks existed in the original signal can be survived after the random selection (method 1)

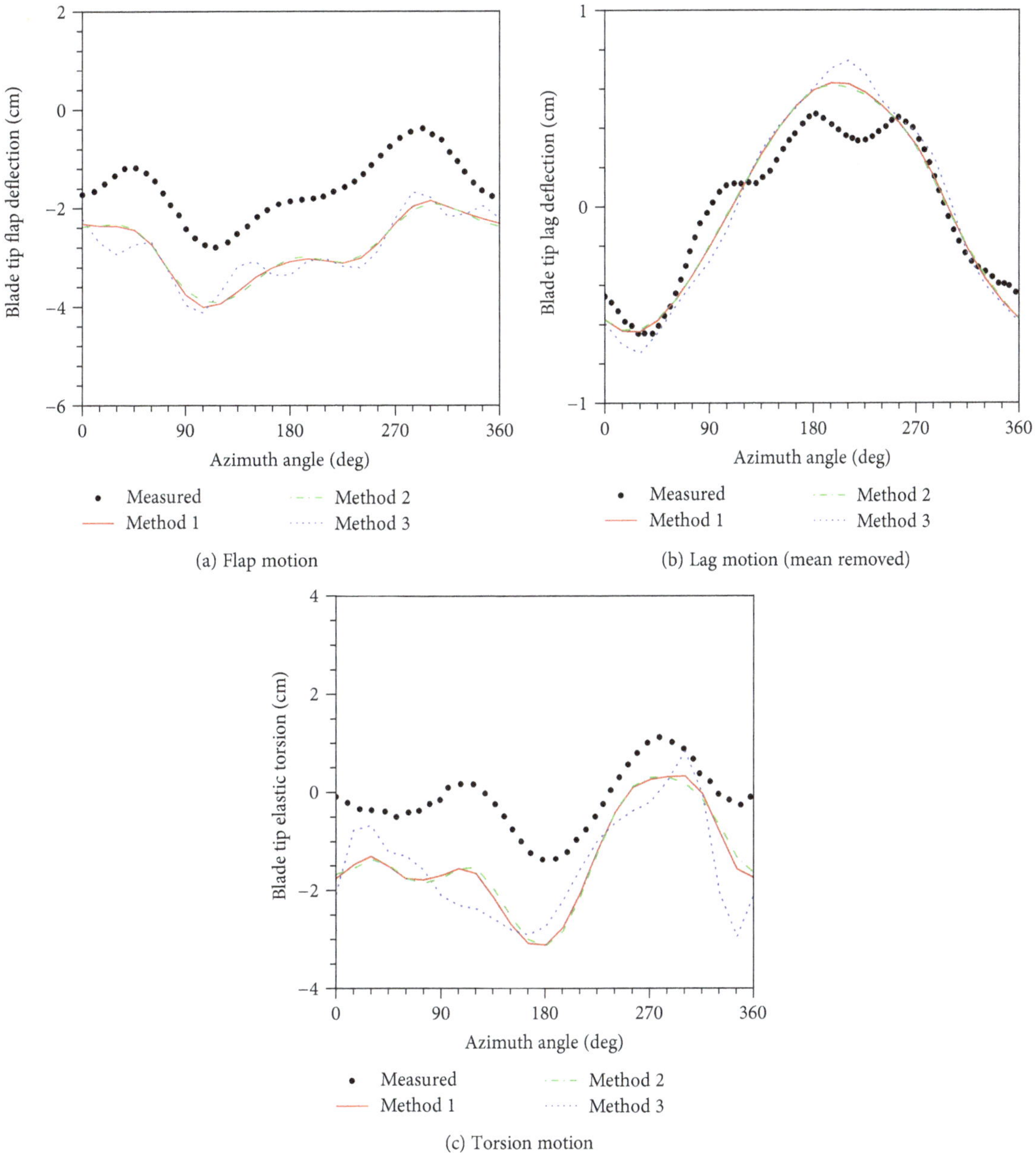

(a) Flap motion

(b) Lag motion (mean removed)

(c) Torsion motion

Figure 12: Comparison of blade tip elastic deflections.

combined with the smoothening process (method 2), and these could be tossed to the consecutive coupling cycles for more intensified oscillations (which is nonphysical), as is observed in Figure 7(b).

Besides the mixture case, method 1 or method 2 indicates reasonable correlation with the measured data. For clarity, enlarged views on the advancing and retreating sides of the section normal forces predicted at 87% rotor radial station are compared against the measured data in Figure 10. Most of the BVI events are seen to be predicted correctly by method 1 or method 2, in terms of the number of peak oscillation counts, peak-to-peak magnitudes, and phases of the measured signals. It should be mentioned that both methods indicate no significant deviations on the airloads predictions, despite the fundamental differences in interpolating the data. Figure 11 presents the comparison of higher harmonic components of the section airloads M^2C_n after removing up to the lowest 10/rev harmonic contents. Once again, both results by methods 1 and 2 show no noticeable deviations with each other. The predicted phase responses show excellent correlation against the measured data with underpredictions in magnitudes around the azimuth angle of 60 deg. The elastic deformations for flap, lag, and torsion predicted at the blade tip by the three data transfer methods are compared with the measured data in Figure 12. Similar to the results found on airloads, both methods 1 and 2 show nearly identical blade motion results, whereas more oscillatory response with higher harmonics is seen in method 3 which should be caused by the stronger peaks combined with the varied mean of the airloads predictions. Based on the observations,

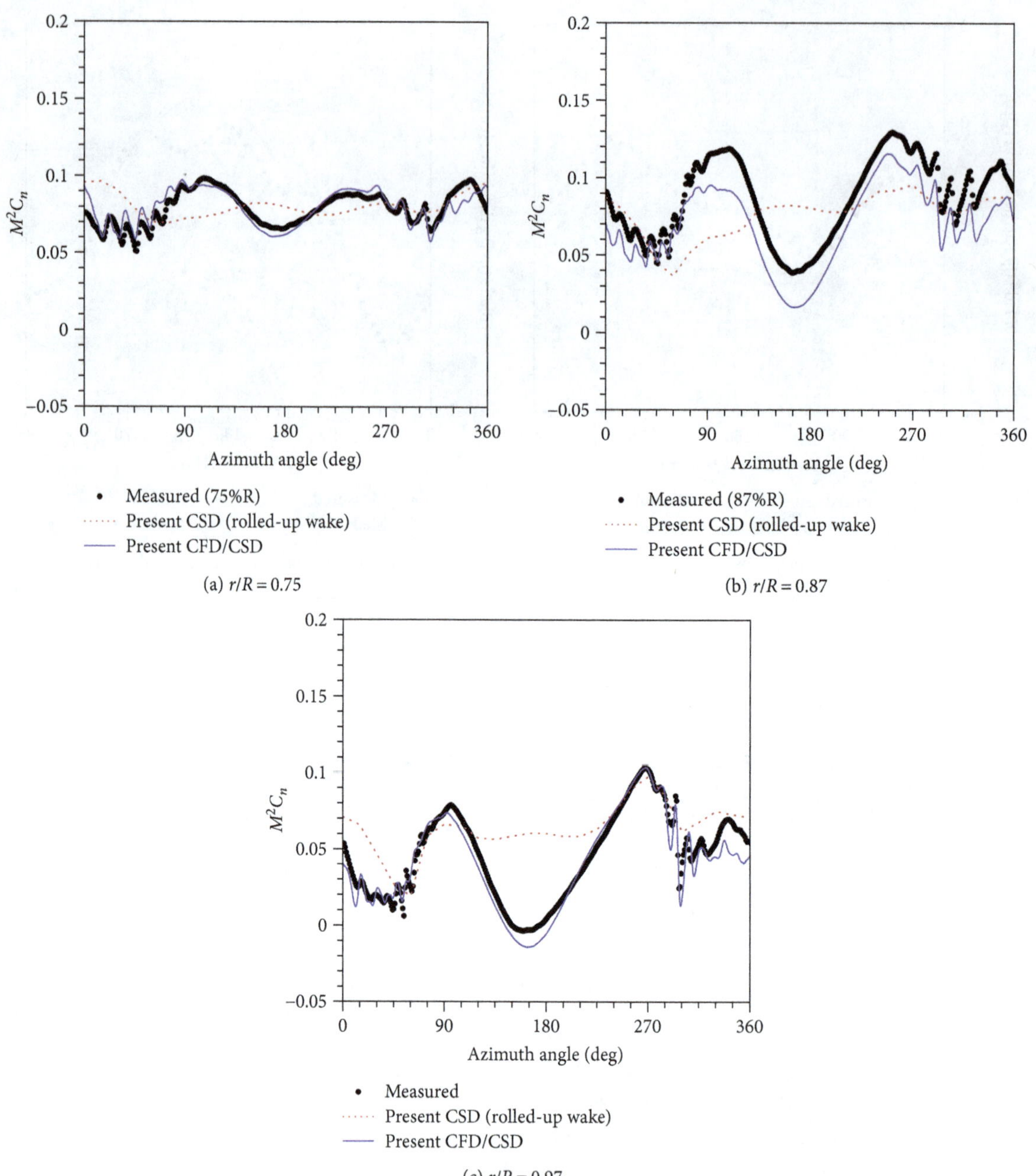

(a) $r/R = 0.75$

(b) $r/R = 0.87$

(c) $r/R = 0.97$

FIGURE 13: Correlation of section normal forces M^2C_n with radial stations.

method 1 is adopted in synthesizing the data for CFD/CSD coupling for the validation of the measured HART data, which will be presented in the next section.

4.2. Validation of HART Rotor. The measured airloads data of HART rotor in descent at different radial locations ($r/R = 0.75$, 0.87, and 0.97) are compared in Figure 13, against the predicted results by CFD/CSD coupling and CSD approach with a rolled-up free-wake model. The free-wake representation is modeled using an initial core size of $0.5c$ with a square root growth over four rotor revolutions. The timewise resolution has 15 deg in the CSD analysis. The CSD predictions on section airloads show only fair correlations with the measured data. Specifically, the typical down-up pattern seen near the front edge of the rotor disk in the measured signal is missed wholly. The CFD/CSD coupling improves the correlation significantly to match the measured data in terms of both waveforms and amplitudes. The means are predicted accurately by CFD/CSD coupling particularly at $r/R = 0.75$ and 0.97 with a slight underestimation at $r/R = 0.87$ as compared to the measured data. It should be mentioned that the correct predictions of phase response as well as the improved BVI signals in the advancing side are due to the incorporation of the fuselage model in the CFD/CSD approach, as observed previously by Jung et al. [14]. The vibratory airloads components keeping 3/rev and higher are compared next in Figure 14 according to the three radial stations. The vibratory components are useful

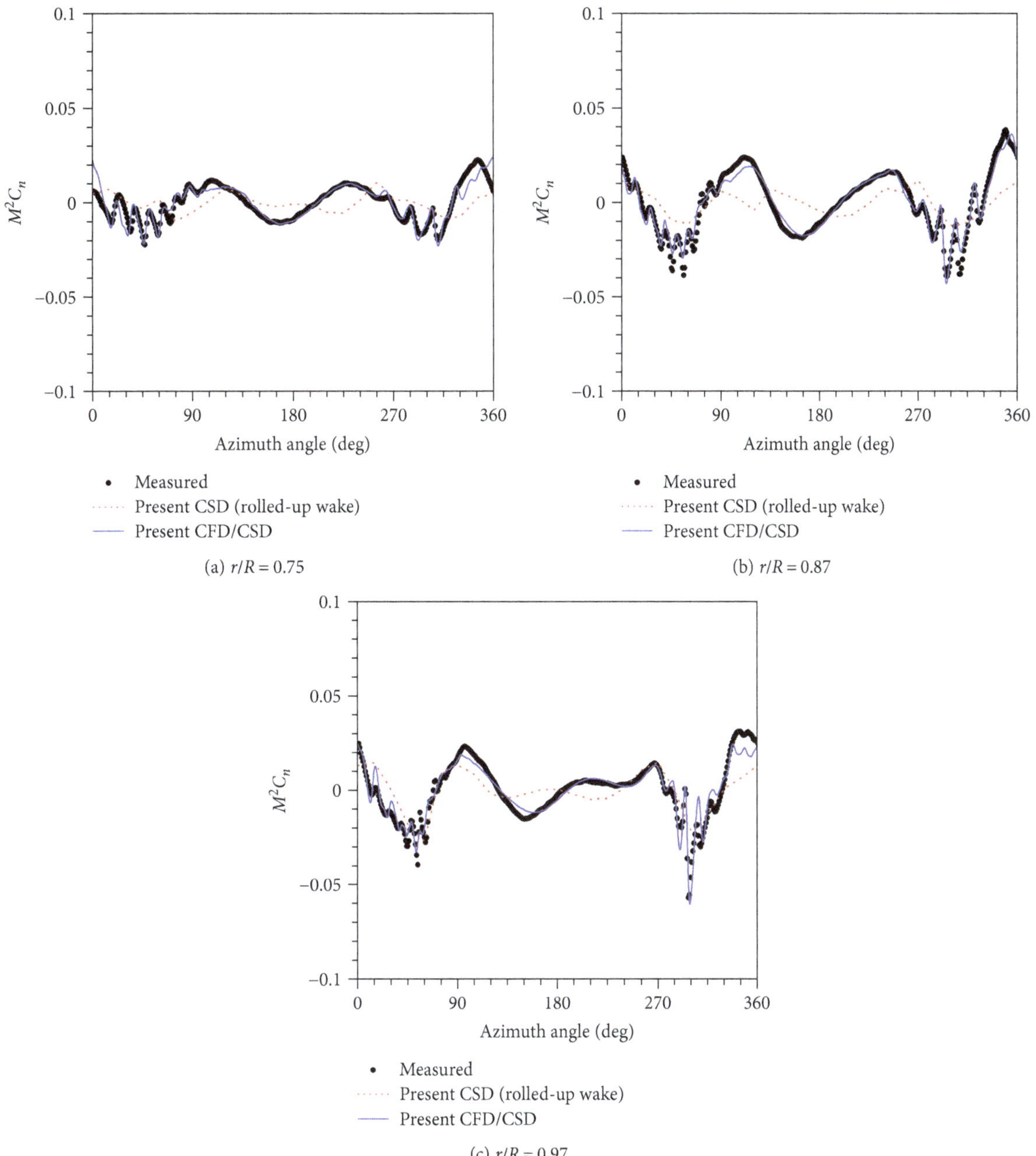

FIGURE 14: Correlation of vibratory components (3/rev and higher) of M^2C_n with radial stations.

to determine the hub vibration characteristic of a rotor with finite number of blades. As can be seen, the predicted airloads exhibit excellent correlations with the measured data particularly on BVI oscillatory peaks in advancing and retreating sides, down-up pattern near the front edge, and peak-to-peak magnitudes and phases of the timewise airloads signal. It is noted again that the CSD results show limited prediction capability in the computation of vibratory airloads components.

To gain more insights on BVI airloads, the gradient of the section normal forces with respect to the time, $d(M^2C_n)/d\psi$, is investigated in Figure 15. The timewise gradient signals are closely related with the acoustic noise emission since the sound radiation involves the partial derivative of air loading (acoustic pressure) with time. The predicted CSD results miss all BVI events in rotor advancing and retreating sides due to the limited modeling capability of its own. The CFD-/CSD-coupled predictions indicate good agreements on the number of BVI counts and the phases while some of peak amplitudes are underestimated compared to the measured data. In predicting BVI peak strengths, the correlation is generally excellent in the retreating side and less satisfactory in the advancing side. The neglect of hub model as well as the limitation of RANS model in approximating the flow equations for such a highly fluctuating BVI event is the possible source of the discrepancy. Despite the limits, however, the computed CFD/CSD results demonstrate the current level of prediction for HART rotor.

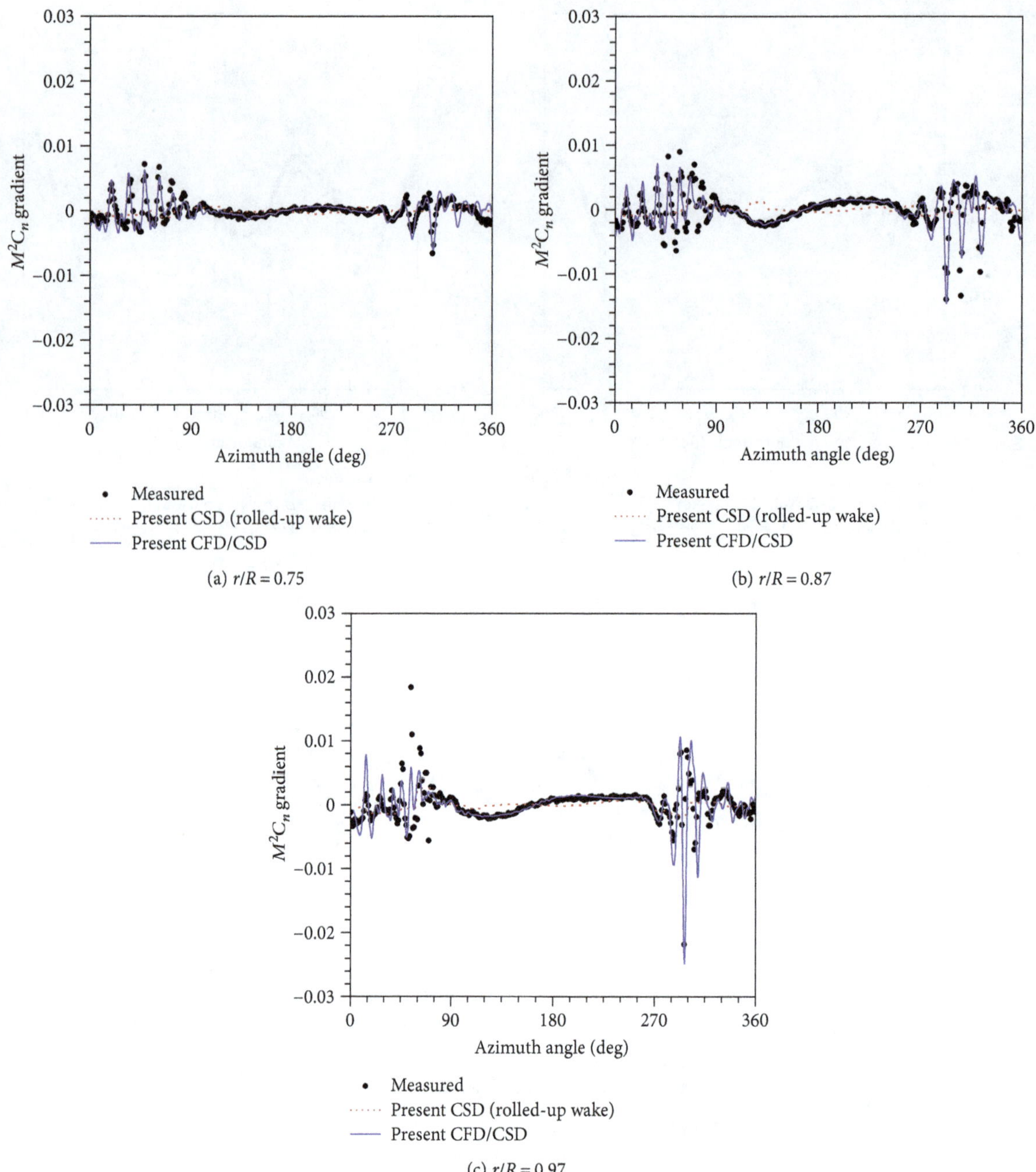

(a) $r/R = 0.75$

(b) $r/R = 0.87$

(c) $r/R = 0.97$

FIGURE 15: Correlation of the gradient of section normal forces $d(M^2C_n)/d\psi$ with radial stations.

The predicted blade tip elastic deformations on flap, lag, and torsion motions are compared with the measured data in Figure 16. Other than the previous airloads comparison, less satisfactory correlation is obtained by CFD/CSD predictions while comparable results are obtained by CSD approach at least in terms of the mean response. However, both the waveform and the phase response are better predicted by CFD/CSD coupling. Specially, the convex pattern around the azimuth angles of 180 deg in the measured flap motion as well as an up-down pattern seen in the second quadrant of the measured torsion is captured nicely with the CFD/CSD coupling. Despite the improved predictions with the CFD/CSD coupling, there remains an apparent offset on flap and twist deformations between CFD/CSD predictions and the measured data. One possible reason of the discrepancy is the blade-to-blade dissimilarity of HART rotor. It should be mentioned that the instrumented blade (number 1 blade) is reported to be heavier by about 6% than the rest of the blades [16]. In addition, the blade property measurement has not been made with the instrumented blade (H1Y) but using number 3 blade (H1B) and a spare HART blade (H1S) [15]. In addition, sophisticated RTB (rotor track and balance) process is skipped during the HART measurement campaign. Even though no measurement record is found for displacements of the individual HART blades, one can estimate the order of magnitudes

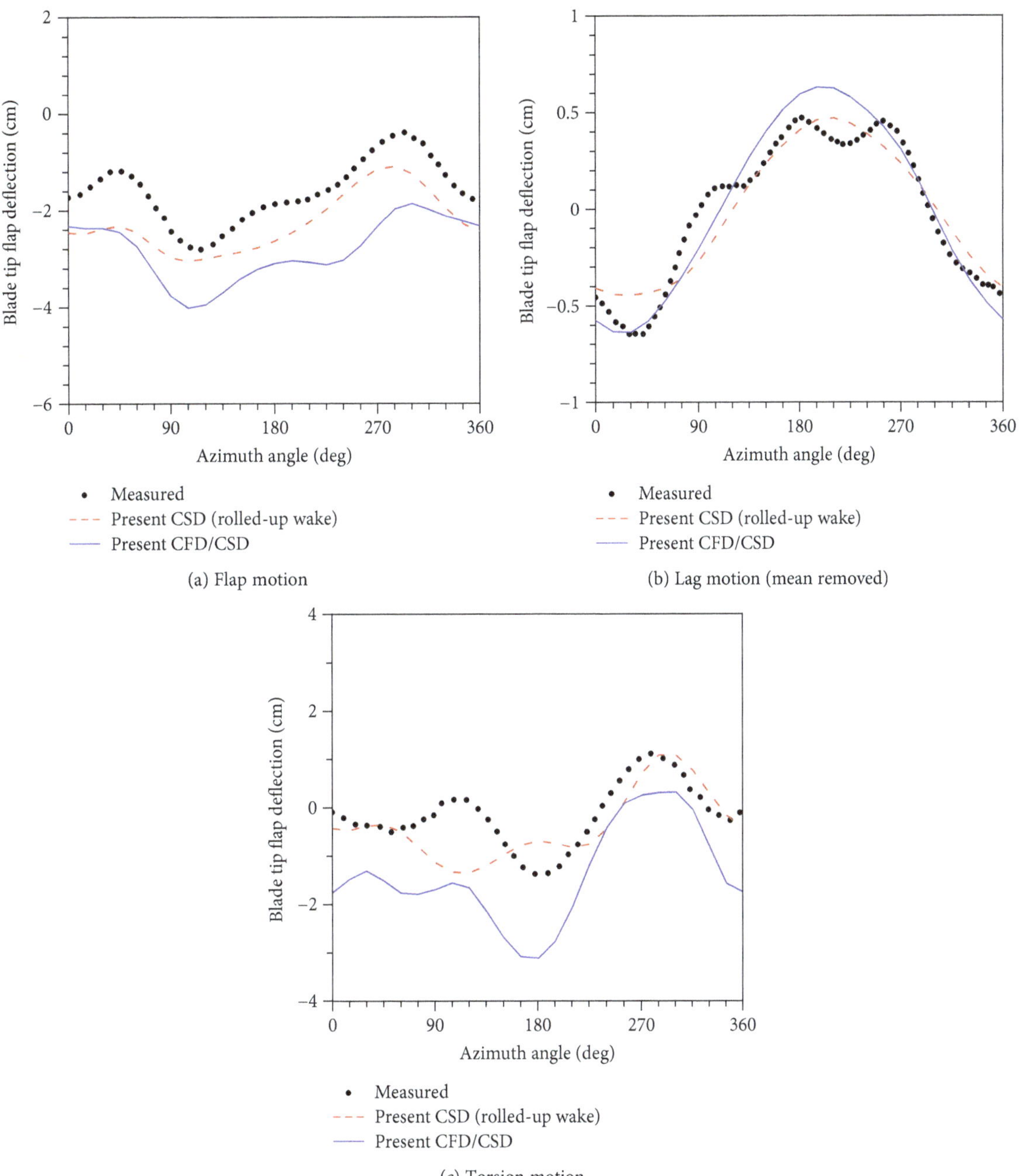

Figure 16: Correlation of blade tip elastic deflections.

considering HART II data. It is found that the maximum deviation between the blades of HART II rotor is about 2% for the tip flap deflections [12]. Taking these facts into account, reasonable agreements appear to be reached using the present CFD/CSD predictions. Finally, in Figure 17, the structural moments at a blade inboard station $r/R = 0.144$ are compared between the present CFD/CSD results, CSD predictions, and the measured data. The mean is removed for comparison. Fair-to-good correlation is observed with the present predictions. The CFD/CSD coupling improves the correlation significantly in terms of the peak-to-peak amplitudes and the waveforms of the measured flap bending and torsion moment signals. The phase response in the retreating side of the flap bending moment as well as the torsion moment in the second quadrant is better predicted using the CFD/CSD coupling analysis than the CSD approach, even though a phase lead is observed in the predicted flap bending moments as compared with the measured data. Use of the blade structural properties other than the instrumented blade may cause the error in the predicted results.

5. Conclusions

In this work, a numerical study is performed for the validation of measured HART data as well as the validity of data

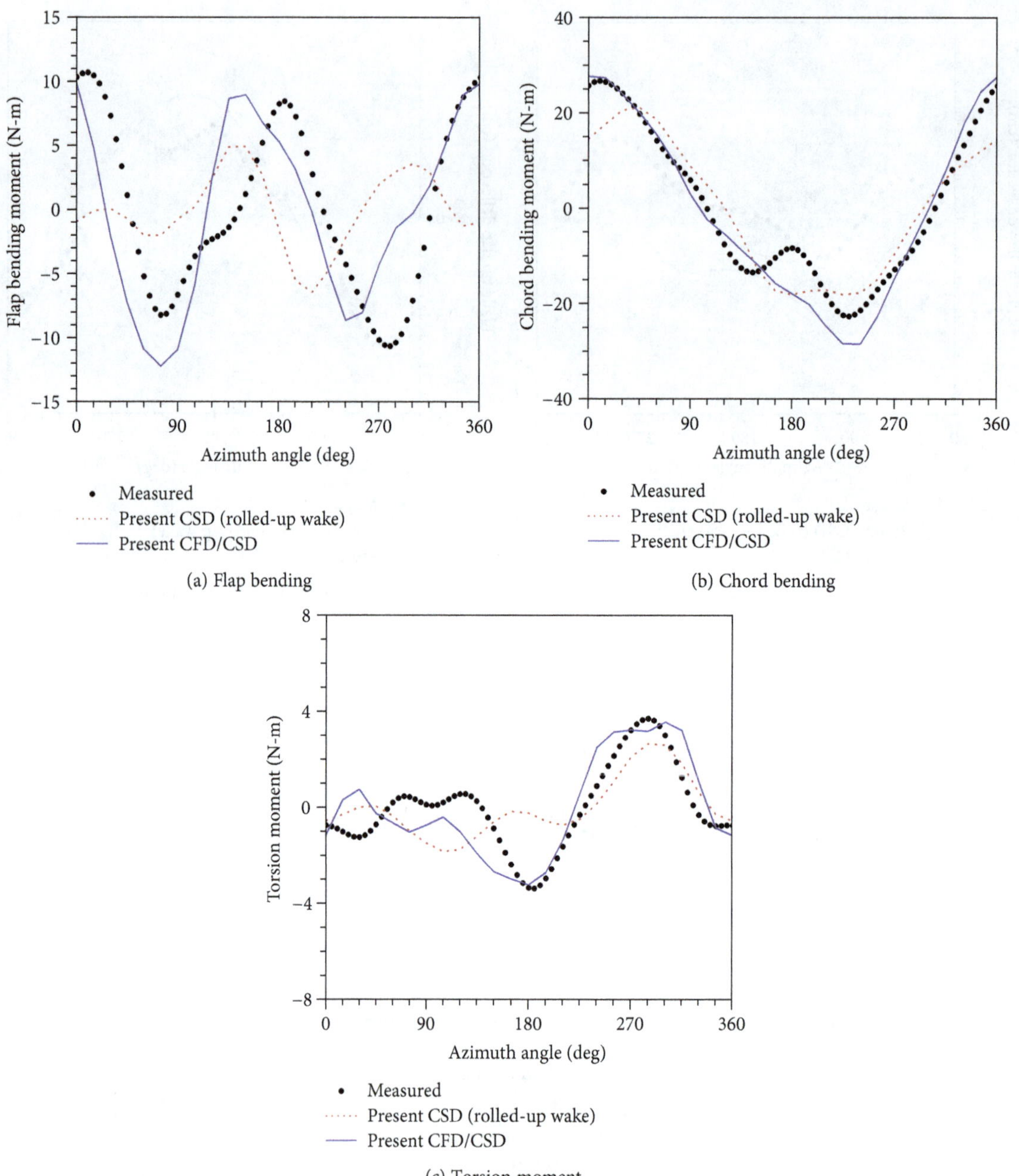

FIGURE 17: Correlation of structural moments at a blade inboard station $r/R = 0.144$ (mean removed).

transfer methods using CFD-/CSD-coupled approach. The data transfer methods include a random data selection matched with the timewise resolution of CSD analysis and/or a low-pass harmonic filter, leading to three different methods (methods 1 to 3). A 3D compressible RANS flow solver KFOW is coupled with a CSD analysis code CAMRAD II for this purpose. Based on the observations, the following conclusions are drawn:

(1) Good convergence characteristic is reached on airloads and delta airloads for all the data transfer methods after six to seven coupling iteration cycles are marched. This verifies the robustness of the delta airloads technique in CFD-/CSD-coupled approach.

(2) In both CFD and CSD analyses, any consistent use of data synthesis techniques (e.g., method 1 or 2) is desired to benefit the coupling approach. A mixture rule (method 3) should be prohibited especially when high-frequency signals such as BVI oscillations are present in the airloads data. It is found that the delta airloads become more oscillatory as some of the kinks and spikes existed in airloads data get survived and intensified in strengths with the coupling iterations, leading to a poor correlation for airloads and blade elastic motions as compared with the measured data.

(3) The data transfer scheme either method 1 or method 2 demonstrates nearly identical results on airloads

and elastic blade motions despite the differences in processing the airloads data. Either method leads to good agreements with the measured HART data. The incorporation of the fuselage model in the CFD analysis contributes to the close correlation, particularly on the phase response around the front edge of the rotor and improved BVI predictions in the advancing side of the rotor disk.

(4) The present CFD/CSD coupling shows good-to-excellent agreements on airloads as well as the time-wise gradient signals $d(M^2C_n)/d\psi$ of HART rotor obtained at the three measurement stations distributed along the blade. The number of BVI oscillatory events and the phase response of the airloads signals are accurately captured with the CFD-/CSD-coupling approach while underestimating some of the gradient signals of BVI peaks in the advancing side of the rotor. The neglect of hub model as well as the use of RANS model in approximating the flow equations around the rotor is the possible source of the discrepancy for such a highly fluctuating BVI event.

Conflicts of Interest

The authors declare that there are no conflicts of interest regarding the publication of this paper.

Acknowledgments

This work was supported by the KEIT Research grant of 2016 (no. 10053155). This paper resulted from the Konkuk University research support program. The authors thank HART team for the test data.

References

[1] C. Tung, F. X. Caradonna, and W. Johnson, "The prediction of transonic flows on an advancing rotor," *Journal of the American Helicopter Society*, vol. 31, no. 4, pp. 4–11, 1986.

[2] P. Beaumier, "A coupling procedure between a rotor dynamics code and a 3D unsteady full potential code," in *American Helicopter Society Aeromechanics Specialists' Conference*, pp. 19–21, San Francisco, CA, USA, 1994.

[3] M. Potsdam, H. Yeo, and W. Johnson, "Rotor airloads prediction using loose aerodynamic/structural coupling," *Journal of Aircraft*, vol. 43, no. 3, pp. 732–742, 2006.

[4] Y. H. Yu, B. Gmelin, H. Heller, J. J. Philippe, E. Mercker, and J. S. Preisser, "HHC aeroacoustics rotor test at the DNW - the joint German/French/US HART project," in *Proceedings of the 20th European Rotorcraft Forum*, pp. 4–7, Amsterdam, Netherlands, 1994.

[5] W. Johnson, "Milestones in rotorcraft aeromechanics Alexander A. Nikolsky honorary lecture," *Journal of the American Helicopter Society*, vol. 56, no. 3, pp. 1–24, 2011.

[6] Y. H. Yu, C. Tung, B. G. van der Wall et al., "The HART-II test: rotor wakes and aeroacoustics with higher-harmonic pitch control (HHC) inputs - the joint German/French/Dutch/US project," in *American Helicopter Society 58th Annual Forum Proceedings*, pp. 11–13, Montreal, Canada, 2002.

[7] C. Tung, J. M. Gallman, R. Kube et al., "Prediction and measurement of blade - vortex interaction loading," in *1st CEAS/AIAA Aeroacoustics Conference Proceedings*, pp. 12–15, Munich, Germany, 1995.

[8] J. W. Lim, Y. H. Yu, and W. Johnson, "Calculation of rotor blade-vortex interaction airloads using a multiple-trailer free-wake model," *Journal of Aircraft*, vol. 40, no. 6, pp. 1123–1130, 2003.

[9] J. W. Lim, T. A. Nygaard, R. Strawn, and M. Potsdam, "Blade-vortex interaction airloads prediction using coupled computational fluid and structural dynamics," *Journal of the American Helicopter Society*, vol. 52, no. 4, pp. 318–328, 2007.

[10] H. Yeo and W. Johnson, "Assessment of comprehensive analysis calculation of airloads on helicopter rotors," *Journal of Aircraft*, vol. 42, no. 5, pp. 1218–1228, 2005.

[11] H. Yeo and W. Johnson, "Prediction of rotor structural loads with comprehensive analysis," *Journal of the American Helicopter Society*, vol. 53, no. 2, pp. 193–209, 2008.

[12] B. G. van der Wall, J. W. Lim, M. J. Smith et al., "The HART II international workshop: an assessment of the state-of-the-art in comprehensive code prediction," *CEAS Aeronautical Journal*, vol. 4, no. 3, pp. 223–252, 2013.

[13] M. J. Smith, J. W. Lim, B. G. van der Wall et al., "The HART II international workshop: an assessment of the state of the art in CFD/CSD prediction," *CEAS Aeronautical Journal*, vol. 4, no. 4, pp. 345–372, 2013.

[14] S. N. Jung, J. H. Sa, Y. H. You, J. S. Park, and S. H. Park, "Loose fluid-structure coupled approach for a rotor in descent incorporating fuselage effects," *Journal of Aircraft*, vol. 50, no. 4, pp. 1016–1026, 2013.

[15] S. N. Jung and B. H. Lau, *Determination of HART I Blade Structural Properties by Laboratory Testing*, 2012, NASA CR 2012-216039.

[16] S. N. Jung, Y. H. You, B. H. Lau, W. Johnson, and J. Lim, "Evaluation of rotor structural and aerodynamic loads using measured blade properties," *Journal of the American Helicopter Society*, vol. 58, no. 4, 2013.

[17] W. Johnson, *CAMRAD II: Comprehensive Analytical Model of Rotorcraft Aerodynamics and Dynamics*, Johnson Aeronautics, Palo Alto, CA, USA, 1992.

[18] J. W. Kim, S. H. Park, and Y. H. Yu, "Euler and Navier-Stokes simulations of helicopter rotor blade in forward flight using an overlapped grid solver," in *19th AIAA CFD Conference*, San Antonio, TX, 2009.

[19] B. G. van der Wall, *Mode Identification and Data Synthesis of HART II Blade Deflection Data*, Institute Report, IB-111-2007/28, Braunschweig, 2007.

[20] J. H. Sa, J. W. Kim, S. H. Park, J. S. Park, S. N. Jung, and Y. H. Yu, "KFLOW results of airloads on HART-II rotor blades with prescribed blade deformation," *International Journal of Aeronautical and Space Sciences*, vol. 10, no. 2, pp. 52–62, 2009.

[21] D. H. Na, Y. H. You, and S. N. Jung, "Comprehensive aeromechanics predictions on air and structural loads of HART I rotor," *International Journal of Aeronautical & Space Sciences*, vol. 18, no. 1, pp. 165–173, 2017.

Analytical Study on Deformation and Structural Safety of Parafoil

Longfang Wang and Weiliang He

School of Astronautics, Beihang University, Beijing 100191, China

Correspondence should be addressed to Weiliang He; heweiliang@buaa.edu.cn

Academic Editor: Filippo Berto

This study focuses on the cell bump distortion and bearing capacity of parafoil structure. Based on the mechanical properties of the membrane structure, the spanwise model of parafoil inflation was established and verified by comparing with the fluid-structure interaction (FSI) results. Because the internal pressure is very low, the chordwise stiffness is mainly generated by suspending lines. The chordwise model of inflated parafoil was established in consideration of elastic force and aerodynamic force. The results show that the cell is slenderer; the canopy surface is smoother; the aerodynamic load has a light effect on the shrinkage and bump ratios; when the cell width is constant, the critical dynamic pressure reduces k times with the k times increasing in parafoil area; and the design parameters of the first-row line OA have significant effects on the structural stiffness of inflated parafoil. The analytical model is useful for the weakening deformation design and the safety discussion of large parafoil for rocket booster recovery.

1. Introduction

The parafoil is a flexible wing maintaining the aerodynamic shape with ram air in cells and decelerating the payload drop with aerodynamic lift. The aerodynamic force on canopy can be changed by pulling down the steering ropes to achieve steerable flight. As a kind of steerable aerodynamic decelerator, parafoil gets extensive attention in the rocket booster recovery. The parafoil is made by flexible textile material and presents large deformation during flight. When used in booster recovery, the parafoil has a tremendous area and bears a large wing load. This puts forward higher requirements for the bearing capacity of parafoil structure. Therefore, it is necessary to investigate the deformation and structural stiffness of large parafoils to improve their performance and safety.

The deformed surface geometry of a parafoil was presented by a video-based photogrammetry during tethered testing in a low-speed tunnel [1, 2]. Due to the fact that this kind of test presents a high cost, numerical simulation has become a popular way to analyze the deformation of parafoils. Kalro et al. set the shape of deformed canopy according to the drop test and analyzed the motion of opening process [3]. Eslambolchi and Johari extracted the inflated canopy geometry from close-up images of the MC-4 canopy during a flight and computed the flow field around the deformed canopy [4]. Ibos et al. simulated the fluid-structure interaction problem of a parafoil using SINPA software [5]. Kalro and Tezduyar calculated the shape of a parafoil in a steady flight using the parallel coupling algorithm for fluid-structure coupling using finite element methods [6]. Fogell et al. analyzed the fluid-structure interaction problem of a single-cell parafoil model [7]. Altmann studied the deformation of canopy by the potential flow theory and the cable finite element [8, 9]. Peralta and Johari investigated the geometry of a fully inflated canopy in steady flight using a prescribed pressure distribution [10]. Mosseev developed a series of software to simulate the aerodynamic deformation of a canopy [11]. The results indicate that the projected area of an inflated canopy is 18% smaller than the area of the initial geometry.

The numerical simulation of the fluid-structure interaction in the parafoil is very complicated and time-consuming. It is difficult to carry out a numerical simulation while designing a parafoil. Besides, the trade studies on design variables cannot be clearly discussed only by the analysis of simulation examples. The deformation and safety of the parafoil structure during the flight should be included in the design process in order to improve the flight performance of the parafoil. It is important to find a design theory which is able to guide the manufacture process of large parafoils. The novelty of the current study is to establish the theoretical spanwise and chordwise model for the design of parafoil. Further, the effect of design variables is studied.

2. Spanwise Structure Model

As is shown in Figure 1, the inflating deformation of parafoil is mainly embodied as the spanwise bumps. This is determined by the mechanical properties of the membrane structure. The widthwise projection of the ideal cell is flat, but the lengthwise projection is curved. When the cell is ballooning, the widthwise curvature changes more than the lengthwise curvature to bear pressure load. The failure of spanwise structure is the primary failure mode of wingtip collapse. Hence, the chordwise deformation and stress can be neglected in the two-dimensional deformation model of canopy structure.

The stress of membrane structure is expressed below [12]:

$$\frac{\sigma_1}{r_1} + \frac{\sigma_2}{r_2} = \frac{P}{t}, \tag{1}$$

where σ_1 and σ_2 are the radial and latitudinal stresses, respectively; r_1 and r_2 are the radial and latitudinal radius of curvature, respectively; P is the differential pressure; and t is the thickness. In the two-dimensional model, (1) is simplified:

$$\frac{\sigma}{r} = \frac{P}{t}, \tag{2}$$

where σ is the canopy stress, and r is the widthwise radius of curvature. The widthwise variation of pressure is neglected, and the canopy stress is constant in each cell, so the cells must be inflated to a circular arc.

Figure 2 presents the 2D geometrical configuration of balloon cells. The ribs are simplified as straight lines and the upper and lower surfaces are simplified as a circular arc with radius r. The cell coordinate system is established to analyze the deformation of the cell. The y-axis is parallel to the rib and the x-axis is perpendicular to the rib. When the parafoil is in steady flight, every cell structure is in equilibrium, and the vector sum of the forces acting upon it is zero. Because the lift of the parafoil is mainly provided by the upper surface, the aerodynamic pressure load is entirely applied on the upper surface. The pressure on the lower surface is equal to the stagnation pressure. By analyzing the forces on the rib, upper and lower surface, the equations are listed as follows:

$$T_1 \cos\varphi + T_2 \cos\varphi = P_2 h, \tag{3}$$

$$2T_1 \sin\varphi = P_1 w, \tag{4}$$

$$2T_2 \sin\varphi = P_2 w, \tag{5}$$

FIGURE 1: Photograph of parafoil in flight.

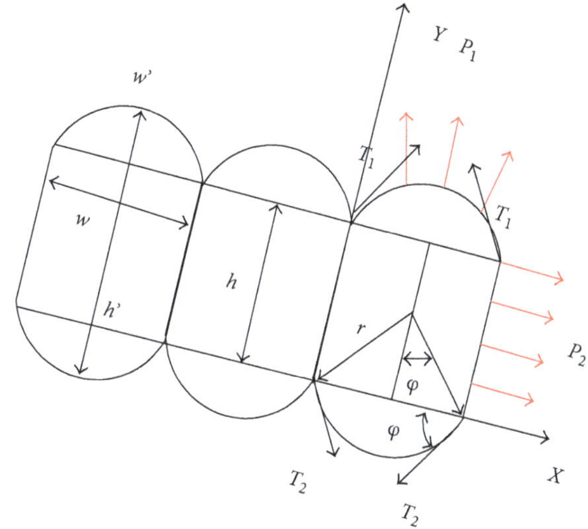

FIGURE 2: Two-dimensional model of balloon cells.

where T_1 and T_2 represent the tensions per unit chord of upper and lower surface, respectively; P_1 and P_2 represent the differential pressure on upper surface and the stagnation pressure, respectively; h and w represent the height and width of the cell, respectively; and φ represents the half of the central angle.

As is mentioned above, the difference between P_1 and P_2 is the aerodynamic pressure load:

$$P_1 = P_2 + \frac{1}{2}\mathrm{CL}'\rho V^2 = \left(1 + \mathrm{CL}'\right)\frac{1}{2}\rho V^2 = \left(1 + \mathrm{CL}'\right)P_2. \tag{6}$$

Here, ρ is the air density; V is the parafoil velocity; and CL' is the lift coefficient defined in the cell coordinate system.

Equation (5) is subtracted from (4) and combined with (6), resulting in (7) below:

$$T_1 = T_2 + \frac{(P_1 - P_2)w}{2\sin\varphi} = T_2 + \mathrm{CL}'\frac{P_2 w}{2\sin\varphi} = \left(1 + \mathrm{CL}'\right)T_2. \tag{7}$$

Equation (7) is substituted into (3), resulting in (8) below:

$$\left(2 + \mathrm{CL}'\right)T_2 \cos\varphi = P_2 h. \tag{8}$$

Equation (9) can be derived by the division operation of (8) and (5):

$$\varphi = \operatorname{atan}\left(\frac{w}{h}\frac{2 + \mathrm{CL}'}{2}\right). \tag{9}$$

According to the geometric relationship shown in Figure 2, the ratio between the arc length w' and the width w can be indicated as follows:

$$r = \frac{w'}{2\varphi} = \frac{w}{2\sin\varphi}, \tag{10}$$

$$\eta_1 = 1 - \frac{w}{w'} = 1 - \frac{\sin\varphi}{\varphi}, \tag{11}$$

$$\eta_2 = \frac{h' - h}{h} = \frac{2r(1-\cos\varphi)}{h}, \tag{12}$$

where η_1 and η_2 are defined as the shrinkage ratio and the bump ratio of the cell, respectively. Because the cell height is mainly concentrated around the rib peak, the cell height of the parafoil is set as the thickness of the rib. The flexible lines can only restrict the relative position of the intersection point of the lines and the connection point of the ribs; therefore, the ribs can move easily on the arc whose center is the intersection point and radius is the line length. The move is similar to the contraction of the accordion and leads to a large difference between the deformed shape and the design shape of the parafoil.

To validate the spanwise model, the fluid-structure interaction (FSI) simulation using the commercial solver ANSYS has been carried out in our previous work [13]. The grid of the parafoil system is shown in Figure 3. The canopy is divided into triangular membrane elements. The two intersection points of the suspending lines are fixed, and in order to avoid the rigid body motion, the symmetrical boundary condition is applied in the middle of the parafoil. The internal and external pressure distributions on the canopy, which were calculated at the angle of attack of 5° using CFD method, were transferred to the canopy structure through mapping interpolation. To avoid divergence, the load was applied on the canopy in a way that it increased linearly with the substeps.

The deformed parafoil obtained from the fluid-structure interaction (FSI) simulation is presented in Figure 4. The deformation is not amplified, and its scale is in accordance with the geometric dimensioning of canopy. The bumps of cells are clearly visible. It can be seen from Figure 4(a) that the inflated canopy reduces the span which is equal to the width of two cells. The actual span in the flight reduced by 13% compared to the designed span. The maximum thickness of the airfoil Clark-Y18 is 18%c, but it will increase to 26%c after the bumps appear.

The cell ratio h/w, stagnation pressure P_2, thickness t, and cell lift coefficient CL' in the case shown in Figure 3 are 0.9, 61.25 Pa, 1 mm, and 0.55, respectively. These parameters are substituted into (9), (10), and (11), then the shrinkage ratio is 14.6% and has little

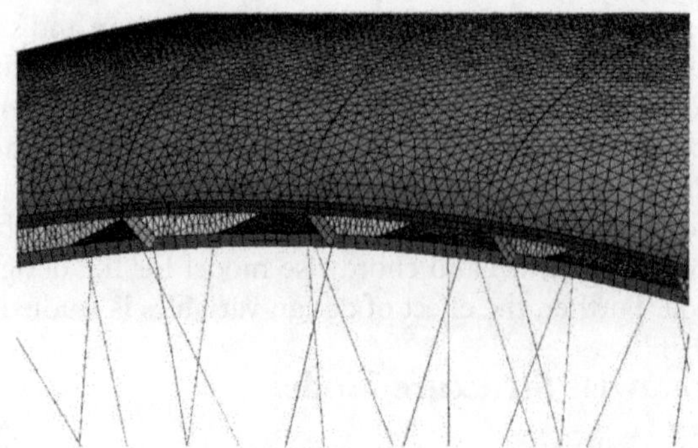

FIGURE 3: Meshes of the parafoil.

difference with the value of 13% presented earlier. The radius of curvature r for the cell is 0.1 m. The bump height h' is 26%c and is equal to the previous value of 26%c. A comparison of the two sets of shrinkage ratio and bump height validates the two-dimensional simplified model. The stress of the lower surface calculated by (2) is 6000 Pa, so the strain of the canopy is very small. The parafoil is a typical large deformation small-strain structure, and the deformation is determined by geometrical configuration rather than material characteristics.

The relationship between the shrinkage/bump ratios and the cell ratio h/w with different CL' values is plotted in Figure 5. When the cell ratio is 1.8, the bump ratio is very close to the wind tunnel test result 18%c to 23%c [7]. As can be seen from the curves, the cell is slenderer, the shrinkage ratio and bump ratio are smaller, and the canopy surface is smoother. This is why some high-performance paragliders are composed of numerous cells. When the cell ratio is greater than 2, the bump ratio changes little, so the design value of 2 is recommended. The cell ratio is decreased, and the shrinkage ratio is increased from the peak to the trailing edge, so the angle of sweepback shows up. The shrinkage and the bump ratios are increased a bit with the increase of CL', and the effect of CL' is great at the medium cell ratio. It means that the aerodynamic shape of the parafoil will change a little with different angles of attack.

The shrinkage and bump ratios are ineradicable. Form rib peak to trailing edge, the cell ratio is monotonously decreasing, consequently, so is the shrinkage ratio. The canopy is made of pieces of cloth. If the pieces of upper and lower surface are designed to be the strips of uniform width, the parafoil is likely to be malformed. The ideal cells of uniform width become wide in the front and narrow in the back after ballooning. This generates the angle of sweepback described previously. In order to guarantee the spanwise appearance, the design of every cell upper and lower surfaces needs to provide margin. A piece of upper or lower surfaces is no longer a rectangular shape. The surcharge is computed by the shrinkage ratio along the chordwise for a CL' value of 0.55. A piece of wider and ideal lower surface for parafoil described previously is shown in Figure 6.

The ribs are made of textile material, so they can only bear tension. The expression for calculating the tension

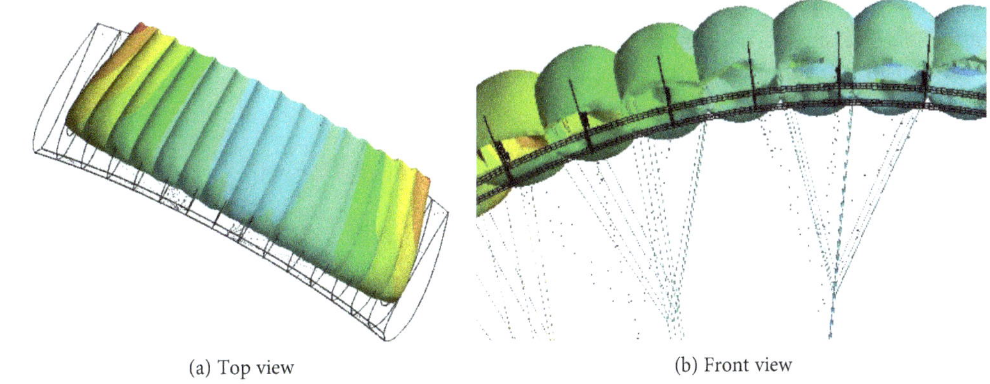

(a) Top view (b) Front view

FIGURE 4: Deformed configuration of a parafoil.

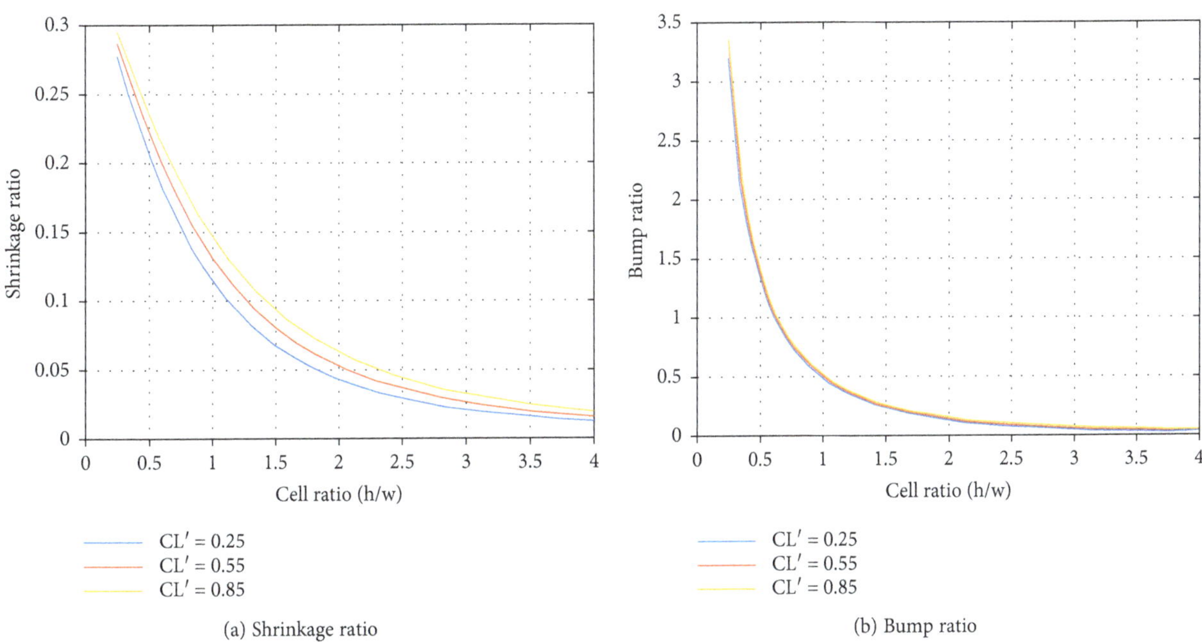

(a) Shrinkage ratio (b) Bump ratio

FIGURE 5: Relationship between shrinkage/bump ratio and cell ratio.

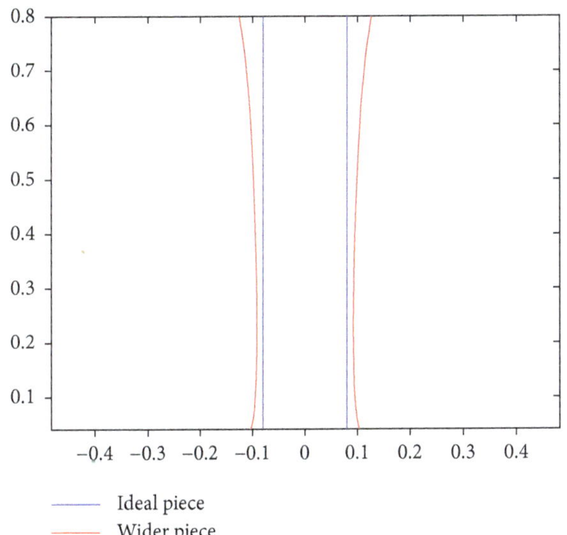

FIGURE 6: Pieces of ideal and wider lower surface.

per unit chord of the rib T_3 can be derived from (4) and (5):

$$T_3 = 2T_1 \sin \varphi - 2T_2 \sin \varphi = (P_1 - P_2)w = \mathrm{CL}' P_2 w. \quad (13)$$

Apparently, the tension of the rib is the lift defined in the cell coordinate system. The physical significance is that the ribs and suspending lines are medium for the lift transmission from the upper surface to payload. Compared with the inflated wing [14], the internal pressure of the parafoil is too small, and the bearing capacity of inflated canopy without the tensile suspending line can be neglected. If the tensions of the ribs in some cells become zero, the ribs and suspending lines will be slacked, and the parafoil structure will collapse. According to (15), the critical pressure on the upper surface is $\mathrm{CL}'P_2$. This critical pressure is not the normal aerodynamic load but the perturbation caused by crosswind. This illustrates that the parafoil with large wing loading has a high ability to resist wingtip collapse. For the flight parameters, the critical pressure is 33.7 Pa for the deformed parafoil shown in Figure 4. This pressure is much lower than the differential pressure of the canopy, which is often considered the critical pressure of wingtip collapse in previous research.

The nonlinear buckling analysis for wingtip collapse of the deformed parafoil shown in Figure 4 was carried out. According to the relationship between the displacement and the load in Figure 7, the critical pressure on the wingtip is 32.9 Pa. The error between this critical pressure and the result above is 2.4%, showing that the two-dimensional model of wingtip collapse can reproduce the phenomena with a low margin of error.

3. Chordwise Structure Model

After the inflating deformation of cell, the parafoil can bear aerodynamic load. The inflated canopy should twist under the action of chordwise aerodynamic moment. Compared with the spanwise bumps, the torsion has a completely different mechanism. It is caused by the aerodynamic loading and varied with flow velocity. As is mentioned above, the bearing capacity of inflated canopy without the tensile suspending line can be neglected. The aerodynamic loads of each cell are directly transmitted to payload by the ribs and suspending lines attached on each cell. The force transmission between cells is negligible, so the analysis of the chordwise torsion can focus on a cell and the suspending lines. Hence, the torsion of parafoil can be simplified to two-dimensional chordwise model.

The two-dimensional chordwise sketch of parafoil is shown in Figure 8. Due to the constraint of line OB, the cell in front of line OB can rotate around point B. When the cell presents an upward rotation angle θ under the aerodynamic moment, the line OA will be stretched and the elongation is $l\theta$; l is the distance between points A and B. Hence, the elastic restoring moment of line OA is $(l^2\theta/L_1)EA$. Here, L_1, E, and A are the length, elasticity modulus, and sectional area of line OA, respectively. The torsional rigidity of the cell is shown below:

$$K_\theta = \frac{l^2}{L_1}EA. \quad (14)$$

The initial angle of attack is α_0, and the free-stream velocity is V. Due to the elasticity of parafoil structure, the new equilibrium at angle of attack $\alpha = \alpha_0 + \theta$ will be established under aerodynamic load. The additional angle θ is defined by torsional rigidity and aerodynamic load, but the aerodynamic load is also affected by θ. According to aerodynamics, the aerodynamic load is composed of the lift force L applied on the aerodynamic center and the nearly invariable moment M_0 about aerodynamic center. The lift L and the aerodynamic moment M about point B are shown below:

$$L = \mathrm{CL}'qS = \frac{\partial \mathrm{CL}'}{\partial \alpha}(\alpha_0 + \theta - \alpha_{zl})qS, \quad (15)$$

$$M = M_0 + Lf = M_0 + \frac{\partial \mathrm{CL}'}{\partial \alpha}(\alpha_0 + \theta - \alpha_{zl})qSf. \quad (16)$$

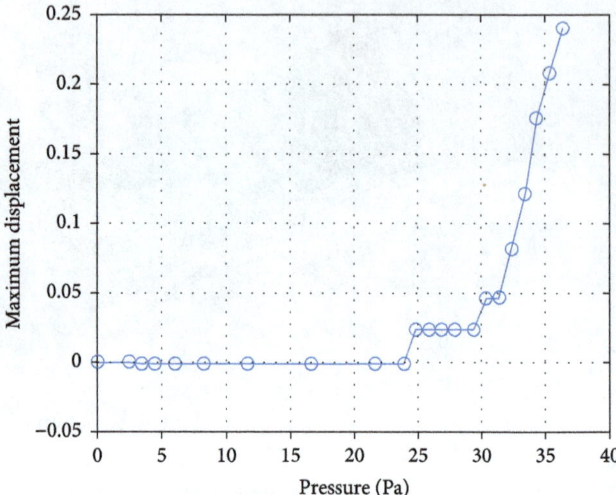

FIGURE 7: Load-displacement curve of deformed canopy.

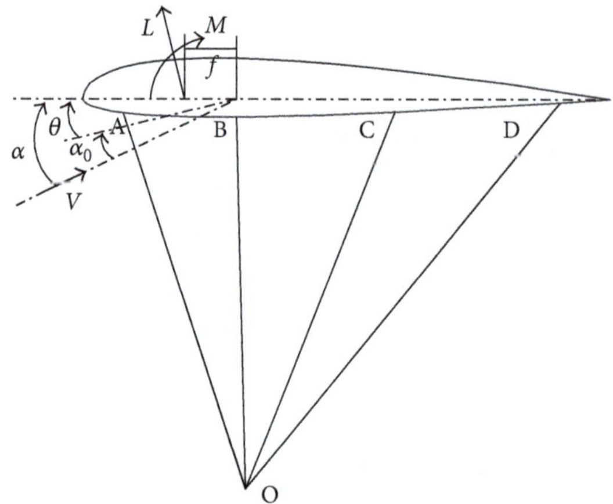

FIGURE 8: Pieces of ideal and wider lower surface.

Here, CL' is the lift coefficient defined in the cell coordinate system; q is the dynamic pressure; S is the reference area of the cell; α_{zl} is the zero lift angle of attack; and f is the distance between the aerodynamic center and point B.

The aerodynamic moment is equal to the elastic moment, so the equilibrium equation can be written as below:

$$K_\theta\theta = M_0 + \frac{\partial \mathrm{CL}'}{\partial \alpha}(\alpha_0 + \theta - \alpha_{zl})qSf. \quad (17)$$

The term θ is moved to the left:

$$\left(K_\theta - \frac{\partial \mathrm{CL}'}{\partial \alpha}qSf\right)\theta = \frac{\partial \mathrm{CL}'}{\partial \alpha}qSf(\alpha_0 - \alpha_{zl}) + M_0. \quad (18)$$

Hence, the expression of θ is shown below:

$$\theta = \frac{\left(\partial \mathrm{CL}'/\partial \alpha\right)qSf(\alpha_0 - \alpha_{zl}) + M_0}{K_\theta - \left(\partial \mathrm{CL}'/\partial \alpha\right)qSf}. \quad (19)$$

When the dynamic pressure q reaches a particular value, the denominator in the equation becomes zero and θ is infinity, and the parafoil structure will crash. The critical dynamic pressure q_D can be written as below:

$$q_D = \frac{K_\theta}{\left(\partial CL'/\partial\alpha\right) Sf} = \frac{l^2}{\left(\partial CL'/\partial\alpha\right) SfL_1} EA. \qquad (20)$$

This critical dynamic pressure is aerodynamic load of incoming flow, rather than the perturbation load for wingtip collapse discussed in Section 2. According to the equation above, when the cell width is constant, the critical dynamic pressure of the parafoil amplified k times reduces k times. The structure stability of a large parafoil needs special consideration. The similar conclusion can be drawn by the wind tunnel test [2].

The calculating example is a large rectangular parafoil used in rocket booster recovery. Its design parameters are presented as follows: the baseline airfoil is Clark-Y18, the span length is 48 m, and the chord length is 16 m. The front projection of the canopy is a quadrant arc and its radius is 32 m. There are four chordwise join points between the ribs and suspending lines on the lower aerofoil. The diameter of the lines is 3.2 mm and its material is Kevlar29. The elastic modulus of this material is 97 GPa. The lift-curve slope $\partial CL'/\partial\alpha$ defined in the cell coordinate system is $0.044\,\text{deg}^{-1}$ and f is 5%c. According to (14), the torsional rigidity of the calculating example is 25,600 Nm. According to (20), the critical dynamic pressure is 5340 Pa. If the maximum dynamic pressure in booster recovery process is larger than 5340 Pa, the recovery mission is likely to fail. Increasing the section area of suspending lines can enlarge the critical dynamic pressure to improve the reliability of recovery.

The large sectional area of suspending lines results in major drag and large weight. The sectional area should be as little as possible under the premise of safety. The relationship between torsion angle θ and dynamic pressure with different torsional rigidity is plotted in Figure 9. The torsion angle of the calculating example is within the acceptable range. In the case of one-tenth sectional area of suspending lines, the torsion angle is too large. Therefore, the design parameters of line OA have significant effects on the structural stiffness of large parafoil.

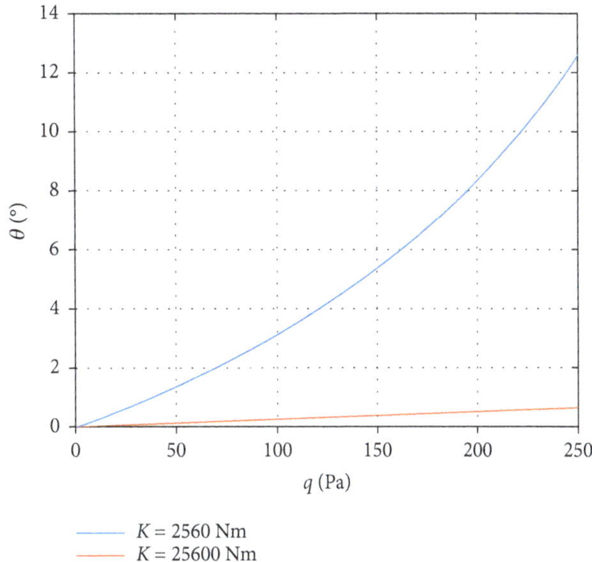

FIGURE 9: Relationship between torsion angle and dynamic pressure.

4. Conclusion

According to the structural characteristic of parafoil, this study has established the spanwise and chordwise structure model of parafoil. The effects of design parameters on deformation and safety of large parafoils are analyzed. This may be useful for the preliminary parafoil design in rocket booster recovery. From the results, the following conclusions are drawn:

(1) Due to the constraint of flexible ropes, after inflation, the parafoil will greatly distort in comparison to its ideal design shape. For the analyzed parafoil with cell ratio 0.9, the actual span of the parafoil reduces by 13% compared to the designed span; the maximum thickness of the airfoil Clark-Y18 increases to 26%c after the bumps appear.

(2) The spanwise model is useful for the weakening deformation design and the margin design of the upper and lower piece of the cell. The cell is slenderer and the canopy surface is smoother. The design value of 2 for the cell ratio is recommended. The aerodynamic load has a bit of effect on the shrinkage and bump ratios.

(3) The critical disturbance load of wingtip collapse is much lower than the difference of pressure inside and outside. The parafoil with large wing loading has high ability to resist wingtip collapse.

(4) When the cell width is constant, the critical dynamic pressure reduces k times with the k times increasing in parafoil area. The design parameters of line OA have significant effects on the structural stiffness of large parafoils. Increasing the section area of line OA can enlarge the critical dynamic pressure to improve the reliability of booster recovery.

Nomenclature

P: Differential pressure
t: Canopy thickness
σ: Canopy stress
r: Widthwise radius of curvature
T_1: Tension per unit chord of upper surface
T_2: Tension per unit chord of lower surface
T_3: Tension per unit chord of the rib
P_1: Pressure on upper surface
P_2: Stagnation pressure

h: Height of cell
w: Width of cell
φ: Half of the central angle
CL': Lift coefficient defined in the cell coordinate system
η_1: Shrinkage ratio of cell
η_2: Bump ratio of cell
l: Distance between points A and B
θ: Upward rotation angle
L_1: Length of line OA
E: Elasticity modulus line OA
A: Sectional area of line OA
K_θ: Torsional rigidity of cell
S: Reference area of the cell
f: Distance between the aerodynamic center and point B.

Conflicts of Interest

The authors declare that there are no conflicts of interest regarding the publication of this paper.

References

[1] C. Matos, R. Mahalingam, G. Ottinger, J. Klapper, R. Funk, and N. Komerath, "Wind tunnel measurements of parafoil geometry and aerodynamics," in *36th AIAA Aerospace Sciences Meeting and Exhibit*, pp. 1–11, Reno, NV, USA, 1998.

[2] R. H. Geiger and W. K. Wailes, "Advanced recovery systems wind tunnel test report," NASA AD-A238157, Moffett Field, CA, USA, 1990.

[3] V. Kalro, S. Aliabadi, W. Garrard, T. Tezduyar, S. Mittal, and K. Stein, "Parallel finite element simulation of large ram-air parachutes," *International Journal for Numerical Methods in Fluids*, vol. 24, no. 12, pp. 1353–1369, 1997.

[4] A. Eslambolchi and H. Johari, "Simulation of flowfield around a ram-air personnel parachute canopy," *Journal of Aircraft*, vol. 50, no. 5, pp. 1628–1636, 2013.

[5] C. Ibos, C. Lacroix, A. Goy, and P. Bordenave, "Fluid-structure simulation of a 3D ram air parachute with SINPA software," in *15th Aerodynamic Decelerator Systems Technology Conference*, pp. 1628–1636, Toulouse, France, 1999, , AIAA-99-1713.

[6] V. Kalro and T. Tezduyar, "A parallel 3D computational method for fluid–structure interactions in parachute systems," *Computer Methods in Applied Mechanics and Engineering*, vol. 190, no. 3-4, pp. 321–332, 2000.

[7] N. A. Fogell, S. Sherwin, C. J. Cotter, L. Iannucci, R. Palacios, and D. J. Pope, "Fluid-structure interaction simulation of the inflated shape of ram-air parachutes," in *AIAA Aerodynamic Decelerator Systems (ADS) Conference*, pp. 1–15, Daytona Beach, FL, USA, 2013.

[8] H. Altmann, "Numerical simulation of parafoil aerodynamics and dynamic behavior," in *20th AIAA Aerodynamic Decelerator Systems Technology Conference and Seminar*, pp. 1–15, Seattle, WA, USA, 2009.

[9] H. Altmann, "Fluid-structure interaction analysis of ram-air parafoil wings," in *23rd AIAA Aerodynamic Decelerator Systems Technology Conference*, pp. 1–10, Daytona Beach, FL, USA, 2015.

[10] R. Peralta and H. Johari, "Geometry of a ram-air parachute canopy in steady flight from numerical simulations," in *23rd AIAA Aerodynamic Decelerator Systems Technology Conference*, pp. 1–13, Daytona Beach, FL, USA, 2015.

[11] Mosseev, "Software tools for the paraglider computer-aided design guide," in *16th AIAA Aerodynamic Decelerator Systems Technology Conference and Seminar*, pp. 52–61, Boston, MA, USA, 2001.

[12] W. Flügge, *Stresses in Shells*, Springer, Berlin, Heidelberg, 1973.

[13] L. F. Wang and W. L. He, "Parafoil aerodynamic deformation simulation based on cable-membrane finite element model," *Journal of Beijing University of Aeronautics and Astronautics*, vol. 43, pp. 47–52, 2017.

[14] Y. C. Gal-Rom and D. E. Raveh, "Simplified aerostructural static model for inflated wings," *AIAA Journal*, vol. 49, no. 6, pp. 1180–1190, 2011.

18

CFD Analysis of Contrarotating Open Rotor Aerodynamic Interactions

Wenbo Shi, Jie Li, Zhao Yang, and Heng Zhang

School of Aeronautics, Northwestern Polytechnical University, Xi'an 710072, China

Correspondence should be addressed to Jie Li; lijieruihao@163.com

Academic Editor: André Cavalieri

High efficiency and low fuel consumption make the contrarotating open rotor (CROR) system a viable economic and environmentally friendly powerplant for future aircraft. While the potential benefits are well accepted, concerns still exist with respect to the vibrations and noise caused by the aerodynamic interactions of CROR systems. In this paper, emphasis is placed on the detailed analysis of the aerodynamic interactions between the front and aft propellers of a puller CROR configuration. For the first step, unsteady Reynolds-averaged Navier-Stokes (URANS) simulations coupled with dynamic patched grid technology are implemented on the isolated single-rotating propeller (SRP) configuration in various operating conditions in order to test the accuracy and feasibility of the numerical approach. The numerical results are verified by a wind tunnel test, showing good agreements with the experimental data. Subsequently, the URANS approach is applied to the CROR configuration. The numerical results obtained through the URANS approach help to improve the understanding of the complex flow field generated by the CROR configuration, and the comparison of SRP flow field and CROR flow field allows for a detailed analysis of the aerodynamic interactions of the front propeller blade wakes and tip vortices with the aft propeller. The main reason of the aerodynamic interactions is the mutual effects of the blade tip vortices, and the aft propeller reduces the strength of the blade tip vortices of the front propeller. Aerodynamic interactions will lead to the periodic oscillations of the aerodynamic forces, and the frequency of the oscillations is linked to the blade numbers. In addition, a CROR has a larger thrust and power coefficient than that of the SRP configuration in the same operating conditions. The URANS approach coupled with a dynamic patched grid method is tested to be an efficient and accurate tool in the analysis of propeller aerodynamic interactions.

1. Introduction

In modern aircrafts, besides supersonic and transonic mainline airliners, regional airliners have the properties of smaller weight and size, lower flight speed and altitude, and good maneuverability; as a result, propeller-driven aircrafts can meet these requirements very well. Due to their shorter take-off distance, larger climbing speed, longer voyage, less fuel consumption, and lower requirements for runways, propeller-driven aircrafts play an important role in both civil and military applications [1–3]. Along with people's attentions to environment issues in recent decades, countries around the world are calling for reducing emissions from aircrafts; therefore, the demands of novel and efficient propeller systems have greatly increased. It is on this premise that CROR systems arouse people's interest as a possible fuel-efficient and environmentally friendly propulsion system for future aircrafts [4, 5].

The efficiency of the CROR system mainly comes from the reduction of the front propeller slipstream swirl by the impingement of the aft propeller slipstream, which will also lead to the extra axial acceleration of the flow downstream when it passes through the aft propeller; therefore, CROR systems can generate greater thrust [6–9]. However, the aerodynamic interactions caused by the front propeller blade tip vortices impinging on the blade tip vortices generated by the aft propeller will result in the noise emissions as well as vibrations [10, 11], making it a main problem for the development of the CROR system in the application of modern aircraft [12–15]. What is more, the mutual interactions between airframe components and the CROR have a remarkable impact on the aerodynamic performance of a CROR. Such installation effects will considerably influence the inflow of the propeller compared to an isolated CROR [16–18].

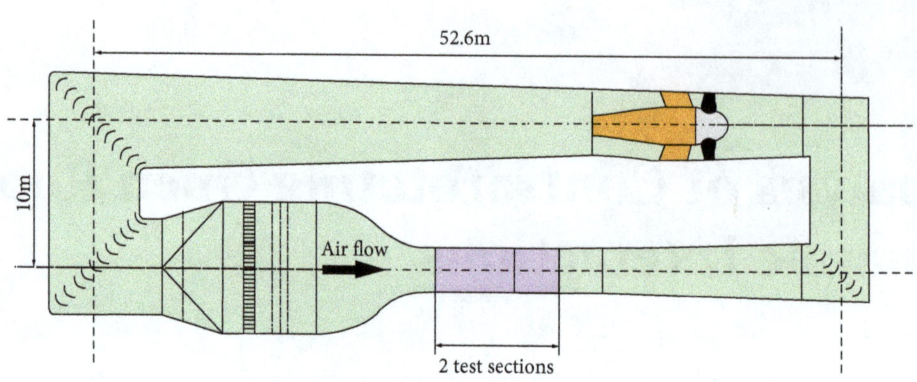

FIGURE 1: Sketch of a wind tunnel.

Some key issues need to be addressed and solved, such as figuring out the flow field around the propeller system and analyzing the slipstream flow interactions between the two propellers in order to design a good CROR system [19, 20].

In recent years, the rapid developments of computer hardware and numerical computation approach have enabled the computational fluid dynamics (CFD) method to become one of the most effective methods to simulate and analyze propeller slipstream flow field [21]. When compared to the traditional wind tunnel test, a CFD method is more economic and time-saving and more importantly, the CFD method can capture the details of flow field while the wind tunnel test cannot [22]. Propeller slipstream flow field is rather complicated due to the strongly perturbed inflow by the blades' periodical affection. Propeller slipstream has unsteady properties; therefore, unsteady numerical approaches should be used in the investigation of propeller slipstream flow field. The URANS calculations have been implemented in the numerical simulations of propeller slipstream, but the mechanism analysis of the complex aerodynamic interactions of CROR systems needs to be further enriched.

In this paper, the URANS code coupled with a dynamic patched grid method is first used on an isolated SRP configuration with experiment verification. The same numerical method is applied to an isolated CROR configuration in different operating conditions with the specific focus on analyzing the mechanism of the complex flow field and the aerodynamic interactions between the front propeller and the aft propeller [23, 24]. Then, some conclusions are drawn with respect to the CROR flow field and aerodynamic interactions.

2. SRP Numerical Simulations with Experiment Verification

2.1. SRP Wind Tunnel Test. An isolated SRP configuration wind tunnel test was conducted in the DNW-LST wind tunnel in Netherland in 2014. The main purpose of the wind tunnel test was to gain access to the thrust and power coefficient as well as the efficiency of the isolated propeller in different rotor speeds under certain blade pitch angles in various operating conditions. The DNW-LST wind tunnel is a low-speed return-flow wind tunnel, which has a 3.0 m (width) × 2.25 m (height) × 8.75 m (length) test section, as shown in Figure 1. The wind tunnel test model was using

FIGURE 2: Isolated single-rotating propeller configuration in the DNW/LST wind tunnel.

TABLE 1: Wind tunnel test settings.

Cases	Mach number	Angle of attack	Rotor speeds	Advance ratio
1	0.2	0.03°	2998 rpm	1.91066
2	0.2	0.03°	3097 rpm	1.8462
3	0.2	0.03°	3198 rpm	1.78863
4	0.2	0.03°	3290 rpm	1.73779

a $1/6^{th}$ scale model that consisted of a nonrotating center shaft and a rotor equipped with 6 blades of the same blade geometry at a fixed diameter of $D = 0.73663$ m with a faring in front of the center shaft. The wind tunnel test model was powered by an air turbine that could drive the rotor speeds to at most 6400 rpm, as shown in Figure 2. The wind tunnel test of the $1/6^{th}$ scale SRP configuration provides a broad scope of database to evaluate the numerical approaches' ability to predict the aerodynamic performance of the SRP configuration as well as the related CROR configuration. Specific emphasis was placed on the flow conditions of the wind tunnel test in order to accommodate the numerical simulations; the flow speed of the wind tunnel test could be set ranging from 1.5 m/s to 80 m/s, and the flow precision can be controlled in the range of 0.2%. The yaw angle of the horizontal and vertical flow is less than 0.1°, while the turbulence level is controlled between 0.02% and 0.03%. The center shaft was connected to a strut, which was connected to an external balance. In this particular wind tunnel test, attentions were paid to aerodynamic

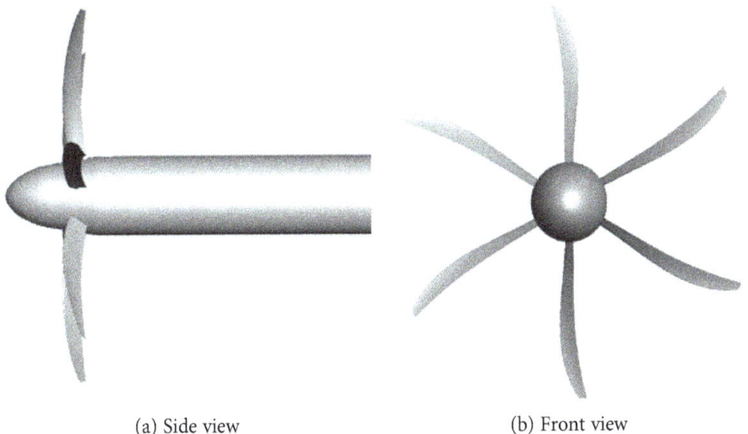

(a) Side view (b) Front view

FIGURE 3: SRP numerical configuration and geometric layout.

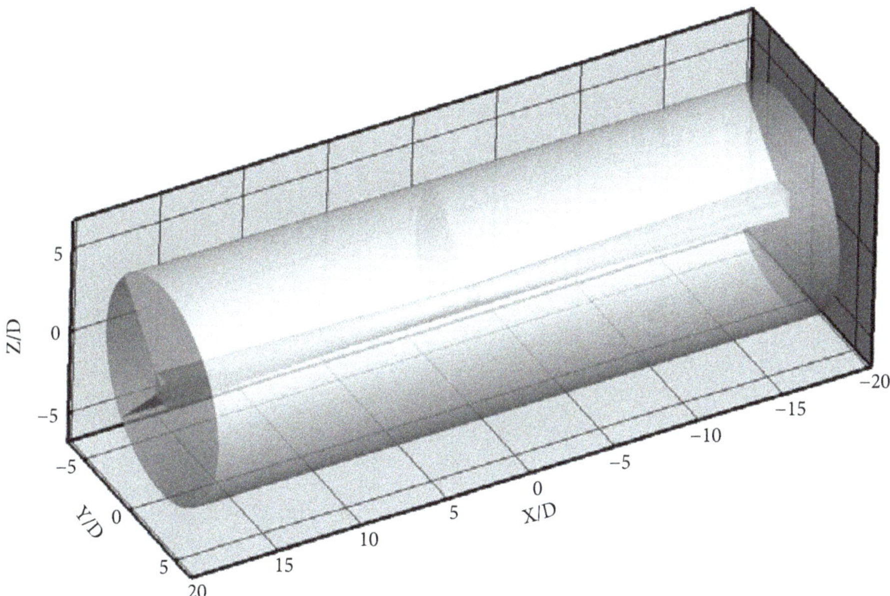

FIGURE 4: Computational domain encompassing the SRP configuration.

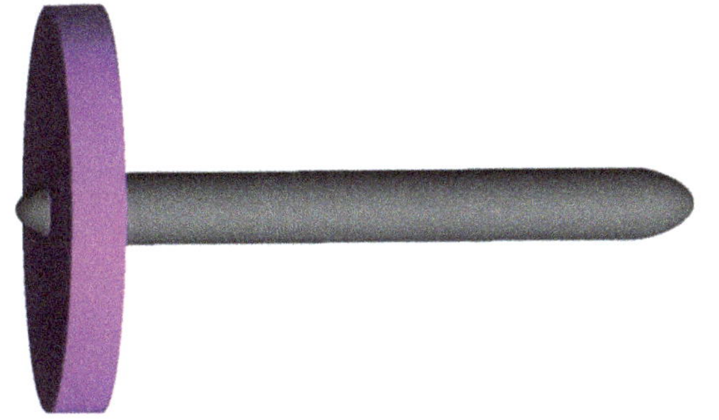

FIGURE 5: Sketch of a dynamic patched grid method of the SRP.

performance by installing a rotated shaft balance inside the faring. Thrust coefficients of different advance ratios are obtained via the rotated shaft balance. The results obtained by the rotated shaft balance were verified through the comparison of the results obtained by the external balance.

The experiment Mach number was set to Ma = 0.2 at an angle of attack $\alpha = 0.03^{\circ}$. Several wind tunnel test conditions were conducted due to various rotor speeds, listed in Table 1. In this wind tunnel test, blade pitch angle was defined as 70% radius blade pitch angle, and in these series of experiments, the blade pitch angle was set as $\beta_{70} = 45.3^{\circ}$.

The propeller thrust coefficient, power coefficient, and efficiency have the following forms:

$$\begin{aligned} C_T &= \frac{T}{\rho n_s^2 D^4}, \\ C_P &= \frac{P}{\rho n_s^3 D^5}, \\ \eta &= \frac{T V_\infty}{P}, \\ P &= 2\pi n_s M, \end{aligned} \quad (1)$$

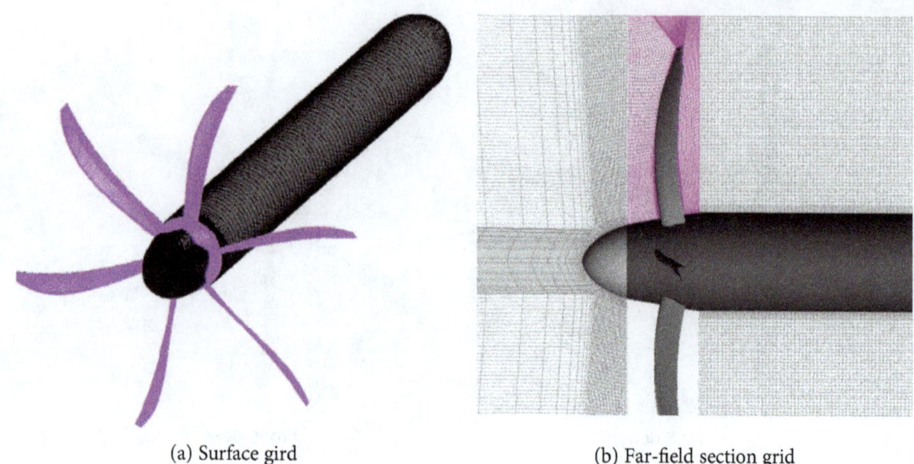

(a) Surface gird (b) Far-field section grid

FIGURE 6: Block-structured grids of an isolated SRP.

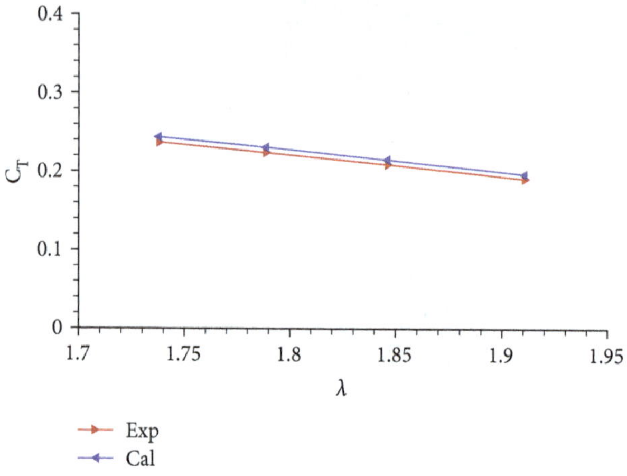

(a) Thrust coefficient

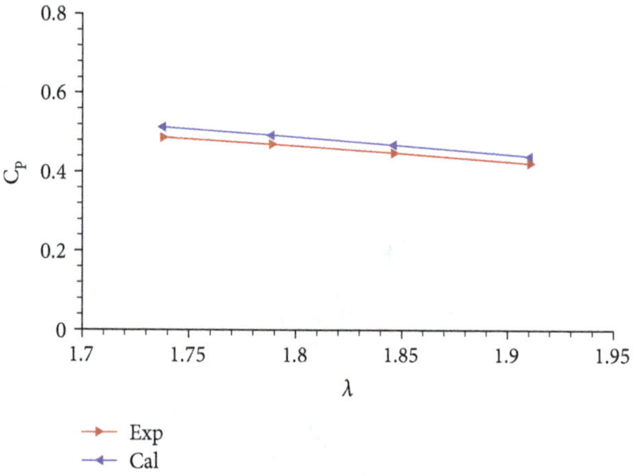

(b) Power coefficient

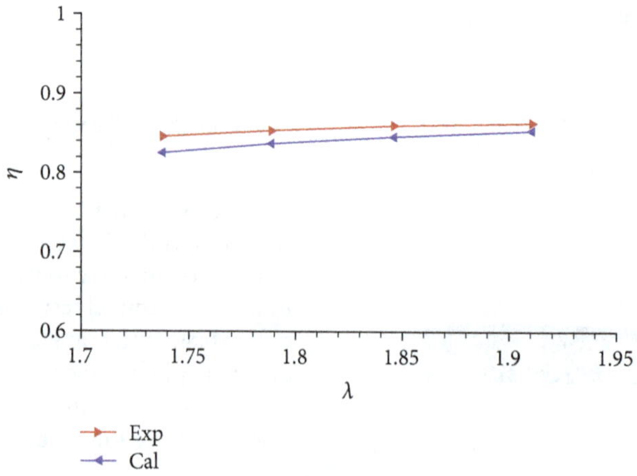

(c) Efficiency

FIGURE 7: Comparison of propeller performance between the wind tunnel test and URANS simulations.

where T is the thrust generated by the propeller and ρ, n_s along with D denote the air density, propeller rotational speed, and propeller diameter, respectively. In addition, P denotes the shaft power and M denotes the torque, whereas V_∞ denotes free stream velocity and η is the efficiency.

2.2. SRP Numerical Simulation Configuration. The SRP numerical simulations mainly focus on the investigation of the propeller flow field and blade aerodynamics, and the most important aspect is the validation of the URANS approach with the wind tunnel test. For the numerical

TABLE 2: Detailed results of the wind tunnel test and URANS numerical simulations.

Cases	C_T (Exp/Cal)	Relative error, %	C_P (Exp/Cal)	Relative error, %
1	0.191/0.197	3.14	0.423/0.441	4.26
2	0.209/0.215	2.87	0.449/0.469	4.45
3	0.224/0.231	3.13	0.470/0.493	4.89
4	0.237/0.244	2.95	0.487/0.513	5.34

studies, a configuration for numerical simulations was approximated as closely to the wind tunnel test model as possible. Modifications were made to the numerical simulation configuration to ensure that there was no adverse impact on the performance of the propeller as well as the flow field in their direct vicinity. The length of the center shaft was chosen to be $2D$ (2 times of the propeller diameter) whereas the strut to which the SRP was connected in the wind tunnel test was removed. The numerical SRP configuration is shown in Figure 3.

2.3. Grid Generation. The accuracy and reliability of solutions obtained by means of CFD have close ties to the density and quality of the computational grids used. A dynamic patched grid method has the primary advantage that the grids of each subdomain can be generated separately without concerning the topology relations among adjacent blocks. One-to-one correspondence of grid points is not required on the two sides of the patched surfaces, and fluxes are interpolated near the patched surfaces in the process of computation; therefore, the flow information of adjacent subdomains is coupled and exchanged. For the numerical simulation described here, the baseline SRP configuration grids have one computational domain and one rotating domain. The computational domain, which is also called far-field, is a cylindrical region that has $20D$ (D is the propeller diameters) long upstream and $20D$ long downstream of the propeller, and the diameter of the computational domain is approximately $14D$, as shown in Figure 4.

The entire grids are included in the computational domain. Except the grids in the rotating domain, the other grids in the computational domain remain static. The rotating domain is a cylindrical region that contains the propeller, and the grids are generated in the cylindrical region, while fluxes are interpolated through the cylinder surfaces, which are patched surfaces, between the grids in the computational domains and rotating domain. The sketch map of the patched grid approach is shown in Figure 5. The rotating cylindrical domain is marked in purple.

The block-structured grids of this SRP configuration have been created with the software ANSYS ICEM CFD. The whole computational grids consist of two parts: the blocks around the rotating propeller and the outer blocks around the static center shaft. Attentions should be paid to the grid generation of the blade passages to make sure that each of the blade passages is created as equal as possible to allow for a reliable spatial resolution for the blades of the propeller. Refined grids are applied in the boundary layer and propeller slipstream region to improve blade wake and tip vortex resolution. The surface grid of the SRP configuration is shown in Figure 6(a). The grid shown in grey in Figure 6(b) is an enlarged image of a far-field section grid. The grid patched in the rotating domain is the propeller section grid, and the grid shown in pink is the propeller section grid. The complete block-structured grids of the isolated SRP configuration consist of 6,664,206 nodes.

2.4. URANS Approach. For the simulations described here, URANS calculations are applied to the isolated SRP configuration in order to testify the feasibility and accuracy of this numerical method, spatial discretization of the convective fluxes is done using a third-order upstream MUSCL scheme, and the viscous fluxes are discretized with central differences. A Spalart–Allmaras (SA) model is chosen as the turbulence model. A fully implicit LU-SGS method with subiteration in pseudotime is implemented as a time-marching method, and the well-established dual-time approach is used to enhance the precision of time discretization. In the course of the URANS simulations, the time step size is set equivalent to a propeller rotation of $\Delta\varphi = 0.25$ deg per physical time step and the numerical results are obtained after 8 full revolutions of the SRP propeller. The numerical approach is implemented on 4 cases, whose numerical parameters are in accordance with the wind tunnel test. The SRP numerical simulations are considered converged if the propeller thrust coefficients show no notable differences. The SRP simulations were run on 64 processors of Inspur TS10000 Cluster in Center for High Performance Computing, Northwestern Polytechnical University. For each case, the approximate total wall-clock runtime for the simulation of the 8 full revolutions was 1 day.

2.5. Result Comparison. The comparison of the SRP wind tunnel test and URANS approach in terms of thrust coefficient C_T and power coefficient C_P is shown in Figure 7, and the results obtained through numerical simulations show a good agreement with the wind tunnel test.

The detailed results of the wind tunnel test and numerical simulations are listed in Table 2. The relative errors of propeller thrust and power coefficient of the 4 operating conditions are quite small, which means that the grid generation and URANS numerical approaches are validated to be accurate in the investigation of the SRP configuration.

3. CROR Numerical Simulations

3.1. CROR Numerical Simulation Configuration. In order to demonstrate the capability of previously used URANS simulations in the analysis of aerodynamic phenomena and interactions of a CROR, an isolated 6×6 puller CROR configuration is put forward built on the isolated SRP configuration mentioned above by installing another propeller of the same blade geometry and blade numbers behind the propeller of the SRP configuration, as shown in Figure 8. This CROR configuration is only fictitious and has never actually been built. The geometric parameters of the CROR are consistent with the SRP configuration. The length of the center shaft is $L_1 = 2D$, while the distance between the two

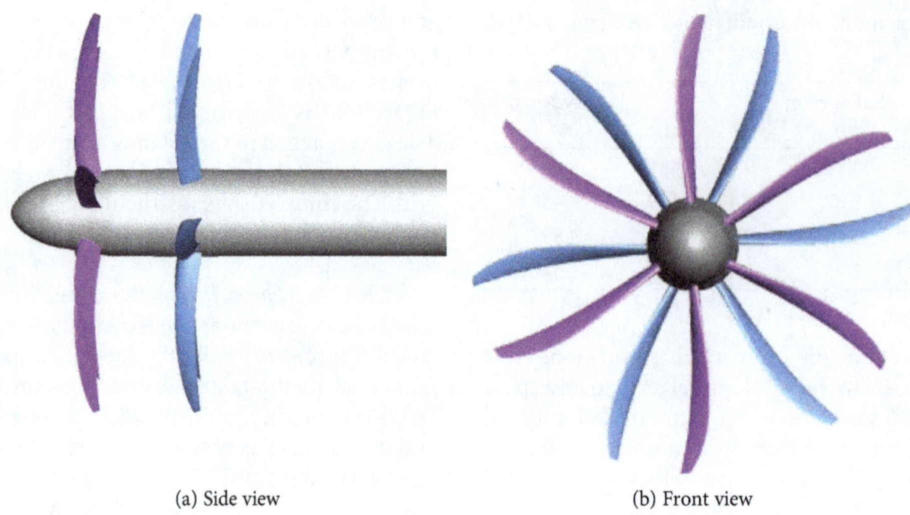

(a) Side view

(b) Front view

FIGURE 8: CROR numerical configuration and geometric layout.

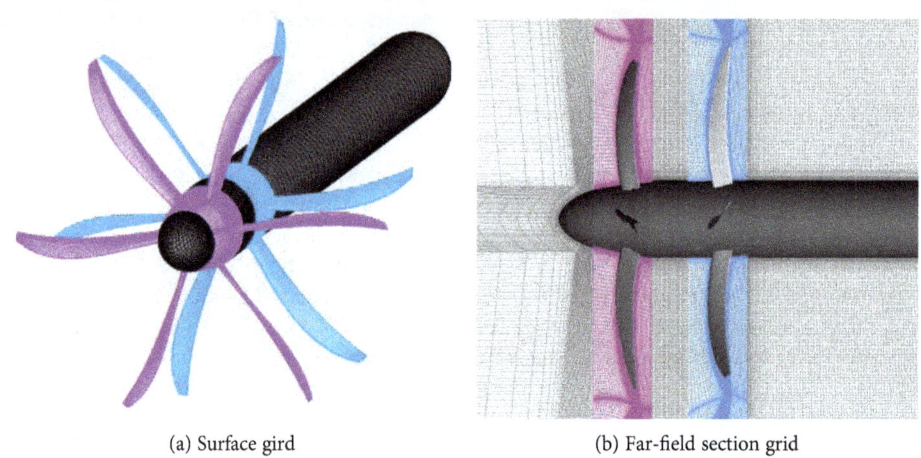

(a) Surface gird

(b) Far-field section grid

FIGURE 9: Block-structured grids of an isolated CROR.

TABLE 3: CROR numerical settings.

Cases	Mach number	Angle of attack	Rotor speeds	Advance ratio
1	0.2	0.03°	2998/2998 rpm	1.91066
2	0.2	0.03°	3097/3097 rpm	1.8462
3	0.2	0.03°	3198/3198 rpm	1.78863
4	0.2	0.03°	3290/3290 rpm	1.73779

propellers is $L_2 = 0.25D$. The blade pitch angles of the front and the aft propellers are both set as $\beta_{70} = 45.3^{\circ}$. The rotational speed is set to be identical for both propellers, and the two propellers rotate in the opposite directions, counterclockwise for the front propeller and clockwise for the aft one. The main reason for using the same blade numbers for a CROR with the identical diameters and blade pitch angles is that the CROR configuration is built on the SRP configuration. The mechanism of aerodynamic interactions of a CROR should be compared with the SRP. The impacts of blade numbers, blade diameters, and blade pitch angles as well as rotor speeds on the aerodynamic interactions have been researched, and the research papers will be released later [25].

3.2. Grid Generation. In the investigation of a CROR, the grid generation approach placed primary emphasis on the adequate resolution of the flow phenomena that played a significant role in the interactions between the front and aft propellers. Thus, the main effort goes into obtaining high-quality and high-resolution propeller grids to address requirements for good aerodynamic analysis results, which are a significant aspect for the design of a high efficient CROR system. The surface grid and section grid of the CROR configuration are shown in Figure 9. The rotational domains of the front and aft propellers are created of the same size to ensure identical grid densities and qualities. Blade passages are created as equal as possible in order to capture the periodic nature of the CROR configuration. The CROR numerical simulations mainly focus on the investigation of the propeller flow field and aerodynamic interactions between the two propellers. Refined grids are applied in the propeller slipstream region of the CROR configuration and

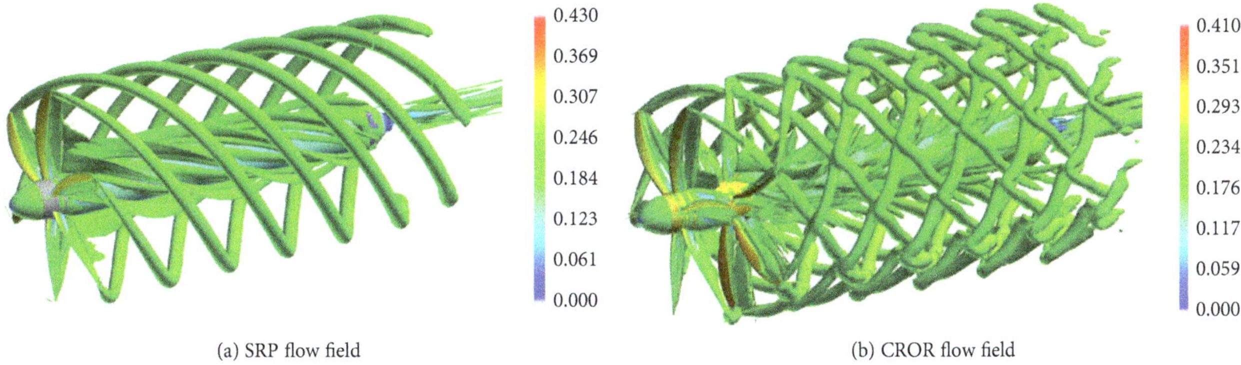

(a) SRP flow field

(b) CROR flow field

FIGURE 10: Slipstream flow field; $J = 1.73779$.

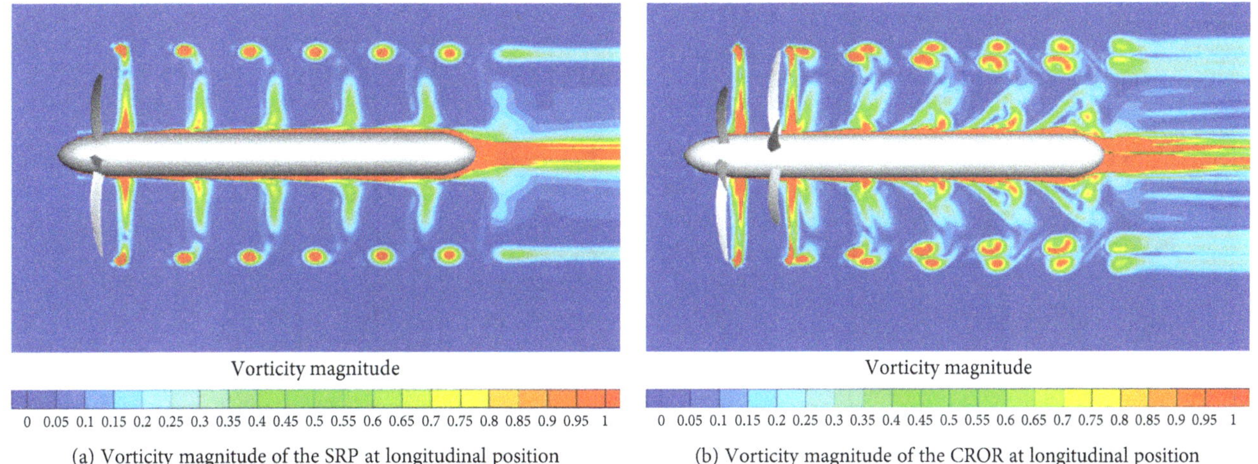

(a) Vorticity magnitude of the SRP at longitudinal position

(b) Vorticity magnitude of the CROR at longitudinal position

FIGURE 11: Cross section vorticity magnitude; $J = 1.73779$.

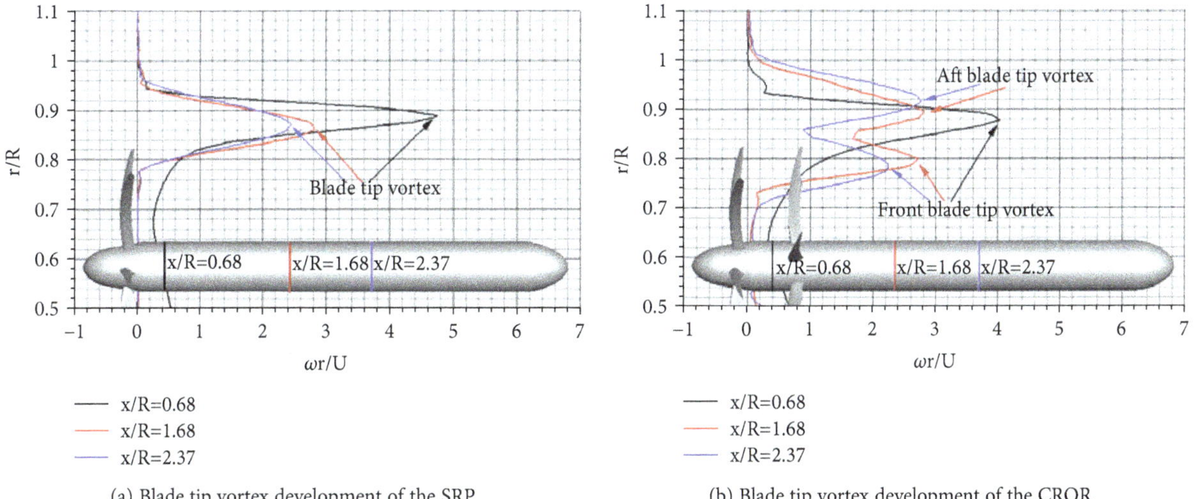

(a) Blade tip vortex development of the SRP

(b) Blade tip vortex development of the CROR

FIGURE 12: Blade tip vortex development at axial positions of $x/R = 0.68$, $x/R = 1.68$, and $x/R = 2.37$.

designed to capture the flow features especially the interactions between the front and aft propellers in terms of blade wake and tip vortices during the process of numerical simulations. The grids used in this investigation are block-structured and generated through ANSYS ICEM CFD and consist of 17,723,511 nodes.

3.3. URANS Approach. In the investigation of the CROR configuration, all the numerical settings are aligned with the settings implemented in the SRP simulations. The CROR flow field is obtained after 12 full propeller revolutions, and the simulations are considered converged if the periodic fluctuations of the blade loadings for the 4 test cases show no

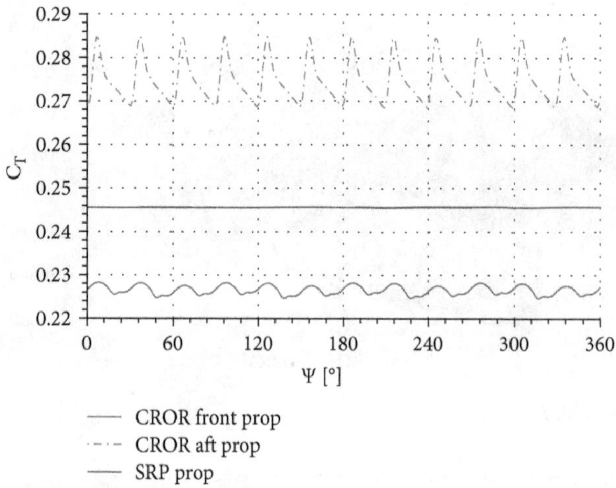

FIGURE 13: Comparison of SRP and CROR thrust coefficient during one full rotation; $J = 1.73779$.

obvious changes. The settings of CROR numerical simulation are listed in Table 3. The CROR simulations were run on 64 processors of Inspur TS10000 Cluster in Center for High Performance Computing, Northwestern Polytechnical University. For each case, the approximate total wall-clock runtime for the simulation of the 12 full revolutions was 5 days.

4. CROR Aerodynamic Interactions

4.1. Flow Field Analysis. The isosurfaces of the Q-criterion of the isolated SRP configuration and CROR configuration are shown in Figure 10 at the advance ratio $J = 1.73779$, which demonstrate numerical simulations of the configuration flow field characterized by Mach number. The vorticity distributions on both sides of the patched surfaces did not break or dislocate; hence, the blade tip vortices and blade wakes smoothly got through the patched surfaces, which indicated the proper applications of the dynamic patched technology in the grid generation of SRP and CROR configurations and validated the numerical approach in terms of URANS used in the investigation. For the SRP configuration, as shown in Figure 10(a), the free stream passed through the propeller, then twisted by the rotating blades, and formed a helical vortex-tube downstream. While for the CROR configuration, the two propellers rotate in the opposite directions, the blade wakes and tip vortices of the front propeller are "chopped" by the blade wakes, and tip vortices generated by the aft propeller and an interwoven net-structured vortex system were formed as shown in Figure 10(b), which lead to periodic unsteady oscillations. Iso-Q surfaces reveal the interactions of the front propeller blade tip vortices and blade wakes with the aft propeller. The front blade wakes played a dominant role in the interaction with the aft blades, so the unsteady flow structures are mainly characterized by the wake impingement of the front propeller on the aft propeller. What should be emphasized is that the blade tip vortices of both propellers are almost perpendicular which happened in each of the test cases regardless of the propeller rotational speeds; the blade tip vortices of the front propeller progressively merged with the tip vortices produced by the aft propeller yielding the net-structured pattern observed downstream.

4.2. Blade Tip Vortex Development. In order to investigate and analyze the vortex systems in depth, instantaneous value of the vorticity magnitude of the cross section is plotted, as shown in Figure 11. The position of the blade tip vortices can be seen clearly for both SRP and CROR configurations by the red spots of vorticity contours. Clearly the blade wakes of each propeller are visible in the slipstream as well as their deformation with respect to the flow acceleration. Visible in Figure 11(a), the blade tip vortices generated on the SRP shows no obvious dissipation as they develop downstream and the diameter of slipstream remains almost the same. However, contraction of slipstream can be seen from Figure 11(b), as the tip vortex emanating from the front rotor interacts with the vortex from the second rotor with the flow develops downstream. According to the law of conservation of mass, the mass flow passing through the slipstream tube remained unchanged and the flow velocity increased, which lead to the decrease of the radius of the slipstream tube and hence the contraction of slipstream. Visible in the plot of vorticity magnitude contour in Figure 11(b), compared to the aft propeller, the front propeller shows a much more rapid decay in the blade tip vortices. Generally, the aerodynamic interactions between the two propellers result in the obvious contraction of the slipstream and the rapid decay of the blade tip vortices of the front propeller. The vorticity magnitude greatly increased in the hub downstream region. As the flow moves downstream, with an increasing axial distance from the propellers, blade wakes and tip vortices are gradually dissipating.

The blade tip vortex development of SRP and CROR configurations is plotted in terms of vorticity magnitude at axial positions of $x/R = 0.68$, $x/R = 1.68$, and $x/R = 2.37$ as shown in Figure 12. As shown in Figure 12(a), the vorticity level is gradually decreasing at axial positions of $x/R = 0.68$, $x/R = 1.68$, and $x/R = 2.37$ (marked as black, red, and blue solid lines) with the peak vorticity locates approximately in the radial position of $r/R = 0.88$. While in Figure 12(b), the black solid line shows the interception of the front blade tip vortex before it impinges on the blade tip vortex generated by the aft propeller at axial position of $x/R = 0.68$, while the red solid line demonstrates the interception of the front and aft blade tip vortices at axial position of $x/R = 1.68$. At $x/R = 0.68$, the peak vorticity level of the front blade tip vortex locates in the radial position of $r/R = 0.88$, and when the front blade tip vortex interacts with the aft blade tip vortex at axial position of $r/R = 1.68$, the peak vorticity of front blade tip vortex locates in the radial position of $r/R = 0.8$. The peak vorticity of front blade tip vortex at $r/R = 1.68$ is lower than that of the front blade tip vortex at axial position of $r/R = 0.68$ (this is due to the reduction in the front blade tip vortex strength by the aft blade tip vortex) and validates the contraction of slipstream. Also, visible in Figure 12(b), the peak vorticity level of the aft blade tip

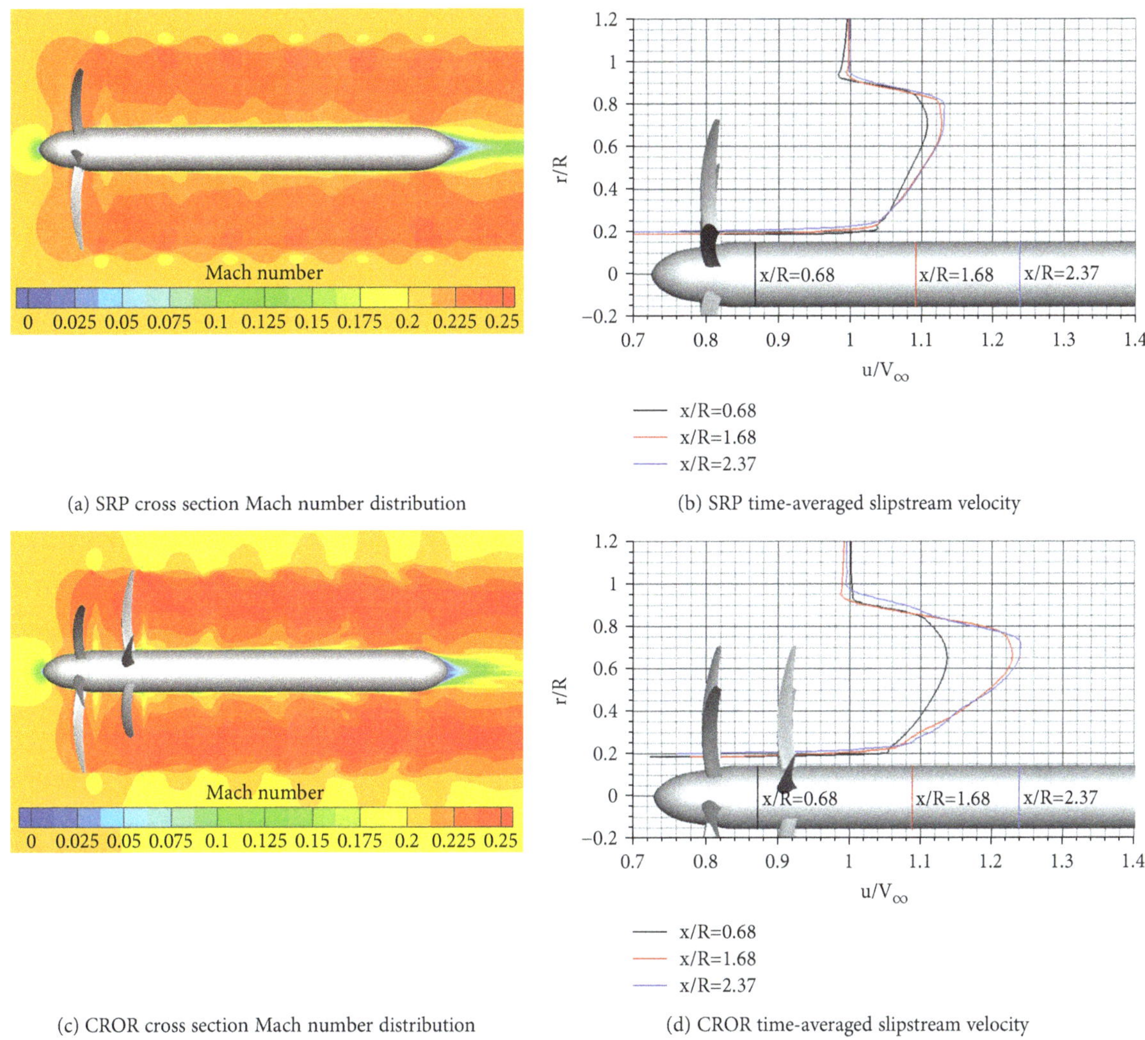

(a) SRP cross section Mach number distribution

(b) SRP time-averaged slipstream velocity

(c) CROR cross section Mach number distribution

(d) CROR time-averaged slipstream velocity

FIGURE 14: Slipstream acceleration effect; $J = 1.73779$.

vortex locates in the radial position of $r/R = 0.9$ and is lower than that of the front blade tip vortex at $x/R = 0.68$; this means that the aft propeller has a function of tip vortex dissipation.

4.3. Aerodynamic Characteristics. Unsteady thrust coefficients are plotted in order to analyze the aerodynamic interactions between the two propellers, as shown in Figure 13. These unsteady blade loadings can be easily obtained from numerical simulations. Figure 13 shows the front and aft propeller thrust coefficient development during one full revolution with the rotational speed of 3290/3290 rpm marked as solid purple and dash-dotted blue lines, respectively. Visible in Figure 13, the thrust coefficients of the two propellers have unsteady but periodic oscillations with respect to aerodynamic interactions between the front and aft propellers. During one full revolution, the thrust coefficient of each propeller will fluctuate twelve times and the thrust coefficient of the two propellers reached maximum or minimum almost at the same time. Due to the 6×6 puller CROR configuration studied here, the oscillations are periodic, recurring every 30 deg of propeller rotation. Strong oscillations of the propeller performance are observed for the aft propeller. For the front propeller, the unsteady loadings are a result of the aft propeller flow field; as for the aft propeller, the much stronger oscillation is caused by the impingement of the front blade wakes and tip vortices. The front propeller delivers a mean thrust coefficient of $C_{T,\text{front}} = 0.226$, while the aft propeller delivers a larger mean thrust coefficient of $C_{T,\text{aft}} = 0.274$. Compared to the CROR configuration, the propeller of the SRP configuration has a relatively constant blade loading in terms of thrust coefficient, which delivers a value of $C_T = 0.245$ marked as red a solid line in Figure 13.

Figure 14 shows the time-averaged plots during one full rotation of the unsteady numerical simulations in terms of Mach number and axial velocity magnitude with the particular focus on the investigation of flow acceleration. As shown in Figure 14(a), the value of Mach number barely increased when the flow passes through the propeller of the SRP configuration. The plot of the time-averaged axial velocity of one full rotation, normalized with free stream Mach number at certain axial positions of $x/R = 0.68$, $x/R = 1.68$, and $x/R = 2.73$, indicates that slipstream is accelerated slightly as it develops downstream, as shown in Figure 14(b). As shown

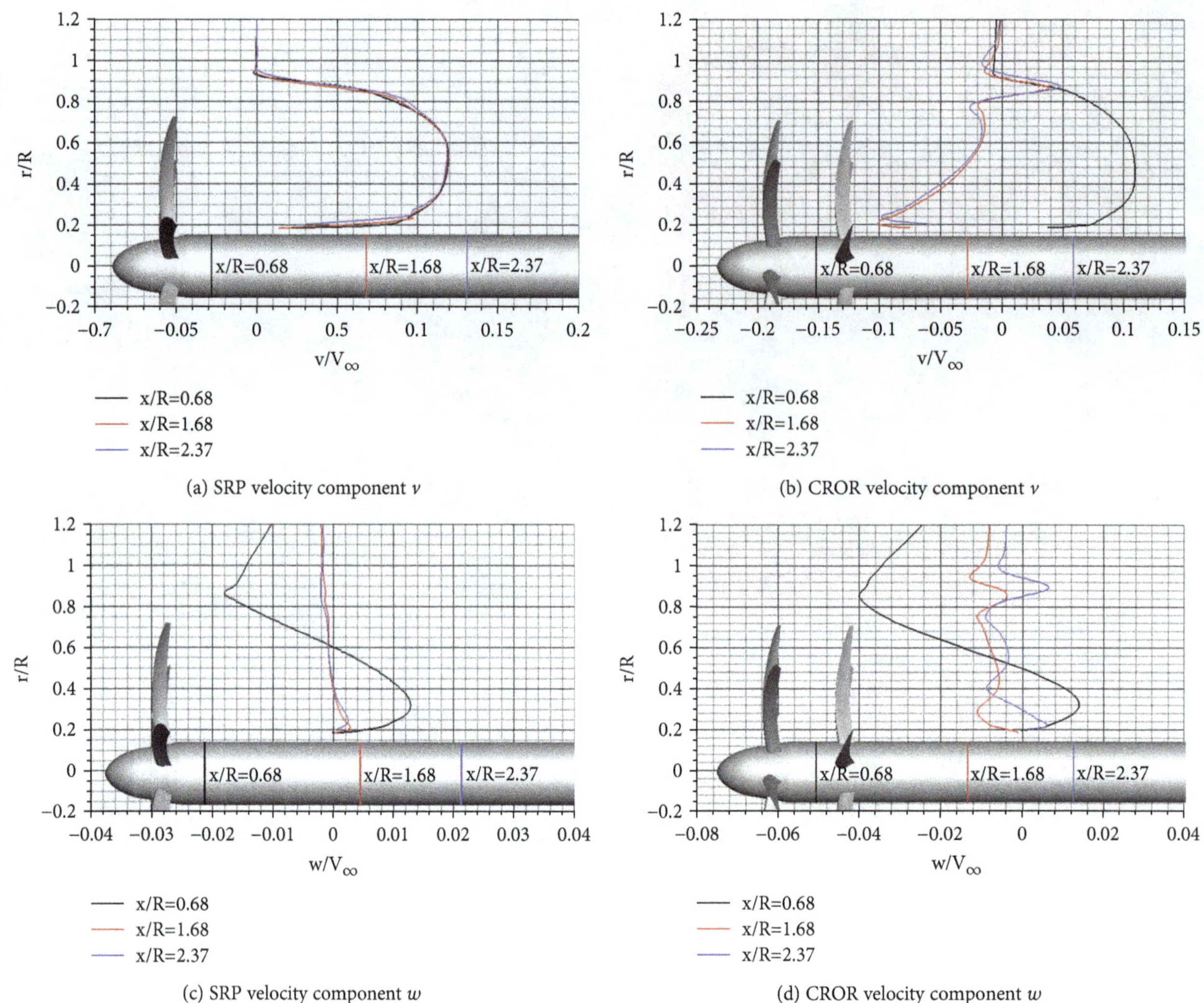

(a) SRP velocity component v

(b) CROR velocity component v

(c) SRP velocity component w

(d) CROR velocity component w

FIGURE 15: Time-averaged slipstream-normalized velocity profiles; $J = 1.73779$.

in Figure 14(c), the value of Mach number increased when the flow passes through the front propeller, and Mach number increases a second time when the flow passes through the aft propeller of the CROR configuration. The acceleration of the flow by the front propeller as well as the further subsequent increase in the Mach number of the aft propeller can be seen clearly from Figure 14(c). Figure 14(d) shows the plot of time-averaged axial velocity magnitude for one complete revolution, normalized with free stream Mach number, at axial positions of $x/R = 0.68$, $x/R = 1.68$, and $x/R = 2.73$ downstream of the front propeller. The black solid line indicates that the flow is first accelerated when it passes through the front propeller in terms of the increase in the axial velocity, while the red solid line shows that the flow is further accelerated when it passes through the aft propeller. The slipstream will be accelerated downstream of the front propeller which is demonstrated by the blue solid line. In addition, with the increasing distance downstream of the front propeller, the slipstream acceleration will gradually weaken. The plot of time-averaged axial velocity magnitude clearly shows that slipstream acceleration is most notable between radial positions of $r/R = 0.4$ and $r/R = 0.9$.

In the numerical investigation of SRP and CROR configurations, the figures of velocity profiles are plotted in terms of time-averaged slipstream-normalized velocity components of one complete revolution. Figures 15(a) and 15(b) show the tangential velocity v at the same axial positions of $x/R = 0.68$, $x/R = 1.68$, and $x/R = 2.37$. For the SRP, the propeller rotates counterclockwise in a right-handed coordinate system aligned with the x-axis, the values of tangential velocity at the three axial positions are positive between radial positions of $r/R = 0.2$ and $r/R = 0.9$, and the three curves are almost coincided, which means that the tangential velocities remain nearly unchanged at each axial position. For the CROR, the front propeller rotates counterclockwise and the aft propeller rotates clockwise in a right-handed coordinate system aligned with the x-axis. This brings about the fact that the tangential velocity is positive at the axial position of $x/R = 0.68$ between radial positions of $r/R = 0.2$ and $r/R = 0.9$, while the tangential velocity v is negative at $x/R = 1.68$ and $x/R = 2.37$ between radial positions of $r/R = 0.2$ and $r/R = 0.8$. The aerodynamic

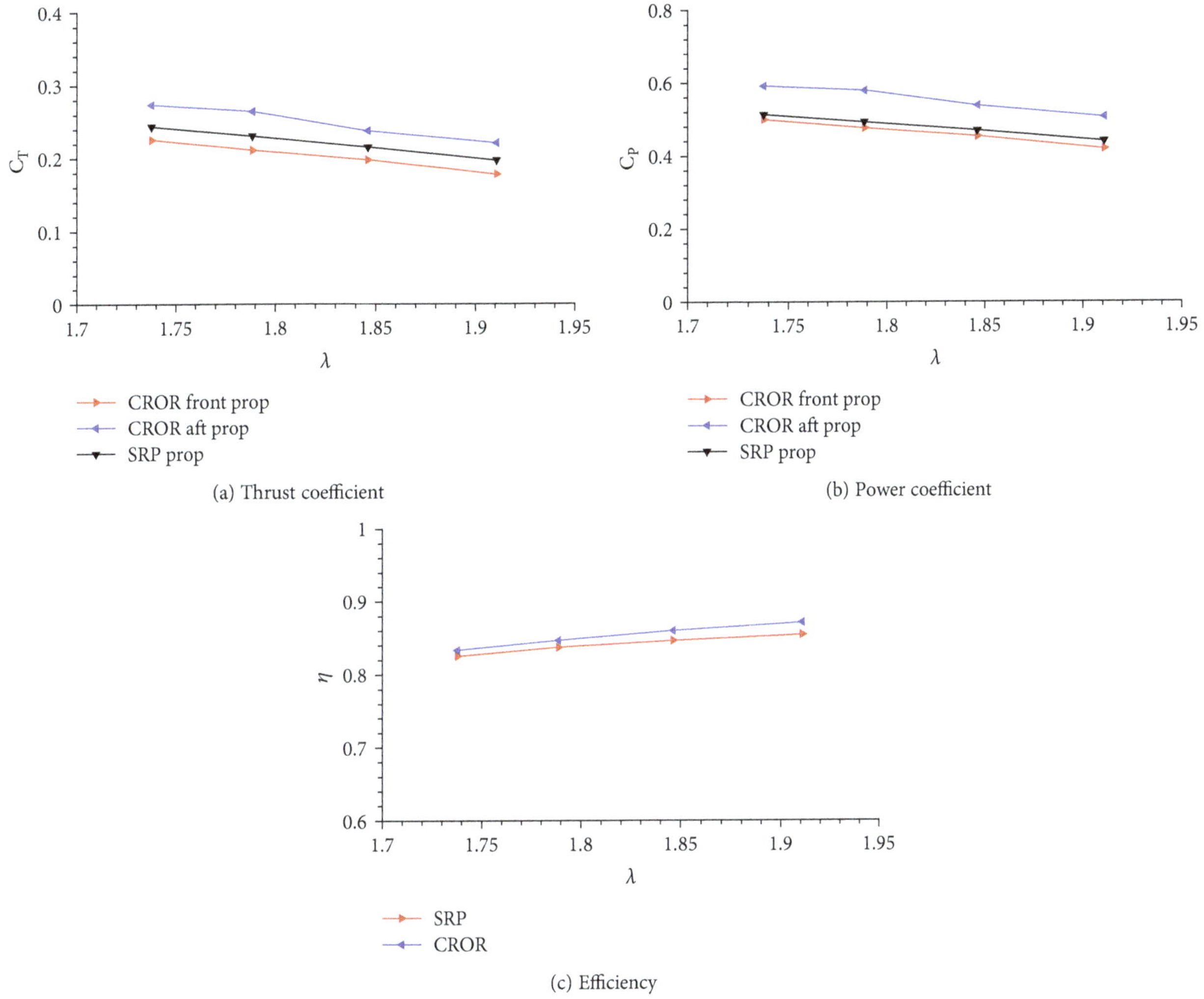

(a) Thrust coefficient

(b) Power coefficient

(c) Efficiency

FIGURE 16: Comparison of CROR and SRP performance.

interactions lead to the decrease of the values between radial positions of $r/R = 0.2$ and $r/R = 0.8$ at $x/R = 1.68$ and $x/R = 2.37$ compared to the absolute value at $x/R = 0.68$; this is because the front blade tip vortex encounters the aft blade tip vortex and vortex strength is reduced. The tangential velocity v becomes positive near the blade tips between $r/R = 0.8$ and $r/R = 0.82$ at axial positions of $x/R = 1.68$ and $x/R = 2.37$, because the strength of the mutual interactions of blade tip vortices has weakened and this indicates that the front propeller of the CROR plays a dominant role in the aerodynamic interactions between the two propellers.

Figure 15(c) shows the plot of time-averaged slipstream-nomalized radial velocity w of the SRP, slipstream contraction results in the negative values of the radial velocity component outboard of the radial position of $r/R = 0.6$ at the axial position of $x/R = 0.68$. The radial velocity component inboard of $r/R = 0.6$ shows a positive value because of the higher pressure near the center shaft. At further downstream positions of $x/R = 1.68$ and $x/R = 2.37$, the radial velocity component decreases close to zero as the flow tends to be stable, which represents the completion of slipstream contraction.

Figure 15(d) is the plot of time-averaged slipstream-nomalized radial velocity w of a CROR, the maximum radial velocity component outboard of the radial position of $r/R = 0.6$ at the axial position of $x/R = 0.68$ is nearly two times as close as that of the SRP due to the increase of Mach number in the region between the two blades. When it comes to the downstream positions of $x/R = 1.68$ and $x/R = 2.37$, the radial velocity component shows an obvious fluctuation along the radial direction because of the strong interference and the slipstream shows a more intense contraction.

Figure 16 shows the comparison of CROR and SRP thrust and power coefficient in regard to advance ratios. As shown in Figure 16, the SRP thrust coefficient and power coefficient are larger than that of the CROR front propeller but smaller than that of the CROR aft propeller in the same advance ratio; this is due to the recovery of the swirl flow behind the front propeller into axial momentum. Naturally, in the given operating conditions, the CROR can generate much larger thrust than SRP does. The efficiency of a CROR is higher than that of the SRP due to the reduction of blade tip vortices by the aft propeller.

5. Conclusions

An analysis of aerodynamic interactions was conducted for a 6 × 6 puller CROR configuration by using a dynamic patched grid method-based URANS numerical simulation approach. The URANS approach has been applied to an isolated SRP configuration with wind tunnel test verification, and this numerical approach is validated to be an efficient and accurate tool in the investigation of propeller slipstream and aerodynamic characteristics. The results obtained through the URANS approach help to improve the understanding of the CROR slipstream flow field and aerodynamic interactions between the two propellers.

The isocontours of vorticity magnitude characterize the vortex system of CROR slipstream flow field, which shows the complex interaction of the front propeller blade wakes and tip vortices with the aft propeller. The mutual interactions between the two propellers result in unsteady periodic blade loading oscillations during one full rotation. Due to the equal diameter and rotational speed of the two propellers, the frequency of blade loading oscillations depends on the blade numbers. The instantaneous contours of vorticity magnitude of the CROR blade tip vortex development reveal the vortex dissipation caused by the aerodynamic interactions. Time-averaged three-component velocity profiles at certain axial positions have been analyzed, and the results show that the freestream is accelerated twice when it passes through the two propellers. With the slipstream developing downstream, the diameter of the slipstream will decrease until the completion of the contraction. The upstream front propeller blade wakes play a dominant role in the interaction with the aft propeller, and these wakes are then ingested by the wakes of the aft propeller and have a strong impact on the vortex system. Generally, a CROR will produce much larger thrust than that of a SRP under the same circumstances, and the higher the rotational speed is, the greater the power consumption will be.

Conflicts of Interest

The authors declare that no conflict of interest regarding the publication of this paper.

Acknowledgments

This study is supported by the State Key Development Program for Basic Research of China (no. 2015CB755800).

References

[1] F. Moens and P. Gardarein, "Numerical simulation of the propeller/wing interactions for transport aircraft," in *19th AIAA Applied Aerodynamics Conference*, Anaheim, CA, USA, June 2001AIAA Paper 2001-2404.

[2] D. Keller and R. Rudnik, "Numerical investigation of engine effects on a transport aircraft with circulation control," *Journal of Aircraft*, vol. 52, no. 2, pp. 421–438, 2015.

[3] P. C. M. van den Borne and J. van Hengst, "Investigation of propeller slipstream effects on the Fokker 50 through in-flight pressure measurements," in *Flight Simulation Technologies Conference and Exhibit*, Portland, OR, USA, August 1990AIAA Paper 90-3084.

[4] M. J. Czech and R. H. Thomas, "Experimental studies of open rotor installation effects," in *3rd AIAA Atmospheric Space Environments Conference*, Honolulu, Hawaii, June 2011AIAA Paper 2011-4047.

[5] T. Sinnige, J. J. A. van Kuijk, K. P. Lynch, D. Ragni, G. Eitelberg, and L. L. M. Veldhuis, "The effects of swirl recovery vanes on single-rotation propeller aerodynamics and aeroacoustics," in *21st AIAA/CEAS Aeroacoustics Conference*, Dallas, TX, USA, June 2015AIAA Paper 2015-2358.

[6] E. W. M. Roosenboom, A. Stürmer, and A. Schröder, "Comparison of PIV measurements with uRANS calculations in a propeller slipstream," in *27th AIAA Applied Aerodynamics Conference*, San Antonio, TX, USA, June 2009AIAA Paper 2009-3626.

[7] E. W. M. Roosenboom, A. Stürmer, and A. Schröder, "Advanced experimental and numerical validation and analysis of propeller slipstream flows," *Journal of Aircraft*, vol. 47, no. 1, pp. 284–291, 2010.

[8] R. Schnell, J. Yin, S. Funke, and H. Siller, "Aerodynamic and basic acoustic optimization of a counter rotating open rotor with experimental verification," in *18th AIAA/CEAS Aeroacoustics Conference (33rd AIAA Aeroacoustics Conference)*, Colorado Springs, CO, USA, June 2012AIAA Paper 2012-2127.

[9] A. Stürmer and R. A. D. Akkermans, "Multidisciplinary analysis of CROR propulsion systems: DLR activities in the JTI SFWA project," *CEAS Aeronautical Journal*, vol. 5, no. 3, pp. 265–277, 2014.

[10] Z. Tang, P. Liu, Y. Chen, and H. Guo, "Experimental study of counter-rotating propellers for high-altitude airships," *Journal of Propulsion and Power*, vol. 31, no. 5, pp. 1491–1496, 2015.

[11] Z. Tang, P. Liu, J. Sun, Y. Chen, H. Guo, and G. Li, "Performance of contra-rotating propellers for stratospheric airships," *International Journal of Aeronautical and Space Sciences*, vol. 16, no. 4, pp. 485–492, 2015.

[12] S. Read and T. Hynes, "Effect of a winglet on open rotor aerodynamics and tip vortex interaction," in *49th AIAA/ASME/SAE/ASEE Joint Propulsion Conference*, San Jose, CA, USA, July 2013AIAA Paper 2012-4039.

[13] D. E. Van Zante and M. P. Wernet, "Tip vortex and wake characteristics of a counterrotating open rotor," in *48th AIAA/ASME/SAE/ASEE Joint Propulsion Conference & Exhibit*, Atlanta, GA, USA, 2012AIAA Paper 2012-4039.

[14] A. Stürmer, C. O. Marquez Gutierrez, E. W. M. Roosenboom et al., "Experimental and numerical investigation of a contra rotating open-rotor flowfield," *Journal of Aircraft*, vol. 49, no. 6, pp. 1868–1877, 2012.

[15] E. Envia, "Contra-rotating open rotor tone noise prediction," in *20th AIAA/CEAS Aeroacoustics Conference*, Atlanta, GA, USA, June 2014AIAA Paper 2014-2606.

[16] R. A. D. Akkermans, M. Pott-Pollenske, H. Buchholz, J. W. Delfs, and D. Almoneit, "Installation effects of a propeller mounted on a high-lift wing with a Coanda flap. Part I: aeroacoustic experiments," in *20th AIAA/CEAS Aeroacoustics Conference*, Atlanta, GA, USA, June 2014AIAA Paper 2014-2606.

[17] J. Dierke, R. A. D. Akkermans, J. W. Delfs, and R. Ewert, "Installation effects of a propeller mounted on a wing with Coanda flap. Part II: numerical investigation and experimental validation," in *20th AIAA/CEAS Aeroacoustics Conference*, Atlanta, GA, USA, June 2014AIAA Paper 2014-2606.

[18] A. Stuermer, J. Yin, and R. Akkermans, "Progress in aerodynamic and aeroacoustic integration of CROR propulsion systems," *The Aeronautical Journal*, vol. 118, no. 1208, pp. 1137–1158, 2014.

[19] L. Soulat, I. Kernemp, M. Sanjose, S. Moreau, and R. Fernando, "Assessment and comparison of tonal noise models for counter-rotating open rotors," in *19th AIAA/CEAS Aeroacoustics Conference*, Berlin, Germany, May 2013AIAA paper 2013-2201.

[20] T. D. Economon, F. Palacios, and J. J. Alonso, "Optimal shape design for open rotor blades," in *30th AIAA Applied Aerodynamics Conference*, New Orleans, LA, USA, June 2012AIAA paper 2012-3018.

[21] R. Boisard and G. Delatre, "HPC capabilities of the elsA CFD software applied to a counter rotating open rotor test rig," in *30th AIAA Applied Aerodynamics Conference*, New Orleans, LA, USA, June 2012AIAA paper 2012-2125.

[22] T. Deconinck, A. Capron, V. Barbieux, C. Hirsch, and G. Ghorbaniasl, "Sensitivity study on computational parameters for the prediction of noise generated by counter-rotating open rotors," in *17th AIAA/CEAS Aeroacoustics Conference (32nd AIAA Aeroacoustics Conference)*, Portland, OR, USA, June 2011AIAA Paper 2011-2765.

[23] J. A. Housman and C. C. Kiris, "Structured overlapping grid simulations of contra-rotating open rotor noise," in *54th AIAA Aerospace Sciences Meeting*, San Diego, CA, USA, January 2016AIAA Paper 2016-0814.

[24] V. N. Vatsa and B. W. Wedan, "Development of a multigrid code for 3-D Navier-Stokes equations and its application to a grid-refinement study," *Computers & Fluids*, vol. 18, no. 4, pp. 391–403, 1990.

[25] R. A. D. Akkermans, J. W. Delfs, C. O. Márquez et al., "Aeroacoustic and aerodynamic importance of a CROR propulsion system with unequal rotor rotation speeds,," in *New Results in Numerical and Experimental Fluid Mechanics IX*, Springer, Stuttgart, Germany, 2012.

Permissions

All chapters in this book were first published in IJAE, by Hindawi Publishing Corporation; hereby published with permission under the Creative Commons Attribution License or equivalent. Every chapter published in this book has been scrutinized by our experts. Their significance has been extensively debated. The topics covered herein carry significant findings which will fuel the growth of the discipline. They may even be implemented as practical applications or may be referred to as a beginning point for another development.

The contributors of this book come from diverse backgrounds, making this book a truly international effort. This book will bring forth new frontiers with its revolutionizing research information and detailed analysis of the nascent developments around the world.

We would like to thank all the contributing authors for lending their expertise to make the book truly unique. They have played a crucial role in the development of this book. Without their invaluable contributions this book wouldn't have been possible. They have made vital efforts to compile up to date information on the varied aspects of this subject to make this book a valuable addition to the collection of many professionals and students.

This book was conceptualized with the vision of imparting up-to-date information and advanced data in this field. To ensure the same, a matchless editorial board was set up. Every individual on the board went through rigorous rounds of assessment to prove their worth. After which they invested a large part of their time researching and compiling the most relevant data for our readers.

The editorial board has been involved in producing this book since its inception. They have spent rigorous hours researching and exploring the diverse topics which have resulted in the successful publishing of this book. They have passed on their knowledge of decades through this book. To expedite this challenging task, the publisher supported the team at every step. A small team of assistant editors was also appointed to further simplify the editing procedure and attain best results for the readers.

Apart from the editorial board, the designing team has also invested a significant amount of their time in understanding the subject and creating the most relevant covers. They scrutinized every image to scout for the most suitable representation of the subject and create an appropriate cover for the book.

The publishing team has been an ardent support to the editorial, designing and production team. Their endless efforts to recruit the best for this project, has resulted in the accomplishment of this book. They are a veteran in the field of academics and their pool of knowledge is as vast as their experience in printing. Their expertise and guidance has proved useful at every step. Their uncompromising quality standards have made this book an exceptional effort. Their encouragement from time to time has been an inspiration for everyone.

The publisher and the editorial board hope that this book will prove to be a valuable piece of knowledge for researchers, students, practitioners and scholars across the globe.

List of Contributors

Du Siliang
Faculty of Mechanical & Material Engineering, Huaiyin Institute of Technology, Huaian 223003, China

Tang Zhengfei
National Key Laboratory of Rotorcraft Aeromechanics, Nanjing University of Aeronautics and Astronautics, Nanjing 210016, China

Zhang Yakui
Aeronautics and Astronautics Engineering College, Air Force Engineering University, Xi'an 710038, China

Guo Shuxiang
Department of Mathematics and Physics College, Air Force Engineering University, Xi'an 710038, China

Mario Rosario Chiarelli and Salvatore Bonomo
DICI, University of Pisa, Largo Lucio Lazzarino 1, 56122 Pisa, Italy

Zuodong Mu, Guiping Lin, Xiaobin Shen, Xueqin Bu and Ying Zhou
Laboratory of Fundamental Science on Ergonomics and Environmental Control, Beihang University, Beijing 100191, China

Si-Yuan Chen and Qun-Li Xia
School of Aerospace Engineering, Beijing Institute of Technology, Beijing 100081, China

Mohammad Mahdi Heydari and Nooredin Ghadiri Massoom
Malek Ashtar University of Technology, Babaei Highway, Lavizan, Tehran 15875-1774, Iran

Hiroshi Yokoyama, Hiroshi Odawara and Akiyoshi Iida
Department of Mechanical Engineering, Toyohashi University of Technology, 1-1 Hibarigaoka, Tempaku, Aichi 441-8580, Japan

Xiaopeng Wang and Zhengyin Ye
Northwestern Polytechnical University, Xi'an 710072, China

Chen'an Zhang, Wen Liu and Famin Wang
State Key Laboratory of High-Temperature Gas Dynamic, Institute of Mechanics, Chinese Academy of Sciences, Beijing 100190, China

Jie Hou, Ji-Hong Zhu and Qing Li
Engineering Simulation and Aerospace Computing (ESAC),The Key Laboratory of Contemporary Design and Integrated Manufacturing Technology, Northwestern Polytechnical University, Xi'an 710072, China

Muhammad Zubair Zahid and Shahid Ikramullah Butt
School of Mechanical and Manufacturing Engineering, National University of Sciences and Technology, Islamabad, Pakistan

Tauqeer Iqbal
Université Catholique de Louvain, Louvain-la-Neuve, Belgium

Syed Zohaib Ejaz
Birmingham City University, Birmingham, UK

Zhang Faping
School of Mechanical Engineering, Beijing Institute of Technology, Beijing, China

Soumya Ranjan Nanda, Sumit Agarwal, Vinayak Kulkarni and Niranjan Sahoo
Department of Mechanical Engineering, Indian Institute of Technology Guwahati, Guwahati 781 039, India

Kuan-Huang Lee and Keh-Chin Chang
Institute of Aeronautics and Astronautics, National Cheng Kung University, Tainan, Taiwan

Kung-Ming Chung
Aerospace Science and Technology Research Centre, National Cheng Kung University, Tainan, Taiwan

Haojie Liu and Yonghui Zhao
State Key Laboratory of Mechanics and Control of Mechanical Structures, Nanjing University of Aeronautics and Astronautics, 210016 Nanjing, China

Adel Ghenaiet
Laboratory of Energetic Mechanics and Conversion Systems, Faculty of Mechanical Engineering, University of Sciences and Technology Houari Boumediene, BP32 El-Alia, Bab-Ezzouar, 16111 Algiers, Algeria

Tsuneyoshi Matsuoka, Kyohei Kamei and Yuji Nakamura
Department of Mechanical Engineering, Toyohashi University of Technology, Toyohashi, Japan

Harunori Nagata
Division of Mechanical and Space Engineering, Hokkaido University, Sapporo, Japan

Young H. You, Deokhwan Na and Sung N. Jung
Department of Aerospace Information Engineering, Konkuk University, Seoul 143-701, Republic of Korea

Longfang Wang and Weiliang He
School of Astronautics, Beihang University, Beijing 100191, China

Wenbo Shi, Jie Li, Zhao Yang and Heng Zhang
School of Aeronautics, Northwestern Polytechnical University, Xi'an 710072, China

Index